CAMBRIDGE LIBRARY COLLECTION

Books of enduring scholarly value

Mathematical Sciences

From its pre-historic roots in simple counting to the algorithms powering modern desktop computers, from the genius of Archimedes to the genius of Einstein, advances in mathematical understanding and numerical techniques have been directly responsible for creating the modern world as we know it. This series will provide a library of the most influential publications and writers on mathematics in its broadest sense. As such, it will show not only the deep roots from which modern science and technology have grown, but also the astonishing breadth of application of mathematical techniques in the humanities and social sciences, and in everyday life.

The Collected Mathematical Papers

Arthur Cayley (1821-1895) was a key figure in the creation of modern algebra. He studied mathematics at Cambridge and published three papers while still an undergraduate. He then qualified as a lawyer and published about 250 mathematical papers during his fourteen years at the Bar. In 1863 he took a significant salary cut to become the first Sadleirian Professor of Pure Mathematics at Cambridge, where he continued to publish at a phenomenal rate on nearly every aspect of the subject, his most important work being in matrices, geometry and abstract groups. In 1882 he spent five months at Johns Hopkins University, and in 1883 he became president of the British Association for the Advancement of Science. Publication of his Collected Papers - 967 papers in 13 volumes plus an index volume - began in 1889 and was completed after his death under the editorship of his successor in the Sadleirian Chair. Volume 3 contains 64 papers first published between 1857 and 1862.

The Collected Mathematical Papers

VOLUME 3

ARTHUR CAYLEY

CAMBRIDGE
UNIVERSITY PRESS

CAMBRIDGE UNIVERSITY PRESS

Cambridge New York Melbourne Madrid Cape Town Singapore São Paolo Delhi

Published in the United States of America by Cambridge University Press, New York

www.cambridge.org
Information on this title: www.cambridge.org/9781108004954

This edition first published 1890
This digitally printed version 2009

ISBN 978-1-108-00495-4

MATHEMATICAL PAPERS.

London: C. J. CLAY & SONS,
CAMBRIDGE UNIVERSITY PRESS WAREHOUSE,
AVE MARIA LANE.

Cambridge: DEIGHTON, BELL AND CO.
Leipzig: F. A. BROCKHAUS.

THE COLLECTED

MATHEMATICAL PAPERS

OF

ARTHUR CAYLEY, Sc.D., F.R.S.,

SADLERIAN PROFESSOR OF PURE MATHEMATICS IN THE UNIVERSITY OF CAMBRIDGE.

VOL. III.

CAMBRIDGE:
AT THE UNIVERSITY PRESS.
1890

CAMBRIDGE:

PRINTED BY C. J. CLAY, M.A. AND SONS,
AT THE UNIVERSITY PRESS.

ADVERTISEMENT.

THE present volume contains 64 papers numbered 159 to 222 originally published in the years 1857 to 1862: the chronological order is slightly departed from for the sake of including a series of papers in the Memoirs and the Monthly Notices of the Royal Astronomical Society: there are thus some earlier papers which have been left over for the next volume.

As papers are referred to by their Numbers only it will be convenient to give in each volume a Table such as the following:

Vol.	I.	Numbers	1	to	100.
,,	II.	,,	101	,,	158.
,,	III.	,,	159	,,	222.

CONTENTS.

[An Asterisk denotes that the Paper is not printed in full.]

CLASSIFICATION.

ANALYSIS

159.

ON SOME INTEGRAL TRANSFORMATIONS.

[From the *Quarterly Mathematical Journal*, vol. I. (1857), pp. 4—6.]

SUPPOSE that x, a, b, c and x', a', b', c' have the same anharmonic ratios, or what is the same thing, let these quantities satisfy the equation

$$\begin{vmatrix} 1, & 1, & 1, & 1 \\ x, & a, & b, & c \\ x', & a', & b', & c' \\ xx', & aa', & bb', & cc' \end{vmatrix} = 0;$$

this equation may be represented under a variety of different forms, which are obtained without difficulty; thus, if for shortness

$$K = a(b'-c')(x'-a') + b\,(c'-a')(x'-b') + c\,(a'-b')(x'-c'),$$

then

$$Kx = -\{bc(b'-c')(x'-a') + ca(c'-a')(x'-b') + ab(a'-b')(x'-c')\},$$

$$K(x-a) = (c-a)(a-b)(b'-c')(x'-a'),$$

$$K(x-b) = (a-b)(b-c)(c'-a')(x'-b'),$$

$$K(x-c) = (b-c)(c-a)(a'-b')(x'-b').$$

Consider x, x' as variables; then

$$K^2\,dx = (b-c)\;(c-a)\;(a-b)\,(b'-c')(c'-a')(a'-b')\,dx';$$

let, d, d' be any corresponding values of x, x'; then

$$\begin{vmatrix} 1, & 1, & 1, & 1 \\ a, & b, & c, & d \\ a', & b', & c', & d' \\ aa', & bb', & cc', & dd' \end{vmatrix} = 0$$

and we have

$$K(x-d)=D(x'-d');$$

where

$$D=(c'-a')(a'-b')(b'-c')\Lambda,$$

and

$$\Lambda=\frac{(a-d)(b-c)}{(a'-d')(b'-c')}=\frac{(b-d)(c-a)}{(b'-d')(c'-a')}=\frac{(c-d)(a-b)}{(c'-d')(a'-b')}.$$

Suppose $\alpha+\beta+\gamma+\delta=-2$; then

$$(x-a)^\alpha(x-b)^\beta(x-c)^\gamma(x-d)^\delta\,dx=J(x'-a')^\alpha(x'-b')^\beta(x'-c')^\gamma(x'-d')^\delta\,dx,$$

where

$$J=(b-c)^{\beta+\gamma+1}(c-a)^{\gamma+\alpha+1}(a-b)^{\alpha+\beta+1}(b'-c')^{\alpha+\delta+1}(c'-a')^{\beta+\delta+1}(a'-b')^{\gamma+\delta+1}D^\delta.$$

We may in particular take for a', b', c', d' the systems b, a, d, c : c, d, a, b and d, c, b, a respectively; this gives, writing successively y, z, w instead of x',

$$\begin{aligned}&(x-a)^\alpha(x-b)^\beta(x-c)^\gamma(x-d)^\delta\,dx\\&=M(y-a)^\beta(y-b)^\alpha(y-c)^\delta(y-d)^\gamma\,dy\\&=N(z-a)^\gamma(z-b)^\delta(z-c)^\alpha(z-d)^\beta\,dz\\&=P(w-a)^\delta(w-b)^\gamma(w-c)^\beta(w-d)^\alpha\,dw,\end{aligned}$$

where

$$\begin{aligned}M&=-(-)^{\gamma+\delta}(a-c)^{\alpha+\gamma+1}(a-d)^{\alpha+\delta+1}(b-c)^{\beta+\gamma+1}(b-d)^{\beta+\delta+1},\\N&=\ \ (-)^{\gamma+\delta}(a-b)^{\alpha+\beta+1}(a-d)^{\alpha+\delta+1}(b-c)^{\beta+\gamma+1}(c-d)^{\gamma+\delta+1},\\P&=(a-b)^{\alpha+\beta+1}(a-c)^{\alpha+\gamma+1}(b-d)^{\beta+\delta+1}(c-d)^{\gamma+\delta+1};\end{aligned}$$

the relations between the variables x, y, z, w being

$$\begin{aligned}x&=\frac{(c+d)ab-(a+b)cd-(ab-cd)y}{ab-cd-(a+b-c-d)y}\\&=\frac{(b+d)ac-(a+c)bd-(ac-bd)z}{ac-bd-(a+c-b-d)z}\\&=\frac{(b+c)ad-(a+d)bc-(ad-bc)w}{ad-bc-(a+d-b-c)w}:\end{aligned}$$

these are, in fact, the formulæ in my note, "On an Integral Transformation," *Camb. and Dubl. Math. Jour.* t. III. (1848), p. 286 [62], which was suggested to me by Gudermann's transformation for elliptic functions, (*Crelle*, t. XXIII. (1846), p. 330).

Suppose now that the values of a', b', c', d' are 0, 1, ∞, ζ, we have in this case

$$x=\frac{a(b-c)+c(a-b)y}{(b-c)+(a-b)y},$$

and representing the denominator by K, then

$$\begin{aligned} K(x-a) &= -(a-b)(a-c)\,y, \\ K(x-b) &= (a-b)(b-c)(1-y), \\ K(x-c) &= (a-c)(b-c), \\ K(x-d) &= (a-b)(c-d)(y-\zeta), \end{aligned}$$

where

$$\zeta = \frac{(a-d)(b-c)}{(a-b)(c-d)},$$

and we have

$$K^2\,dx = -(a-b)(a-c)(b-c)\,dy,$$

whence

$$(x-a)^\alpha (x-b)^\beta (x-c)^\gamma (x-d)^\delta\,dx = \\ -(-)^\alpha (a-b)^{\alpha+\beta+\delta+1}(a-c)^{\alpha+\gamma+1}(b-c)^{\beta+\gamma+1}(c-d)^\delta\, y^\alpha (1-y)^\beta (y-\zeta)^\delta\,dy.$$

It is easy, by means of this equation, to generalise a remarkable formula given by M. Serret in his memoir, "Sur la Représentation géométrique des Fonctions elliptiques et ultrà-elliptiques," *Liouville*, t. XI. and XII. [1846 and 1847], and *Recueil des Savans étrangers*, t. XI. [1851].(1) In fact, suppose that the indices α, β, γ, δ are integers, and that two of these indices, e.g. γ, δ, are negative, the remaining two indices being positive, then writing $-\gamma$, $-\delta$ instead of γ, δ the integral

$$\int \frac{(x-a)^\alpha (x-b)^\beta\,dx}{(x-c)^\gamma (x-d)^\delta},$$

where $\gamma+\delta=\alpha+\beta+2$, depends on the integral

$$\int \frac{y^\alpha (1-y)^\beta\,dy}{(y-\zeta)^\delta}.$$

Suppose that the fraction under the integral sign is resolved into simple fractions, each of these fractions will be integrable algebraically, except the fraction having for its denominator the simple power $y-\zeta$, the integral of which is a logarithm. The coefficient of this fraction is at once found by writing in the numerator $\zeta+(y-\zeta)$ for y; and expanding in ascending powers of $y-\zeta$ and equating this coefficient to zero, we have

$$\left(\frac{d}{d\zeta}\right)^{\delta-1} \zeta^\alpha (1-\zeta)^\beta = 0,$$

which [observing that $(\gamma-1)+(\delta-1)=\alpha+\beta$] is easily seen to be equivalent to

$$\left(\frac{d}{d\zeta}\right)^{\gamma-1} \zeta^\beta (1-\zeta)^\alpha = 0.$$

[1] M. Serret has reproduced the theorem in his very interesting and instructive treatise, "Cours d'Algèbre supérieure," deuxième édition, Paris, 1854, [quatrième édition, Paris, 1877].

Hence if the function $\zeta = \dfrac{(a-d)(b-c)}{(a-b)(c-d)}$ satisfy this condition, the indefinite integral

$$\int \frac{(x-a)^{\alpha}(x-b)^{\beta}\,dx}{(x-c)^{\gamma}(x-d)^{\delta}},$$

where $\gamma + \delta = \alpha + \beta + 2$, will be expressible as a rational algebraical fraction.

It may be noticed, that in the general case, observing that $x = a$, $x = b$ give $y = 0$, $y = 1$, the integral

$$\int_a^b (x-a)^{\alpha}(x-b)^{\beta}(x-c)^{\gamma}(x-d)^{\delta}\,dx$$

depends on

$$\int_0^1 y^{\alpha}(1-y)^{\beta}(\zeta - y)^{\delta}\,dy,$$

or, putting $\zeta = \dfrac{1}{u}$, upon

$$\int_0^1 y^{\alpha}(1-y)^{\beta}(1-uy)^{\delta}\,dy,$$

which is expressible by means of a hypergeometric series having u for its argument or fourth element.

2, *Stone Buildings, Lincoln's Inn, Feb.* 1854.

160.

ON A THEOREM RELATING TO RECIPROCAL TRIANGLES.

[From the *Quarterly Mathematical Journal*, vol. I. (1857), pp. 7—10.]

THE following theorem is, I assume, known; but the analytical demonstration of it depends upon a formula in determinants which is not without interest. The theorem referred to may be thus stated:

"A triangle and its reciprocal are in perspective;" where by the reciprocal of a triangle is meant the triangle the sides of which are the polars of the angles of the first-mentioned triangle with respect to a conic; and triangles are in perspective when the three lines forming the corresponding angles meet in a point, or what is the same thing, when the three points of intersection of the corresponding sides lie in a line.

Let the equation of the conic be

$$x^2+y^2+z^2=0,$$

and take (α, β, γ), $(\alpha', \beta', \gamma')$, $(\alpha'', \beta'', \gamma'')$ for the coordinates of the angles of the triangle, then if K be the determinant, and (A, B, C) (A', B', C') (A'', B'', C'') the inverse system, i.e. if

$$\begin{aligned}
KA\ &=(\beta'\gamma''-\beta''\gamma'), & KB\ &=\gamma'\alpha''-\gamma''\alpha', & KC\ &=\alpha'\beta''-\alpha''\beta',\\
KA' &=(\beta''\gamma-\beta\gamma''), & KB' &=\gamma''\alpha-\gamma\alpha'', & KC' &=\alpha''\beta-\alpha\beta'',\\
KA'' &=(\beta\gamma'-\beta'\gamma), & KB'' &=\gamma\alpha'-\gamma'\alpha, & KC'' &=\alpha\beta'-\alpha'\beta,
\end{aligned}$$

equations which may be represented in the notation of matrices by the single equation

$$\left|\begin{matrix}\alpha, & \beta, & \gamma\\ \alpha', & \beta', & \gamma'\\ \alpha'', & \beta'', & \gamma''\end{matrix}\right|^{-1}=\left|\begin{matrix}A, & A', & A''\\ B, & B', & B''\\ C, & C', & C''\end{matrix}\right|,$$

then the equations of the sides of the triangle are

$$A\,x+B\,y+C\,z=0,$$
$$A'\,x+B'\,y+C'\,z=0,$$
$$A''x+B''y+C''z=0,$$

and the coordinates of the angles of the reciprocal triangle may be taken to be (A, B, C) (A', B', C') (A'', B'', C''); the equations of the lines joining the corresponding angles of the two triangles are therefore

$$(B\gamma\ \ -C\,\beta\ )\,x+(C\,\alpha\ -A\,\gamma\ )\,y+(A\beta\ \ \ -B\,\alpha\ )\,z=0,$$
$$(B'\gamma'\ -C'\beta'\,)\,x+(C'\alpha'-A'\gamma')\,y+(A'\beta'\ -B'\alpha')\,z=0,$$
$$(B''\gamma''-C''\beta'')\,x+(C''\alpha''-A''\gamma'')\,y+(A''\beta''-B''\alpha'')\,z=0;$$

the condition that these lines may meet in a point is therefore

$$\begin{vmatrix} B\,\gamma-C\,\beta\,, & C\,\alpha\,-A\,\gamma\,, & A\,\beta-B\,\alpha \\ B'\gamma'-C'\beta'\,, & C'\alpha'-A'\gamma', & A'\beta'-B'\alpha' \\ B''\gamma''-C''\beta'', & C''\alpha''-A''\gamma'', & A''\beta''-B''\alpha'' \end{vmatrix}=0,$$

an equation which is satisfied identically when A, B, C; A', B', C'; A'', B'', C'' are replaced by their values. To prove this I transform the different quantities which enter into the determinant as follows: putting

$$F=\alpha'\alpha''+\beta'\beta''+\gamma'\gamma'',$$
$$G=\alpha''\alpha\ +\beta''\beta\ +\gamma''\gamma,$$
$$H=\alpha\,\alpha'\ +\beta\,\beta'\ +\gamma\,\gamma';$$

we have

$$\begin{aligned} K\,(B\gamma-C\beta) &= \gamma\ \ (\gamma'\alpha''-\gamma''\alpha')-\beta\,(\alpha\beta''-\alpha''\beta') \\ &= \alpha''\,(\beta\beta'+\gamma\gamma')\ -\alpha'\,(\beta\beta''+\gamma\gamma''\,) \\ &= \alpha''\,(\alpha\alpha'+\beta\beta'+\gamma\gamma')-\alpha'\,(\alpha\alpha''+\beta\beta''+\gamma\gamma'') \\ &= \alpha''H-\alpha'G, \\ &\text{\&c.}; \end{aligned}$$

and the equation becomes

$$\begin{vmatrix} \alpha''H-\alpha'G, & \beta''H-\beta'G, & \gamma''H-\gamma'G \\ \alpha\,F-\alpha''H, & \beta\,F-\beta''H, & \gamma\,F-\gamma''H \\ \alpha'G-\alpha\,F, & \beta'G-\beta\,F, & \gamma'G-\gamma\,F \end{vmatrix}=0.$$

Now the minor $(\beta F-\beta''H)(\gamma'G-\gamma F)-(\gamma F-\gamma''H)(\beta'G-\beta F)$ is equal to

$$GH\,(\beta'\gamma''-\beta''\gamma')+HF\,(\beta''\gamma-\beta\gamma'')+FG\,(\beta\gamma'-\beta'\gamma),$$

i.e. to

$$K\,(GHA+HFA'+FGA'');$$

and expressing the other minors in a similar form, the equation to be proved is

$$\left.\begin{aligned}&(GHA+HFA'+FGA'')(B\gamma-C\beta)\\ &+(GHB+HFB'+FGB'')(C\alpha-A\gamma)\\ &+(GHC+HFC'+FGC'')(A\beta-B\alpha)\end{aligned}\right\}=0,$$

i. e.

$$HF\begin{vmatrix} A', & B', & C' \\ A, & B, & C \\ \alpha, & \beta, & \gamma \end{vmatrix} + FG\begin{vmatrix} A'', & B'', & C'' \\ A, & B, & C \\ \alpha, & \beta, & \gamma \end{vmatrix} = 0.$$

The first determinant is

$$-\{\alpha(BC'-B'C)+\beta(CA'-C'A)+\gamma(AB'-A'B)\}=-\frac{1}{K}(\alpha\alpha''+\beta\beta''+\gamma\gamma'')=-\frac{1}{K}G,$$

and the second determinant is

$$\{\alpha(B''C-BC'')+\beta(C''A-CA'')+\gamma(A''B-AB'')\}=\frac{1}{K}(\alpha\alpha'+\beta\beta'+\gamma\gamma')=\frac{1}{K}H,$$

and we have therefore identically

$$HF(-G)+FG(H)=0.$$

The corresponding theorem in geometry of three dimensions is that a tetrahedron and its reciprocal have to each other a certain relation, viz. the four lines joining the corresponding angles are generating lines of a hyperboloid, or, what is the same thing, the four lines of intersection of corresponding faces are generating lines of a hyperboloid. The demonstration would show how the theorem in determinants is to be generalised.

2, *Stone Buildings, Lincoln's Inn, February*, 1855.

161.

A PROBLEM IN PERMUTATIONS.

[From the *Quarterly Mathematical Journal*, vol. I. (1857), p. 79.]

THE game called Mousetrap gives rise to a singular problem in permutations. A set of cards, ace, two, three, &c., say up to thirteen, are arranged in a circle with their faces upwards—you begin at any card, and count one, two, three, &c., and if upon counting suppose the number five, you arrive at the card five, that card is thrown out; and beginning again with the next card, you count one, two, three, &c., throwing out if the case happen a new card as before, and so on until you have counted up to thirteen, without coming to a card which ought to be thrown out. It is easy to see that, whatever the number of the cards is, they may be so arranged as to be all thrown out in the order of their numbers; but that it is not possible in general to arrange the cards so that all the cards, or any specified cards, may be thrown out in a given order. Thus, if all the cards are to be thrown out in the order of their numbers, the arrangements in the case of a single card, two, three, &c. cards, are

1

1 2

1 3 2

1 4 2 3

1 3 2 5 4

1 4 2 5 6 3

1 5 2 7 4 3 6

1 6 2 4 5 3 7 8

&c.

It is required to investigate the general theory.

162.

TWO LETTERS ON CUBIC FORMS.

[From the *Quarterly Mathematical Journal,* vol. I. (1857), pp. 85—87 and 90—91.]

CHER MONS. HERMITE,

Il y a longtemps que j'ai voulu vous écrire, mais j'en ai été empêché je ne sais comment; j'ai assez à vous dire par rapport aux covariants, mais à présent je vais vous parler des formes cubiques à deux indéterminées. Il me semble que l'on peut simplifier la théorie de Eisenstein, et l'étendre au cas d'un déterminant négatif quelconque, de la manière que voici.

Soit $(a, b, c, d \between x, y)^3$ une forme cubique, je représente par Hessn. $(a, b, c, d \between x, y)^3$ la forme quadratique dérivée $(ac - b^2, \tfrac{1}{2}(ad - bc), bd - c^2 \between x, y)^2$. Cela étant, soit (A, B, C) une forme représentative (réduite et proprement primitive) au déterminant $-D$; à moins que $(A, B, C)^2 = (A, -B, C)$, c'est-à-dire, à moins que (A, B, C) ne soit une forme laquelle par sa triplication produit la forme principale, il n'existe pas de forme cubique (a, b, c, d) telle que $-$ Hessn. $(a, b, c, d \between x, y)^3 = (A, B, C \between x, y)^2$, ou, si l'on veut, telle que $b^2 - ac = A$, $bc - ad = 2B$, $c^2 - bd = C$; mais en supposant que l'on ait $(A, B, C)^2 = (A, -B, C)$ on peut trouver une seule forme cubique qui satisfait à l'équation dont il s'agit. J'écarte, cela va sans dire, l'une ou l'autre des deux formes (a, b, c, d) et $(-a, -b, -c, -d)$.

En effet on a identiquement

$$(b^2 - ac, -\tfrac{1}{2}(bc - ad), c^2 - bd \between bxx' + cxy' + cx'y + dyy', axx' + bxy' + bx'y + cyy')^2$$
$$= (b^2 - ac, \tfrac{1}{2}(bc - ad), c^2 - bd \between x, y)^2 \times (b^2 - ac, \tfrac{1}{2}(bc - ad), c^2 - bd \between x', y')^2,$$

donc, en supposant que $b^2 - ac = A$, $bc - ad = 2B$, $c^2 - bd = C$, il s'ensuit que $(A, B, C)^2 = (A, -B, C)$.

Je suppose donc $(A, B, C)^2 = (A, -B, C)$, et je dis qu'il ne peut pas y avoir deux formes cubiques, (a, b, c, d) et $(a_{,}, b_{,}, c_{,}, d_{,})$, qui aient la propriété dont il s'agit; car, en écrivant

$$\left.\begin{aligned}\xi &= bxx' + cxy' + cx'y + dyy',\\ \eta &= axx' + bxy' + bx'y + cyy',\end{aligned}\right\} \qquad \left.\begin{aligned}\xi_{,} &= b_{,}xx' + c_{,}xy' + c_{,}x'y + d_{,}yy',\\ \eta_{,} &= a_{,}xx' + b_{,}xy' + b_{,}x'y + c_{,}yy',\end{aligned}\right\}$$

on trouverait

$$(A, -B, C \between \xi, \eta)^2 = (A, -B, C \between \xi_{,}, \eta_{,})^2,$$

ce qui implique d'abord que $\xi_{,}, \eta_{,}$ soient des fonctions linéaires de ξ, η. Mais $(A, -B, C)$ étant une forme réduite et proprement primitive au déterminant $-D$, il n'existe pas de transformation de la forme quadratique en elle-même, hormis $\xi_{,} = \xi$, $\eta_{,} = \eta$. Le cas $D = 1$ doit se traiter à part; dans ce cas particulier il n'y a que la forme cubique $(0, 1, 0, 1)$. Donc &c. Enfin, si $(A, B, C)^2 = (A, -B, C)$, il existe une forme cubique (a, b, c, d) telle que $b^2 - ac = A$, &c.; car en cherchant par la méthode de Gauss les valeurs des coefficients p, p', p'', p''' et q, q', q'', q''' qui donnent cette transformation, on obtient d'abord $p' = p''$, $q' = q''$. On peut donc représenter ces coefficients par $b_{,}, c_{,}, c_{,}, d_{,}$; a, b, b, c; savoir, on peut trouver $a, b, c, b_{,}, c_{,}, d_{,}$ de manière que

$$(A, -B, C \between b_{,}xx' + c_{,}xy' + c_{,}x'y + d_{,}yy', \quad axx' + bxy' + bx'y + cyy')^2$$
$$= (A, B, C \between x, y)^2 . (A, B, C \between x', y')^2.$$

Cela étant, les équations de Gauss donnent

$$\begin{aligned} A &= bb_{,} - ac_{,}, & A &= b^2 \ - ac,\\ 2B &= cb_{,} - ad_{,}, & -2B &= cb_{,} + ad_{,} - 2bc_{,},\\ C &= cc_{,} - bd_{,}, & C &= c_{,}^2 \ - b_{,}d_{,}, \end{aligned}$$

et de là on obtient

$$\begin{aligned} b\ (b - b_{,}) - a\ (c - c_{,}) &= 0,\\ c\ (b - b_{,}) - b\ (c - c') &= 0,\\ c_{,}(b - b_{,}) - b_{,}(c - c_{,}) &= 0,\\ d_{,}(b - b_{,}) - c_{,}(c - c_{,}) &= 0; \end{aligned}$$

c'est-à-dire, ou $\dfrac{a}{b} = \dfrac{b}{c} = \dfrac{b_{,}}{c_{,}} = \dfrac{c_{,}}{d_{,}}$, ce qui n'est pas vrai (car cela donnerait $A = 0$, $B = 0$, $C = 0$), ou $b - b_{,} = 0$, $c - c_{,} = 0$. Donc $b_{,} = b$, $c_{,} = c$; et en écrivant d au lieu de $d_{,}$, on voit que l'équation de transformation devient

$$(A, -B, C \between bxx' + cxy' + cx'y + dyy', \quad axx' + bxy' + bx'y + cyy')^2$$
$$= (A, B, C \between x, y)^2 . (A, B, C \between x, y')^2,$$

où

$$A = b^2 - ac, \quad 2B = bc - ad, \quad C = c^2 - bd. \quad \text{C. q. f. à d.}$$

Je ne sais pas si je vous ai mentionné que j'ai calculé les formes quadratiques pour les treize numéros -307, &c. aux déterminants irréguliers. Pour $D=-307$ ces formes sont:

Ordre P. P., 1 genre, 9 classes, c'est-à-dire:

Composition.						
1	δ	δ^2	m	(1, 0, 307),	(7, 1, 44),	(7, −1, 44),
$\delta_{,}$	$\delta\delta_{,}$	$\delta^2\delta_{,}$	$\overline{307}$	(4, 1, 77),	(11, −1, 28),	(17, 4, 19),
$\delta_{,}^2$	$\delta\delta_{,}^2$	$\delta^2\delta_{,}^2$	+	(4, −1, 77),	(17, −4, 19),	(11, 1, 28),

où $\delta^3=1$, $\delta_{,}^3=1$.

Ordre I. P., 1 genre, 3 classes, c'est-à-dire:

$$\sigma,\quad \sigma\delta,\quad \sigma\delta^2 \quad | \; + \; | \; (2,\ 1,\ 154),\quad (14,\ 1,\ 22),\quad (14,\ -1,\ 22).$$

A chaque forme de l'ordre P. P. il corresponde donc une forme et une seule forme cubique au déterminant -1228. Ces formes sont

$$\begin{array}{lll} (0,\ 1,\ \ 0,\ -307), & (1,\ \ 1,\ -6,\ \ 8), & (1,\ -1,\ -6,\ -8), \\ (0,\ 2,\ \ 1,\ -\ \ 38), & (1,\ -3,\ -2,\ \ 8), & (4,\ \ 1,\ -4,\ -3), \\ (0,\ 2,\ -1,\ -\ \ 38), & (4,\ -1,\ -4,\ \ 3), & (1,\ \ 3,\ -2,\ \ 8). \end{array}$$

Je serai bien aise d'avoir de vos nouvelles, et je vous prie de me croire votre très-dévoué

A. Cayley.

Cher Mons. Hermite,

On démontre sans peine la proposition avancée dans ma dernière lettre, savoir qu'en supposant

$$\begin{aligned} \xi &= b\,xx' + c\,xy' + c\,x'y + d\,yy', \\ \eta &= a\,xx' + b\,xy' + b\,x'y + c\,yy', \\ \xi_{,} &= b_{,}xx' + c_{,}xy' + c_{,}x'y + d_{,}yy', \\ \eta_{,} &= a_{,}xx' + b_{,}xy' + c_{,}x'y + d_{,}yy', \end{aligned}$$

$$(A,\ -B,\ C \between \xi,\ \eta)^2 = (A,\ -B,\ C \between \xi,\ \eta)^2,$$

on doit avoir

$$\begin{aligned} \xi_{,} &= \alpha\xi + \beta\eta, \\ \eta_{,} &= \gamma\xi + \delta\eta, \end{aligned}$$

où α, β, γ, δ sont des entiers. En effet en trouve d'abord en éliminant par ex. xy' et $x'y$,

$$\xi_{,} = \alpha\xi + \beta\eta + \lambda xx' + \mu yy',$$
$$\eta_{,} = \gamma\xi + \delta\eta + \nu xx' + \rho yy',$$

où α, β, &c., λ, μ, &c., sont des quantités rationnelles. En substituant ces valeurs de ξ, η, on aura $(A, -B, C)(\lambda, \nu)^2 = 0$, $(A, -B, C)(\mu, \rho)^2 = 0$. Donc le déterminant n'étant pas un carré, $\lambda = 0$, $\nu = 0$, $\mu = 0$, $\rho = 0$ et

$$\xi_{,} = \alpha\xi + \beta\eta,$$
$$\eta_{,} = \gamma\xi + \delta\eta,$$

où α, β, γ, δ sont des quantités rationnelles. Donc

$$b_{,} = \alpha b + \beta a, \quad a_{,} = \gamma b + \delta a,$$
$$c_{,} = \alpha c + \beta b, \quad b_{,} = \gamma c + \delta b,$$
$$d_{,} = \alpha d + \beta c, \quad c_{,} = \gamma d + \delta c,$$

donc α, β seront des quantités rationnelles ayant pour dénominateur l'une quelconque à volonté des trois quantités $b^2 - ac$, $c^2 - bd$, $ad - bc$, c'est-à-dire, des quantités A, B, C; et, puisque (A, B, C) est une forme P. P., cela ne peut arriver à moins que α, β ne soient des entiers; de même, γ, δ seront des entiers. Je remarque de plus les équations

$$a\beta + b(\alpha - \delta) - c\gamma = 0,$$
$$b\beta + c(\alpha - \delta) - d\gamma = 0,$$

lesquelles donnent

$$\begin{aligned} \beta : \alpha - \delta : -\gamma &= bd - c^2 : bc - ad : ac - b^2 \\ &= \quad C \quad : -2B \quad : A, \end{aligned}$$

cela fait voir que la transformation en elle-même de la forme $(A, -B, C)(\xi, \eta)^2$ à moyen de $\xi_{,} = \alpha\xi + \beta\eta$, $\eta_{,} = \gamma\xi + \delta\eta$ est une transformation propre. J'aime cependant mieux la manière dont vous vous êtes servi pour déduire d'une forme cubique donnée toutes les autres formes cubiques qui correspondent à la même forme quadratique. Il serait facile de la même manière, étant donnée une transformation quelconque d'une forme quadratique dans le produit de deux autres formes quadratiques, d'en déduire toutes les autres transformations; car il y a pour la fonction $axx'x'' + \&c\ \ldots$ un covariant qui corresponde au cubicovariant de la fonction $ax^3 + \&c.\ldots$ Je suis votre très-dévoué

A. CAYLEY.

2, Stone Buildings, Lincoln's Inn, 6 Mars, 1855.

163.

ON HANSEN'S LUNAR THEORY.

[From the *Quarterly Mathematical Journal*, vol. I. (1857), pp. 112—125.]

THE following paper was written in order to exhibit, in as clear a form as may be, the investigation of the remarkable equations for the motion of the moon established in Hansen's "Fundamenta Nova Investigationis Orbitæ veræ quam Luna perlustrat," &c., Gothæ, 1838. I have availed myself for this purpose of the remarks in Jacobi's two letters in answer to a letter of Hansen's, *Crelle*, t. XLII. [1851], p. 12; it may be convenient to remark that the quantity there represented by Λ, and which does not occur in Hansen's own investigation, is in this paper represented by Θ.

The position of the moon referred to the earth as centre is determined by

r, the radius vector,
L, the longitude,
Λ, the latitude.

Suppose, moreover, that the attractive force at distance unity, $=\kappa(M+E)$, is represented by n^2a^3, then the principal function will be $V=-\frac{n^2a^3}{r}$, and the disturbing function R may be represented by $n^2a^3\Omega$; the expression for the half of the vis viva is

$$T=\tfrac{1}{2}\left\{\left(\frac{dr}{dt}\right)^2+r^2\cos^2\Lambda\left(\frac{dL}{dt}\right)^2+r^2\left(\frac{d\Lambda}{dt}\right)^2\right\},$$

and the equations of motion are therefore

$$\frac{d}{dt}\frac{dr}{dt}-r\cos^2\Lambda\left(\frac{dL}{dt}\right)^2-r\left(\frac{d\Lambda}{dt}\right)^2+\frac{n^2a^3}{r^2}=n^2a^3\frac{d\Omega}{dr},$$

$$\frac{d}{dt}\left(r^2\cos^2\Lambda\frac{dL}{dt}\right)=n^2a^3\frac{d\Omega}{dL},$$

$$\frac{d}{dt}\left(r^2\frac{d\Lambda}{dt}\right)+r^2\cos\Lambda\sin\Lambda\left(\frac{dL}{dt}\right)^2=n^2a^3\frac{d\Omega}{d\Lambda},$$

where Ω is considered as a function of r, L, Λ.

Consider the orbit as an ellipse, then putting

a, the mean distance,

n, the mean motion $= \sqrt{\dfrac{\kappa(M+E)}{a^3}}$,

e, the excentricity,

c, the mean anomaly at epoch,

ω, the distance of perigee from node,

θ, the longitude of node,

i, the inclination,

Φ, the distance from node,

Ψ, the reduced distance from node, $= L - \theta$,

U, the excentric anomaly,

f, the true anomaly, $= \Phi - \omega$,

the elements of the orbit are a, e, c, ω, θ, i, and we have from the theory of elliptic motion,

$$nt + c = U - e\sin U,$$

$$f = \tan^{-1}\frac{\sqrt{(1-e^2)}\sin U}{\cos U - e},$$

$$r = a(1 - e\cos U) = \frac{a(1-e^2)}{1 + e\cos f}.$$

Moreover i is the angle at the base of a right-angled spherical triangle, the base, perpendicular, and hypothenuse of which are Ψ, Λ, Φ, hence

$$\tan\Psi = \cos i\tan\Phi,$$

$$\sin\Lambda = \sin i\sin\Phi,$$

$$\sin\Psi = \cot i\tan\Lambda,$$

$$\cos\Phi = \cos\Lambda\cos\Psi.$$

Considering the elements as constant, we have

$$\frac{dr}{dt} = \frac{nae\sin f}{\sqrt{(1-e^2)}},$$

$$\frac{df}{dt} = \frac{na^2\sqrt{(1-e^2)}}{r^2},$$

$$\frac{d\Phi}{dt} = \frac{na^2\sqrt{(1-e^2)}}{r^2},$$

$$\frac{d\Psi}{dt} = \frac{\cos i}{\cos^2\Lambda}\,\frac{na^2\sqrt{(1-e^2)}}{r^2},$$

$$\frac{dL}{dt} = \frac{\cos i}{\cos^2\Lambda}\,\frac{na^2\sqrt{(1-e^2)}}{r^2},$$

$$\frac{d\Lambda}{dt} = \sin i\cos\Psi\,\frac{na^2\sqrt{(1-e^2)}}{r^2}.$$

Hence also

$$\frac{d}{dt}\left(\frac{dr}{dt}\right) = \frac{n^2a^3e\cos f}{r^2},$$

$$\frac{d}{dt}\left(r^2\cos^2\Lambda\frac{dL}{dt}\right) = 0,$$

$$\frac{d}{dt}\left(r^2\frac{d\Lambda}{dt}\right) = -\frac{\cos^2 i\sin\Lambda}{\cos^3\Lambda}\frac{n^2a^4(1-e^2)}{r^2},$$

$$\frac{d}{dt}\left(\frac{df}{dt}\right) = -\frac{2n^2a^3}{r^3}e\sin f.$$

The foregoing values show that the equations of motion, neglecting the terms which involve the disturbing functions, are satisfied by the elliptic values of r, L, Λ: and in order to satisfy the actual equations of motion, we have only to consider the elements as variable and to write

$$dr = 0,$$

$$dL = 0,$$

$$d\Lambda = 0,$$

$$d\frac{dr}{dt} = n^2a^3\frac{d\Omega}{dr}dt,$$

$$d\left(r^2\cos^2\Lambda\frac{dL}{dt}\right) = n^2a^3\frac{d\Omega}{dL}dt,$$

$$d\left(r^2\frac{d\Lambda}{dt}\right) = n^2a^3\frac{d\Omega}{d\Lambda}dt,$$

where the differentiations relate only to the elements, or, what is the same thing, to t in so far only as it enters through the variable elements: the system is at once transformed into

$$dr = 0,$$

$$dL = 0,$$

$$d\Lambda = 0,$$

$$d\frac{nae\sin f}{\sqrt{(1-e^2)}} = n^2a^3\frac{d\Omega}{dr}dt,$$

$$d\,na^2\sqrt{(1-e^2)}\cos i = n^2a^3\frac{d\Omega}{dL}dt,$$

$$d\,na^2\sqrt{(1-e^2)}\sin i\cos\Psi = n^2a^3\frac{d\Omega}{d\Lambda}dt.$$

Now $\Psi = L - \theta$, or (supposing, as before, that the differentiations relate to t, only in so far as it enters through the variable elements) $d\Psi = -d\theta$, and thence $d\theta = \frac{\tan\Phi}{\sin i} di$; we have also $d\Phi = -\cos i\, d\theta$. The equations containing $\frac{d\Omega}{dL}$ and $\frac{d\Omega}{d\Lambda}$ give

$$\cos i\, d\, na^2\sqrt{(1-e^2)} - na^2\sqrt{(1-e^2)}\sin i\, di = n^2a^3\frac{d\Omega}{dL}dt,$$

$$\cos\Psi\sin i\, d\, na^2\sqrt{(1-e^2)} + na^2\sqrt{(1-e^2)}(\cos\Psi\cos i\, di + \sin\Psi\sin i\, d\theta) = n^2a^3\frac{d\Omega}{d\Lambda}dt;$$

or, expressing $d\theta$ by means of di and reducing, the second of these equations becomes

$$\sin i\, d\, na^2\sqrt{(1-e^2)} + na^2\sqrt{(1-e^2)}\frac{\cos i}{\cos^2\Lambda\cos^2\Psi}di = \frac{1}{\cos\Psi}n^2a^3\frac{d\Omega}{d\Lambda}dt,$$

and combining this with the first of the two equations, and observing that

$$\frac{\cos^2 i}{\cos^2\Lambda\cos^2\Psi} + \sin^2 i = \frac{1}{\cos^2\Psi},$$

we find

$$d\, na^2\sqrt{(1-e^2)} = n^2a^3\left(\frac{\cos i}{\cos^2\Lambda}\frac{d\Omega}{dL} + \sin i\cos\Psi\frac{d\Omega}{d\Lambda}\right)dt,$$

$$di = \frac{na^2}{\sqrt{(1-e^2)}}\left(-\sin i\cos^2\Psi\frac{d\Omega}{dL} + \cos i\cos\Psi\frac{d\Omega}{d\Lambda}\right)dt.$$

Now, considering Ω as a function of r, θ, i, Φ, then Λ, L are given as functions of θ, i, Φ by the equations $\sin\Lambda = \sin i\sin\Phi$, $\tan\Psi = \cos i\tan\Phi$, $\Psi = L - \theta$, and after some simple reductions,

$$\frac{d\Omega}{dr} = \frac{d\Omega}{dr},$$

$$\frac{d\Omega}{d\theta} = \frac{d\Omega}{dL},$$

$$\frac{d\Omega}{di} = \tan\Phi\left(-\sin i\cos^2\Psi\frac{d\Omega}{dL} + \cos i\cos\Psi\frac{d\Omega}{d\Lambda}\right),$$

$$\frac{d\Omega}{d\Phi} = \left(\frac{\cos i}{\cos^2\Lambda}\frac{d\Omega}{dL} + \sin i\cos\Psi\frac{d\Omega}{d\Lambda}\right);$$

whence also

$$\frac{d\Omega}{d\theta} = \cos i\frac{d\Omega}{d\Phi} - \sin i\cot\Phi\frac{d\Omega}{di}.$$

We have therefore

$$
\begin{aligned}
d\, na^2\sqrt{(1-e^2)} &= n^2a^3\frac{d\Omega}{d\Phi}dt,\\
di &= \frac{na\cot\Phi}{\sqrt{(1-e^2)}}\frac{d\Omega}{di}dt,\\
d\theta &= \frac{na}{\sqrt{(1-e^2)}\sin i}\frac{d\Omega}{di}dt,\\
d\Phi &= \frac{-na\cot i}{\sqrt{(1-e^2)}}\frac{d\Omega}{di}dt,\\
dr &= 0,\\
d\frac{nae\sin f}{\sqrt{(1-e^2)}} &= n^2a^3\frac{d\Omega}{dr}dt.
\end{aligned}
$$

Suppose now that we have

ρ, a radius vector, τ for t,

ϕ, a true anomaly, do.,

v, an excentric anomaly, do.,

i.e. let ρ, ϕ, v, be what the radius vector, the true anomaly and the excentric anomaly become when the time t, in so far as it enters directly, and not through the variable elements, is replaced by a new variable τ. We have

$$
\begin{aligned}
n\tau + c &= v - e\sin v,\\
\phi &= \tan^{-1}\frac{\sqrt{(1-e^2)}\sin v}{\cos v - e},\\
\rho &= a(1-e\cos v) = \frac{a(1-e^2)}{1+e\cos\phi};
\end{aligned}
$$

and of course the differential coefficients of ρ, ϕ with respect to τ may be at once deduced from the corresponding expressions for the differential coefficients of r, f with respect to t, the elements being considered as constant. Now, using l to denote a logarithm, and supposing that the differentiations affect only the elements, we have

$$
\begin{aligned}
dl\rho &= \frac{da}{a} - \frac{2ede}{1-e^2} - \frac{\rho\cos\phi de}{a(1-e^2)} + \frac{\rho e\sin\phi d\phi}{a(1-e^2)},\\
dlr &= \frac{da}{a} - \frac{2ede}{1-e^2} - \frac{r\sin f de}{a(1-e^2)} + \frac{re\sin f df}{a(1-e^2)};
\end{aligned}
$$

and putting for shortness

$$
\begin{aligned}
X_, &= dl\rho - \frac{\rho e\sin\phi d\phi}{a(1-e^2)},\\
X &= dlr - \frac{re\sin f df}{a(1-e^2)},
\end{aligned}
$$

we find

$$
X_, - \frac{\rho\sin\phi}{r\sin f}X = \left(1-\frac{\rho\sin\phi}{r\sin f}\right)\left(\frac{da}{a} - \frac{2ede}{1-e^2}\right) - \frac{\rho}{a(1-e^2)}\left(\cos\phi - \frac{\sin\phi\cos f}{\sin f}\right)de,
$$

or reducing

$$X_{,} - \frac{\rho \sin \phi}{r \sin f} X =$$

$$\frac{1}{\sin f(1 + e\cos\phi)}\left\{\left(\sin f - \sin\phi + e\sin(f-\phi)\right)\left(\frac{da}{a} - \frac{2ede}{1-e^2}\right) - \sin(f-\phi)\,de\right\}.$$

Write for a moment

$$P = na^2\sqrt{(1-e^2)},\quad Q = \frac{nae\sin f}{\sqrt{(1-e^2)}},\quad R = \frac{a(1-e^2)}{1+e\cos f},$$

so that

$$a(1-e^2) = \frac{P^2}{n^2a^3},$$

$$e^2 = \frac{1}{n^4a^6}P^2Q^2 + \frac{1}{n^4a^6}\frac{P^4}{R^2} - \frac{2}{n^2a^3}\frac{P^2}{R} + 1.$$

We have therefore

$$\frac{da}{a} - \frac{2ede}{1-e^2} = \frac{2dP}{P} = \frac{2}{na^2(1-e^2)}dP,$$

$$ede = \left(\frac{1}{n^4a^6}PQ^2 - \frac{2}{n^4a^6}\frac{P^3}{R^2} - \frac{2}{n^2a^3}\frac{P}{R}\right)dP + \frac{1}{n^4a^6}P^2QdQ + \left(-\frac{1}{n^4a^6}\frac{P^4}{R^3} + \frac{2}{n^2a^3}\frac{P^2}{R^2}\right)dR,$$

which, after reduction, becomes

$$de = \frac{1}{na^2(1-e^2)}\left(e\sin^2 f + 2(\cos f + e\cos^2 f)\right)dP + \frac{1}{na}\sqrt{(1-e^2)}\sin f\,dQ - \frac{(1+e\cos f)^2\cos f}{a(1-e^2)}dR,$$

and substituting these values,

$$X_{,} - \frac{\rho\sin\phi}{r\sin f}X = \frac{1}{1+e\cos\phi}\left\{\left(2 - 2\cos(f-\phi) + e\sin f\sin(f-\phi)\right)\frac{1}{na^2\sqrt{(1-e^2)}}dP\right.$$

$$- a(1-e^2)\sin(f-\phi)\ \frac{1}{na^2\sqrt{(1-e^2)}}dQ$$

$$\left. + \frac{(1+e\cos f)^2\cot f\sin(f-\phi)}{\sqrt{(1-e^2)}}\ \frac{1}{na^2\sqrt{(1-e^2)}}dR\right\},$$

or substituting for X, $X_{,}$, P, Q, R, their values

$$dl\rho - \frac{\rho\sin\phi}{r\sin f}dlr + \frac{\rho e\sin\phi}{a(1-e^2)}(df - d\phi) =$$

$$\frac{\rho}{a(1-e^2)}\left(2 - 2\cos(f-\phi) + e\sin f\sin(f-\phi)\right)\frac{1}{na^2\sqrt{(1-e^2)}}\,dna^2\sqrt{(1-e^2)}$$

$$- \rho\sin(f-\phi)\frac{1}{na^2\sqrt{(1-e^2)}}\,d\,\frac{nae\sin f}{\sqrt{(1-e^2)}}$$

$$+ \frac{\rho a\sqrt{(1-e^2)}}{r^2}\cot f\sin(f-\phi)\frac{1}{na^2\sqrt{(1-e^2)}}dr.$$

Now

$$-\left\{\frac{\rho}{r}\cos(f-\phi)-1+\frac{\rho}{a(1-e^2)}\left(\cos(f-\phi)-1\right)\right\}$$

$$=\frac{\rho}{a(1-e^2)}\left(2-2\cos(f-\phi)+e\sin f\sin(f-\phi)\right);$$

therefore

$$dl\rho-\frac{\rho\sin\phi}{r\sin f}dlr+\frac{\rho e\sin\phi}{a(1-e^2)}(df-d\phi)=$$

$$-\left\{\frac{\rho}{r}\cos(f-\phi)-1+\frac{\rho}{a(1-e^2)}\left(\cos(f-\phi)-1\right)\right\}\frac{1}{na^2\sqrt{(1-e^2)}}\,dna^2\sqrt{(1-e^2)}$$

$$-\rho\sin(f-\phi)\frac{1}{na^2\sqrt{(1-e^2)}}\,d\,\frac{nae\sin f}{\sqrt{(1-e^2)}}$$

$$+\frac{\rho a\sqrt{(1-e^2)}}{r^2}\cot f\sin(f-\phi)\frac{1}{na^2\sqrt{(1-e^2)}}\,dr.$$

So far the variations of the elements have, in fact, been treated as independent; but if we substitute for $dna^2\sqrt{(1-e^2)}$, $d\dfrac{nae\sin f}{\sqrt{(1-e^2)}}$, dr, their values in the disturbed motion, the equation becomes

$$dl\rho+\frac{\rho e\sin\phi}{a(1-e^2)}(df-d\phi)=-\left\{\frac{\rho}{r}\cos(f-\phi)-1+\frac{\rho}{a(1-e^2)}\left(\cos(f-\phi)-1\right)\right\}\frac{na}{\sqrt{(1-e^2)}}\frac{d\Omega}{d\Phi}dt$$

$$-\rho\sin(f-\phi)\frac{na}{\sqrt{(1-e^2)}}\frac{d\Omega}{dr}dt.$$

Consider now the point in which the orbit is intersected by any orthogonal trajectory to the successive positions of the orbit, or to fix the ideas, the orthogonal trajectory passing through ♈, the point in question may, for want of a recognised name, be called the "departure point;" and the angular distances in the orbit measured from this point may be termed "departures;" the expression, "the departure," is to be understood as meaning the departure of the moon. Write now

χ, the departure of the perigee,

$v_{,}$, the departure, $=f+\chi$,

σ, the departure of the node, $=\chi-\omega$,

Θ, the longitude in orbit of departure point, $=\theta-\sigma$.

It should be remarked that χ is not properly an element, i.e. it is not a function of a, e, c, ϖ, θ, i without t, and in like manner σ and Θ (which depend upon χ) are not elements.

We have from the geometrical definition

$$d\chi=d\omega+\cos i\,d\theta;$$

and therefore

$$d\sigma = \cos i \, d\theta,$$
$$d\Theta = (1 - \cos i) \, d\theta.$$

Moreover $v_, = \Phi + \sigma$, which gives (assuming that the differentiations are performed with respect to t, only in so far as it enters through the variable elements) $dv_, = d\Phi + d\sigma = d\Phi + \cos i \, d\theta$, i.e. $dv_, = 0$, an equation which might have been assumed for the purpose of defining the departure point; the equation, in fact, expresses that the departure $v_,$ is measured from a point not actually fixed, but such that the increment of $v_,$ in the interval of time dt is the angular distance between two consecutive positions of the moon.

We have, as above noticed, $d\sigma = \cos i \, d\theta$, and thence and from what has preceded

$$di = \frac{na \cot \Phi}{\sqrt{(1 - e^2)}} \frac{d\Omega}{di},$$

$$d\sigma = \frac{na \cot i}{\sqrt{(1 - e^2)}} \frac{d\Omega}{di}.$$

Now the position of the moon can be determined by means of the quantities r, $v_,$, Θ, σ, i; hence Ω (which has been considered as a function of r, Φ, θ, i) may, if we please, be considered as a function of r, $v_,$, Θ, σ, i and from the differential relations

$$dr = dr,$$
$$dv_, = d\Phi + \cos i \, d\theta,$$
$$d\Theta = (1 - \cos i) \, d\theta,$$
$$d\sigma = d\omega + \cos i \, d\theta,$$
$$di = di,$$

we find

$$\frac{d\Omega}{dr} = \frac{d\Omega}{dr},$$

$$\frac{d\Omega}{d\Phi} = \frac{d\Omega}{dv_,},$$

$$\frac{d\Omega}{d\theta} = \cos i \left(\frac{d\Omega}{dv_,} + \frac{d\Omega}{d\sigma}\right) + \left(1 - \cos i\right) \frac{d\Omega}{d\Theta},$$

$$\frac{d\Omega}{di} = \frac{d\Omega}{di};$$

we have therefore

$$\frac{d\Omega}{d\sigma} + \frac{1 - \cos i}{\cos i} \frac{d\Omega}{d\Theta} = -\frac{d\Omega}{d\Phi} + \frac{1}{\cos i} \frac{d\Omega}{d\theta},$$

or in virtue of a preceding equation

$$\frac{d\Omega}{d\sigma} + \frac{1 - \cos i}{\cos i} \frac{d\Omega}{d\Theta} = - \tan i \cot \Phi \frac{d\Omega}{di};$$

and effecting the substitutions, and collecting the results,

$$d\ na^2\sqrt{(1-e^2)} = n^2a^3\frac{d\Omega}{dv_{,}}\,dt,$$

$$dr = 0,$$

$$d\ \frac{nae\sin f}{\sqrt{(1-e^2)}} = n^2a^3\frac{d\Omega}{dr}\,dt,$$

$$di = -\frac{na\cot i}{\sqrt{(1-e^2)}}\left(\frac{d\Omega}{d\sigma} + \frac{1-\cos i}{\cos i}\frac{d\Omega}{d\Theta}\right)dt,$$

$$d\sigma = \frac{na\cot i}{\sqrt{(1-e^2)}}\frac{d\Omega}{di}\,dt,$$

where Ω is considered as a function of r, $v_{,}$, Θ, σ, i.

Instead of σ, i we may introduce the new quantities p, q defined by the equations

$$p = \sin i \sin \sigma,$$

$$q = \sin i \cos \sigma;$$

this gives $\sin^2 i = p^2 + q^2$, $\sigma = \tan^{-1}\frac{p}{q}$ and retaining in the formulæ the sine and cosine of i, to avoid the introduction of irrational functions of p^2+q^2, we have

$$d\Theta = (1-\cos i)\,d\theta = \frac{1-\cos i}{\cos i}\,d\sigma = \frac{1-\cos i}{\cos i \sin^2 i}.(qdp - pdq),$$

i.e.

$$d\Theta = \frac{qdp - pdq}{\cos i\,(1+\cos i)},$$

which determines Θ by means of p and q. We have moreover

$$dp = \sin i \cos \sigma\, d\sigma + \cos i \sin \sigma\, di,$$

$$dq = -\sin i \sin \sigma\, d\sigma + \cos i \cos \sigma\, di,$$

$$\frac{d\Omega}{dp} = \frac{\sin\sigma}{\cos i}\frac{d\Omega}{di} + \frac{\cos\sigma}{\sin i}\frac{d\Omega}{d\sigma},$$

$$\frac{d\Omega}{dq} = \frac{\cos\sigma}{\cos i}\frac{d\Omega}{di} - \frac{\sin\sigma}{\sin i}\frac{d\Omega}{d\sigma},$$

from which equations and the foregoing values of di and $d\sigma$ we find the values of dp and dq; the other equations of the system remain unaltered, and we have therefore

$$d\ na^2\sqrt{(1-e^2)} = n^2a^3\frac{d\Omega}{dv_{,}}\,dt,$$

$$dr = 0,$$

$$d\ \frac{nae\sin f}{\sqrt{(1-e^2)}} = n^2a^3\frac{d\Omega}{dr}\,dt,$$

$$dp = \frac{na\cos^2 i}{\sqrt{(1-e^2)}}\left(\frac{d\Omega}{dq} - \frac{p}{\cos i\,(1+\cos i)}\,\frac{d\Omega}{d\Theta}\right)dt,$$

$$dq = -\frac{na\cos^2 i}{\sqrt{(1-e^2)}}\left(\frac{d\Omega}{dp} + \frac{q}{\cos i\,(1+\cos i)}\,\frac{d\Omega}{d\Theta}\right)dt,$$

where Ω is considered as a function of r, $v_{,}$, Θ, p, q. The symbols $\frac{d\Omega}{dp}$, $\frac{d\Theta}{dq}$, as employed by Hansen, mean that the differentiations are to be performed as if Θ was a function of p, q, such that

$$d\Theta = \frac{d\Theta}{dp}\,dp + \frac{d\Theta}{dq}\,dq = \frac{qdp - pdq}{\cos i\,(1+\cos i)};$$

the last two equations being therefore nothing else than what Hansen represents by

$$dp = \frac{na\cos^2 i}{\sqrt{(1-e^2)}}\,\frac{d\Omega}{dq}\,dt,$$

$$dq = \frac{na\cos^2 i}{\sqrt{(1-e^2)}}\,\frac{d\Omega}{dp}\,dt.$$

Write now

λ, the departure, τ for t,

i.e. λ is what $v_{,}$ becomes when t, in so far as it enters explicitly, and not through the variable elements, is replaced by the new variable τ; so that, in fact $\lambda = \phi + \chi$. The values of r, $v_{,}$ could be at once found from those of ρ, λ by changing τ into t; and to determine the values of ρ, λ, Hansen proceeds as follows:

writing

$$\lambda = \Pi\,(\zeta,\ t),$$

$$l\rho = \Gamma\,(\zeta,\ t) + \beta,$$

where ζ and β are new variables functions of τ and t, and Π, Γ are arbitrary functional symbols; so that if z, w are what ζ, β become when τ is changed into t, we should have

$$v_{,} = \Pi\,(z,\ t),$$

$$lr = \Gamma\,(z,\ t) + w;$$

then the foregoing equations give

$$\frac{d\lambda}{d\tau} = \Pi'\,(\zeta,\ t)\,\frac{d\zeta}{d\tau},$$

$$\frac{d\lambda}{dt} = \Pi'\,(\zeta,\ t)\,\frac{d\zeta}{dt} + \Pi_{,}\,(\zeta,\ t),$$

$$\frac{dl\rho}{d\tau} = \Gamma'\,(\zeta,\ t)\,\frac{dl\rho}{d\tau} + \frac{d\beta}{d\tau},$$

$$\frac{dl\rho}{dt} = \Gamma'\,(\zeta,\ t)\,\frac{dl\rho}{dt} + \Gamma_{,}\,(\zeta,\ t) + \frac{d\beta}{dt};$$

[where the accents and strokes denote differentiation in regard to ζ, t respectively].

Hence eliminating $\Pi'(\zeta, t)$ and $\Gamma'(\zeta, t)$ we have

$$\frac{d\lambda}{d\tau}\frac{d\zeta}{dt} - \frac{d\lambda}{dt}\frac{d\zeta}{d\tau} = -\Pi_{,}(\zeta, t)\frac{d\zeta}{d\tau},$$

$$\frac{dl\rho}{d\tau}\frac{d\zeta}{dt} - \frac{dl\rho}{dt}\frac{d\zeta}{d\tau} = \frac{d\beta}{d\tau}\frac{d\zeta}{dt} - \frac{d\beta}{dt}\frac{d\zeta}{d\tau} - \Gamma_{,}(\zeta, t)\frac{d\zeta}{dt},$$

or, what is the same thing,

$$\frac{\frac{d\zeta}{dt}}{\frac{d\zeta}{d\tau}} = \frac{\frac{d\lambda}{dt}}{\frac{d\lambda}{d\tau}} - \Pi_{,}(\zeta, t)\frac{1}{\frac{d\lambda}{d\tau}},$$

$$\frac{d\beta}{dt} - \frac{d\beta}{d\tau}\frac{\frac{d\zeta}{dt}}{\frac{d\zeta}{d\tau}} = \frac{dl\rho}{dt} - \frac{\frac{d\lambda}{dt}}{\frac{d\lambda}{d\tau}}\frac{dl\rho}{d\tau} + \Pi_{,}(\zeta, t)\frac{\frac{dl\rho}{d\tau}}{\frac{d\lambda}{d\tau}} - \Gamma_{,}(\zeta, t),$$

or writing

$$T = \frac{d}{d\tau}\frac{\frac{d\zeta}{dt}}{\frac{d\zeta}{d\tau}}, \quad R = \frac{d\beta}{dt} - \frac{d\beta}{d\tau}\frac{\frac{d\zeta}{dt}}{\frac{d\zeta}{d\tau}},$$

we have

$$T = \frac{1}{\frac{d\lambda}{d\tau}}\frac{d^2\lambda}{dtd\tau} - \frac{\frac{d^2\lambda}{d\tau^2}}{\left(\frac{d\lambda}{d\tau}\right)^2}\frac{d\lambda}{dt} + \Pi_{,}(\zeta, t)\frac{\frac{d^2\lambda}{d\tau^2}}{\left(\frac{d}{d\tau}\right)^2} - \Pi_{,}(\zeta, t)\frac{\frac{d\zeta}{d\tau}}{\frac{d\lambda}{d\tau}},$$

$$R = \frac{dl\rho}{dt} - \frac{\frac{dl\rho}{d\tau}}{\frac{d\lambda}{d\tau}}\frac{d\lambda}{d\tau} + \Pi_{,}(\zeta, t)\frac{\frac{dl\rho}{d\tau}}{\frac{d\lambda}{d\tau}} - \Gamma_{,}(\zeta, t).$$

Now from the equation $\lambda = \phi + \chi$ where χ is independent of τ, the differential coefficients of λ and ρ with respect to τ are at once deduced from those of f, r with respect to t, and we have,

$$\frac{d^2\lambda}{d\tau^2} = -\frac{2n^2a^3}{\rho^3}e\sin\phi,$$

$$\frac{d\lambda}{d\tau} = \frac{na^2\sqrt{(1-e^2)}}{\rho^2},$$

$$\frac{dl\rho}{d\tau} = \frac{nae\sin\phi}{\rho\sqrt{(1-e^2)}}.$$

Consequently

$$T = \frac{\rho^2}{na^2\sqrt{(1-e^2)}} \frac{d}{dt} \frac{na^2\sqrt{(1-e^2)}}{\rho^2} + \frac{2\rho e \sin\phi}{a(1-e^2)} \frac{d\lambda}{dt}$$

$$- \frac{2\rho e \sin\phi}{a(1-e^2)} \Pi_{,}(\zeta,\ t) - \frac{\rho^2}{na^2\sqrt{(1-e^2)}} \Pi'_{,}(\zeta,\ t) \frac{d\zeta}{dt},$$

$$= -2\frac{dl\rho}{dt} + \frac{2\rho e\sin\phi}{a(1-e^2)}\frac{d\lambda}{dt} + \frac{1}{na^2\sqrt{(1-e^2)}}\frac{d}{dt}na^2\sqrt{(1-e^2)}$$

$$- \frac{2\rho e \sin\phi}{a(1-e^2)} \Pi_{,}(\zeta,\ t) - \frac{\rho^2}{na^2\sqrt{(1-e^2)}} \Pi'_{,}(\zeta,\ t) \frac{d\zeta}{dt},$$

$$R = \frac{dl\rho}{dt} - \frac{\rho e \sin\phi}{a(1-e^2)}\frac{d\lambda}{dt} + \frac{\rho e\sin\phi}{a(1-e^2)}\Pi_{,}(\zeta,\ t) - \Gamma_{,}(\zeta,\ t),$$

and substituting in these equations the values of

$$\frac{dl\rho}{dt} - \frac{\rho e\sin\phi}{a(1-e^2)}\frac{d\lambda}{dt} \text{ and } \frac{d}{dt}na^2\sqrt{(1-e^2)}$$

we find

$$T = \left\{2\frac{\rho}{r}\cos(v_{,}-\lambda) - 1 + \frac{2\rho}{a(1-e^2)}\left(\cos(v_{,}-\lambda)-1\right)\right\}\frac{na}{\sqrt{(1-e^2)}}\frac{d\Omega}{dv_{,}} + 2\frac{\rho}{r}\sin(v_{,}-\lambda)\frac{na}{\sqrt{(1-e^2)}}r\frac{d\Omega}{dr}$$

$$-2\Pi_{,}(\zeta,\ t)\frac{\rho e\sin\phi}{a(1-e^2)} - \Pi'_{,}(\zeta,\ t)\frac{\rho^2}{na^2\sqrt{(1-e^2)}}\frac{d\zeta}{dt}$$

$$R = -\left\{\frac{\rho}{r}\cos(v_{,}-\lambda) - 1 + \frac{\rho}{a(1-e^2)}\left(\cos(v_{,}-\lambda)-1\right)\right\}\frac{na}{\sqrt{(1-e^2)}}\frac{d\Omega}{dv_{,}} - \frac{\rho}{r}\sin(v_{,}-\lambda)\frac{na}{\sqrt{(1-e^2)}}r\frac{d\Omega}{dr}$$

$$+\Pi_{,}(\zeta,\ t)\frac{\rho e\sin\phi}{a(1-e^2)} - \Gamma_{,}(\zeta,\ t),$$

which are Hansen's values, except that $\frac{d\lambda}{d\tau}$ in the coefficient of $\Pi'_{,}(\zeta,\ t)$ has been replaced by its value.

2, *Stone Buildings*, 31*st March*, 1855.

164.

ON GAUSS' METHOD FOR THE ATTRACTION OF ELLIPSOIDS.

[From the *Quarterly Mathematical Journal*, vol. I. (1857), pp. 162—166.]

THE following is the method employed in Gauss' Memoir "Theoria Attractionis Corporum Sphæroidicorum ellipticorum homogeneorum methodo novo tractata," 1813. *Comm. Gott. recent.*, t. II. [and *Werke* t. VI. pp. 1—22]. I have somewhat developed the geometrical considerations upon which the method depends.

The attraction of the ellipsoid is found by means of the following theorems, which apply generally to the case of a homogeneous solid bounded by a closed surface:— M denotes the attracted point, P a point of the surface, PQ is the normal (lying outside the surface) at the point P, dS is the element of the surface at this point, MQ, QX, and MX denote angles at the point P, viz. MQ the $\angle MPQ$, and QX and MX the inclinations of QP and MP respectively to a line PX drawn in a direction assumed as that of the axis of X, MP denotes the distance between the points M and P. And X is the attraction in the direction opposite to that of the axis of x; the integrations extend over the entire surface.

THEOREM. The integral

$$\iint \frac{dS \cos MQ}{\overline{MP}^2}$$

has for its value

$$0, \quad -2\pi, \quad \text{or} \quad -4\pi,$$

according as M is exterior to, upon, or interior to, the surface.

This is obviously a purely geometrical theorem.

THEOREM. The attraction is given by the formula

$$X = \iint \frac{dS \cos QX}{\overline{MP}}.$$

THEOREM. The attraction is also given by the formula

$$X = -\iint \frac{dS \cos MQ \cos MX}{MP}.$$

Consider now an ellipsoid, the semi-axes of which are A, B, C, and putting a, b, c for the coordinates of the attracted point M, and x, y, z for the coordinates of P, assume

$$x = A\xi,$$
$$y = B\eta,$$
$$z = C\zeta,$$

so that ξ, η, ζ are the coordinates of a point P' on a sphere, radius unity, corresponding in a definite manner to the point P on the ellipsoid; and let $d\sigma$ be the corresponding element of the spherical surface, we have

$$dS = \frac{ABC}{p}\, d\sigma,$$

where p denotes the perpendicular let fall from the centre of the ellipsoid upon the tangent plane at P.

Moreover,

$$\cos QX = \frac{p\xi}{A}, \quad \cos MX = \frac{a-x}{MP} = \frac{a - A\xi}{MP};$$

and therefore

$$dS \cos QX = \frac{BC\xi d\sigma}{MP}.$$

The second theorem gives therefore

$$A\,\frac{X}{ABC} = \iint \frac{\xi d\sigma}{MP},$$

where the integration is extended over the surface of the sphere, and the third theorem gives

$$\frac{X}{ABC} = -\iint \frac{(a - A\xi)\cos MQ dS}{MP^2},$$

where the integration is extended over the surface of the ellipsoid.

Suppose now a confocal ellipsoid, the semi-axes of which are $A + \delta A$, $B + \delta B$, $C + \delta C$, and let $P_{,}$ be the point on this new ellipsoid which *corresponds* to the point P on the original ellipsoid, i.e. let $P_{,}$ be the point whose coordinates are $(A + \delta A)\,\xi$, $(B + \delta B)\,\eta$, $(C + \delta C)\,\zeta$; the decrement of MP will be equal to the normal distance δN between the two ellipsoids at the point P, multiplied into the cosine of the angle MQ, and we have, by a property of confocal ellipsoids, $A\delta A = B\delta B = C\delta C = p\delta N$; we have therefore

$$\delta\,\overline{MP} = -\frac{A\delta A \cos MQ}{p},$$

which gives

$$d\sigma \,.\, \delta \overline{MP} = \frac{A\delta A}{ABC} dS \cos MQ.$$

Now from the equation

$$A \frac{X}{ABC} = \iint \frac{\xi d\sigma}{\overline{MP}},$$

we find

$$A\delta \frac{X}{ABC} + \frac{X}{ABC} \delta A = \iint \frac{\xi d\sigma \delta \overline{MP}}{\overline{MP}^2}$$

$$= \frac{\delta A}{ABC} \iint \frac{A\xi \cos MQdS}{\overline{MP}^2}.$$

But

$$-\frac{X}{ABC} \delta A = \frac{\delta A}{ABC} \iint \frac{(a - A\xi) \cos MQdS}{\overline{MP}^2},$$

and consequently

$$A\delta \,.\, \frac{X}{ABC} = \frac{a\delta A}{ABC} \iint \frac{\cos MQdS}{\overline{MP}^2}.$$

Hence, by the first theorem:

In the case of an exterior point, we have

$$\delta \,.\, \frac{X}{ABC} = 0,$$

i.e. the attractions, in the directions of the axes, of confocal ellipsoids vary as the masses; which is Maclaurin's theorem for the attractions of ellipsoids upon an exterior point.

In the case of an interior point, we have

$$\delta \,.\, \frac{X}{ABC} = -4\pi \frac{a\delta A}{A^2BC};$$

or, taking α, β, γ as the semi-axes of an ellipsoid confocal with the ellipsoid (A, B, C), but exterior to it, and supposing that (X) refers to the ellipsoid (α, β, γ), we have

$$\delta \frac{(X)}{\alpha\beta\gamma} = -4\pi a \frac{\delta\alpha}{\alpha^2\beta\gamma}.$$

Now introducing instead of α the new variable θ, such that $\alpha^2 = A^2 + \theta$, we have $\dfrac{\delta\alpha}{\alpha^2} = \dfrac{\delta\theta}{(A^2+\theta)^{\frac{3}{2}}}$, $\beta = (B^2+\theta)^{\frac{1}{2}}$, $\gamma = (C^2+\theta)^{\frac{1}{2}}$, and consequently writing d for δ,

$$d\frac{(X)}{\alpha\beta\gamma} = -4\pi a \frac{d\theta}{(A^2+\theta)^{\frac{3}{2}} (B^2+\theta)^{\frac{1}{2}} (C^2+\theta)^{\frac{1}{2}}},$$

and thence, effecting the integration,

$$\frac{X_,}{(A^2+\theta_,)^{\frac{1}{2}}(B^2+\theta_,)^{\frac{1}{2}}(C^2+\theta_,)^{\frac{1}{2}}}-\frac{X}{ABC}=-4\pi a\int_0^{\theta_,}\frac{d\theta}{(A^2+\theta)^{\frac{3}{2}}(B^2+\theta)^{\frac{1}{2}}(C^2+\theta)^{\frac{1}{2}}},$$

where $X_,$ refers to the ellipsoid whose semi-axes are $(A^2+\theta_,)^{\frac{1}{2}}$, $(B^2+\theta_,)^{\frac{1}{2}}$, $(C^2+\theta_,)^{\frac{1}{2}}$. In the case where $\theta_,=\infty$, we have

$$\frac{X_,}{(A^2+\theta_,)^{\frac{1}{2}}(B^2+\theta_,)^{\frac{1}{2}}(C^2+\theta_,)^{\frac{1}{2}}}=0,$$

and consequently

$$X=4\pi a\,ABC\int_0^{\infty}\frac{d\theta}{(A^2+\theta)^{\frac{3}{2}}(B^2+\theta)^{\frac{1}{2}}(C^2+\theta)^{\frac{1}{2}}},$$

which is the expression for the attraction, in the direction opposite to that of the axis of x, of the ellipsoid (A, B, C) upon an interior point, the coordinates of which are (a, b, c).

In the case of an exterior point, let $A_,$, $B_,$, $C_,$ be the semi-axes of the confocal ellipsoid passing through the attracted point; so that putting $A_,=\sqrt{(A^2+\eta)}$, $B_,=\sqrt{(B^2+\eta)}$, $C_,=\sqrt{(C^2+\eta)}$, we have

$$\frac{a^2}{A^2+\eta}+\frac{b^2}{B^2+\eta}+\frac{c^2}{C^2+\eta}=1,$$

the attraction is equal to $\dfrac{ABC}{A_,B_,C_,}\times$ attraction of the ellipsoid which passes through the point, i.e.

$$X=4\pi a\,ABC\int_0^{\infty}\frac{d\theta}{(A_,^2+\theta)^{\frac{3}{2}}(B_,^2+\theta)^{\frac{1}{2}}(C_,^2+\theta)^{\frac{1}{2}}};$$

or, putting $\theta-\eta$ instead of θ,

$$X=4\pi a\,ABC\int_{\eta}^{\infty}\frac{d\theta}{(A^2+\theta)^{\frac{3}{2}}(B^2+\theta)^{\frac{1}{2}}(C^2+\theta)^{\frac{1}{2}}},$$

which is the expression for the attraction upon an exterior point. The formulæ coincide, as they ought to do, in the case of a point upon the surface.

2, *Stone Buildings*, 9*th April*, 1855.

165.

ON SOME GEOMETRICAL THEOREMS RELATING TO A TRIANGLE CIRCUMSCRIBED ABOUT A CONIC.

[From the *Quarterly Mathematical Journal*, vol. I. (1857), pp. 169—175.]

THE following investigations were suggested to me by Sir F. Pollock's interesting paper "On a Geometrical Theorem relating to an Equilateral Triangle circumscribed about a Circle," [*Quart. Math. Jour.* t. I. (1857), pp. 167—169].

If on the sides of a triangle ABC, there be taken points α, β, γ, such that $A\alpha$, $B\beta$, $C\gamma$ meet in a point O; and if on each side of the triangle there be taken two points forming with the two angles on the same side an involution having the first-mentioned point on the same side for a double point; then if three of the six points lie in a line, the two lines are said to be harmonically related with respect to the triangle ABC and point O. Call these the lines (r), (s).

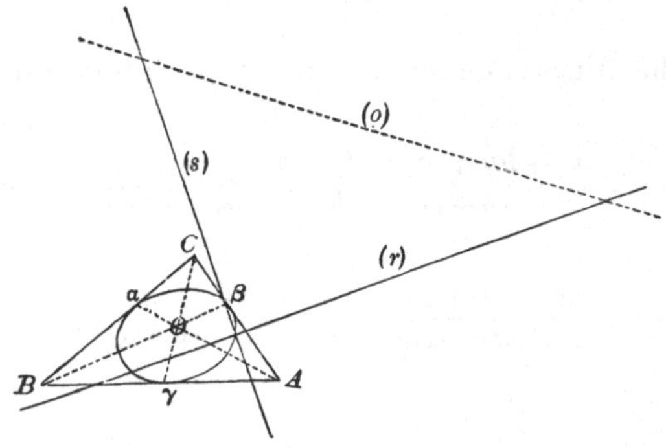

The triangle ABC and point O give rise to a determinate conic; viz., the conic touching the sides at the points α, β, γ.

The harmonic of the point O with respect to the triangle is a line (o), which is also the polar of O with respect to the conic. The conic meets this line in two points (as the figure is drawn, imaginary ones), I, J.

Suppose now that the harmonic lines (r), (s) are harmonically related to the points I, J, (i.e. let the lines (r), (s) and the lines through their point of intersection and I, J, be a harmonic pencil, (or, what is the same thing, let I, J, and the points in which the line of junction meets the lines (r), (s) be a harmonic range), then, *Theorem*, the point of intersection of the lines (r), (s) lies on the conic.

In order to prove this, take

$$x=0,\ y=0,\ z=0, \text{ for the equations of } BC,\ CA,\ AB,$$

$$x=y=z, \text{ for the point } O,$$

the equations of the harmonic lines will be

$$ax+by+cz=0,$$

$$\frac{x}{a}+\frac{y}{b}+\frac{z}{c}=0,$$

the equation of the conic is

$$x^2+y^2+z^2-2yz-2zx-2xy=0\,;$$

the equation of the line (o) is

$$x+y+z=0.$$

The coordinates of the points I, J, are

$$x:y:z=1:\omega:\omega^2,$$

and

$$x:y:z=1:\omega^2:\omega,$$

where ω is an imaginary cube root of unity.

The equations of the lines joining the point of intersection of (r), (s) with the points I, J, are

$$\frac{ax+by+cz}{a+b\omega+c\omega}=\frac{bcx+cay+abz}{bc+ca\omega+ab\omega^2},$$

and

$$\frac{ax+by+cz}{a+b\omega^2+c\omega}=\frac{bcx+cay+abz}{bc+ca\omega^2+ab\omega},$$

and these will form with the lines

$$ax+by+cz=0,$$

$$bcx+cay+abz=0,$$

a harmonic pencil, if

$$(a+b\omega\ +c\omega^2)(bc+ca\omega^2+ab\omega\ )$$
$$+(a+b\omega^2+c\omega\ )(bc+ca\omega\ +ab\omega^2)=0\,;$$

i.e., if

$$6abc-bc^2-b^2c-ca^2-c^2a-ab^2-a^2b=0\,;$$

which is, therefore, the condition in order that (r), (s), I, J may be harmonically related.

The coordinates of the point of intersection of (r), (s) are

$$x : y : z = a(b^2-c^2) : b(c^2-a^2) : c(a^2-b^2)\,;$$

and substituting these values in

$$x^2+y^2+z^2-2yz-2zx-2xy,$$

we see, *à priori*, that the result contains the factor

$$2abc+bc^2+b^2c+ca^2+c^2a+ab^2+a^2b=(b+c)(c+a)(a+b)\,;$$

in fact $b+c=0$ gives $x : y : z = 0 : b(b^2-a^2) : b(b^2-a^2)$, i.e. $x=0$, $y-z=0$, which makes $x^2+y^2+z^2-2yz-2zx-2xy$ vanish; and so for the other factors.

Effecting the substitution, we find

$$x^2+y^2+z^2-2yz-2zx-xy=(2abc+bc^2+b^2c+ca^2+c^2a+ab^2+a^2b)$$
$$\times(-6abc+bc^2+b^2c+ca^2+c^2a+ab^2+a^2b),$$

which equals 0, on account of the second factor.

Hence the point of intersection of the lines (r), (s) lies upon the conic. Q.E.D.

Consider, next, a side of the triangle: then, *Theorem*,—the following four points on a side of the triangle, viz. the point of contact, the point of intersection with the line (o), and the points of intersection with the harmonic lines (r), (s), form a harmonic range.

In fact, for the side $x=0$ the points in question are given by means of the equations

$$y-z=0,\quad y+z=0,\quad by+cz+0,\quad \frac{y}{b}+\frac{z}{c}=0,$$

and forming the equations

$$y^2-z^2=0,$$
$$y^2+\left(\frac{b}{c}+\frac{c}{b}\right)yz+z^2=0\,;$$

the theorem is at once seen to be true. Q.E.D.

It is interesting to give the analytical expressions for some other of the points and lines of the figure. The harmonic lines (*r*), (*s*), meet in a point of the conic given by the equations

$$\xi : \eta : \zeta = a(b^2 - c^2) : b(c^2 - a^2) : c(a^2 - b^2).$$

This point may be spoken of as the common point of intersection of the harmonic lines with the conic. The foregoing values give

$$\xi^2 - \eta^2 - \zeta^2 = \eta\zeta\left(\frac{b}{c} + \frac{c}{b}\right),$$

$$\eta^2 - \zeta^2 - \xi^2 = \zeta\xi\left(\frac{c}{a} + \frac{a}{c}\right),$$

$$\zeta^2 - \xi^2 - \eta^2 = \xi\eta\left(\frac{a}{b} + \frac{b}{a}\right)$$

which may also be written

$$\xi^2 - (\eta - \zeta)^2 = \eta\zeta\frac{1}{bc}(b + c)^2 = QR,$$

$$\eta^2 - (\zeta - \xi)^2 = \zeta\xi\frac{1}{ca}(c + a)^2 = RP,$$

$$\zeta^2 - (\xi - \eta)^2 = \xi\eta\frac{1}{ab}(a + b)^2 = PQ,$$

if for shortness

$$-\xi + \eta + \zeta = P,$$

$$\xi - \eta + \zeta = Q,$$

$$\xi + \eta - \zeta = R,$$

formulæ which will be presently useful.

Each harmonic line meets the conic in another point, which may be termed the simple point of intersection. For the line $ax + by + cz = 0$, the coordinates of the simple point of intersection are given by $x : y : z = \eta\zeta(b + c)^2 : \zeta\xi(c + a)^2 : \zeta\eta(a + b)^2$; or, what is the same thing, by

$$x : y : z = \frac{1}{Pa} : \frac{1}{Qb} : \frac{1}{Rc},$$

and, in like manner, the coordinates of the simple point of intersection of the line $\frac{x}{a} + \frac{y}{b} + \frac{z}{c} = 1$ are given by

$$x : y : z = \frac{a}{P} : \frac{b}{Q} : \frac{c}{R}.$$

Hence the equation of the line joining the two simple points of intersection is

$$\begin{vmatrix} x\ , & y\ , & z \\ \frac{1}{Pa}, & \frac{1}{Qb}, & \frac{1}{Rc} \\ \frac{a}{P}, & \frac{b}{Q}, & \frac{c}{R} \end{vmatrix} = 0,$$

or expanding and reducing

$$P\xi x + Q\eta y + R\zeta z = 0.$$

The equation of the tangent to the conic at the common point of intersection is evidently

$$Px + Qy + Rz = 0.$$

The last-mentioned lines, together with the harmonic lines (r), (s), viz. the lines

$$ax + by + cz = 0,$$

$$\frac{x}{a} + \frac{y}{b} + \frac{z}{c} = 0,$$

may be considered as the sides of an inscribed quadrilateral; the equation of the conic must therefore be expressible in a form in which this is put in evidence; to do this, I first form the equation

$$\left(ax + by + cz\right)\left(\frac{x}{a} + \frac{y}{b} + \frac{z}{c}\right) = x^2 + y^2 + z^2 - 2\left(1 - \frac{QR}{\eta\zeta}\right) yz - \left(1 - \frac{RP}{\zeta\xi}\right) zx - 2\left(1 - \frac{PQ}{\zeta\eta}\right) xy,$$

which may also be written

$$\left(ax + by + cz\right)\left(\frac{x}{a} + \frac{y}{b} + \frac{z}{c}\right) = x^2 + y^2 + z^2 - 2\lambda yz - 2\mu zx - 2\nu xy,$$

where

$$\lambda = \frac{\eta^2 + \zeta^2 - \xi^2}{2\eta\zeta},$$

$$\mu = \frac{\zeta^2 + \xi^2 - \eta^2}{2\zeta\xi},$$

$$\nu = \frac{\xi^2 + \eta^2 - \zeta^2}{2\xi\eta},$$

and then putting

$$\Delta = \xi^2 + \eta^2 + \zeta^2 - 2\eta\zeta - 2\zeta\xi - 2\xi\eta,$$

an equation which gives

$$\begin{aligned} \Delta &= P^2 - 4\eta\zeta \\ &= Q^2 - 4\zeta\xi \\ &= R^2 - 4\xi\eta, \end{aligned}$$

and

$$-PQR-8\xi\eta\zeta=(\xi+\eta+\zeta)\,\Delta,$$

it is easy by means of these relations to verify the identical equation

$$3\xi\eta\zeta(x^2+y^2+z^2-2yz-2zx-2xy)+\xi\eta\zeta(ax+by+cz)\left(\frac{x}{a}+\frac{y}{b}+\frac{z}{c}\right)$$

$$-(P\xi x+Q\eta y+R\zeta z)(Px+Qy+Rz)=\Delta\{-\xi x^2-\eta y^2-\zeta z^2+(\xi+\eta+\zeta)(yz+zx+xy)\},$$

or, writing for Δ its value $\Delta=0$, the equation of the conic takes the form

$$\xi\eta\zeta(ax+by+cz)\left(\frac{x}{a}+\frac{y}{b}+\frac{z}{c}\right)-(P\xi x+Q\eta y+R\zeta z)(Px+Qy+Rz)=0,$$

which is as it should be.

It may be added, that the common point of intersection and the points in which the harmonic lines (r), (s) meet a side of the triangle lie in a conic passing through the points I, J, such that with respect to this conic the point of contact is the pole of the line (o). Thus for the side $x=0$ the equation of the conic in question is

$$\theta x(x+y+z)+\left(1-\frac{b}{c}-\frac{c}{b}\right)x^2+(by+cz)\left(\frac{y}{b}+\frac{z}{c}\right)=0,$$

where

$$\theta=\frac{a(b-c)^2(b+c)}{bc(a-b)(c-a)}=\frac{(b+c)^2}{2bc},$$

and similarly for the other two conics.

In the case where the triangle is an equilateral triangle and the line (o) is the line at ∞, the conic becomes a circle, the points I, J, are the circular points at infinity, and lines harmonic in respect to the points I, J, become lines at right angles to each other: the foregoing results agree, therefore, with Sir F. Pollock's Theorem.

166.

NOTE ON THE HOMOLOGY OF SETS.

[From the *Quarterly Mathematical Journal*, vol. I. (1857), p. 178.]

LET L denote a set of any four elements a, b, c, d, and in like manner Λ, L_1 &c. sets of the four elements $\alpha, \beta, \gamma, \delta$; $a_{,}, b_{,}, c_{,}, d_{,}$, &c.; then we may establish a relation of homology between four sets L, L_1, L_2, L_3, and four other sets $\Lambda, \Lambda_1, \Lambda_2, \Lambda_3$; viz., considering the corresponding anharmonic ratios of the different sets, we may suppose a relation of homology between these ratios. Thus considering the set to L, write

$$x=(a-b)(c-d),$$
$$y=(a-c)(d-b),$$
$$z=(a-d)(b-c),$$

then $x+y+z=0$ and the anharmonic ratios of the set are $x : y : z$—we may, if we please, take $x : y$ as the anharmonic ratio of the set. And in like manner taking $\xi : \eta$ as the anharmonic ratio of the set $\alpha, \beta, \gamma, \delta$, &c., the assumed relation between the sets L, L_1, L_2, L_3 and the sets $\Lambda, \Lambda_1, \Lambda_2, \Lambda_3$ will be

$$\begin{vmatrix} x\xi, & x\eta, & y\xi, & y\eta \\ x_1\xi_1, & x_1\eta_1, & y_1\xi_1, & y_1\eta_1 \\ x_2\xi_2, & x_2\eta_2, & y_2\xi_2, & y_2\eta_2 \\ x_3\xi_3, & x_3\eta_3, & y_3\xi_3, & y_3\eta_3 \end{vmatrix}=0;$$

and it is to be observed, that this relation is independent of the particular ratio $x : y$ which has been chosen as the anharmonic ratio of the set; in fact, if we write $x=-y-z$, $\xi=-\eta-\zeta$, &c., then reducing the result by means of an elementary property of determinants, the equation will preserve its original form, but will contain the ratios $y : z$; $\eta : \zeta$, &c., instead of the ratios $x : y$; $\xi : \eta$, &c.

167.

APROPOS OF PARTITIONS.

[From the *Quarterly Mathematical Journal*, vol. I. (1857), pp. 183—184.]

LET $\Pi(1-x^a) = (1-x^a)\,(1-x^b)\ldots(\kappa \text{ factors})$ and assume that $\left[\dfrac{1}{\Pi(1-x^a)}\right]_{1-x}$ is the part of $\dfrac{1}{\Pi(1-x^a)}$ which involves negative powers of $1-x$, then

$$\text{Coefficient } x^q \text{ in } \left[\frac{1}{\Pi(1-x^a)}\right]_{1-x} = \text{coefficient } z^{\kappa-1} \text{ in } \left(\frac{1}{(1-z)^{q+1}}\,\Pi\,\frac{z}{1-(1-z)^a}\right),$$

which suggests the question of the expansion in powers of z, of the function

$$\frac{1}{(1-z)^{q+1}}\,\Pi\,\frac{z}{1-(1-z)^a}.$$

Now by the definition of Bernoulli's numbers

$$\frac{1}{e^t-1} = \frac{1}{t} - \frac{1}{2} + B_1\frac{t}{1.2} - B_2\frac{t^3}{1.2.3.4} + B_3\frac{t^5}{1.2.3.4.5.6} - \&c.$$

from which it is easy to deduce

$$\frac{t}{1-e^{-t}} = e^{\frac{1}{2}t - B_1\frac{t^2}{1.2^2} + B_2\frac{t^4}{1.2.3.4^2} - B_3\frac{t^6}{1.2.3.4.5.6^2} + \&c.};$$

and, writing in this formula $t = -a\log(1-z)$, we have

$$\frac{-a\log(1-z)}{1-(1-z)^a} = e^{-\frac{1}{2}\log(1-z)\,a - B_1\frac{\log^2(1-z)}{1.2^2}a^2 + B_2\frac{\log^4(1-z)}{1.2.3.4^2}a^4 + \&c.},$$

i.e.

$$\frac{z}{1-(1-z)^a} = \frac{1}{a}\left(\frac{-z}{\log(1-z)}\right) e^{-\frac{1}{2}\log(1-z)\,a - \&c.} = \frac{1}{a} e^{\log\left(\frac{-z}{\log(1-z)}\right) - \frac{1}{2}\log(1-z)\,a - \&c.},$$

and putting p_κ for $abc\ldots$ and S_1, $S_2\ldots$ for the sums of the powers, we have, taking the product

$$\Pi \frac{z}{1-(1-z)^a} = \frac{1}{p_\kappa} e^{\kappa\log\left(\frac{-z}{\log(1-z)}\right) - \frac{1}{2}\log(1-z)\,S_1 - B_1\frac{\log^2(1-z)}{1.2^2}S_2 + B_2\frac{\log_4(1-z)}{1.2.3.4^2}S_4 - \&c.},$$

whence also

$$\frac{1}{(1-z)^{q+1}}\Pi\frac{z}{1-(1-z)^a} = \frac{1}{p_\kappa} e^{\kappa\log\left(\frac{-z}{\log(1-z)}\right) - (q+1+\frac{1}{2}S_1)\log(1-z) - B_1\frac{\log^2(1-z)}{1.2^2}S_2 + B_2\frac{\log^4(1-z)}{1.2.3.4^2}S_4 - \&c.}$$

from which the development may be found.

The index of e is

$$\begin{aligned} &(q+1-\tfrac{1}{2}\kappa+\tfrac{1}{2}S_1)\,z \\ &+(\tfrac{1}{2}q+\tfrac{1}{2}-\tfrac{5}{24}\kappa+\tfrac{1}{4}S_1-\tfrac{1}{24}S_2)\,z^2 \\ &+(\tfrac{1}{3}q+\tfrac{1}{3}-\tfrac{1}{8}\kappa+\tfrac{1}{6}S_1-\tfrac{1}{24}S_2)\,z^3 \\ &+\&c. \end{aligned}$$

and developing the exponential,

1°. The coefficient of z is

$$q+\tfrac{1}{2}S_1-\tfrac{1}{2}(\kappa-2).$$

2°. The coefficient of z^2 is

$$\tfrac{1}{2}q^2+q\,\{\tfrac{1}{2}S_1-\tfrac{1}{2}(\kappa-3)\}+\tfrac{1}{8}S_1^2-\tfrac{1}{24}S_2-\tfrac{1}{4}(\kappa-3)\,S_1+(\kappa-3)\,(\tfrac{1}{8}\kappa-\tfrac{1}{3}),$$

and so on.

The peculiarity is the appearance of the factors $\kappa-2$, $\kappa-3$, &c. If we neglect these terms, and consider as well q as $a, b, c\ldots$ to be each of them of the dimension unity, the coefficients will be homogeneous.

168.

A DEMONSTRATION OF THE FUNDAMENTAL PROPERTY OF GEODESIC LINES.

[From the *Quarterly Mathematical Journal*, vol. I. (1857), pp. 185—6.]

This is the translation of a paragraph in M. Bertrand's Memoir "Démonstration géométrique de quelques théorèmes relatifs à la théorie des Surfaces," Liouv. t. XIII. (1848), p. 73, in which he shows that the theorem that the geodesic lines on a surface have at each point their osculating plane normal to the surface is an immediate consequence of Meunier's theorem.

169.

EISENSTEIN'S GEOMETRICAL PROOF OF THE FUNDAMENTAL THEOREM FOR QUADRATIC RESIDUES. (*Translated from the Original Memoir*, Crelle, t. XXVIII. (1844), *with an Addition, by* A. CAYLEY.)

[From the *Quarterly Mathematical Journal*, vol. I. (1857), pp. 186—191.]

LET p be a positive odd prime, a the aggregate of all the even numbers $< p$ and > 0, viz. $a = 2, 4, 6, \ldots p-1$; and let q be any integer not divisible by the modulus p; then if r denote the general term of the residues of the multiples qa in respect to the modulus p, it is clear that the numbers of the series the general term of which is $(-)^r r$ will coincide to multiples of p près with the numbers of the series a; so that we shall have the two congruences

$$q^{\frac{1}{2}(p-1)}\,\Pi a \equiv \Pi r \text{ (mod. } p\text{), and } \Pi a \equiv (-)^{\Sigma r}\,\Pi r \text{ (mod. } p\text{)},$$

from which it follows that

$$q^{\frac{1}{2}(p-1)} \equiv (-)^{\Sigma r}\,1 \text{ (mod. } p\text{), and therefore } \left(\frac{p}{q}\right) = (-)^{\Sigma r}\,1.$$

Let $E\left(\frac{qa}{p}\right)$ denote the greatest whole number contained in the fraction $\frac{qa}{p}$, then it is clear that $\Sigma qa = p\Sigma E\left(\frac{qa}{p}\right) + \Sigma r$; and since all the a's are even, and $p \equiv 1$ (mod. 2) it follows that $\Sigma r \equiv \Sigma E\left(\frac{qa}{p}\right)$ (mod. 2); and we have, therefore,

$$\left(\frac{q}{p}\right) = (-)^{\Sigma E\left(\frac{qa}{p}\right)}$$

When $q=2$, the formula gives at once the value of $\left(\frac{2}{p}\right)$; when, on the other hand, q is odd, and therefore $q-1$ even, we find, by a simple transformation,

$$\Sigma E\left(\frac{qa}{p}\right) \equiv -E\left(\frac{q}{p}\right)+E\left(\frac{2q}{p}\right)-E\left(\frac{3q}{p}\right)\ldots \pm E\left(\frac{\frac{1}{2}(p-1)q}{p}\right)$$

$$\equiv \quad E\left(\frac{q}{p}\right)+E\left(\frac{2q}{p}\right)+E\left(\frac{3q}{p}\right)\ldots +E\left(\frac{\frac{1}{2}(p-1)q}{p}\right), \text{ (mod. 2)};$$

and if the last sum be represented by μ, then also

$$\left(\frac{q}{p}\right)=(-)^{\mu}\,1.$$

Imagine now in a plane, a rectangular system of coordinates (x, y) and the whole plane divided by lines parallel to the axes at distances $=1$ from each other into squares of the dimension $=1$. And let the angles which do not lie on the axes of coordinates be called "lattice points."

Take, now, upon any vertical parallel a point corresponding to the ordinate y, then $E(y)$ will denote the number of lattice points which lie between this point and the horizontal axis; and, in like manner, taking upon any horizontal parallel a point

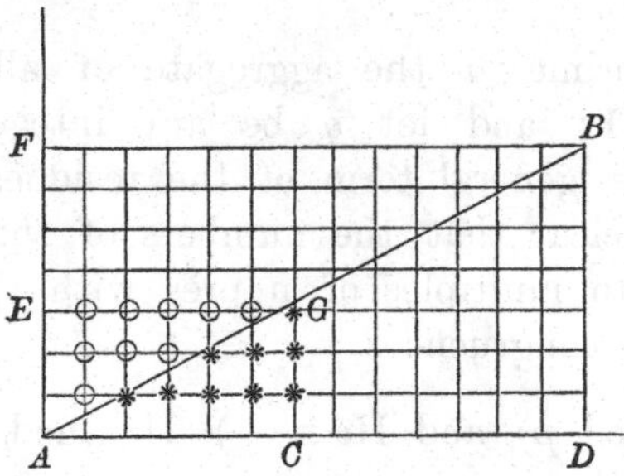

corresponding to the abscissa x, then $E(x)$ will denote the number of lattice points which lie between this point and the vertical axis. If, therefore, we draw in the plane a curve the equation of which is $y=\phi x$, then the sum

$$E\phi 1+E\phi 2+E\phi 3+E\phi 4+\&\text{c.},$$

will denote the number of lattice points which lie between the curve and the axis of x, including any lattice points which lie upon the curve.

Suppose now, to return to the subject, that AB represents the straight line which has for its equation $y=\frac{q}{p}x$, where p and q are now assumed to be both of them positive odd primes; and let $AD=FB$ be $=p$, $AF=DB=q$, $AC=EG=\frac{1}{2}(p-1)$, $AE=CG=\frac{1}{2}(q-1)$. Then if μ denote the number of the lattice points between AB and AD as far as the ordinate CG inclusively (these lattice points are distinguished in the figure by the mark [*]), by what precedes $\left(\frac{q}{p}\right)=(-)^{\mu}1$. But since the equation of the straight line AB may also be written $x=\frac{p}{q}y$, we have in the same way, if ν denote the number of the lattice points which lie between AB and AF as far as the abscissa EG inclusively (these are distinguished in the figure by the mark [∘]), $\left(\frac{p}{q}\right)=(-)^{\nu}1$. But the lattice points marked with [*] and [∘] taken together, i.e. all the lattice points to the right, and all the lattice points to the left of AB, make up the entire system of lattice points of the rectangle $AEGC$, the number of which is $\frac{1}{2}(p-1).\frac{1}{2}(q-1)$; consequently, $\mu+\nu=\frac{1}{2}(p-1).\frac{1}{2}(q-1)$, and therefore

$$\left(\frac{q}{p}\right)\left(\frac{p}{q}\right)=(-)^{\mu+\nu}1=(-)^{\frac{1}{2}(p-1)\cdot\frac{1}{2}(q-1)}1$$

which is the theorem in question.

It may be noticed that the foregoing transformation $\Sigma E\left(\frac{qa}{p}\right)\equiv\mu \pmod 2$ may itself be proved by the simple geometrical consideration that $\Sigma E\left(\frac{qa}{p}\right)$ is nothing else than the number of lattice points which lie on the even ordinates (those corresponding to $x=2, 4, 6, \ldots p-1$) between AB and AD as far as BD, and that each ordinate, between the axes AD and BF exclusively, contains $(q-1)$, i.e. an even number of lattice points; and besides, that the two triangles BAD and ABF are congruent, and that the latter of them stands in the same relation to BF and BD as the former does to AD and AF; the completion of which is left to the reader.

Remark. There are figures for which simple formulæ may be obtained for the number of the interior lattice points. Imagine, for example, a circle, the centre of which lies in the axis of coordinates, and the radius of which is $\sqrt{m}$, then the number S of the lattice points which lie within the circle, including those which lie on the axes, is given by the following formula,

$$S=1+4\{E(m)-E(\tfrac{1}{3}m)+E(\tfrac{1}{5}m)-\&c.\}$$

continued until the series stops of itself. It is easy to see that the equation expresses a relation between the number of lattice points of the circle and the number of lattice points of a segment included between two rectangular hyperbolas. Writing in the formula

$$\frac{1}{m}S=\frac{1}{m}+4\left\{\frac{1}{m}Em-\frac{1}{m}E(\tfrac{1}{3}m)+\frac{1}{m}E(\tfrac{1}{5}m)-\&c.\right\}$$

$m=\infty$, the left-hand side becomes equal to π, and the right-hand side becomes equal to $4(1-\frac{1}{3}+\frac{1}{5}-\&c.)$, which gives the formula of Leibnitz. There are similar formulæ for the number of lattice points of a system of sectors of ellipses or hyperbolas; and similar relations exist also in space, and in cases of more than three dimensions. We shall return to this important subject, which has the closest connection with the properties of the higher forms, upon another occasion.

(*Addition by the Translator.*) Eisenstein is now, alas! dead; too soon for the complete development of his various and profound researches in elliptic functions and the theory of numbers; and the promise at the conclusion of the foregoing memoir has not, I believe, been fulfilled. The formula in the Remark must, I think, have been established by geometrical considerations, and would have served to give the number of decompositions of a number into the sum of two squares; but, as I do not perceive how this is to be done, I shall follow a reverse course, and establish the theorem from considerations founded on the theory of numbers. I remark, first, that the number of lattice points in a quadrant of the circle, inclusive of those on the vertical axis, but exclusive of those on the horizontal axis and of the centre, is equal to $E\sqrt{m} + E\sqrt{(m-1)} + E\sqrt{(m-4)} + E\sqrt{(m-9)} + \&c.$; and that this sum, multiplied by four and increased by unity, gives precisely the number of lattice points of the circle, including those on the horizontal and vertical axes. The formula to be established is therefore

$$E\sqrt{m} + E\sqrt{(m-1)} + E\sqrt{(m-4)} + E\sqrt{(m-9)} + \&c. = E(m) - E(\tfrac{1}{3}m) + E(\tfrac{1}{5}m) - E(\tfrac{1}{7}m) + \&c.$$

where, as already noticed, the left-hand side denotes the number of lattice points of a quadrant of the circle, inclusive of those on the vertical axis, but exclusive of those on the horizontal axis and of the centre.

Let Xn be the number of ways in which the integer number n can be expressed as the sum of two squares. {If $n = \alpha^2 + \beta^2$, then if α and β are unequal, and neither of them is zero, this counts as two decompositions, viz. $x = \alpha$, $y = \beta$ or $x = \beta$, $y = \alpha$; but if $\alpha = \beta$, this counts only as a single decomposition; or if either of the numbers α, β, e.g. α, is zero, then, since $y = 0$ is excluded, this counts as a single decomposition, $x = 0$, $y = \beta$.} Xn will denote the number of lattice points *on* the quadrant of the circle radius $\sqrt{n}$. Suppose also that $E'\left(\frac{1}{k}n\right)$ stands for unity or zero, according as $\frac{1}{k}n$ is or is not an integer. Then as m passes through the integer number n, i.e. from a value between $n-1$ and n to a value between n and $n+1$, the left-hand side of the equation is increased by Xn, and the right-hand side of the equation is increased by $E'(n) - E'(\tfrac{1}{3}n) + E'(\tfrac{1}{5}n) - \&c.$ We ought therefore to have

$$Xn = E'(n) - E'(\tfrac{1}{3}n) + E'(\tfrac{1}{5}n) - E'(\tfrac{1}{7}n) + \&c.,$$

and conversely from this equation, the original equation will at once follow. The right-hand side denotes, it should be observed, the number of factors of n of the form $\equiv 1 \pmod{4}$, less the number of factors of the form $\equiv 3 \pmod{4}$. Let $n = 2^k f^\alpha f'^{\alpha'} \ldots g^\beta g'^{\beta'} \ldots$ where $f, f' \ldots$ are odd primes $\equiv 1 \pmod{4}$ and $g, g' \ldots$ are odd primes $\equiv 3 \pmod{4}$. Consider first the factors $g, g' \ldots$, and forming the product

$$(1 + g \ldots + g^\beta)(1 + g + \ldots g'^{\beta'}) \ldots,$$

it is easy to see that if all or any one or more of the indices β, β'... are odd, then the number of terms of the product which are $\equiv 1$ (mod. 4) is equal to the number of terms of the product which are $\equiv 3$ (mod. 4); but if all the indices β, β'... are even, then the number of terms of the first form is greater by unity than the number of terms of the second form. Now the terms of the product $(1+f+\ldots f^{\alpha})(1+f'+\ldots f'^{\alpha'})\ldots$ are all $\equiv 1$ (mod. 4), and the number of terms is $(1+\alpha)(1+\alpha')\ldots$. Hence if all or any one or more of the indices β, β'... are odd, then the number of factors of n of the first form *less* the number of factors of the second form is zero; but if the indices β, β'... are all even, the number of factors of the first form *less* the number of factors of the second form is $(1+\alpha)(1+\alpha')\ldots$, i.e. we have

$$E'(n) - E'(\tfrac{1}{3}n) + E'(\tfrac{1}{5}n) - \&c. = 0, \text{ or } = (1+\alpha)(1+\alpha')\ldots,$$

according as β, β'... are all or any one or more of them odd, or according as they are all of them even. Now it is well known that the number $n = 2^k f^{\alpha} f'^{\alpha'} \ldots g^{\beta} g'^{\beta'} \ldots$ does not, in the case of all or any one or more of the indices β, β'... being odd, admit of decomposition into two squares, i.e. in this case $Xn = 0$; but if the indices β, β'... are all even, then the number n will admit of precisely as many decompositions into two squares as the number $n' = 2^k f^{\alpha} f'^{\alpha'} \ldots$ (in fact, the only decompositions of n are those obtained from the decompositions of n' by multiplying the roots into the common factor $g^{\frac{1}{2}\beta} g'^{\frac{1}{2}\beta'} \ldots$), and the number of decompositions of n' is moreover equal to the number of decompositions of $n'' = f^{\alpha} f'^{\alpha'} \ldots$, which last number is in fact the product of the numbers of decompositions of f^{α}, $f'^{\alpha'}$...; the number of decompositions of n into two squares (estimated according to the foregoing convention) is thus shown to be, in the case of β, β'..., all of them even, equal to $(1+\alpha)(1+\alpha')\ldots$. And we have therefore in every case

$$Xn = E'(n) - E'(\tfrac{1}{3}n) + E'(\tfrac{1}{5}n) - \&c.,$$

and the principal theorem is thus shown to be true.

170.

ON SCHELLBACH'S SOLUTION OF MALFATTI'S PROBLEM.

[From the *Quarterly Mathematical Journal*, vol. I. (1857), pp. 222—226.]

THE following elegant solution of Malfatti's Problem as applied to spherical triangles is given by Dr Schellbach (*Crelle*, t. XLV. (1853), p. 186); for the reason which will be mentioned I have made a change of notation.

In a spherical triangle ABC to describe three small circles, each of them touching the other two, and also two sides of the triangle.

Let the sides of the triangle be a, b, c, and let x, y, z, be the distances of the points of contact from the adjacent angles of the triangle. Then writing

$$a + b + c = 2s,$$

$$a - \tfrac{1}{2}s = l,\ b - \tfrac{1}{2}s = m,\ c - \tfrac{1}{2}s = n,$$

whence

$$l + m + n = \tfrac{1}{2}s,$$

and putting also

$$\tfrac{1}{2}s - x = \xi,\ \tfrac{1}{2}s - y = \eta,\ \tfrac{1}{2}s - z = \zeta,$$

it is easy to obtain

$$\left\{\begin{array}{l} \dfrac{\cos l\ \cos\eta\cos\zeta}{\cos\frac{1}{2}s} - \dfrac{\sin l\ \sin\eta\sin\zeta}{\sin\frac{1}{2}s} = 1, \\[2ex] \dfrac{\cos m\cos\zeta\cos\xi}{\cos\frac{1}{2}s} - \dfrac{\sin m\sin\zeta\sin\xi}{\sin\frac{1}{2}s} = 1, \\[2ex] \dfrac{\cos n\ \cos\xi\cos\eta}{\cos\frac{1}{2}s} - \dfrac{\sin n\sin\xi\sin\eta}{\sin\frac{1}{2}s} = 1, \end{array}\right.$$

from which equations the unknown quantities ξ, η, ζ, are to be determined. And the equations may be solved without assuming the existence of the relation $l + m + n = \frac{1}{2}s$.

To solve the equations, let the subsidiary angles λ, μ, ν, be determined by the conditions

$$\left\{\begin{aligned}
&\frac{\cos\lambda\cos m\cos n}{\cos\frac{1}{2}s}+\frac{\sin\lambda\sin m\sin n}{\sin\frac{1}{2}s}=1,\\
&\frac{\cos\mu\cos n\cos l}{\cos\frac{1}{2}s}+\frac{\sin\mu\sin n\sin l}{\sin\frac{1}{2}s}=1,\\
&\frac{\cos\nu\cos l\cos m}{\cos\frac{1}{2}s}+\frac{\sin\nu\sin l\sin m}{\sin\frac{1}{2}s}=1,
\end{aligned}\right.$$

then it may be shown that

$$\left\{\begin{aligned}
&\cos(\eta+\zeta)=\frac{\cos\frac{1}{2}(s+\lambda-\zeta)}{\cos\frac{1}{2}(\lambda+l)},\ \cos(\eta-\zeta)=\frac{\cos\frac{1}{2}(s-\lambda+l)}{\cos\frac{1}{2}(\lambda+l)},\\
&\cos(\zeta+\xi)=\frac{\cos\frac{1}{2}(s+\mu-m)}{\cos\frac{1}{2}(\mu+m)},\ \cos(\zeta-\xi)=\frac{\cos\frac{1}{2}(s-\mu+m)}{\cos\frac{1}{2}(\mu+m)},\\
&\cos(\xi+\eta)=\frac{\cos\frac{1}{2}(s+\nu-n)}{\cos\frac{1}{2}(\nu+n)},\ \cos(\xi-\eta)=\frac{\cos\frac{1}{2}(s-\nu+n)}{\cos\frac{1}{2}(\nu+n)}.
\end{aligned}\right.$$

If we write

$$\tan\phi=\tan m\tan n\cot\tfrac{1}{2}s,$$
$$\tan\chi=\tan n\tan l\cot\tfrac{1}{2}s,$$
$$\tan\psi=\tan l\tan m\cot\tfrac{1}{2}s,$$

then

$$\cos(\lambda-\phi)=\frac{\cos\frac{1}{2}s\cos\phi}{\cos m\cos n},$$

$$\cos(\mu-\chi)=\frac{\cos\frac{1}{2}s\cos\chi}{\cos n\cos l},$$

$$\cos(\nu-\psi)=\frac{\cos\frac{1}{2}s\cos\psi}{\cos l\cos m},$$

equations which give the values of λ, μ, ν, from which ξ, η, ζ are determined as above.

If we suppose that the sides become indefinitely small, we have the case of a plane triangle, and the equations then are

$$\eta^2+\zeta^2+\frac{4l}{s}\eta\zeta=\tfrac{1}{4}s^2-l^2,$$

$$\zeta^2+\xi^2+\frac{4m}{s}\zeta\xi=\tfrac{1}{4}s^2-m^2,$$

$$\xi^2+\eta^2+\frac{4n}{s}\xi\eta=\tfrac{1}{4}s^2-n^2.$$

We have here

$$(\eta+\zeta)^2=\tfrac{1}{4}\{(s+\lambda-l)^2-(\lambda+l)^2\}\ =(\tfrac{1}{2}s+\lambda)(\tfrac{1}{2}s-l),$$
$$(\eta-\zeta)^2=\tfrac{1}{4}\{(s-\lambda+l)^2-(\lambda+l)^2\}\ =(\tfrac{1}{2}s-\lambda)(\tfrac{1}{2}s+l),$$

and consequently

$$\eta^2+\zeta^2=\tfrac{1}{2}s^2-2\lambda l, \quad \eta\zeta=s(\lambda-l),$$

where if

$$\phi=\frac{2mn}{s}, \ (\lambda-\phi)^2=\tfrac{1}{4}s^2+\phi^2-m^2-n^2,$$

i.e.

$$\lambda=\frac{2mn}{s}+\sqrt{\tfrac{1}{4}s^2+\frac{4m^2n^2}{s^2}-m^2-n^2}$$

$$=\frac{2mn}{s}+\frac{1}{2s}\sqrt{(s^2-4m^2)(s^2-4n^2)}$$

Hence

$$\eta^2+\zeta^2=\tfrac{1}{2}s^2-\frac{4lmn}{s}-\frac{l}{s}\sqrt{s^2-4m^2}\sqrt{s^2-4n^2},$$

$$\eta\zeta=2mn-sl+\tfrac{1}{2}\sqrt{s^2-4m^2}\sqrt{s^2-4n^2},$$

$$\zeta^2+\xi^2=\tfrac{1}{2}s^2-\frac{4lmn}{s}-\frac{m}{s}\sqrt{s^2-4n^2}\sqrt{s^2-4l^2},$$

$$\zeta\xi=2nl-sm+\tfrac{1}{2}\sqrt{s^2-4n^2}\sqrt{s^2-4l^2},$$

$$\xi^2+\eta^2=\tfrac{1}{2}s^2-\frac{4lmn}{s}-\frac{n}{s}\sqrt{s^2-4l^2}\sqrt{s^2-4m^2},$$

$$\xi\eta=2lm-sn+\tfrac{1}{2}\sqrt{s^2-4l^2}\sqrt{s^2-4m^2},$$

which is in fact at once deducible from the formulæ in my paper "On a System of Equations connected with Malfatti's Problem and on another Algebraical System," (*Camb. and Dubl. Math. Journ.* t. IV. (1849), p. 270 [79]).

Write now for l, m, n, ξ, η, ζ, their values in terms of a, b, c, x, y, z. We have

$$(\tfrac{1}{2}s-y)^2+(\tfrac{1}{2}s-z)^2-\frac{4}{s}(\tfrac{1}{2}s-a)(\tfrac{1}{2}s-y)(\tfrac{1}{2}s-z)=\tfrac{1}{4}s^2-(\tfrac{1}{2}s-a)^2,$$

i.e.

$$y^2+z^2-\frac{4}{s}(\tfrac{1}{2}s-a)yz-2a(y+z)+a^2=0,$$

or reducing

$$(y+z-a)^2-4\left(1-\frac{a}{s}\right)yz=0,$$

and we have thus the system

$$y+z+2\sqrt{1-\frac{a}{s}}\sqrt{yz}=a,$$

$$z+x+2\sqrt{1-\frac{b}{s}}\sqrt{zx}=b,$$

$$x+y+2\sqrt{1-\frac{c}{s}}\sqrt{xy}=c,$$

which are given by Schellbach in the same volume, p. 29; and it was for the sake of facilitating the comparison that the notation has been altered in the case of the spherical triangle. To solve the system, Schellbach writes

$$a = s\sin^2\phi,\ b = s\sin^2\chi,\ c = s\sin^2\psi,$$

reducing the equations to

$$y + z + 2\sqrt{yz}\cos\phi = s\sin^2\phi,$$
$$z + x + 2\sqrt{zx}\cos\chi = s\sin^2\chi,$$
$$x + y + 2\sqrt{xy}\cos\psi = s\sin^2\psi,$$

whence, putting

$$\phi + \chi + \psi = 2\sigma,$$

the equations are satisfied by

$$x = s\sin^2(\sigma-\phi),\ y = s\sin^2(\sigma-\chi),\ z = s\sin^2(\sigma-\psi),$$

which leads to a simple geometrical construction. And if we substitute for ϕ, χ, ψ, σ, their values, it is easy to obtain

$$x = \tfrac{1}{2}s\left\{1 - \sqrt{\left(1-\frac{a}{s}\right)\left(1-\frac{b}{s}\right)\left(1-\frac{c}{s}\right)} + \sqrt{1-\frac{a}{s}}\sqrt{\frac{b}{s}}\sqrt{\frac{c}{s}} - \sqrt{1-\frac{b}{s}}\sqrt{\frac{c}{s}}\sqrt{\frac{a}{s}} - \sqrt{1-\frac{c}{s}}\sqrt{\frac{a}{s}}\sqrt{\frac{b}{s}}\right\},$$

$$y = \tfrac{1}{2}s\left\{1 - \sqrt{\left(1-\frac{a}{s}\right)\left(1-\frac{b}{s}\right)\left(1-\frac{c}{s}\right)} - \sqrt{\left(1-\frac{a}{s}\right)}\sqrt{\frac{b}{s}}\sqrt{\frac{c}{s}} + \sqrt{1-\frac{b}{s}}\sqrt{\frac{c}{s}}\sqrt{\frac{a}{s}} - \sqrt{1-\frac{c}{s}}\sqrt{\frac{a}{s}}\sqrt{\frac{b}{s}}\right\},$$

$$z = \tfrac{1}{2}s\left\{1 - \sqrt{\left(1-\frac{a}{s}\right)\left(1-\frac{b}{s}\right)\left(1-\frac{c}{s}\right)} - \sqrt{1-\frac{a}{s}}\sqrt{\frac{b}{s}}\sqrt{\frac{c}{s}} - \sqrt{1-\frac{b}{s}}\sqrt{\frac{c}{s}}\sqrt{\frac{a}{s}} + \sqrt{1-\frac{c}{s}}\sqrt{\frac{a}{s}}\sqrt{\frac{b}{s}}\right\},$$

$$yz = \tfrac{1}{2}s\left\{\sqrt{1-\frac{b}{s}}\sqrt{1-\frac{c}{s}} - \sqrt{1-\frac{a}{s}} + \sqrt{\frac{b}{s}}\sqrt{\frac{c}{s}}\right\},$$

$$zx = \tfrac{1}{2}s\left\{\sqrt{1-\frac{c}{s}}\sqrt{1-\frac{a}{s}} - \sqrt{1-\frac{b}{s}} + \sqrt{\frac{c}{s}}\sqrt{\frac{a}{s}}\right\},$$

$$xy = \tfrac{1}{2}s\left\{\sqrt{1-\frac{a}{s}}\sqrt{1-\frac{b}{s}} - \sqrt{1-\frac{c}{s}} + \sqrt{\frac{a}{s}}\sqrt{\frac{b}{s}}\right\},$$

values which are also at once obtained from the formula in my paper above referred to. It may be remarked that the above equations for the determination of x, y, z (the distances of the points of contact from the adjacent angles of the triangle) are very similar in form to those given in the same paper for the determination of X, Y, Z, the radii of the inscribed circles.

171.

NOTE ON MR SALMON'S EQUATION OF THE ORTHOTOMIC CIRCLE.

[From the *Quarterly Mathematical Journal*, vol. I. (1857), pp. 242—244.]

LET $U_1=0$, $U_2=0$, $U_3=0$ be the equations of three circles, and let V be the functional determinant of U_1, U_2, U_3, the functions being in the first instance made homogeneous by the introduction of a variable z, which is ultimately replaced by unity; then the equation of the circle cutting at right angles the three given circles, or, as it may be called, the orthotomic circle, is $V=0$. This elegant theorem of Mr Salmon's is connected with the theory developed by Hesse in the memoir, "Ueber die Wendepuncte der Curven dritter Ordnung," *Crelle*, t. XXVIII. (1844), p. 97).

In fact, let $U_1=0$, $U_2=0$, $U_3=0$ be the equations of three conics, the locus of a point such that its polars with respect to each of these conics, or indeed with respect to any conic having for its equation $\lambda U_1+\mu U_2+\nu U_3=0$ (where λ, μ, ν are arbitrary), pass through the same point, is a curve of the third order $V=0$, where V is the functional determinant of U_1, U_2, U_3.

Conversely, if the curve of the third order $V=0$ be given, and U be a function of the third order, such that the functional determinant of $\frac{dU}{dx}$, $\frac{dU}{dy}$, $\frac{dU}{dz}$, or, what is the same thing, the "Hessian" of the function U is equal to V, a condition which may be written $V=H(U)$, then we may take for the conics any three conics the equations of which are of the form $\lambda\frac{dU}{dx}+\mu\frac{dU}{dy}+\nu\frac{dU}{dz}=0$. The equation $V=H(U)$ affords the means of determining U; in fact, we shall have $U=aV+bH(V)$, where a and b are constants to be determined. This gives $H(U)=H(aV+bH(V))=AV+BH(V)$, where A and B are given functions of a, b (a practical method of determining these functions was first given in Aronhold's memoir, "Zur Theorie der homogenen Functionen dritten Grades von zwei Variabeln," *Crelle*, t. XXXIX. (1850), pp. 140—159); and we have therefore

$V=AV+BH(V)$, i.e. $A=1$, $B=0$: the latter equation determines, what is alone important, the ratio $b : a$; the equation is of the third order, so that there are in general three distinct solutions $U=aV+BH(V)=0$.

In the particular case in which the curves of the third order $V=0$ is made up of a line $P=0$ and a conic $W=0$, i.e. where $V=PW=0$, the curve $H(V)=0$ is made up of the same line $P=0$ and of a conic having double contact with the conic $W=0$ at the point of intersection with the line $P=0$, i.e. $H(PW)=P(lW+mP^2)$. And $U=aPW+bH(PW)$ is consequently a function of the same form, i.e. the cubic $U=0$ is made up of the line $P=0$ and of a conic having double contact with the conic $W=0$ at its points of intersection with the line $P=0$. We may therefore write $U=P(fW+gP^2)$, and forming with this value the equation $PW=P(FW+GP^2)$, it may be noticed that, owing to the occurrence of a special factor which may be rejected, the resulting equation $G=0$ gives only a single value for the ratio $f : g$. Forming from the value $U=P(fW+gP^2)$, the equation $\lambda\frac{dU}{dx}+\mu\frac{dU}{dy}+\nu\frac{dU}{dz}=0$, the equation thus obtained will be of the form $W+PQ=0$, which is the equation of a conic passing through the points of intersection of the line and conic $P=0$, $W=0$, and besides intersecting the conic $W=0$ in two other points. And it may also be shown that the four points of intersection, (i.e. the points given by the equations $W=0$, $W+PQ=0$), the pole of the line $P=0$ with respect to the conic $W=0$, and the pole of this same line with respect to the conic $W+PQ=0$, lie all six in the same conic. We see, therefore, that, given a curve of the third order, the aggregate of a line $P=0$ and a conic $W=0$, as the locus of the point such that its polars, with respect to three several conics (or a system depending on three conics), meet in a point, each conic of the system is a conic passing through the points of intersection of the line and conic $P=0$, $W=0$, and, moreover, such that the four points of intersection with the conic $W=0$ and the poles of the line $P=0$, with respect to the conic of the system, and with respect to the conic $W=0$, lie all six in the same conic. In the particular case in which the line and conic $P=0$, $W=0$ are the line at ∞ and a circle, each conic of the system is a circle such that its points of intersection with the circle $W=0$ and the centres of the two circles lie in a circle, i.e. the conics are circles cutting at right angles the circle $W=0$, which agrees with Mr Salmon's theorem.

To verify the assumed theorems in the case of the curve of the third order $V=PW=0$, we may take

$$P=\alpha x+\beta y+\gamma z=0$$

for the equation of the line, and

$$W=x^2+y^2+z^2=0$$

for the equation of the conic. I write, for greater convenience, $U=P(\tfrac{1}{2}fW+\tfrac{1}{6}gP^2)$; the Hessian of this is

$$\begin{vmatrix} f(P+2\alpha x)+g\alpha^2 P, & f(\beta x+\alpha y)+g\alpha\beta P, & f(\alpha z+\gamma x)+g\gamma\alpha P \\ f(\beta x+\alpha y)+g\alpha\beta P, & f(P+2\beta y)+g\beta^2 P, & f(\gamma y+\beta z)+g\beta\gamma P \\ f(\alpha z+\gamma x)+g\gamma\alpha P, & f(\gamma y+\beta z)+g\beta\gamma P, & f(P+2\gamma z)+g\gamma^2 P \end{vmatrix},$$

which is equal to

$$f^3P\left(4P^2-(\alpha^2+\beta^2+\gamma^2)\,W\right)+f^2gP\,(\alpha^2+\beta^2+\gamma^2)\,P^2,$$

i.e. we must have $4f+g\,(\alpha^2+\beta^2+\gamma^2)=0$, or putting $g=-24$ and $\therefore f=6\,(\alpha^2+\beta^2+\gamma^2)$, we have

$$U=P\left(3\,(\alpha^2+\beta^2+\gamma^2)\,W-4P^2\right).$$

Forming the function $\lambda\dfrac{dU}{dx}+\mu\dfrac{dU}{dy}+\nu\dfrac{dU}{dz}$, and dividing by the constant factor $3\,(\alpha^2+\beta^2+\gamma^2)(\lambda\alpha+\mu\beta+\nu\gamma)$, we have for the equation of any one of the conics

$$W+P\left\{\frac{2\,(\lambda x+\mu y+\nu z)}{\lambda\alpha+\mu\beta+\nu\gamma}-\frac{4P}{\alpha^2+\beta^2+\gamma^2}\right\}=0,$$

which may be written under the form $W+2PQ=0$, where

$$Q=ax+by+cz=\frac{\lambda x+\mu y+\nu z}{\lambda\alpha+\mu\beta+\nu\gamma}-\frac{2\,(\alpha x+\beta y+\gamma z)}{\alpha^2+\beta^2+\gamma^2}.$$

We have therefore $a\alpha+b\beta+c\gamma=1-2=-1$, i.e. $a\alpha+b\beta+c\gamma+1=0$. And there is no difficulty in showing that, given the two conics

$$x^2+y^2+z^2=0,$$

$$x^2+y^2+z^2+2\,(\alpha x+\beta y+\gamma z)(ax+by+cz)=0,$$

the condition in order that the four points of intersection and the poles, with respect to each conic, of the line $\alpha x+\beta y+\gamma z=0$, may lie in a conic is precisely this equation $a\alpha+b\beta+c\gamma+1=0$.

172.

NOTE ON THE LOGIC OF CHARACTERISTICS.

[From the *Quarterly Mathematical Journal*, vol. I. (1857), pp. 257—259.]

THE conditions in order that an equation of the sixth degree

$$(a,\ b,\ c,\ d,\ e,\ f,\ g),\ (x,\ y)^6 = 0$$

may have five of its roots equal are

$$\begin{aligned}
A &= ae - 4bd + 3c^2 = 0,\\
B &= af - 3be + 2cd = 0,\\
C &= bf - 4ce + 3d^2 = 0,\\
D &= ag - 9ce + 8d^2 = 0,\\
E &= bg - 3cf + 2de = 0,\\
F &= cg - 4df + 3e^2 = 0,
\end{aligned}$$

equivalent of course to four relations between the coefficients: among the connections of these equations are

$$fA - \ eB - \qquad bF + \ cE = 0,$$

$$(3e^2 - 2df)\,A - 2deB + ecD - AF - 2cdE + (3c^2 - 2bd)\,F = 0.$$

The system is one of the tenth order. To verify this, I write first

$$(A,\ B,\ C,\ F) = (A,\ B,\ C,\ F,\ cE) = (A,\ B,\ C,\ F,\ c) + (A,\ B,\ C,\ F,\ E),$$

i.e. the equations $A=0$, $B=0$, $C=0$, $F=0$ imply (by the first of the connectives) the additional equation $cE=0$, viz. the system $A=0$, $B=0$, $C=0$, $F=0$, $cE=0$, or what is the same thing, one of the systems $A=0$, $B=0$, $C=0$, $F=0$, $c=0$ and $A=0$, $B=0$, $C=0$, $F=0$, $E=0$.

We have in like manner

$$(A,\ B,\ C,\ F,\ E) = (A,\ B,\ C,\ F,\ E,\ ceD) = (A,\ B,\ C,\ F,\ E,\ D)$$

since (A, B, C, F, E, c) and (A, B, C, F, E, e) respectively vanish as being each of them equivalent not to four but to five relations, and therefore as not adding to the order of the system.

Again,

$$\begin{aligned}(A,\ B,\ C,\ c) &= (A,\ B,\ C,\ c,\ bF)\\ &= (A,\ B,\ C,\ c,\ b) + (A,\ B,\ C,\ c,\ F)\\ &= (ae,\ af,\ d^2,\ b,\ c) + (A,\ B,\ C,\ c,\ F).\end{aligned}$$

But here

$$\begin{aligned}(ae,\ af,\ d^2,\ b,\ c) &= (a,\ af,\ d^2,\ b,\ c) + (e,\ af,\ d^2,\ b,\ c)\\ &= (a,\ af,\ d^2,\ b,\ c),\end{aligned}$$

(for (e, af, d^2, b, c) vanishes as being equivalent to five relations, and therefore as not adding to the order of the system),

$$\begin{aligned}&= (a,\ a,\ d^2,\ b,\ c) + (a,\ f,\ d^2,\ b,\ c)\\ &= (a,\ d^2,\ b,\ c),\end{aligned}$$

since, (a, f, d^2, b, c) vanishes for the above-mentioned reason.

Hence

$$(A,\ B,\ C,\ c) = (a,\ d^2,\ b,\ c) + (A,\ B,\ C,\ c,\ F).$$

We have consequently

$$\begin{aligned}(A,\ B,\ C,\ D,\ E,\ F) &= (A,\ B,\ C,\ E,\ F)\\ &= (A,\ B,\ C,\ F) - (A,\ B,\ C,\ F,\ c)\\ &= (A,\ B,\ C,\ F) - \{(A,\ B,\ C,\ c) - (a,\ b,\ c,\ d^2)\},\end{aligned}$$

which may be thus interpreted:—the system (A, B, C, c) contains the system (a, b, c, d^2), or what is the same thing, contains twice-over the system (a, b, c, d). Discarding this contained system, the remainder of the system (A, B, C, c) is contained in the system (A, B, C, F), and the residue of the last-mentioned system is the system (A, B, C, D, E, F), i.e. the system represented by the equations which express the equality of five roots of the given equation of the sixth degree.

It follows immediately that the order of the system (A, B, C, D, E, F) is $16-(8-2)=10$, i.e. that the system is, as above stated, one of the tenth order. The preceding is, I think, a good example of the kind of reasoning to be employed in what Mr Sylvester has most happily termed the Logic of Characteristics.

173.

ON LAPLACE'S METHOD FOR THE ATTRACTION OF ELLIPSOIDS.

[From the *Quarterly Mathematical Journal*, vol. I. (1857), pp. 285—300.]

[THE method referred to is that given in Book III. of the Mécanique Céleste, Ed. I. 1798.] Let the equation of the surface of the ellipsoid be

$$lx^2 + my^2 + nz^2 - k = 0,$$

and take a, b, c as the coordinates of the attracted point, which is supposed to be exterior to the surface. Imagine an indefinitely thin cone, having the attracted point for vertex, and intersecting the surface of the ellipsoid; let ξ, η, ζ be the direction-cosines of the cone, and dS its spherical angle. Then if for a moment r denote the radius vector corresponding to a point within the ellipsoid, the element of the mass within the cone is $r^2 dr dS$; and we may thence, by an integration with respect to r, find the attractions parallel to the axes (and tending towards the centre) and the potential of the mass within the cone, viz. if r', r'' be the values of r at the surface of the ellipsoid, the attractions are $(r'' - r')\,\xi dS$, $(r'' - r')\,\eta dS$, $(r'' - r')\,\zeta dS$, and the potential is $\frac{1}{2}(r''^2 - r'^2)\,dS$. And putting for shortness

$$\begin{aligned}
L &= l\xi^2 + m\eta^2 + n\zeta^2, \\
I &= la\xi + mb\eta + nc\zeta, \\
P &= la^2 + mb^2 + nc^2 - 1, \\
R &= I^2 - PL,
\end{aligned}$$

then r', r'' will be the roots of the equation $Lr^2 - 2Ir + P = 0$, and we have consequently $r' = \dfrac{I - \sqrt{R}}{L}$, $r'' = \dfrac{I + \sqrt{R}}{L}$. We have, therefore, for the attractions parallel to the axes (and tending towards the centre) and for the potential of the entire ellipsoid the well-known expressions

$$A = 2\int dS \frac{\xi\sqrt{R}}{L},$$

$$B = 2\int dS \frac{\eta\sqrt{R}}{L},$$

$$C = 2\int dS \frac{\zeta\sqrt{R}}{L},$$

and

$$V = 2\int \frac{I\sqrt{R}dS}{L^2},$$

where the integrations extend over the spherical angle of the circumscribed cone $R=0$.

We have moreover, by the general theory of attractions,

$$A = -\frac{dV}{da},\ B = -\frac{dV}{db},\ C = -\frac{dV}{dc}.$$

Since at the limits of the integration the quantity under the integral sign vanishes, it is easy to see that the first differentials of V, A, B, C with respect to any of the quantities a, b, c, l, m, n, k, may be found by simply differentiating under the integral sign without its being necessary to pay attention to the variation of the limits, so that, for instance, $\frac{dV}{da} = 2\int dS \frac{d}{da}\frac{I\sqrt{R}}{L}$. It is proper to remark that the expressions thus obtained for the attractions A, B, C, are of a different form from the foregoing expressions for the same quantities. Laplace writes

$$F = aA + bB + cC,$$

and he remarks that it may be shown by differentiation that the quantities, B, C, F, V, are connected by an equation which (writing k for k^2, and $\frac{m}{l}$, $\frac{n}{l}$ for m, n, the equation of the ellipsoid being with Laplace $x^2+my^2+nz^2=k^2$) becomes, in the notation of the present memoir,

$$\begin{aligned}&\{l(a^2+b^2+c^2)-k\}\left(\frac{dV}{dk}-\frac{dF}{dk}\right)+V-F\\&+\left(1-\frac{l}{m}\right)b\left(\frac{dF}{db}-\tfrac{1}{2}\frac{dV}{db}-B\right)\\&+\left(1-\frac{l}{n}\right)c\left(\frac{dF}{dc}-\tfrac{1}{2}\frac{dV}{dc}-C\right)\\&-(m-l)\frac{dF}{dm}-(n-l)\frac{dF}{dn}=0.\end{aligned}$$

I write this equation in the form

$$
\begin{aligned}
&(la^2+mb^2+nc^2-k)\left(\frac{dV}{dk}-\frac{dF}{dk}\right)+V-F\\
&\quad+(l\ -l)\left\{-a^2\left(\frac{dV}{dk}-\frac{dF}{dk}\right)+\frac{a}{l}\left(\frac{dF}{da}-\tfrac{1}{2}\frac{dV}{da}-A\right)-\frac{dF}{dl}\right\}\\
&\quad+(m-l)\left\{-b^2\left(\frac{dV}{dk}-\frac{dF}{dk}\right)+\frac{b}{m}\left(\frac{dF}{db}-\tfrac{1}{2}\frac{dV}{db}-B\right)-\frac{dF}{dm}\right\}\\
&\quad+(n\ -l)\left\{-c^2\left(\frac{dV}{dk}-\frac{dF}{dk}\right)+\frac{c}{n}\left(\frac{dF}{dc}-\tfrac{1}{2}\frac{dV}{dc}-C\right)-\frac{dF}{dn}\right\}\\
&\quad=0\,;
\end{aligned}
$$

and I remark that this equation may be broken up into two equations, each of which separately is satisfied; these equations are

$$
\begin{aligned}
&-k\left(\frac{dV}{dk}-\frac{dF}{dk}\right)+(V-F)-l\frac{dF}{dl}-m\frac{dF}{dm}-n\frac{dF}{dn}\\
&\qquad+a\left(\frac{dF}{da}-\tfrac{1}{2}\frac{dV}{da}-A\right)+b\left(\frac{dF}{db}-\tfrac{1}{2}\frac{dV}{db}-B\right)+c\left(\frac{dF}{dc}-\tfrac{1}{2}\frac{dV}{dc}-C\right)=0,
\end{aligned}
$$

and

$$
\begin{aligned}
&(a^2+b^2+c^2)\left(\frac{dV}{dk}-\frac{dF}{dk}\right)-\frac{dF}{dl}-\frac{dF}{dm}-\frac{dF}{dn}\\
&\qquad+\frac{a}{l}\left(\frac{dF}{da}-\tfrac{1}{2}\frac{dV}{da}-A\right)+\frac{b}{m}\left(\frac{dF}{db}-\tfrac{1}{2}\frac{dV}{db}-B\right)+\frac{c}{n}\left(\frac{dF}{dc}-\tfrac{1}{2}\frac{dV}{dc}-C\right)=0.
\end{aligned}
$$

It may be added that the functions under the integral signs, and consequently the integrals, are all of them homogeneous of the degree zero in l, m, n, k. The thing to be verified is that the foregoing two equations are satisfied by the functions under the integral signs, independently of the integrations, in fact by the values

$$
\left\{
\begin{aligned}
&A=\frac{\xi\sqrt{R}}{L},\ B=\frac{\eta\sqrt{R}}{L},\ C=\frac{\zeta\sqrt{R}}{L},\\
&V=\frac{I\sqrt{R}}{L},\\
&F=\frac{(a\xi+b\eta+c\zeta)\sqrt{R}}{L}.
\end{aligned}
\right.
$$

We find by differentiating these values, and after a few obvious reductions,

$$
\begin{aligned}
\frac{dV}{da}&=l\xi\left(\frac{\sqrt{R}}{L^2}+\frac{I^2}{L^2\sqrt{R}}\right)+la\left(\frac{-I}{L\sqrt{R}}\right),\\
\frac{dV}{dl}&=a\xi\left(\frac{\sqrt{R}}{L^2}+\frac{I^2}{L^2\sqrt{R}}\right)+a^2\left(\frac{-I}{2L\sqrt{R}}\right)+\xi^2\left(-\frac{3I\sqrt{R}}{2L^3}-\frac{I^3}{2L^3\sqrt{R}}\right),\\
\frac{dV}{dk}&=\qquad\qquad\qquad\qquad\frac{I}{2L\sqrt{R}},
\end{aligned}
$$

$$\frac{dF}{da} - \frac{\xi\sqrt{R}}{L} = (a\xi + b\eta + c\zeta)\left\{l\zeta\left(\frac{I}{L\sqrt{R}}\right) + la\left(-\frac{1}{\sqrt{R}}\right)\right\},$$

$$\frac{dF}{dl} = (a\xi + b\eta + c\zeta)\left\{a\xi\left(\frac{I}{L\sqrt{R}}\right) + a^2\left(\frac{-1}{2\sqrt{R}}\right) + \xi^2\left(-\frac{\sqrt{R}}{2L^2} - \frac{I^2}{2L^2\sqrt{R}}\right)\right\},$$

$$\frac{dF}{dk} = (a\xi + b\eta + c\zeta)\left\{\frac{1}{2\sqrt{R}}\right\},$$

values which give, as they should do,

$$l\frac{dF}{dl} + m\frac{dF}{dm} + n\frac{dF}{dn} + k\frac{dF}{dk} = 0.$$

Hence forming the values of the different parts of the first equation,

$$-k\left(\frac{dV}{dk} - \frac{dF}{dk}\right) + V - F - l\frac{dF}{dl} - m\frac{dF}{dm} - n\frac{dF}{dn}$$

$$= \frac{I\sqrt{R}}{L^2} - \frac{kI}{2L\sqrt{R}} + (a\xi + b\eta + c\zeta)\left\{-\frac{\sqrt{R}}{L} + \frac{k}{\sqrt{R}}\right\},$$

$$a\left(\frac{dF}{da} - A\right) + b\left(\frac{dF}{db} - B\right) + c\left(\frac{dF}{dc} - C\right) = (a\xi + b\eta + c\zeta)\left\{\frac{\sqrt{R}}{L} - \frac{k}{\sqrt{R}}\right\},$$

$$-\tfrac{1}{2}\left(a\frac{dV}{da} + b\frac{dV}{db} + c\frac{dV}{dc}\right) = -\frac{I\sqrt{R}}{L^2} + \frac{kI}{2L\sqrt{R}},$$

which satisfy the first equation.

And in like manner for the second equation,

$$-(a^2+b^2+c^2)\left(\frac{dV}{dk} - \frac{dF}{dk}\right) = -(a^2+b^2+c^2)\frac{I}{2L\sqrt{R}} + (a^2+b^2+c^2)(a\xi + b\eta + c\zeta)\frac{1}{2\sqrt{R}},$$

$$\frac{a}{l}\left(\frac{dF}{da} - A\right) + \frac{b}{m}\left(\frac{dF}{db} - B\right) + \frac{c}{n}\left(\frac{dF}{dc} - C\right)$$

$$= (a\xi + b\eta + c\zeta)^2\frac{I}{L\sqrt{R}} - (a^2+b^2+c^2)(a\xi + b\eta + c\zeta)\frac{1}{\sqrt{R}},$$

$$-\tfrac{1}{2}\left(\frac{a}{l}\frac{dV}{da} + \frac{b}{m}\frac{dV}{db} + \frac{c}{n}\frac{dV}{dc}\right) = (a^2+b^2+c^2)\frac{I}{2L\sqrt{R}} - (a\xi + b\eta + c\zeta)\left(\frac{\sqrt{R}}{2L^2} + \frac{I^2}{2L^2\sqrt{R}}\right),$$

$$-\frac{dF}{dl} - \frac{dF}{dm} - \frac{dF}{dn} = -(a\xi + b\eta + c\zeta)^2\frac{I}{2L\sqrt{R}} + (a^2+b^2+c^2)(a\xi + b\eta + c\zeta)\frac{1}{2\sqrt{R}}$$

$$+ (a\xi + b\eta + c\zeta)\left(\frac{\sqrt{R}}{2L^2} + \frac{I^2}{2L^2\sqrt{R}}\right),$$

which satisfy the second equation

Now considering V, F, A, B, C as standing for the definite integrals, we may replace A, B, C by the differential coefficients of V, and retaining for shortness F to stand for

$$F = -a\frac{dV}{da} - b\frac{dV}{db} - c\frac{dV}{dc},$$

the first equation becomes

$$-k\left(\frac{dV}{dk} - \frac{dF}{dk}\right) + V - F - l\frac{dF}{dl} - m\frac{dF}{dm} - n\frac{dF}{dn}$$
$$+ a\left(\frac{dF}{da} + \tfrac{1}{2}\frac{dV}{da}\right) + b\left(\frac{dF}{db} + \tfrac{1}{2}\frac{dV}{db}\right) + c\left(\frac{dF}{dc} + \tfrac{1}{2}\frac{dV}{dc}\right) = 0;$$

and the second equation becomes

$$-(a^2+b^2+c^2)\left(\frac{dV}{dk} - \frac{dF}{dk}\right) - \frac{dF}{dl} - \frac{dF}{dm} - \frac{dF}{dn}$$
$$+ \frac{a}{l}\left(\frac{dF}{da} + \tfrac{1}{2}\frac{dV}{da}\right) + \frac{b}{m}\left(\frac{dF}{db} + \tfrac{1}{2}\frac{dV}{db}\right) + \frac{c}{n}\left(\frac{dF}{dc} + \tfrac{1}{2}\frac{dV}{dc}\right) = 0.$$

If we put as before $V = \frac{I\sqrt{R}}{L^2}$, the preceding values of the differential coefficients of V give $F = \frac{2I\sqrt{R}}{L^2} + \frac{kI}{L\sqrt{R}}$, or as we may write it $F = -2V + W$, where $W = \frac{kI}{L\sqrt{R}}$. I put then for the moment

$$\begin{cases} V = \dfrac{I\sqrt{R}}{L^2},\ W = \dfrac{kI}{L\sqrt{R}}, \\ F = -2V + W. \end{cases}$$

It should be remarked that there is nothing in what has preceded which tends to show that these values must satisfy the differential equations. The definite integrals must, of course, satisfy as before the equations, but it does not follow that the equations are satisfied by the elements separately. And in fact only the first equation is so satisfied; the second equation is not satisfied. To verify this I form the differential coefficients

$$\frac{dW}{da} = l\xi\left(\frac{k}{L\sqrt{R}} - \frac{kI^2}{LR\sqrt{R}}\right) + la\left(\frac{kI}{R\sqrt{R}}\right),$$

$$\frac{dW}{dk} = \qquad \frac{I}{L\sqrt{R}} - \frac{kI}{2R\sqrt{R}},$$

$$\frac{dW}{dl} = a\xi\left(\frac{k}{L\sqrt{R}} - \frac{kI^2}{LR\sqrt{R}}\right) + \xi^2\left(\frac{-3kI}{2L^2\sqrt{R}} + \frac{kI^3}{2L^2R\sqrt{R}}\right) + a^2\frac{kI}{2R\sqrt{R}}.$$

We have as before

$$l\frac{dF}{dl}+m\frac{dF}{dm}+n\frac{dF}{dn}+k\frac{dF}{dk}=0,$$

and forming the expressions for the different parts of the first equation,

$$-k\left(\frac{dV}{dk}-\frac{dF}{dk}\right)+V-F-l\,\frac{dF}{dl}-m\,\frac{dF}{dm}-n\,\frac{dF}{dn}$$

$$=-\,5k\frac{dV}{dk}+2k\frac{dW}{dk}+3V-W \qquad = \frac{3I\sqrt{R}}{L^2}-\frac{3kI}{2L\sqrt{R}}-\frac{k^2I}{R\sqrt{R}},$$

$$a\left(\frac{dF}{da}+\tfrac{1}{2}\frac{dV}{da}\right)+\text{\&c.}=$$

$$-\tfrac{3}{2}\left(a\,\frac{dV}{da}+b\,\frac{dV}{db}+c\,\frac{dV}{dc}\right)+\left(a\,\frac{dW}{da}+b\,\frac{dW}{db}+c\,\frac{dW}{dc}\right),$$

and

$$-\tfrac{3}{2}\left(a\,\frac{dV}{da}+b\,\frac{dV}{db}+c\,\frac{dV}{dc}\right) \qquad =-\frac{3I\sqrt{R}}{L^2}+\frac{3kI}{2L\sqrt{R}},$$

$$\left(a\,\frac{dW}{da}+b\,\frac{dW}{db}+c\,\frac{dW}{dc}\right) \qquad = \qquad \frac{k^2I}{R\sqrt{R}},$$

which show that the first equation is satisfied.

Proceeding in the same manner with the second equation,

$$-\,(a^2+b^2+c^2)\left(\frac{dV}{dk}-\frac{dF}{dk}\right)=$$

$$-\,(a^2+b^2+c^2)\left(3\,\frac{dV}{dk}-\frac{dW}{dk}\right)=(a^2+b^2+c^2)\left(-\frac{I}{L\sqrt{R}}-\frac{kI}{2R\sqrt{R}}\right)$$

$$\frac{a}{l}\left(\frac{dF}{da}+\tfrac{1}{2}\frac{dV}{da}\right)+\text{\&c.}=$$

$$-\tfrac{3}{2}\left(\frac{a}{l}\frac{dV}{da}+\frac{b}{m}\frac{dV}{db}+\frac{c}{n}\frac{dV}{dc}\right)+\left(\frac{a}{l}\frac{dW}{da}+\frac{b}{m}\frac{dW}{db}+\frac{c}{n}\frac{dW}{dc}\right);$$

and

$$-\tfrac{3}{2}\left(\frac{a}{l}\frac{dV}{da}+\frac{b}{m}\frac{dV}{db}+\frac{c}{n}\frac{dV}{dc}\right)=(a^2+b^2+c^2)\frac{3I}{2L\sqrt{R}}+(a\xi+b\eta+c\zeta)\left(-\frac{3\sqrt{R}}{2L^2}-\frac{3I^2}{2L^2\sqrt{R}}\right),$$

$$\left(\frac{a}{l}\frac{dW}{da}+\frac{b}{m}\frac{dW}{db}+\frac{c}{n}\frac{dW}{dc}\right)=(a^2+b^2+c^2)\frac{kI}{R\sqrt{R}}+(a\xi+b\eta+c\zeta)\left(\frac{k}{L\sqrt{R}}-\frac{kI^2}{LR\sqrt{R}}\right);$$

$$-\frac{dF}{dl}-\frac{dF}{dm}-\frac{dF}{dn}=2\left(\frac{dV}{dl}+\frac{dV}{dm}+\frac{dV}{dn}\right)-\left(\frac{dW}{dl}+\frac{dW}{dm}+\frac{dW}{dn}\right),$$

and

$$2\left(\frac{dV}{dl}+\frac{dV}{dm}+\frac{dV}{dn}\right)=(a^2+b^2+c^2)\left(\frac{-1}{L\sqrt{R}}\right)+(a\xi+b\eta+c\zeta)\left(\frac{2\sqrt{R}}{L^2}+\frac{2I^2}{L^2\sqrt{R}}\right)-\frac{3I\sqrt{R}}{L^3}-\frac{I^3}{L^3\sqrt{R}}$$

$$-\left(\frac{dW}{dl}+\frac{dW}{dm}+\frac{dW}{dn}\right)=(a^2+b^2+c^2)\left(\frac{-kI}{2R\sqrt{R}}\right)+(a\xi+b\eta+c\zeta)\left(\frac{-k}{L\sqrt{R}}+\frac{kI^2}{LR\sqrt{R}}\right)$$

$$+\frac{3kI}{2L^2\sqrt{R}}-\frac{kI^3}{2L^2R\sqrt{R}};$$

and the value of the left-hand side of the second equation is therefore,

$$(a\xi+b\eta+c\zeta)\left(\frac{\sqrt{R}}{2L^2}+\frac{3I^2}{2L^2\sqrt{R}}\right)-\frac{3I\sqrt{R}}{L^3}-\frac{I^3}{2L^3\sqrt{R}}+\frac{3kI}{2L^2\sqrt{R}}-\frac{kI^3}{2L^2R\sqrt{R}},$$

which is not equal to zero, or the second equation is not satisfied.

I consider again V, F as denoting the definite integrals, and I eliminate $\frac{dV}{dk}$, $\frac{dF}{dk}$ by means of the equations

$$l\frac{dV}{dl}+m\frac{dV}{dm}+n\frac{dV}{dn}+k\frac{dV}{dk}=0,$$

$$l\frac{dF}{dl}+m\frac{dF}{dm}+n\frac{dF}{dn}+k\frac{dF}{dk}=0.$$

The first equation thus becomes

$$l\frac{dV}{dl}+m\frac{dV}{dm}+n\frac{dV}{dn}-2\left(l\frac{dF}{dl}+m\frac{dF}{dm}+n\frac{dF}{dn}\right)+V-F$$

$$+\tfrac{1}{2}\left(a\frac{dV}{da}+b\frac{dV}{db}+b\frac{dV}{dc}\right)+\left(a\frac{dF}{da}+b\frac{dF}{db}+c\frac{dF}{dc}\right)=0,$$

and the second equation becomes

$$(a^2+b^2+c^2)\frac{1}{k}\left(l\frac{dV}{dl}+m\frac{dV}{dm}+n\frac{dV}{dn}\right)-(a^2+b^2+c^2)\frac{1}{k}\left(l\frac{dF}{dl}+m\frac{dF}{dm}+n\frac{dF}{dn}\right)$$

$$+\tfrac{1}{2}\left(\frac{a}{l}\frac{dV}{da}+\frac{bd}{m}\frac{V}{db}+\frac{c}{n}\frac{dV}{dc}\right)+\left(\frac{a}{l}\frac{dF}{da}+\frac{b}{m}\frac{dF}{db}+\frac{c}{n}\frac{dF}{dc}\right)-\left(\frac{dF}{dl}+\frac{dF}{dm}+\frac{dF}{dn}\right)=0;$$

and it will be remembered that in these equations

$$F=-a\frac{dV}{da}-b\frac{dV}{db}-c\frac{dV}{dc}.$$

The first equation (it is easy to perceive) shows merely that V is made up of terms separately homogeneous in a, b, c, and in l, m, n, and such that the degrees in the two sets respectively being κ, λ, then $\lambda-\frac{1}{2}(\kappa-2)$. In fact V being a function of the form in question, if we attend only to the term the degrees of which are κ, λ, then, by the properties of homogeneous functions, $F = -V$, and the first equation is satisfied if only

$$\lambda + 2\lambda\kappa + 1 + \kappa + \tfrac{1}{2}\kappa - \kappa^2 = 0,$$

i.e. if $\lambda(2\kappa+1) = \kappa^2 - \tfrac{3}{2}\kappa - 1 = \tfrac{1}{2}(\kappa-2)(2\kappa+1)$ or $\lambda = \tfrac{1}{2}(\kappa - 2)$. Or, what is the same thing, we may say that the first equation shows that V is made up of terms the degrees of which in a, b, c and in l, m, n are $-2i-1$, and $-i-\tfrac{3}{2}$ respectively. Attending henceforth only to the second equation, I write $\frac{l}{k} = \frac{1}{\alpha+\theta}$, $\frac{m}{k} = \frac{1}{\beta+\theta}$, $\frac{n}{k} = \frac{1}{\gamma+\theta}$; so that $\alpha+\theta$, $\beta+\theta$, $\gamma+\theta$ denote the squared semiaxes of the ellipsoid. We have

$$\frac{d}{dl} = -\frac{(\alpha+\theta)^2}{k}\frac{d}{d\alpha}, \text{ \&c.,}$$

and the equation becomes

$$\begin{aligned}
&-(a^2+b^2+c^2)\left((\alpha+\theta)\frac{dV}{d\alpha} + (\beta+\theta)\frac{dV}{d\beta} + (\gamma+\theta)\frac{dV}{d\gamma}\right)\\
&+(a^2+b^2+c^2)\left((\alpha+\theta)\frac{dF}{d\alpha} + (\beta+\theta)\frac{dF}{d\beta} + (\gamma+\theta)\frac{dF}{d\gamma}\right)\\
&+\tfrac{1}{2}\left(a(\alpha+\theta)\frac{dV}{da} + b(\beta+\theta)\frac{dV}{db} + c(\gamma+\theta)\frac{dV}{dc}\right)\\
&+\left(a(\alpha+\theta)\frac{dF}{da} + b(\beta+\theta)\frac{dF}{db} + c(\gamma+\theta)\frac{dF}{dc}\right)\\
&+\left((\alpha+\theta)^2\frac{dF}{d\alpha} + (\beta+\theta)^2\frac{dF}{d\beta} + (\gamma+\theta)^2\frac{dF}{d\gamma}\right) = 0.
\end{aligned}$$

Put for shortness $\Theta = (\alpha+\theta)(\beta+\theta)(\gamma+\theta)$, and write

$$V = v\sqrt{\Theta},\quad F = f\sqrt{\Theta},$$

($\sqrt{\Theta}$ is to a constant factor *près* the mass of the ellipsoid) then v, f are connected by the equation

$$f = -a\frac{dv}{da} - b\frac{dv}{db} - c\frac{dv}{dc};$$

and observing that

$$(\alpha+\theta)\frac{d\sqrt{\Theta}}{d\alpha} + (\beta+\theta)\frac{d\sqrt{\Theta}}{d\beta} + (\gamma+\theta)\frac{d\sqrt{\Theta}}{d\gamma} = 3\Theta\frac{1}{2\sqrt{\Theta}} = \tfrac{3}{2}\sqrt{\Theta},$$

$$\begin{aligned}
&(\alpha+\theta)^2\frac{d\sqrt{\Theta}}{d\alpha} + (\beta+\theta)^2\frac{d\sqrt{\Theta}}{d\beta} + (\gamma+\theta)^2\frac{d\sqrt{\Theta}}{d\gamma}\\
&\qquad = (\alpha+\beta+\gamma+3\theta)\,\Theta\frac{1}{2\sqrt{\Theta}} = \tfrac{1}{2}(\alpha+\beta+\gamma+3\theta)\sqrt{\theta},
\end{aligned}$$

the equation becomes

$$-(a^2+b^2+c^2)\left((\alpha+\theta)\frac{dv}{d\alpha}+(\beta+\theta)\frac{dv}{d\beta}+(\gamma+\theta)\frac{dv}{d\gamma}\right)$$

$$+(a^2+b^2+c^2)\left((\alpha+\theta)\frac{df}{d\alpha}+(\beta+\theta)\frac{df}{d\beta}+(\gamma+\theta)\frac{df}{d\gamma}\right)$$

$$-\tfrac{3}{2}(a^2+b^2+c^2)\,v+\tfrac{3}{2}(a^2+b^2+c^2)f$$

$$+\tfrac{1}{2}\left(a(\alpha+\theta)\frac{dv}{da}+b(\beta+\theta)\frac{dv}{db}+c(\gamma+\theta)\frac{dv}{dc}\right)$$

$$+\ \left(a(\alpha+\theta)\frac{df}{da}+b(\beta+\theta)\frac{df}{db}+c(\gamma+\theta)\frac{df}{dc}\right)$$

$$+\left((\alpha+\theta)^2\frac{df}{d\alpha}+(\beta+\theta)^2\frac{df}{d\beta}+(\gamma+\theta)^2\frac{df}{d\gamma}\right)$$

$$+\tfrac{1}{2}(\alpha+\beta+\gamma+3\theta)f=0.$$

Now $v\left(=\frac{1}{\sqrt{\Theta}}V\right)$ may be expanded in the form

$$v=U_0+U_1+U_2\ldots+U_i\ldots,$$

where U_i is of the degree $2i$ in the semiaxes $\sqrt{(\alpha+\theta)}$, $\sqrt{(\beta+\theta)}$, $\sqrt{(\gamma+\theta)}$, and of the degree $-2i-1$ in the coordinates a, b, c. And it is easy to see that the first term of the expansion is

$$U_0=\frac{4\pi}{3}\,\frac{1}{\sqrt{(a^2+b^2+c^2)}}.$$

The preceding value of v gives

$$f=U_0+3U_1+5U_2\ldots+(2i+1)\,U_i+\ldots,$$

and substituting the values of v, f in the differential equation, and attending only to the terms which are of the same dimensions as $(a^2+b^2+c^2)\,U_{i+1}$, we have

$$\left\{-(i+1)+(i+1)(2i+3)-\tfrac{3}{2}+\tfrac{3}{2}(2i+3)\right\}(a^2+b^2+c^2)\,U_{i+1}$$

$$+\left\{a\alpha\frac{dU_i}{da}+b\beta\frac{dU_i}{d\beta}+c\gamma\frac{dU_i}{d\gamma}-(2i+1)\,\theta U_i\right\}\left(\tfrac{1}{2}+2i+1\right)$$

$$+\left\{\alpha^2\frac{dU_i}{d\alpha}+\beta^2\frac{dU_i}{d\beta}+\gamma^2\frac{dU_i}{d\gamma}+2i\theta U_i-\theta^2\left(\frac{dU_i}{d\alpha}+\frac{dU_i}{d\beta}+\frac{dU_i}{d\gamma}\right)\right\}(2i+1)$$

$$+\tfrac{1}{2}(\alpha+\beta+\gamma+3\theta)(2i+1)\,U_i=0;$$

or, reducing,

$$(i+1)(2i+5)(a^2+b^2+c^2)\,U_{i+1}$$

$$+(2i+\tfrac{3}{2})\left(a\alpha\frac{dU_i}{da}+b\beta\frac{dU_i}{db}+c\gamma\frac{dU_i}{dc}\right)$$

$$+(2i+1)\left(\alpha^2\frac{dU_i}{d\alpha}+\beta^2\frac{dU_i}{d\beta}+\gamma^2\frac{dU_i}{d\gamma}+\tfrac{1}{2}(\alpha+\beta+\gamma)\,U_i\right)$$

$$-(2i+1)\left(\frac{dU_i}{d\alpha}+\frac{dU_i}{d\beta}+\frac{dU_i}{d\gamma}\right)=0.$$

Now U_i is a function of $\alpha+\theta$, $\beta+\theta$, $\gamma+\theta$; if, therefore, for any particular value of i, U_i is independent of θ, it is clear that U_i must be a function of the differences of these quantities, and we shall have $\frac{dU_i}{d\alpha}+\frac{dU_i}{d\beta}+\frac{dU_i}{d\gamma}=0$; and this being so, the equation of differences, and consequently U_{i+1}, will be independent of θ. But U_0 is independent of θ, hence U_1, U_2, &c. are all of them independent of θ, or v is independent of θ; i.e. for ellipsoids having the same foci for their principal sections, and acting on the same external point, the potentials, and therefore the attractions, are proportional to the masses, which is Maclaurin's theorem for the attractions of ellipsoids upon an external point.

The foregoing equation, omitting the term which vanishes, gives

$$U_{i+1}=\frac{-(2i+\tfrac{3}{2})\left(a\alpha\frac{dU_i}{da}+b\beta\frac{dU_i}{db}+c\gamma\frac{dU_i}{dc}\right)-(2i+1)\left(\alpha^2\frac{dU_i}{d\alpha}+\beta^2\frac{dU_i}{d\beta}+\gamma^2\frac{dU_i}{d\gamma}+\tfrac{1}{2}(\alpha+\beta+\gamma)\,U_i\right)}{(i+1)(2i+5)(a^2+b^2+c^2)}.$$

It may be remarked that this equation, with the assistance of the equation

$$\frac{dU_i}{d\alpha}+\frac{dU_i}{d\beta}+\frac{dU_i}{d\gamma}=0,$$

gives

$$(i+1)(2i+5)\left(\frac{dU_{i+1}}{d\alpha}+\frac{dU_{i+1}}{d\beta}+\frac{dU_{i+1}}{d\gamma}\right)=-(2i+\tfrac{3}{2})\left(a\frac{dU_i}{da}+b\frac{dU_i}{db}+c\frac{dU_i}{dc}\right)$$

$$-(2i+1)\left(2\left(\alpha\frac{dU_i}{d\alpha}+\beta\frac{dU_i}{d\beta}+\gamma\frac{dU_i}{d\gamma}\right)+\tfrac{3}{2}U_i\right)=\left((2i+\tfrac{3}{2})(2i+1)-(2i+1)(2i+\tfrac{3}{2})\right)U_i=0,$$

which is as it should be.

Write

$$U_i=\frac{Q_i}{(a^2+b^2+c^2)^{2i+\frac{1}{2}}},$$

where U_i is of the degree i in α, β, γ, and of the degree $2i$ in a, b, c. We have

$$a\alpha\frac{dU_i}{da}+b\beta\frac{dU_i}{db}+c\gamma\frac{dU_i}{dc}=(a^2+b^2+c^2)^{-2i-\frac{1}{2}}\left(a\alpha\frac{dQ_i}{da}+b\beta\frac{dQ_i}{db}+c\gamma\frac{dQ_i}{dc}\right)$$

$$-(4i+1)(a^2+b^2+c^2)^{-2i-\frac{3}{2}}(a^2\alpha+b^2\beta+c^2\gamma)\,Q_i,$$

$$\alpha^2\frac{dU_i}{d\alpha}+\beta^2\frac{dU_i}{d\beta}+\gamma^2\frac{dU_i}{d\gamma}+\tfrac{1}{2}(\alpha+\beta+\gamma)\,U_i$$

$$=(a^2+b^2+c^2)^{-2i-\frac{1}{2}}\left(\alpha^2\frac{dQ_i}{d\alpha}+\beta^2\frac{dQ_i}{d\beta}+\gamma^2\frac{dQ_i}{d\gamma}\right)+\tfrac{1}{2}(\alpha+\beta+\gamma)\,Q_i.$$

Hence, putting in like manner

$$(i+1)(2i+5)\,Q_{i+1}=-(2i+\tfrac{3}{2})\left\{(a^2+b^2+c^2)\left(a\alpha\frac{dQ_i}{da}+b\beta\frac{dQ_i}{db}+c\gamma\frac{dQ_i}{dc}\right)\right.$$

$$\left.-(4i+1)(a^2\alpha+b^2\beta+c^2\gamma)\,Q_i\right\}$$

$$-(2i+1)(a^2+b^2+c^2)\left(\alpha^2\frac{dQ_i}{d\alpha}+\beta^2\frac{dQ_i}{d\beta}+\gamma^2\frac{dQ_i}{d\gamma}+\tfrac{1}{2}(\alpha+\beta+\gamma)\,Q_i\right),$$

i.e.

$$(i+1)(2i+5)\,Q_{i+1}=-(a^2+b^2+c^2)\left\{(2i+\tfrac{3}{2})\left(a\alpha\frac{dQ_i}{da}+d\beta\frac{dQ_i}{db}+c\gamma\frac{dQ_i}{dc}\right)\right.$$

$$\left.+(2i+1)\left(\alpha^2\frac{dQ_i}{d\alpha}+\beta^2\frac{dQ_i}{d\beta}+\gamma^2\frac{dQ_i}{d\gamma}+\tfrac{1}{2}(\alpha+\beta+\gamma)\,Q_i\right)\right\}+(2i+\tfrac{3}{2})(4i+1)(a^2\alpha+b^2\beta+c^2\gamma)\,Q_i,$$

from which the functions Q_i may be calculated successively.

We may, it is clear, write

$$U_i=\frac{4\pi}{3}\cdot\frac{1}{2^i\,1\,.\,2\,.\,3\,..\,i\;5\,.\,7\,..\,2i+3}K_i,$$

and we shall then have

$$(a^2+b^2+c^2)\,K_{i+1}+(4i+3)\left(a\alpha\frac{dK_i}{da}+b\beta\frac{dK_i}{db}+c\gamma\frac{dK_i}{dc}\right)$$

$$+(4i+2)\left(\alpha^2\frac{dK_i}{d\alpha}+\beta^2\frac{dK_i}{d\beta}+\gamma^2\frac{dK_i}{d\gamma}\right)$$

$$+(2i+1)\;(\alpha+\beta+\gamma)\,K_i=0,$$

where

$$K_0=\frac{1}{\sqrt{(a^2+b^2+c^2)}}.$$

I proceed to show that

$$K_i=\left(\alpha\frac{d^2}{da^2}+\beta\frac{d^2}{db^2}+\gamma\frac{d^2}{dc^2}\right)^i\frac{1}{\sqrt{(a^2+b^2+c^2)}}.$$

In fact, assuming this equation for any particular value of i, we find first

$$\left(\alpha^2 \frac{d}{d\alpha} + \beta^2 \frac{d}{d\beta} + \gamma^2 \frac{d}{d\gamma}\right) K_i = i\left(\alpha \frac{d^2}{da^2} + \beta \frac{d^2}{db^2} + \gamma \frac{d^2}{dc^2}\right)^{i-1} \left(\alpha^2 \frac{d^2}{da^2} + \beta^2 \frac{d^2}{db^2} + \gamma^2 \frac{d^2}{dc^2}\right) \frac{1}{\sqrt{(a^2+b^2+c^2)}}$$

$$= i\left(\alpha^2 \frac{d^2}{da^2} + \beta^2 \frac{d^2}{db^2} + \gamma^2 \frac{d^2}{dc^2}\right) \left(\alpha \frac{d^2}{da^2} + \beta \frac{d^2}{db^2} + \gamma \frac{d^2}{dc^2}\right)^{i-1} \frac{1}{\sqrt{(a^2+b^2+c^2)}}$$

$$= i\left(\alpha^2 \frac{d^2}{da^2} + \beta^2 \frac{d^2}{db^2} + \gamma^2 \frac{d^2}{dc^2}\right) K_{i-1}.$$

Now putting for shortness

$$\Delta = \alpha \frac{d^2}{da^2} + \beta \frac{d^2}{db^2} + \gamma \frac{d^2}{dc^2},$$

we find, replacing ΔK_{i+1} and ΔK_i by their values K_{i+2} and K_{i+1},

$$\begin{aligned}
\Delta(a^2+b^2+c^2)K_{i+1} = {} & (a^2+b^2+c^2)K_{i+2} \\
& + 4\left(a\alpha \frac{d}{da} + b\beta \frac{d}{db} + c\gamma \frac{d}{dc}\right) K_{i+1} \\
& + 2(\alpha+\beta+\gamma)K_{i+1}, \\
\Delta\left(a\alpha \frac{d}{da} + b\beta \frac{d}{db} + c\gamma \frac{d}{dc}\right) K_i = {} & \left(a\alpha \frac{d}{da} + d\beta \frac{d}{db} + c\gamma \frac{d}{dc}\right) K_{i+1} \\
& + 2\left(\alpha^2 \frac{d^2}{da^2} + \beta^2 \frac{d^2}{db^2} + \gamma^2 \frac{d^2}{dc^2}\right) K_i, \\
\Delta\left(\alpha^2 \frac{d}{d\alpha} + \beta^2 \frac{d}{d\beta} + \gamma^2 \frac{d}{d\gamma}\right) K_i = {} & \left(\alpha^2 \frac{d}{d\alpha} + \beta^2 \frac{d}{d\beta} + \gamma^2 \frac{d}{d\gamma}\right) K_{i+1} \\
& - \left(\alpha^2 \frac{d^2}{da^2} + \beta^2 \frac{d^2}{db^2} + \gamma^2 \frac{d^2}{dc^2}\right) K_i, \\
\Delta(\alpha+\beta+\gamma)K_i = {} & (\alpha+\beta+\gamma)K_{i+1};
\end{aligned}$$

hence operating on the equation of differences with the symbol Δ, we obtain

$$\begin{aligned}
(a^2+b^2+c^2)K_{i+2} + (4i+7)\left(a\alpha \frac{d}{da} + b\beta \frac{d}{db} + c\gamma \frac{d}{dc}\right) K_{i+1} & \\
+ (4i+2)\left(\alpha^2 \frac{d}{d\alpha} + \beta^2 \frac{d}{d\beta} + \gamma^2 \frac{d}{d\gamma}\right) K_{i+1} & \\
+ \{2(4i+3) - (4i+2)\}\left(\alpha^2 \frac{d^2}{da^2} + \beta^2 \frac{d^2}{db^2} + \gamma^2 \frac{d^2}{dc^2}\right) K_i & \\
+ (2i+3)(\alpha+\beta+\gamma)K_{i+1} = 0, &
\end{aligned}$$

the third line of which is

$$4(i+1)\left(\alpha^2\frac{d^2}{da^2}+\beta^2\frac{d^2}{db^2}+\gamma^2\frac{d^2}{dc^2}\right)K_i, =4\left(\alpha^2\frac{d}{d\alpha}+\beta^2\frac{d}{d\beta}+\gamma^2\frac{d}{d\gamma}\right)K_{i+1},$$

by a foregoing equation, and the assumed equation of difference thus leads to

$$\begin{aligned}(a^2+b^2+c^2)K_{i+2}+(4i+7)\left(a\alpha\frac{d}{da}+b\beta\frac{d}{db}+c\gamma\frac{d}{dc}\right)K_{i+1}\\ +(4i+6)\left(\alpha^2\frac{d}{d\alpha}+\beta^2\frac{d}{d\beta}+\gamma^2\frac{d}{d\gamma}\right)K_{i+1}\\ +(2i+3)\left(\alpha+\beta+\gamma\right)K_{i+1}=0,\end{aligned}$$

which is the assumed equation, writing $i+1$ instead of i. The equation, if true for i, is therefore true for $i+1$, and it is easily seen to be true for $i=0$; hence it is true generally, or the value

$$K_i=\left(\alpha\frac{d^2}{da^2}+\beta\frac{d^2}{db^2}+\gamma\frac{d^2}{dc^2}\right)^i\frac{1}{\sqrt{(a^2+b^2+c^2)}}$$

satisfies the equation obtained by Laplace's method, and gives, moreover, the proper value for K_0. We have thus the value of K_i; and remembering that

$$V=\sqrt{(\alpha+\theta)(\beta+\theta)(\gamma+\theta)}\,v,$$

and observing that the symbol Δ may be replaced by

$$\Delta=(\alpha+\theta)\frac{d^2}{da^2}+(\beta+\theta)\frac{d^2}{db^2}+(\gamma+\theta)\frac{d^2}{dc^2},$$

the value of V is

$$V=\frac{4\pi}{3}\sqrt{(\alpha+\theta)(\beta+\theta)(\gamma+\theta)}\,S_{i_0}^{\infty}\left(\frac{1}{2^i\,1\,.\,2\ldots i\,.\,5\,.\,7\ldots\,2i+3}\Delta^i\frac{1}{\sqrt{(a^2+b^2+c^2)}}\right);$$

which is in fact the value which I have found by a much more simple method in the *Cambridge Mathematical Journal*, t. III. p. 69 [2].

174.

ON THE OVAL OF DESCARTES.

[From the *Quarterly Mathematical Journal*, vol. I. (1857), pp. 320—328.]

This is an extract of the passage in the Geometry (La Géométrie de M. Descartes, 12° Paris, 1705, pp. 79—92) in which he applies his method of coordinates to this now well-known curve: the marginal reference is "Explication de quatre nouveaux genres d'Ovales qui servent à l'Optique." The paper was intended to be introductory to a discussion in some detail of the history and properties of the Curve, but no continuation was written.

175.

ON THE PORISM OF THE IN-AND-CIRCUMSCRIBED TRIANGLE.

[From the *Quarterly Mathematical Journal*, vol. I. (1857), pp. 344—354.]

THE porism of the in-and-circumscribed triangle in its most general form relates to a triangle the angles of which lie in fixed curves, and the sides of which touch fixed curves; but at present I consider only the case in which the angles lie in one and the same fixed curve, which for greater simplicity I assume to be a conic. We have therefore a triangle ABC, the angles of which lie in a fixed conic $\mathfrak{S}$, and the sides of which touch the fixed curves $\mathfrak{A}$, $\mathfrak{B}$, $\mathfrak{C}$; the points of contact may be represented by α, β, γ. And if we consider the conic $\mathfrak{S}$ and the curves $\mathfrak{A}$, $\mathfrak{B}$ as given, the curve $\mathfrak{C}$ will be the envelope of the side AB; to construct this side we have only to take at pleasure a point C on the curve $\mathfrak{S}$ and to draw through this point tangents to the curve $\mathfrak{B}$, $\mathfrak{A}$ respectively meeting the conic $\mathfrak{S}$ in the points A and B; the line joining these points is the required side AB. I may notice that in the case supposed of the curve $\mathfrak{S}$ being a conic, the lines $A\alpha$, $B\beta$, $C\gamma$ meet in a point; which gives at once a construction for γ, the point of contact of AB with the curve C. For the sake however of exhibiting the reasoning in a form which may be modified so as to be applicable to a curve $\mathfrak{S}$ of any order, instead of the conic $\mathfrak{S}$, I shall dispense with the employment of the property just mentioned, which is peculiar to the case of the conic.

Suppose for a moment (figs. 1 and 1 *bis*) that the curves $\mathfrak{A}$, $\mathfrak{B}$ are points, and let the line through $\mathfrak{A}$, $\mathfrak{B}$ meet the conic $\mathfrak{S}$ in the points M, N. If we take the point N for the angle C of the triangle, the points A, B will each of them coincide with M, and the side AB will be the tangent at M to the conic $\mathfrak{S}$: call this tangent T. Consider next a point C in the neighbourhood of N, we shall have two points A, B in the neighbourhood of M, and the point in which AB intersects T will be the point of contact γ of T with the curve $\mathfrak{C}$. To find this point, suppose that $M\mathfrak{A}=a$, $M\mathfrak{B}=b$, $N\mathfrak{A}=a'$, $N\mathfrak{B}=b'$ and let the distance of C from N be ds; the distances parallel to T of A, B from the line MN will be proportional to $\frac{b}{b'}$, ds, $\frac{a}{a'}ds$, and the perpendicular distances of these points from T are consequently proportional to $\frac{b^2}{b'^2}ds^2$, $\frac{a^2}{a'^2}ds^2$. The inclination of AB to T is therefore proportional to

$$\left(\frac{b^2}{b'^2}-\frac{a^2}{a'^2}\right)ds^2 \div \left(\frac{b}{b}-\frac{a}{a'}\right)ds, \text{ i.e. to } \left(\frac{b}{b'}+\frac{a}{a'}\right)ds,$$

which is of the same order as ds, and it is at once seen that the point γ will

Fig. 1.

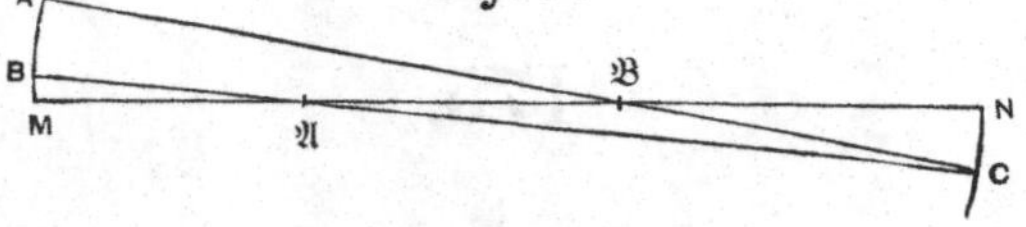

Fig. 1. bis.

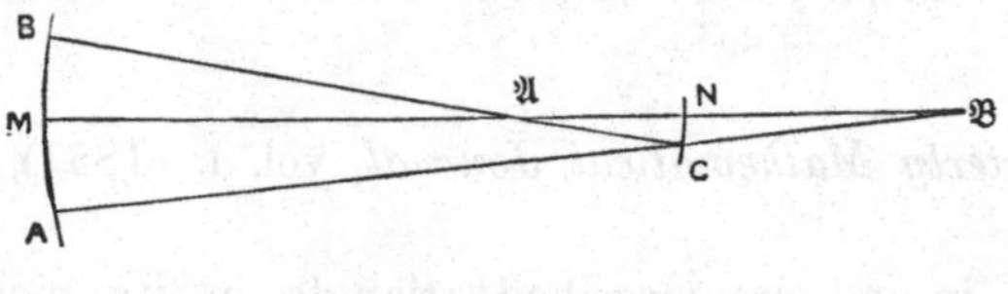

coincide with M, i.e. that the curve $\mathfrak{C}$ touches the conic $\mathfrak{S}$ at the point M. If, however, $\frac{a}{a'}+\frac{b}{b'}=0$, i.e. if the points M, N are harmonically related to $\mathfrak{A}$, $\mathfrak{B}$, then the inclination is in general of the order ds^2, and the point γ will be at a finite distance from M; moreover T is in this case a stationary tangent (i.e. a tangent at a point of inflexion) of the curve $\mathfrak{C}$. Now reverting to the general case of $\mathfrak{A}$, $\mathfrak{B}$ being any two curves, then if there be a common tangent touching these curves in the points α, β, and meeting the curve $\mathfrak{S}$ in the points M, N, the like reasonings apply to this case. Hence

FIRST LEMMA. If a common tangent to the curves $\mathfrak{A}$, $\mathfrak{B}$ touch these curves in α, β and meet the conic $\mathfrak{S}$ in the points M, N; the point N gives rise to a branch of the curve $\mathfrak{C}$ which (except in the case after mentioned) touches the conic $\mathfrak{S}$ at the point M. If however M, N are harmonically related with respect to α, β, then the branch of the curve $\mathfrak{C}$ does *not* pass through M, but it has for a *stationary* tangent the tangent M to the conic $\mathfrak{S}$.

Suppose again that the curve $\mathfrak{A}$ (fig. 2) is a point, and let the curve $\mathfrak{B}$ intersect the conic $\mathfrak{S}$ at the point M, and let the tangent to $\mathfrak{B}$ at M meet $\mathfrak{S}$ in a point R, and $R\mathfrak{A}$ meet $\mathfrak{S}$ in a point Q. Then taking the point R for the angle C of the triangle, we shall have A, B coinciding with M, Q respectively, and thence MQ a tangent to the curve $\mathfrak{C}$. To find the point of contact, take C in the neighbourhood of R at a distance from it ds. B will be a point in the neighbourhood of Q and distant from it by an infinitesimal of the order ds, A will be in the neighbourhood of M and distant from it by an infinitesimal of the order ds^2. Hence AB will intersect MQ at the point M, or the curve $\mathfrak{C}$ will pass through M. Reverting to the general case where $\mathfrak{A}$ is a curve, we have only to consider RQ as a tangent to the curve $\mathfrak{A}$ at a point α, and the like reasoning will apply to this case: hence

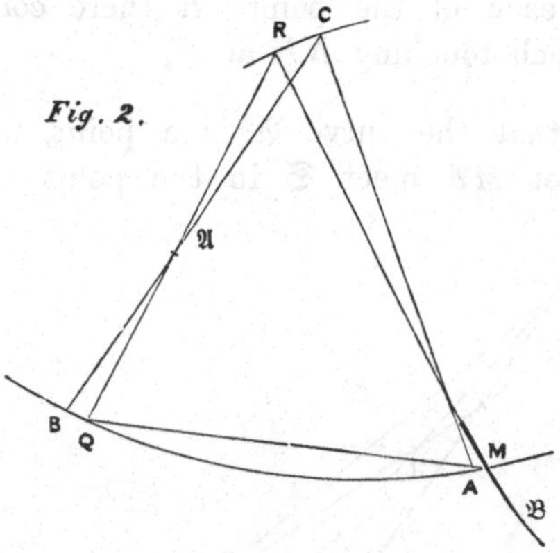

Fig. 2.

SECOND LEMMA. If the curve $\mathfrak{B}$ intersect the conic $\mathfrak{S}$ at a point M, and the tangent to $\mathfrak{B}$ at M intersect $\mathfrak{S}$ in R, if moreover a tangent through R to the curve $\mathfrak{A}$ intersect $\mathfrak{S}$ in Q; then to each of the points Q there corresponds a branch through M of the curve $\mathfrak{C}$, viz. a branch touching MQ at the point M.

Suppose as before (fig. 3) that $\mathfrak{A}$ is a point, and let the curve $\mathfrak{B}$ intersect the

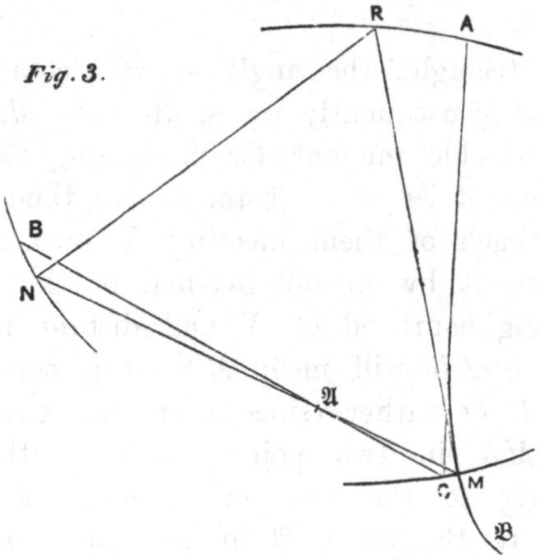

Fig. 3.

conic $\mathfrak{S}$ at the point M, and the tangent to $\mathfrak{B}$ at M meet $\mathfrak{S}$ in the point R. And let $M\mathfrak{A}$ meet $\mathfrak{S}$ in the point N. Then taking the point M for the angle C of the triangle we shall have A, B coinciding with R, N respectively, and thence NR a tangent to the curve $\mathfrak{C}$. To find the point of contact, take A in the neighbourhood of R at a distance from it ds, then C will be in the neighbourhood of M and distant from it by an infinitesimal of the order ds^2, and B will be in the neighbourhood of N and distant from it by an infinitesimal of the same order ds^2, and consequently AB will intersect NR at the point N, or the curve $\mathfrak{C}$ passes through N. Reverting to the general case where $\mathfrak{A}$ is a curve, we have only to consider MN as a tangent to the curve $\mathfrak{A}$ at a point α and the like reasoning applies: hence,

THIRD LEMMA. If the curve $\mathfrak{B}$ intersect the conic $\mathfrak{A}$ at a point M, and the tangent to $\mathfrak{B}$ at M intersect $\mathfrak{S}$ in R, if moreover a tangent through M to the curve

𝔄 meet 𝔖 in N; then to each of the points R there corresponds a branch through N of the curve 𝔈, viz. a branch touching NR at N.

Suppose again (fig. 4) that the curve 𝔄 is a point, and let the curve 𝔅 *touch* 𝔖 at the point M. And let M𝔄 meet 𝔖 in the point N. Then taking the point

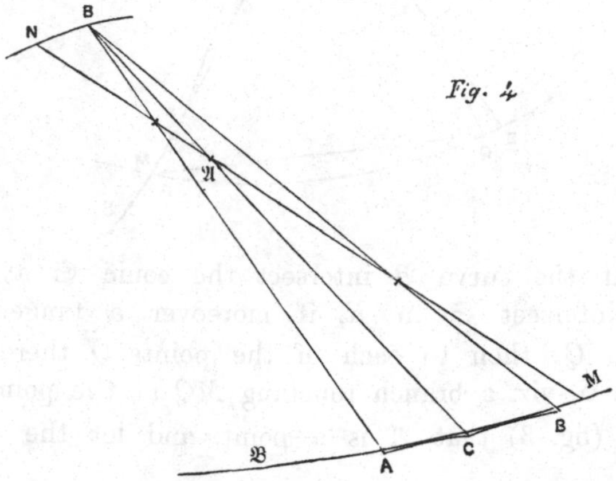

Fig. 4

M for the angle C of the triangle, the angle A will coincide with M and the angle B will coincide with N, and consequently we shall have MN a tangent to the curve 𝔈. And MN is in fact a double tangent, for proceeding to find the point of contact, take C in the neighbourhood of M at a distance ds; then from C we may draw to the curve 𝔅 two tangents each of them meeting 𝔖 in a point 𝔄 in the neighbourhood of M and distant from it by an infinitesimal of the order ds; again CA meets 𝔖 in a point B in the neighbourhood of N and distant from it by an infinitesimal of the same order ds; hence AB will meet MN at a point which will be in general at a finite distance from M, or rather (since there are two positions of the point A) the lines AB will meet MN in two points, each of them in general at a finite distance from M. Reverting to the case of 𝔄 being a curve, we must as before consider MN as a tangent to the curve 𝔄 at the point α, and the same reasoning applies: hence

FOURTH LEMMA. If the curve 𝔅 *touch* the conic 𝔖 at the point M, and if a tangent through M to the curve 𝔄 meet 𝔖 in N, then MN is a double tangent of the curve 𝔈, viz. the line MN has two distinct points of contact with the curve 𝔈.

Suppose (fig. 5) that the curves 𝔄 and 𝔅 meet the conic 𝔖 in one and the same point M, and let the tangent at M to the curve 𝔄 meet 𝔖 in the point P, and the tangent at M to the curve 𝔅 meet 𝔖 in the point N; then if we take the point M for the angle C of the triangle, the angles A and B of the triangle will coincide with N, P respectively, and NP will be a tangent of the curve 𝔈. But NP will be a double tangent, for proceeding to find the point of contact, take C in the neighbourhood of M, and distant from it by an infinitesimal of the second order ds^2; then since from the point C there may be drawn two tangents touching

𝔄 in the neighbourhood of M, and two tangents touching 𝔅 in the neighbourhood of M, there will be two points A in the neighbourhood of N and distant from it by infinitesimals of the order ds, and in like manner two points B in the neighbourhood of P and distant from it by infinitesimals of the order ds. Call these points A, A';

Fig. 5.

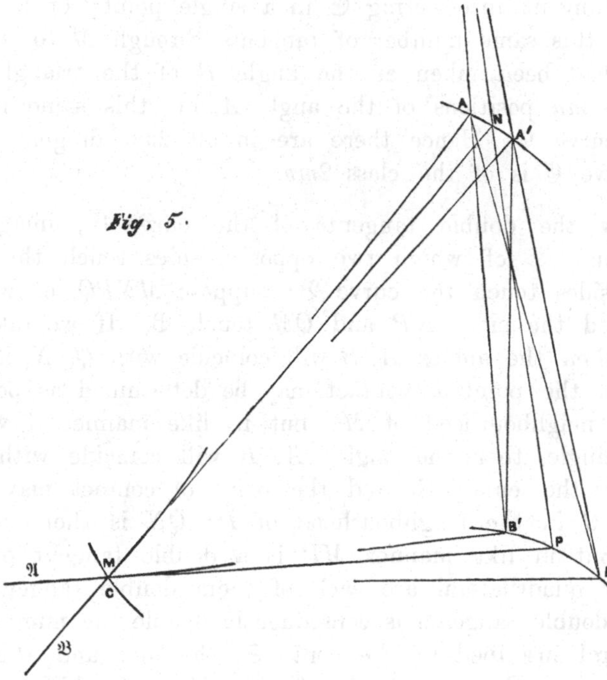

B, B'; then AB', $A'B'$ will meet NP in one and the same point, and AB', $A'B'$ will in like manner meet NP in one and the same point; these two points, which will be in general at finite distances from N, P, will be points of contact of NP with the curve ℭ: hence,

FIFTH LEMMA. If the curves 𝔄, 𝔅 meet the conic 𝔖 in one and the same point M, and if the tangents at M to the curves 𝔅 and 𝔄 respectively meet 𝔖 in the points N and P, then, joining these points, the line NP will be a double tangent to the curve ℭ, viz. there will be two distinct points of contact of the line NP with the curve ℭ.

The double tangents of the fourth and fifth lemmas exist by virtue of the particular relations assumed between the curves 𝔄, 𝔅 and the conic 𝔖, viz. from one or both of the curves 𝔄, 𝔅 touching the conic 𝔖, or from the two curves having a common point of intersection or common points of intersection with 𝔖; there are besides double tangents which exist independently of any such relations, and the theory of which will be presently investigated, but it will be convenient in the first instance to find the class of ℭ.

The class of $\mathfrak{C}$ is at once determinable from the classes (which I represent by m, n) of the curves $\mathfrak{A}$, $\mathfrak{B}$. In fact, take any point M on the conic $\mathfrak{S}$ for the angle A of the triangle. Through the point M we may draw N tangents to $\mathfrak{B}$, each of which will intersect the conic $\mathfrak{S}$ in a single point, or there will be N positions of the point C, and since from each of these we may draw M tangents to the curve $\mathfrak{A}$, each tangent intersecting $\mathfrak{S}$ in a single point; or we have mn positions of the angle B, i.e. this same number of tangents through M to the curve $\mathfrak{C}$. But if the same point M had been taken as the angle B of the triangle, there would have been in like manner mn positions of the angle A, i.e. this same number of tangents through M to the curve $\mathfrak{C}$. Hence there are in all $2mn$ tangents through M to the curve $\mathfrak{C}$, or the curve $\mathfrak{C}$ is of the class $2mn$.

Considering now the double tangents of the curve $\mathfrak{C}$; imagine a quadrilateral inscribed in the conic $\mathfrak{S}$ of which two opposite sides touch the curve $\mathfrak{A}$, and the other two opposite sides touch the curve $\mathfrak{B}$: suppose $MNPQ$, of which the sides MN and PQ touch $\mathfrak{A}$ and the sides NP and QM touch $\mathfrak{B}$. If we take M for the angle C of the triangle, then the angles A, B will coincide with Q, N, i.e. NQ is a tangent to the curve $\mathfrak{C}$, and the point of contact may be determined as before by considering a point C in the neighbourhood of M; but in like manner if we take P for the angle C of the triangle, then the angles A, B will coincide with N, Q, i.e. QN is again a tangent to the curve $\mathfrak{C}$, and the point of contact may be determined by considering a point C in the neighbourhood of P; QN is therefore a double tangent of the curve $\mathfrak{C}$. But in like manner MP is a double tangent of the curve $\mathfrak{C}$, i.e. the diagonals of the quadrilateral are each of them double tangents of the curve $\mathfrak{C}$, and the number of double tangents is consequently double the number of quadrilaterals. Imagine a pentilateral inscribed in the conic $\mathfrak{S}$, the first and third sides of which touch the curve $\mathfrak{A}$ and the second and fourth sides of which touch the curve $\mathfrak{B}$, the fifth or closing side will envelope a curve $\mathfrak{C}'$, and in the cases in which the curve $\mathfrak{C}'$ and the conic $\mathfrak{S}$ have a common tangent, the fifth side will vanish, and the pentilateral becomes a quadrilateral of the kind before referred to. Now as $\mathfrak{C}$ was shown to be of the class $2mn$, so it may be shown that $\mathfrak{C}'$ is of the class $2m^2n^2$, hence $\mathfrak{C}'$ and $\mathfrak{S}$ have $4m^2n^2$ common tangents. The quadrilateral may reduce itself to a common tangent of the curves $\mathfrak{A}$ and $\mathfrak{B}$: this gives rise to $2mn$ points of contact of $\mathfrak{C}'$ and $\mathfrak{S}$, and the common tangent at a point of contact reckons as two common tangents; the number of the remaining common tangents is therefore $4m^2n^2 - 4mn$. And these are in fact tangents at points of contact of $\mathfrak{C}'$ and $\mathfrak{S}$, i.e. $\mathfrak{C}'$ and $\mathfrak{S}$ touch in $2m^2n^2 - 2mn$ points. And since each angle of the quadrilateral may be taken as the first angle, the number of quadrilaterals is one fourth of this, or $\frac{1}{2}(m^2n^2 - mn)$, and the number of double tangents of the curve $\mathfrak{C}$ from the before-mentioned cause is therefore $m^2n^2 - mn$.

But there is another way in which double tangents arise; we may have a quadrilateral $MNPQ$ inscribed in the conic $\mathfrak{S}$, such that two adjacent sides MN, NP touch the curve $\mathfrak{A}$, and the other two adjacent sides PQ, QM touch the curve $\mathfrak{B}$. In fact in this case one of the diagonals, viz. NQ, is a double tangent of the curve $\mathfrak{C}$; the number of double tangents is therefore equal to the number of quadrilaterals.

Consider a pentilateral inscribed in the conic, and such that the first and second sides touch the curve $\mathfrak{A}$ and the third and fourth sides touch the curve $\mathfrak{B}$, the fifth or closing side will envelope a curve $\mathfrak{C}''$, and in the cases in which the curve $\mathfrak{C}''$ and the conic $\mathfrak{S}$ have a common tangent, the fifth side will vanish, and the pentilateral will become a quadrilateral of the kind last referred to. The curve $\mathfrak{C}''$ is of the class $2mn(m-1)(n-1)$, hence $\mathfrak{C}''$ and $\mathfrak{S}$ have $4mn(m-1)(n-1)$ common tangents; but these are tangents at points of contact, or the curves touch in $2mn(m-1)(n-1)$ points, and the number of quadrilaterals is $mn(m-1)(n-1)$. The number of double tangents of the curve $\mathfrak{C}$ from the cause last referred to is therefore $mn(m-1)(n-1)$, which is equal to $mn(mn-m-n+1)$. The total number of double tangents of the curve $\mathfrak{C}$ is consequently $mn(2mn-m-n)$. And the curve $\mathfrak{C}$ has not in general any stationary tangents or what is the same thing any inflexions. It has been shown that $\mathfrak{C}$ is of the class $2mn$, it is therefore of the order $2mn(2mn-1)-2mn(2mn-m-n)$, which is equal to $2mn(m+n-1)$. Hence

Theorem. If a triangle ABC be inscribed in a conic $\mathfrak{S}$, and the sides BC, AC touch curves $\mathfrak{A}$, $\mathfrak{B}$ of the classes m, n respectively, the side AB will envelope a curve $\mathfrak{C}$ of the class $2mn$ with in general $mn(2mn-m-n)$ double tangents, but no stationary tangents, and therefore of the order $2mn(m+n-1)$. If the curve $\mathfrak{A}$ touch the conic $\mathfrak{S}$, each point of contact will give rise to n double tangents of the curve $\mathfrak{C}$, and so if the curve $\mathfrak{B}$ touch the conic $\mathfrak{S}$, each point of contact will give rise to m double tangents of the curve $\mathfrak{C}$. And if $\mathfrak{A}$ and $\mathfrak{B}$ intersect on the conic $\mathfrak{S}$, each such intersection will give rise to a double tangent of the curve $\mathfrak{C}$. The curve $\mathfrak{C}$ in general touches the conic $\mathfrak{S}$ in the points in which it is intersected by any common tangent of the curves $\mathfrak{A}$ and $\mathfrak{B}$; but if the points of contact be harmonically situated with respect to the conic $\mathfrak{S}$, then $\mathfrak{C}$ does *not* pass through the points of intersection, but the tangents to $\mathfrak{S}$ at the points of intersection are stationary tangents of $\mathfrak{C}$. There is of course in the above-mentioned special cases a corresponding reduction in the order of $\mathfrak{C}$.

It sometimes happens that the number of double tangents of $\mathfrak{C}$ becomes infinite, i.e. in fact that $\mathfrak{C}$ is made up of two coincident curves; instances of this will be presently mentioned.

Suppose that the curves $\mathfrak{A}$, $\mathfrak{B}$ are points; $\mathfrak{C}$ is in general a conic having double contact with the conic $\mathfrak{S}$ in the points in which it is intersected by the line joining $\mathfrak{A}$, $\mathfrak{B}$. But if the points $\mathfrak{A}$, $\mathfrak{B}$ are harmonically situated with respect to the conic $\mathfrak{S}$, then $\mathfrak{C}$ does *not* pass through the points of intersection, but the tangents to $\mathfrak{S}$ at the points of intersection are stationary tangents of $\mathfrak{C}$. This implies that the curve $\mathfrak{C}$ is made up of two coincident points at the point of intersection of the two tangents of $\mathfrak{S}$: call this the point $\mathfrak{C}$, then $\mathfrak{A}$, $\mathfrak{B}$, $\mathfrak{C}$ are conjugate points of the conic $\mathfrak{S}$, and we have the well-known theorem that in a conic $\mathfrak{S}$ there may be inscribed an infinity of triangles the sides of which pass through three conjugate points of the conic. It should be remarked that for each position of the side AB, there are two positions of the angle $\mathfrak{C}$, i.e. each side AB is properly a double tangent of the curve (point) $\mathfrak{C}$.

Let $\mathfrak{A}$ be a point and $\mathfrak{B}$ be a conic, the curve $\mathfrak{C}$ is in general of the class 4, with two double tangents, and therefore of the order 8. It is proper to remark that the double tangents originate in a quadrilateral of the kind first considered, viz. a quadrilateral of which two opposite sides pass through the point $\mathfrak{A}$ and the other two opposite sides touch the conic $\mathfrak{B}$. It is easy in the present case to construct the quadrilateral: consider the point $\mathfrak{A}$ as a pole, and take its polar with respect to the conic $\mathfrak{S}$, take the pole of this polar with respect to the conic $\mathfrak{B}$. Join the two poles, and from the point in which the joining line meets the polar draw tangents to the conic $\mathfrak{B}$, these tangents will meet the conic $\mathfrak{S}$ in four points, lying two and two in lines passing through the point $\mathfrak{A}$: we have thus the quadrilateral, and the diagonals of the quadrilateral (or the other two lines passing through the four points) are the double tangents of the curve $\mathfrak{C}$.

There are two particular cases to be considered. First, where the conic $\mathfrak{B}$ has double contact with the conic $\mathfrak{S}$. Here the lines joining the point $\mathfrak{A}$ with the points of contact are double tangents of the curve $\mathfrak{C}$, which has therefore in all four double tangents, and being a curve of the fourth class it must break up into two curves of the second class, i.e. into two conics. And these are curves having double contact with the conic $\mathfrak{S}$; for the curve $\mathfrak{C}$ touches the conic $\mathfrak{S}$ in the four points in which $\mathfrak{S}$ is intersected by the tangents through $\mathfrak{A}$ to the conic $\mathfrak{B}$. Secondly, where $\mathfrak{A}$ is one of the conjugate points of the conics $\mathfrak{B}$, $\mathfrak{S}$. The general construction for the quadrilateral shows that if from any point of the common polar of $\mathfrak{A}$ with respect to $\mathfrak{B}$ and with respect to $\mathfrak{S}$, we draw tangents to $\mathfrak{B}$, these will meet $\mathfrak{S}$ in four points, lying two and two in lines passing through $\mathfrak{A}$, i.e. that the number of the double tangents of the curve $\mathfrak{C}$ is indefinite; $\mathfrak{C}$ is therefore made up of two coincident curves of the second class, i.e. of two coincident conics. Moreover, $\mathfrak{C}$ passes through the points of intersection of $\mathfrak{B}$, $\mathfrak{S}$; hence, disregarding one of the two coincident conics, we may say that the curve $\mathfrak{C}$ is a conic passing through the points of intersection of the conics $\mathfrak{B}$, $\mathfrak{S}$.

Next, let the curves $\mathfrak{A}$, $\mathfrak{B}$ be each of them a conic. The curve $\mathfrak{C}$ is of the class 8, with in general 16 double tangents, and therefore of the order 24. But there are two particular cases to be considered: first, where the conics $\mathfrak{A}$, $\mathfrak{B}$ have each of them double contact with the conic $\mathfrak{S}$. Here the tangents drawn from the points of contact of either of the conics $\mathfrak{A}$ or $\mathfrak{B}$ with $\mathfrak{S}$ to the other of the conics, $\mathfrak{B}$ or $\mathfrak{A}$, is a double tangent of the curve $\mathfrak{C}$, i.e. there are 8 new double tangents, or in all 24 double tangents of the curve $\mathfrak{C}$, which is therefore of the order 8; and being of the class 8 with 24 double tangents, it must break up into four curves of the second class, i.e. into four conics. And the curve $\mathfrak{C}$ touches the conic $\mathfrak{S}$ in the points in which $\mathfrak{S}$ is intersected by any one of the four common tangents of the conics $\mathfrak{A}$, $\mathfrak{B}$, viz. 8 points in all; hence each of the four conics has double contact with the conic $\mathfrak{S}$. Attending only to one of the four conics of which $\mathfrak{C}$ is made up, we have thus what (in a restricted sense of the expression, the porism of the in-and-circumscribed triangle) I call the porism (homographic) of the in-and-circumscribed triangle, viz.

If a triangle be inscribed in a conic, and two of the sides touch conics having double contact with the circumscribed conic, then will the third side touch a conic having double contact with the circumscribed conic.

Secondly, the conics $\mathfrak{A}$ and $\mathfrak{B}$ may cut the conic $\mathfrak{S}$ in the same four points. Here it may be seen that there are an infinity of inscribed quadrilaterals of the kind first considered, viz. of which two opposite sides touch the conic $\mathfrak{A}$, and the other two opposite sides touch the conic $\mathfrak{B}$. Hence, the curve $\mathfrak{C}$ is made up of two coincident curves of the class 4. But the curve of the class 4 has in fact 4 double tangents, viz. considering each of the points of intersection of $\mathfrak{A}$, $\mathfrak{B}$, $\mathfrak{S}$, and drawing tangents to $\mathfrak{A}$ and $\mathfrak{B}$ meeting $\mathfrak{S}$ in two new points, the line joining these points is a double tangent of the curve in question, which is therefore of the 4th order, and being of the class 4 with 4 double tangents, it must break up into two curves of the second class, i.e. into two conics. Each of these conics passes through the points of intersection of $\mathfrak{A}$, $\mathfrak{B}$, $\mathfrak{S}$, and touches the four lines last referred to, the conics would of course be completely determined by the condition of passing through the four points and touching one of the four lines. Attending only to one of the two conics, we have thus what I call the porism (allographic) of the in-and-circumscribed triangle, viz.

If a triangle be inscribed in a conic, and two of the sides touch conics meeting the circumscribed conic in the same four points, the remaining side will touch a conic meeting the circumscribed conic in the four points.

The *à posteriori* demonstration of these theorems will form the subject of another paper, [178].

176.

NOTE ON JACOBI'S CANONICAL FORMULÆ FOR DISTURBED MOTION IN AN ELLIPTIC ORBIT.

[From the *Quarterly Mathematical Journal*, vol. I. (1857), pp. 355—356.]

CONSIDER a body (afterwards called the disturbed body) revolving about a central body under the influence of their mutual attraction and of any disturbing forces. Then referring the disturbed body to axes through the central body and parallel to fixed lines in space, write

x, y, z, the coordinates of the disturbed body,

r, , the radius vector, $= \sqrt{(x^2 + y^2 + z^2)}$,

M , the mass of the disturbed body,

M'' , the mass of the central body.

Write also

R, the disturbing function, taken negatively, i.e. the sign of R is taken as in the Mécanique Céleste.

The equations of motion then are

$$\frac{d^2x}{dt^2} = -\frac{(M+M'')\,x}{r^3} - \frac{dR}{dx},$$

$$\frac{d^2y}{dt^2} = -\frac{(M+M'')\,y}{r^3} - \frac{dR}{dy},$$

$$\frac{d^2z}{dt^2} = -\frac{(M+M'')\,z}{r^3} - \frac{dR}{dz},$$

and the motion may be represented by supposing that the body moves in an ellipse with variable elements, and such that the direction and velocity of motion are always the same as in the actual orbit.

We may, to fix the ideas, take the plane of xy to be the plane of the ecliptic and the axis of x to be the line through the first point of Aries, or origin of longitude.

Jacobi's canonical elements may be taken to be,

First, the constant of vis viva or invariable part of the half-square of the velocity, which is equal to the sum of the masses divided by twice the mean distance and with the sign minus.

Secondly, the constant of areas, which is equal to the square root of the sum of the masses into the half of the latus rectum.

Thirdly, the constant of the reduced area (i.e. of the area described on the plane of the ecliptic), which is equal to the square root of the sum of the masses into the half of the latus rectum into the cosine of the inclination.

Fourthly, the constant attached by addition to the time t, or what is the same thing, the epoch or time of pericentric passage, taken with the sign minus.

Fifthly, the angular distance from node (or argument of latitude) of the pericentre.

Sixthly, the longitude of the node.

(The first and fourth elements are taken by Jacobi with the contrary sign, but this difference is not material.)

Representing the preceding system of canonical elements by $\mathfrak{A}$, $\mathfrak{B}$, $\mathfrak{C}$, $\mathfrak{F}$, $\mathfrak{G}$, $\mathfrak{H}$, and observing that Jacobi's disturbing function Ω is the same as the disturbing function R of the Mécanique Céleste, except that the sign is reversed (i.e. $\Omega = -R$), the expressions for the variations of the canonical elements are

$$\frac{d\mathfrak{A}}{dt} = -\frac{dR}{d\mathfrak{F}}, \quad \frac{d\mathfrak{B}}{dt} = -\frac{dR}{d\mathfrak{G}}, \quad \frac{d\mathfrak{C}}{dt} = -\frac{dR}{d\mathfrak{H}},$$

$$\frac{d\mathfrak{F}}{dt} = +\frac{dR}{d\mathfrak{A}}, \quad \frac{d\mathfrak{G}}{dt} = +\frac{dR}{d\mathfrak{B}}, \quad \frac{d\mathfrak{H}}{dt} = +\frac{dR}{d\mathfrak{C}}.$$

In the ordinary case in which the disturbing force is the attraction of a third body, write

x', y', z', the coordinates of the disturbing body,

r' , the radius vector, $= \sqrt{(x'^2 + y'^2 + r'^2)}$,

M' , the mass of the disturbing body.

Then the expression for the disturbing function is

$$R = M' \left\{ \frac{xx' + yy' + zz'}{r'^3} - \frac{1}{\sqrt{(x-x')^2 + (y-y')^2 + (z-z')^2}} \right\}.$$

The preceding formulæ form a convenient standard of reference for the various systems of elements which have been made use of by writers upon Physical Astronomy; any such system may be without difficulty derived from the canonical system by expressing the elements adopted in terms of the above-mentioned canonical elements.

177.

SOLUTION OF A MECHANICAL PROBLEM.

[From the *Quarterly Mathematical Journal*, vol. I. (1857), pp. 405—406.]

A HEAVY plane is supported by parallel elastic strings of small extensibility; and the strings are of the same length and extensibility: required the position of equilibrium.

Imagine the plane horizontal, and let n be the number of strings, (a, b), (a', b'), &c. the coordinates of the points of attachment; ξ, η the coordinates of the centre of gravity of the plane; W the weight; let the equation of the horizontal line about which the plane turns be

$$x\cos\alpha + y\sin\alpha - p = 0;$$

and let $\delta\theta$ be the inclination of the plane in its position of equilibrium to the horizontal plane, and $\omega\delta l$ the force generated by an increase δl in the length of one of the strings.

We have for the conditions of equilibrium

$$\begin{aligned}\Sigma\,(a\cos\alpha + b\sin\alpha - p)\ \ \omega\delta\theta - W\ &= 0,\\ \Sigma\,(a\cos\alpha + b\sin\alpha - p)\ a\omega\delta\theta - W\xi &= 0,\\ \Sigma\,(a\cos\alpha + b\sin\alpha - p)\ b\omega\delta\theta - W\eta &= 0;\end{aligned}$$

or putting $\Sigma a = L$, $\Sigma b = M$, $\Sigma a^2 = A$, $\Sigma ab = H$, $\Sigma b^2 = B$, we have

$$L\cos\alpha + M\sin\alpha - np - \frac{W}{\omega\delta\theta} = 0,$$

$$A\cos\alpha + H\sin\alpha - Lp - \frac{W\xi}{\omega\delta\theta} = 0,$$

$$H\cos\alpha + B\sin\alpha - Mp - \frac{W\eta}{\omega\delta\theta} = 0.$$

Combining with these the equations

$$x \cos \alpha + \eta \sin \alpha - p = 0,$$

and eliminating linearly $\cos \alpha$, $\sin \alpha$, p and W, we have

$$\begin{vmatrix} . & x, & y, & 1 \\ \xi, & A, & H, & L \\ \eta, & H, & B, & M \\ 1, & L, & M, & n \end{vmatrix} = 0$$

for the equation of the required line $x \cos \alpha + y \sin \alpha - p = 0$. Replacing L, M, A, H, B by their values, the equation is readily transformed into

$$\Sigma \left\{ \begin{vmatrix} x, & y, & 1 \\ a, & b, & 1 \\ a', & b', & 1 \end{vmatrix} \times \begin{vmatrix} \xi, & \eta, & 1 \\ a, & b, & 1 \\ a', & b', & 1 \end{vmatrix} \right\} = 0$$

where the summation extends to each pair of points (a, b) and (a', b'). This is, in fact, an extension of the harmonic relation of a point and line with respect to a triangle.

178.

ON THE A POSTERIORI DEMONSTRATION OF THE PORISM OF THE IN-AND-CIRCUMSCRIBED TRIANGLE.

[From the *Quarterly Mathematical Journal*, vol. II. (1858), pp. 31—38.]

In my former paper "On the Porism of the In-and-circumscribed Triangle" (*Journal*, t. I. p. 344 [175]) the two porisms (the homographic and the allographic) were established *à priori*, i.e. by means of an investigation of the order of the curve enveloped by the third side of the triangle. I propose in the present paper to give the *à posteriori* demonstration of these two porisms; first according to Poncelet, and then in a form not involving (as do his demonstrations) the principle of projections. My objection to the employment of the principle may be stated as follows: viz. that in a systematic development of the subject, the theorems relating to a particular case and which are by the principle in question extended to the general case, are not in anywise more simple or easier to demonstrate than are the theorems for the general case; and, consequently that the circuity of the method can and ought to be avoided.

The porism (homographic) of the in-and-circumscribed triangle, viz.

If a triangle be inscribed in a conic, and two of the sides envelope conics having double contact with the circumscribed conic, then will the third side envelope a conic having double contact with the circumscribed conic.

The following is Poncelet's demonstration, the Nos. are those of the *Traité des Propriétés Projectives* [Paris, 1822]:

No. 431. If a triangle be inscribed in a circle and two of the sides are parallel to given lines, then the third side envelopes a concentric circle.

This is evident, for, the angle in the segment subtended by the third side being constant, the length of the third side is constant; hence, the length of the perpendicular from the centre upon the third side is also constant, and the third side envelopes a concentric circle.

Hence, by the principle of projections:

If a triangle be inscribed in a conic and two of the sides pass through given points, the remaining side envelopes a conic having double contact with the circumscribed conic, the line through the two points being the chord of contact.

No. 434. Conversely, if there be a triangle inscribed in a conic and the first side envelope a conic having double contact with the circumscribed conic, and the second side pass through a fixed point in the chord of contact, then will the third side also pass through a fixed point in the chord of contact.

No. 437. In particular, if there be a triangle inscribed in a conic and two of the sides pass through fixed points, then will the third side pass through a fixed point, viz. the point forming with the other two points a conjugate system.

No. 439. It follows that:

If there be a triangle inscribed in a conic and the first side passes through a fixed point, and the second side envelopes a conic having double contact with the circumscribed conic, then will the third side envelope a conic having double contact with the circumscribed conic.

For the chord of contact meets the polar of the fixed point with respect to the circumscribed conic in a point; the line joining this point with the third angle (i.e. the angle opposite the third side) of the triangle meets the conic in a variable point; and joining this variable point with the first and second angles of the triangle we have a new triangle; two of the sides of this new triangle (by Nos. 434 and 437) pass through fixed points; hence the remaining side, i.e. the third side of the original triangle, touches a conic having double contact with the circumscribed conic.

We have thus passed from the case of the two sides passing through fixed points to that of one of the two sides enveloping a conic having double contact with the given conic and the other of them passing through a fixed point; and, by a repetition of the reasoning, Poncelet passes to the general case, viz.

If there be a triangle inscribed in a conic, and two of the sides envelope conics having double contact with the circumscribed conic, then will the third side envelope a conic having double contact with the circumscribed conic.

But it is somewhat more simple to omit the intermediate case of a conic and point, and pass directly, by the reasoning of No. 439, from the case of two points to that of two conics.

In fact, considering the point of intersection of the two chords of contact, the line joining this point with the third angle of the triangle meets the conic in a variable point, and joining this variable point with the first and second angles of the triangle we have a new triangle: two of the sides of this new triangle (by No. 434) pass through fixed points; hence the remaining side, i.e. the third side of the original triangle, envelopes a conic having double contact with the circumscribed conic; and the general case is thus established.

The porism (allographic) of the in-and-circumscribed triangle, viz.

If a triangle be inscribed in a conic and two of the sides envelope conics meeting the circumscribed conic in the same four points, then the third side will touch a conic meeting the circumscribed conic in the four points.

The following is Poncelet's demonstration:

No. 433. In the particular case of the homographic porism, viz.—that in which two of the sides of the triangle pass through fixed points and the remaining side envelopes a conic having double contact with the circumscribed conic—it is easy to see that the lines joining the angles of the triangle with the two fixed points and with the point of contact on the third side, meet in a point; this follows at once by the principle of projection from the case in No. 431, viz. the case of a triangle inscribed in a circle when two of the sides are parallel to given lines and the third side touches a concentric circle. Hence,

No. 531. If there be a triangle inscribed in a conic, and two of the sides envelope fixed curves, and the third side envelopes a certain curve; the lines joining the angles of the triangle with the points of contact meet in a point.

In fact, attending only to the infinitesimal variation of the position of the triangle, the curves enveloped by the first and second sides may be replaced by the points of contact on these sides, and the curve enveloped by the third side may be replaced by a conic having double contact with the circumscribed conic, and the general case thus follows at once from the particular one.

Nos. 162 and 163. LEMMA[1]. If, on the sides of a triangle ABC, there are taken any three points L, M, N in the same line, and the harmonics A', B', C' of these points (i.e. the harmonic of each point with respect to the two vertices on the same side of the triangle), then the lines AA', BB', CC' will meet in a point; and, moreover, if $A'L$, $B'M$, $C'N$ are bisected in F, G, H (or, what is the same thing, if $\overline{FA'}^2 = FB \cdot FC$, $\overline{GB'}^2 = GC \cdot GA$, $\overline{HC'}^2 = HA \cdot HB$), then will the three points F, G, H lie in a line. This is, in fact, the theorem No. 164,—In any complete quadrilateral the middle points of the three diagonals lie in a line.

It is now easy to prove a particular case of the allographic porism, viz.

No. 531. If there be a triangle inscribed in a circle, such that two of the sides envelope circles having a common secant (real or ideal) with the circumscribed circle; then will the third side envelope a circle having the same common secant with the circumscribed circle.

For if the triangle be ABC, and the points of contact of the sides CB, CA with the enveloped circles and the point of contact of the side AB with the enveloped curve, be A', B', C'; if moreover the points of intersection of the circumscribed circle and the two enveloped circles be M, N, and the common secant MN meet the sides

[1] I have not thought it necessary to give the figures; they can be supplied without difficulty.

of the triangle in F, G, H; then F, G, H and A', B', C' are points on the sides of the triangle ABC, such that F, G, H lie in a line, and AA', BB', CC' meet in a point. And by a property of the circle

$$\overline{FA'}^2 = FM \,.\, FN = FB \,.\, FC,$$

$$\overline{GB'}^2 = GM \,.\, GN = GC \,.\, GA\,;$$

whence by the lemma (or rather its converse) $\overline{HC'}^2 = HA \,.\, HB$ and by a property of the circle $HA \,.\, HB = HM \,.\, HN$; and therefore, $\overline{HC'}^2 = HM \,.\, HN$, a property which can only belong to a circle having, with the other circles, the common secant MN: the particular case is thus demonstrated. And the principle of projections leads at once to the general case of the allographic porism.

To exhibit the demonstrations in a form independent of the principle of projections, it will be convenient to enunciate the following three lemmas: the first of them being, in fact, the theorem contained in No. 434, as generalised by No. 531; the second of them a theorem connected with and including the properties of the circle assumed in Poncelet's demonstration of the allographic porism; and the third of them a theorem derivable by the principle of projections from the theorem in Nos. 162 and 163.

LEMMA I. If there be a triangle inscribed in a conic, such that two of the sides envelope given curves and the third side envelopes a curve; then the lines joining the angles of the triangle with the points of contact of the opposite sides meet in a point.

LEMMA II. If there be three conics meeting in the same four points, then any line meets the conic in six points forming a system in involution.

COROLL. 1. If the line be a tangent to one of the conics, then the point of contact is the double or sibi-conjugate point of the involution formed by the points of intersection with the other two conics. And conversely if the curve enveloped by the line is not given, but the preceding property holds for all positions of the tangent line; then the curve enveloped by such line is a conic passing through the points of intersection of the two given conics.

COROLL. 2. If one of the conics be a pair of coincident lines, then the other two conics are conics having double contact, with the line in question for their chord of contact; any line meets the chord of contact in a point which is a double or sibi-conjugate point of the involution formed by the points of intersection with the other two conics; and if the line be a tangent to one of the conics, then the point of contact and the point of intersection with the chord of contact are harmonics with respect to the points of intersection with the other conic. And conversely if every tangent of a curve intersect a line and conic in such manner that the point of contact and the point of intersection with the line are harmonics with respect to the points of intersection with the conic; then the curve is a conic having double contact with the given conic, and the line in question is the chord of contact.

The third lemma is a theorem (first explicitly stated, so far as I am aware, by Steiner, Lehrsätze 24 and 25, Crelle, t. III. [1828], p. 212, and demonstrated by Bauer, t. XIX. [1839], p. 227) which, in a note in the *Phil. Mag.*, Augt. 1853 [118], I have called the *Theorem of the harmonic relation of two lines with respect to a quadrilateral.*

LEMMA III. If on each of the diagonals of a quadrilateral there be taken two points harmonically related with respect to the angles upon this diagonal; then if three of the points lie in a line, the other three points will also lie in a line: the two lines are said to be harmonically related with respect to the quadrilateral.

The relation may be exhibited under a different form. The three diagonals of the quadrilateral form a triangle, the sides of which contain the six angles of the quadrilateral; and considering only three of the six angles (one angle on each diagonal) these three angles are points which either lie in a line, or else are such that the lines joining them with the opposite angles of the triangle meet in a point. Each of the three points is, with respect to the involution formed by the two angles of the triangle and the two points harmonically related thereto, a double or sibi-conjugate point, and we have thus a theorem of the harmonic relation of two lines to a triangle and line, or else to a triangle and point, viz. *Theorem,* If on the sides of a *triangle* there be taken three points which either lie in a *line* or else are such that the lines joining them with the opposite angles of the triangle meet in a *point;* and if on each side of the triangle there be taken two points forming with the two angles on the same side an involution having the first-mentioned point on the same side for a double or sibi-conjugate point; then if three of the six points lie in a *line,* the other three of the six points will also lie in a line; the two lines are said to be harmonically related to the triangle and line, or (as the case may be) to the triangle and point.

The proof of the two porisms is by the preceding lemmas rendered very simple.

Demonstration of the homographic porism.

First, the particular case, where two of the sides pass through fixed points. Lemma I. gives the construction of the point of contact on the third side, and the figure shows that the point of contact and the point in which the third side is intersected by the line through the two given points are harmonics with respect to the points of intersection of the third side with the circumscribed conic. Hence, (Lemma II. Coroll. 2) the curve touched by the third side is a conic having double contact with the circumscribed conic, and the chord of contact is the line joining the two given points; and conversely if one of the sides touch a conic having double contact with the circumscribed conic and another of the sides passes through a fixed point on the chord of contact, then the third side will also pass through a fixed point on the chord of contact. The general case is deduced from the particular one precisely as before, viz. where two of the sides touch conics having double contact with the circumscribed conic, then considering the point of intersection of the two chords of contact, the line joining this point with the third angle of the triangle meets the circumscribed conic in a variable point, and joining this variable point with the first and second angles of the triangle, we have a new triangle, two of the sides of which (by the converse of the particular case) pass through fixed points: hence the remaining side, i.e. the third side of the original triangle, touches a conic having double contact with the circumscribed conic.

Demonstration of the allographic porism.

Let ABC be the triangle, A', B', C' the points of contact on the three sides, then by Lemma I. the lines AA', BB', CC' meet in a point. Take a pair of lines passing through the points of intersection of the circumscribed conic with the two given conics enveloped by the sides CA, CB, and let one of these lines meet the sides of the triangle in the points F, G, H, and the other of them meet the sides of the triangle in the points F', G', H'. Then considering the following three conics, viz. the last-mentioned pair of lines, the circumscribed conic, and the conic enveloped by the side CA; these are conics passing through the same four points, and the side CA is a tangent to one of them: hence by Lemma II. Coroll. 1, G, G', C, A will be an involution having the point B' for a double or sibi-conjugate point, and similarly F, F', G, B are an involution having the point A' for a double or sibi-conjugate point. It follows from Lemma III. that H, H', A, B will be an involution having C' for a double or sibi-conjugate point. Hence by Lemma II. Coroll. 1 (the two given conics being the before-mentioned pair of lines and the circumscribed conic) the curve enveloped by the side AB will be a conic passing through the points of intersection of the pair of lines and the circumscribed conic, or, what is the same thing, the points of intersection of the circumscribed conic and the conics enveloped by the other two sides.

2, *Stone Buildings, Oct.* 2, 1856.

179.

ON CERTAIN FORMS OF THE EQUATION OF A CONIC.

[From the *Quarterly Mathematical Journal*, vol. II. (1858), pp. 44—48.]

To find the general equation of a conic which passes through two given points and touches a given line.

Let the coordinates of the given points be (α, β, γ), $(\alpha', \beta', \gamma')$, and the equation of the given line be $\lambda x + \mu y + \nu z = 0$. Then writing

$$u = \begin{vmatrix} x, & y, & z \\ \alpha, & \beta, & \gamma \\ a, & b, & c \end{vmatrix}, \quad v = s \begin{vmatrix} x, & y, & z \\ \alpha, & \beta, & \gamma \\ \alpha', & \beta', & \gamma' \end{vmatrix}, \quad w = \begin{vmatrix} x, & y, & z \\ \alpha', & \beta', & \gamma' \\ a, & b, & c \end{vmatrix},$$

where a, b, c, s are arbitrary coefficients, the general equation of a conic passing through the two given points will be

$$uw - v^2 = 0.$$

We have identically

$$s \begin{vmatrix} \lambda x + \mu y + \nu z, & x, & y, & z \\ \lambda\alpha + \mu\beta + \nu\gamma, & \alpha, & \beta, & \gamma \\ \lambda\alpha' + \mu\beta' + \nu\gamma', & \alpha', & \beta', & \gamma' \\ \lambda a + \mu b + \nu c, & a, & b, & c \end{vmatrix} = 0;$$

and hence putting

$$\nabla = \begin{vmatrix} \alpha, & \beta, & \gamma \\ \alpha', & \beta', & \gamma' \\ a, & b, & c \end{vmatrix},$$

$$A = \quad (\lambda\alpha' + \mu\beta' + \nu\gamma')\, s,$$
$$B = -(\lambda a + \mu b + \nu c)\, s,$$
$$C = -(\lambda\alpha + \mu\beta + \nu\gamma)\, s,$$

we have

$$(\lambda x + \mu y + \nu z)\, s\nabla + Au + Bv + Cw = 0,$$

and consequently the equation $\lambda x + \mu y + \nu z = 0$ is equivalent to

$$Au + Bv + Cw = 0,$$

and we have only to express that the line represented by this equation touches the conic $uw - v^2 = 0$.

Combining the two equations, we find $Au + Cw + B\sqrt{(uw)} = 0$, that is

$$(Au + Cw)^2 - B^2 uw = 0,$$

an equation which must have equal roots; and the condition for this is obviously $4AC - B^2 = 0$. Or putting the condition under the form $-B + 2\sqrt{(AC)} = 0$ and substituting for A, B, C their values, the condition becomes $\{i = \sqrt{(-1)}$ as usual$\}$

$$\lambda a + \mu b + \nu c + 2is\sqrt{\{(\lambda\alpha + \mu\beta + \nu\gamma)(\lambda\alpha' + \mu\beta' + \nu\gamma')\}} = 0.$$

We have consequently

$$s^2 = -\frac{(\lambda a + \mu b + \nu c)^2}{4(\lambda\alpha + \mu\beta + \nu\gamma)(\lambda\alpha' + \mu\beta' + \nu\gamma')},$$

and the equation of the conic is

$$4(\lambda\alpha+\mu\beta+\nu\gamma)(\lambda\alpha'+\mu\beta'+\nu\gamma')\begin{vmatrix} x, & y, & z \\ \alpha, & \beta, & \gamma \\ a, & b, & c \end{vmatrix}\begin{vmatrix} x, & y, & z \\ \alpha', & \beta', & \gamma' \\ a, & b, & c \end{vmatrix} + (\lambda a+\mu b+\nu c)^2 \begin{vmatrix} x, & y, & z \\ \alpha, & \beta, & \gamma \\ \alpha', & \beta', & \gamma' \end{vmatrix}^2 = 0.$$

But the equation of the conic may be obtained in a different form as follows: we may first write $B^2v^2 = 4ACuw$, and then substituting for $-Bv$ the value

$$(\lambda x + \mu y + \nu z)\, s\nabla + Au + Cw,$$

the equation becomes

$$\{Au + Cw + (\lambda x + \mu y + \nu z)\, s\nabla\}^2 = 4ACuw,$$

or, extracting the root of each side and transposing,

$$\{\sqrt{(Au)} + \sqrt{(Cw)}\}^2 + (\lambda x + \mu y + \nu z)\, s\nabla = 0,$$

and thence

$$\sqrt{(Au)}+\sqrt{(Cw)}+i\sqrt{(s\nabla)}\sqrt{(\lambda x+\mu y+\nu z)}=0,$$

or substituting the values of A, C, ∇, u, w, and omitting the common factor $\sqrt{(s)}$ the equation becomes

$$\sqrt{(\lambda\alpha'+\mu\beta+\nu\gamma')}\sqrt{\left(\begin{vmatrix} x, & y, & z \\ \alpha, & \beta, & \gamma \\ a, & b, & c\end{vmatrix}\right)}+i\sqrt{(\lambda\alpha+\mu\beta+\nu\gamma)}\sqrt{\left(\begin{vmatrix} x, & y, & z \\ \alpha', & \beta', & \gamma' \\ a, & b, & c\end{vmatrix}\right)}$$

$$+i\sqrt{(\lambda x+\mu y+\nu z)}\sqrt{\left(\begin{vmatrix} \alpha, & \beta, & \gamma \\ \alpha', & \beta', & \gamma' \\ a, & b, & c\end{vmatrix}\right)}=0,$$

a form symmetrically related to the three lines

$$\lambda x+\mu y+\nu z=0,\quad \begin{vmatrix} x, & y, & z \\ \alpha, & \beta, & \gamma \\ a, & b, & c\end{vmatrix}=0,\quad \begin{vmatrix} x, & y, & z \\ \alpha', & \beta', & \gamma' \\ a, & b, & c\end{vmatrix}=0.$$

Let it be required to find the conic passing through the two points (α, β, γ), $(\alpha', \beta', \gamma')$, and touching the three lines

$$\lambda_1 x+\mu_1 y+\nu_1 z=0,\quad \lambda_2 x+\mu_2 y+\nu_2 z=0,\quad \lambda_3 x+\mu_3 y+\nu_3 z=0.$$

The constants a, b, c have to be determined in such manner that the equations obtained from the preceding, by writing successively $(\lambda_1, \mu_1, \nu_1)$, $(\lambda_2, \mu_2, \nu_2)$, $(\lambda_3, \mu_3, \nu_3)$ for (λ, μ, ν) may represent one and the same equation; the three equations so obtained will therefore subsist simultaneously, and we may from the equations in question eliminate a, b, c; the resulting equation

$$\begin{vmatrix} \sqrt{(\lambda_1 x+\mu_1 y+\nu_1 z)}, & \sqrt{(\lambda_2 x+\mu_2 y+\nu_2 z)}, & \sqrt{(\lambda_3 x+\mu_3 y+\nu_3 z)} \\ \sqrt{(\lambda_1\alpha+\mu_1\beta+\nu_1\gamma)}, & \sqrt{(\lambda_2\alpha+\mu_2\beta+\nu_2\gamma)}, & \sqrt{(\lambda_3\alpha+\mu_3\beta+\nu_3\gamma)} \\ \sqrt{(\lambda_1\alpha'+\mu_1\beta'+\nu_1\gamma')}, & \sqrt{(\lambda_2\alpha'+\mu_2\beta'+\nu_2\gamma')}, & \sqrt{(\lambda_3\alpha'+\mu_3\beta'+\nu_3\gamma')}\end{vmatrix}=0$$

is the equation of the conic in question; this is in fact evident from other considerations.

To find the condition in order that a conic passing through the points (α, β, γ), $(\alpha', \beta', \gamma')$ may touch the four lines

$$\lambda_1 x+\mu_1 y+\nu_1 z=0,\quad \lambda_2 x+\mu_2 y+\nu_2 z=0,\quad \lambda_3 x+\mu_3 y+\nu_3 z=0,\quad \lambda_4 x+\mu_4 y+\nu_4 z=0.$$

The relation first obtained between a, b, c, s gives four equations from which these quantities may be eliminated, the resulting equation

$$\begin{vmatrix} \lambda_1, & \mu_1, & \nu_1, & \sqrt{\{(\lambda_1\alpha+\mu_1\beta+\nu_1\gamma)\ (\lambda_1\alpha'+\mu_1\beta'+\nu_1\gamma')\}} \\ \lambda_2, & \mu_2, & \nu_2, & \sqrt{\{(\lambda_2\alpha+\mu_2\beta+\nu_2\gamma)\ (\lambda_2\alpha'+\mu_2\beta'+\nu_2\gamma')\}} \\ \lambda_3, & \mu_3, & \nu_3, & \sqrt{\{(\lambda_3\alpha+\mu_3\beta+\nu_3\gamma)\ (\lambda_3\alpha'+\mu_3\beta'+\nu_3\gamma')\}} \\ \lambda_4, & \mu_4, & \nu_4, & \sqrt{\{(\lambda_4\alpha+\mu_4\beta+\nu_4\gamma)\ (\lambda_4\alpha'+\mu_4\beta'+\nu_4\gamma')\}} \end{vmatrix} = 0$$

is the required relation.

The preceding investigations apply directly to the circle, which is a conic passing through two given points. Thus the equation of a circle touching the three lines

$$\begin{aligned} Ax\ +By\ +C\ &= 0, \\ A'x\ +B'y\ +C'\ &= 0, \\ A''x+B''y+C'' &= 0, \end{aligned}$$

is

$$\begin{vmatrix} \sqrt{(Ax+By+C)}, & \sqrt{(A'x+B'y+C')}, & \sqrt{(A''x+B''y+C'')} \\ \sqrt{(A\ +Bi)}, & \sqrt{(A'\ +B'i)}, & \sqrt{(A''\ +B''i)} \\ \sqrt{(A\ -Bi)}, & \sqrt{(A'\ -B'i)}, & \sqrt{(A''\ +B''i)} \end{vmatrix} = 0.$$

Hence also the equation of a circle touching the three lines

$$\begin{aligned} x\cos\alpha+y\sin\alpha-p &= 0, \\ x\cos\beta+y\sin\beta-q &= 0, \\ x\cos\gamma+y\sin\gamma-r &= 0, \end{aligned}$$

is

$$\sin\tfrac{1}{2}(\beta-\gamma)\sqrt{(x\cos\alpha+y\sin\alpha-p)}+\sin\tfrac{1}{2}(\gamma-\alpha)\sqrt{(x\cos\beta+y\sin\beta-q)} \\ +\sin\tfrac{1}{2}(\alpha-\beta)\sqrt{(x\cos\gamma+y\sin\gamma-r)}=0.$$

To rationalise the equation, I remark that an equation $\sqrt{(A)}+\sqrt{(B)}+\sqrt{(C)}=0$ gives in general

$$(1,\ 1,\ 1,\ \bar{1},\ \bar{1},\ \bar{1})\ (A,\ B,\ C)^2=0,$$

and that

$$(1,\ 1,\ 1,\ \bar{1},\ \bar{1},\ \bar{1}\between 2p\sin^2\tfrac{1}{2}(\beta-\gamma),\ \ 2q\sin^2\tfrac{1}{2}(\gamma-\alpha),\ \ 2r\sin^2\tfrac{1}{2}(\alpha-\beta)^2)$$

or as it may also be written

$$(1,\ 1,\ 1,\ \bar{1},\ \bar{1},\ \bar{1}\between p\{1-\cos(\beta-\gamma)\},\ q\{1-\cos(\gamma-\alpha)\},\ r\{1-\cos(\alpha-\beta)\})^2,$$

is identically equal to

$$\begin{aligned} &\{p(\sin\beta-\sin\gamma)+q(\sin\gamma-\sin\alpha)+r(\sin\alpha-\sin\beta)\}^2 \\ +&\{p(\cos\beta-\cos\gamma)+q(\cos\gamma-\cos\alpha)+r(\cos\alpha-\cos\beta)\}^2 \\ -&\{p\sin(\beta-\gamma)+q\sin(\gamma-\alpha)+r\sin(\alpha-\beta)\}^2. \end{aligned}$$

Hence if we replace p, q, r by

$$x\cos\alpha+y\sin\alpha-p,\quad x\cos\beta+y\sin\beta-q,\quad x\cos\gamma+y\sin\beta-r,$$

the last-mentioned expression equated to zero will give the equation of the circle, and we obtain

$$\begin{aligned}&\{\nabla x+p(\sin\beta-\sin\gamma)+q(\sin\gamma-\sin\alpha)+r(\sin\alpha-\sin\beta)\}^2\\ +&\{\nabla y-p(\cos\beta-\cos\gamma)-q(\cos\gamma-\cos\alpha)-r(\cos\alpha-\cos\beta)\}^2\\ -&\{p\sin(\beta-\gamma)+q\sin(\gamma-\alpha)+r\sin(\alpha-\beta)\}^2=0,\end{aligned}$$

where

$$\nabla=\sin(\beta-\gamma)+\sin(\gamma-\alpha)+\sin(\alpha-\beta),$$

and we have thus the equation of the circle in the usual form with the coordinates of the centre and the radius put in evidence.

The condition that there may be a circle touching the four lines

$$\begin{aligned}Ax+By+C&=0,\\ A'x+B'y+C'&=0,\\ A''x+B''y+C''&=0,\\ A'''x+B'''y+C'''&=0,\end{aligned}$$

is by the general formula shown to be

$$\begin{vmatrix} A, & B, & C, & \sqrt{(A^2+B^2)} \\ A', & B', & C', & \sqrt{(A'^2+B'^2)} \\ A'', & B'', & C'', & \sqrt{(A''^2+B''^2)} \\ A''', & B''', & C''', & \sqrt{(A'''^2+B'''^2)} \end{vmatrix}=0,$$

which is in fact obvious from other considerations.

180.

NOTE ON THE REDUCTION OF AN ELLIPTIC ORBIT TO A FIXED PLANE.

[From the *Quarterly Mathematical Journal*, vol. II. (1858), pp. 49—54.]

THE principal object of the present Note is to obtain an expression for the quantity ϵ_0 which I call the modified mean longitude at epoch, viz. taking as the elements the longitude of the node, inclination and any four elements which determine the motion in the plane of the orbit, then the longitude measured in the fixed plane (or reduced longitude) will be a function of the form

$$nt + \epsilon_0 + \text{periodic terms},$$

where ϵ_0 is a determinate function of the elements, and it is proposed to find the expression of this function. But as the corresponding formulæ relating to the eccentricity and longitude of the pericentre are not in general given as part of the theory of elliptic motion, but occur only, so far as I am aware, in works on the lunar theory, I have thought it desirable to include these formulæ and take as the subject of this Note the reduction of an elliptic orbit to a fixed plane. Write

a_1, the semiaxis-major,

e_1, the eccentricity $(=\sin\kappa_1)$,

ϖ_1, the longitude of pericentre in orbit,

ϵ_1, the mean longitude in orbit at epoch,

θ, the longitude of node,

ϕ, the inclination $(=\tan^{-1}\gamma)$,

and moreover

n_1, the mean motion $\left\{=\sqrt{\left(\dfrac{\sigma}{a_1^3}\right)}\right\}$, where by longitude in orbit is to be understood as usual a longitude measured in the fixed plane as far as the node and from the node in the plane of the orbit: the meaning of ϵ_1 is perhaps more clearly fixed by saying that $\epsilon_1 - \varpi_1$ denotes the mean anomaly at epoch.

The elements most nearly corresponding to the above, in the orbit reduced to the fixed plane, are

a_0, the modified semiaxis-major,

e_0, the modified eccentricity,

ϖ_0, the modified longitude of pericentre,

ϵ_0, the modified mean longitude at epoch,

θ, the longitude of node,

ϕ, the inclination ($= \tan^{-1} \gamma$),

and moreover

$$n\text{, the mean motion }\left\{\text{not equal to } \sqrt{\left(\frac{\sigma}{a_0^3}\right)}\right\},$$

where θ, ϕ, n are the same as in the actual orbit, but a_0, e_0, ϖ_0, ϵ_0 are defined as follows: viz. e_0, ϖ_0 are functions of e_1, ϖ_1, θ, ϕ given by the equations

$$\tan(\varpi_0 - \theta) = \sec\phi \tan(\varpi_1 - \theta),$$

$$e_0 = \frac{e_1 \cos(\varpi_1 - \theta)}{\cos(\varpi_0 - \theta)},$$

a_0 is determined by the condition

$$a_0(1 - e_0^2) = a_1(1 - e_1^2),$$

and ϵ_0 is determined so that the reduced longitude may be equal to

$$nt + \epsilon_0 + \text{periodic terms.}$$

It is easy to see that considering the orbit and the fixed plane as great circles of the sphere, and projecting the pericentre upon the fixed plane by an arc *perpendicular to the orbit*, then ϖ_0 denotes the longitude of such projection of the pericentre; and e_0 is equal to e_1 into the secant of the projecting arc. In fact we have a right-angled spherical triangle, of which the projecting arc in question is the perpendicular, and the hypothenuse and base are $\varpi_0 - \theta$ and $\varpi_1 - \theta$ respectively, and the base angle is the inclination ϕ. It is to be remarked that ϖ_0 is *not* the reduced longitude of the pericentre, an expression that would signify the longitude of the projection of the pericentre by an arc perpendicular to the fixed plane; this is the reason why I have throughout used the word *modified* instead of what would at first sight have appeared the natural one, viz. the word *reduced*. The modified semiaxis-major is obviously a semiaxis-major calculated from the latus rectum of the orbit by means of the modified eccentricity e_0.

The relations between e_0, ϖ_0, e_1, ϖ_1 may be written

$$\tan(\varpi_0 - \theta) = \sec\phi \tan(\varpi_1 - \theta),$$
$$e_0 = e_1 \sec\phi \sqrt{\{1 - \sin^2\phi \sin^2(\varpi_1 - \theta)\}},$$

or again

$$\tan(\varpi_1 - \theta) = \cos\phi \tan(\varpi_0 - \theta),$$
$$e_1 = e_0 \sqrt{\{1 - \sin^2\phi \sin^2(\varpi_0 - \theta)\}}.$$

Write now

r_1, the radius vector,

v_1, the longitude in orbit,

λ, the latitude ($=\tan^{-1} s$),

and in like manner

r_0, the reduced radius vector,

v_0, the reduced longitude,

λ, the latitude ($=\tan^{-1} s$).

Then $v_1 - \theta$ and $v_0 - \theta$ are the hypothenuse and base of a right-angled spherical triangle, the perpendicular being λ and the angle at the base being ϕ. We have

$$\tan\lambda = \tan\phi \sin(v_0 - \theta),$$
$$\sin\lambda = \sin\phi \sin(v_1 - \theta),$$
$$\tan(v_0 - \theta) = \cos\phi \tan(v_1 - \theta),$$
$$\cos(v_0 - \theta) = \sec\lambda \cos(v_1 - \theta).$$

We have for the radius vector

$$\frac{1}{r_1} = \frac{1}{a_1(1 - e_1^2)}\{1 + e_1 \cos(v_1 - \varpi_1)\},$$

and the reduced radius vector is thence found as follows: viz. we have $r_0 = r_1 \cos\lambda$, that is

$$\frac{1}{r_0} = \frac{1}{a_1(1 - e_1^2)}\{\sec\lambda + e_1 \sec\lambda \cos(v_1 - \varpi_1)\};$$

but $e_1 \sec\lambda \cos(v_1 - \varpi_1)$

$$= e_1 \sec\lambda \cos\{(v_1 - \theta) - (\varpi_1 - \theta)\},$$
$$= e_1 \sec\lambda \cos(v_1 - \theta)\cos(\varpi_1 - \theta) + e_1 \sec\lambda \sin(v_1 - \theta)\sin(\varpi_1 - \theta),$$
$$= e_1 \sec\lambda \cos(v_1 - \theta)\cos(\varpi_1 - \theta)\{1 + \tan(v_1 - \theta)\tan(\varpi_1 - \theta)\},$$
$$= e_0 \cos(v_0 - \theta)\cos(\varpi_0 - \theta)\{1 + \tan(v_0 - \theta)\tan(\varpi_0 - \theta)\},$$
$$= e_0 \cos(v_0 - \theta)\cos(\varpi_0 - \theta) + e_0 \sin(\varpi_0 - \theta)\sin(\varpi_0 - \theta),$$
$$= e_0 \cos(v_0 - \varpi_0),$$

and by the definition of a_0 we have $a_0(1-e_0^2)=a_1(1-e_1^2)$. Hence

$$\frac{1}{r_0}=\frac{1}{a_0(1-e_0^2)}\{\sec\lambda+e_0\cos(v_0-\varpi_0)\},$$

which, combined with the equation

$$\tan\lambda=\tan\phi\sin(v_0-\theta),$$

determines the position of the body in terms of the modified elements and of the reduced longitude v_0. Introducing into the two equations $s\,(=\tan\lambda)$ and $\gamma\,(=\tan\phi)$ in the place of λ and ϕ, they become

$$\frac{1}{r_0}=\frac{1}{a_0(1-e_0^2)}\{\sqrt{(1+s^2)}+e_0\cos(v_0-\varpi_0)\},$$

$$s=\gamma\sin(v_0-\theta),$$

which is the form in which the equations occur in the lunar theory.

Proceeding now to the formulæ which involve the time, it is to be remarked that the true anomaly and the quotient of the radius vector by the semiaxis-major are given functions of the eccentricity and the mean anomaly, and calling for a moment the last-mentioned quantities e, ξ, I represent the functions in question by

$$\text{elta}\,(e,\ \xi),\ \text{elqr}\,(e,\ \xi),$$

or more simply when the mean anomaly only is attended to by

$$\text{elta}\ \xi,\ \text{elqr}\ \xi.$$

I have found this notation very convenient as a means of dispensing with the introduction of the eccentric anomaly.

The reduced longitude is found in terms of the time by means of the equations

$$\tan(v_0-\theta)=\sec\phi\tan(v_1-\theta),$$

$$v_1-\varpi_1=\text{elta}\,(nt+\epsilon_1-\varpi_1),$$

the former equation gives, as is well known,

$$v_0-\theta=v_1-\theta-\tan^2\tfrac{1}{2}\phi\sin(2v_1-2\theta)+\tfrac{1}{2}\tan^4\tfrac{1}{2}\sin(4v_1-4\theta)-\&c.,$$

(where the successive coefficients are the reciprocals of the natural numbers) we have therefore

$$v_0=v_1-\tan^2\tfrac{1}{2}\phi\sin\{(2v_1-2\varpi_1)+(2\varpi_1-2\theta)\}+\&c.,$$

or, as it may be written,

$$\begin{aligned}v_0=v_1-\ &\tan^2\tfrac{1}{2}\phi\,\{\sin(2v_1-2\varpi_1)\cos(2\varpi_1-2\theta)+\cos(2v_1-2\varpi_1)\sin(2\varpi_1-2\theta)\}\\ &+\tfrac{1}{2}\tan^4\tfrac{1}{2}\phi\,\{\sin(4v_1-4\varpi_1)\cos(4\varpi_1-4\theta)+\cos(4v_1-4\varpi_1)\sin(4\varpi_1-4\theta)\}\\ &-\&c.,\end{aligned}$$

and for the present purpose it is only necessary to attend to the non-periodic part of the function on the right-hand side. Now

$$v_1 - \varpi_1 = \text{elta}\,(nt + \epsilon_1 - \varpi_1),$$

the non-periodic part of which is $nt + \epsilon_1 - \varpi_1$. And the non-periodic part of $\begin{smallmatrix}\cos\\ \sin\end{smallmatrix}\,\mu\,(v_1 - \varpi_1)$ is given by the equation (62) of Hansen's Memoir "Entwickelung des Products u. s. w." *Abhand. der K. Sächs. Gesellschaft zu Leipzig*, t. II. (1853). In fact, Hansen's

$$\beta \text{ is } = \frac{e_1}{1 + \sqrt{(1 - e_1^2)}}, = \tan \tfrac{1}{2}\kappa_1$$

and the formula gives for the non-periodic parts

$$\cos \mu\,(v_1 - \varpi_1) = (-)^\mu \tan^\mu \tfrac{1}{2}\kappa_1\,(1 + \mu \cos \kappa_1),$$
$$\sin \mu\,(v_1 - \varpi_1) = 0.$$

Hence, substituting these values and putting for the non-periodic part of v_0 the assumed value $nt + \epsilon_0$, we find

$$\begin{aligned}\epsilon_0 = \epsilon_1 - {} & \tan^2 \tfrac{1}{2}\phi \tan^2 \tfrac{1}{2}\kappa_1 (1 + 2\cos\kappa_1) \sin (2\varpi_1 - 2\theta) \\ & + \tfrac{1}{2} \tan^4 \tfrac{1}{2}\phi \tan^4 \tfrac{1}{2}\kappa_1 (1 + 4\cos\kappa_1) \sin (4\varpi_1 - 2\theta) \\ & - \&c.\end{aligned}$$

The series on the right-hand side may be summed without difficulty, and we obtain

$$\begin{aligned}\epsilon_0 = \epsilon_1 - \tan^{-1} & \left\{\frac{\tan^2 \frac{1}{2}\phi \tan^2 \frac{1}{2}\kappa_1 \sin (2\varpi_1 - 2\theta)}{1 + \tan^2 \frac{1}{2}\phi \tan^2 \frac{1}{2}\kappa_1 \cos (2\varpi_1 - 2\theta)}\right\} \\ & - 2\cos\kappa_1 \frac{\tan^2 \frac{1}{2}\phi \tan^2 \frac{1}{2}\kappa_1 \sin (2\varpi_1 - 2\theta)}{1 + 2\tan^2 \frac{1}{2}\phi \tan^2 \frac{1}{2}\kappa_1 \cos (2\varpi_1 - 2\theta) + \tan^4 \frac{1}{2}\phi \tan^4 \frac{1}{2}\kappa_1},\end{aligned}$$

in which formula the values of $\tan \frac{1}{2}\phi$, $\tan \frac{1}{2}\kappa_1$ (in terms of γ, e_1) are

$$\frac{\gamma}{1 + \sqrt{(1 + \gamma^2)}}, \quad \frac{e_1}{1 + \sqrt{(1 - e_1^2)}},$$

and that of $\cos \kappa_1$ is $\sqrt{(1 - e_1^2)}$. We have thus the required expression for the modified mean longitude at epoch, and all the modified elements are now expressed in terms of the original elements.

The following investigation leads to a theorem which it is, I think, worth while to notice. We have

$$r_0^2 \frac{dv_0}{dt} = \sqrt{\{\sigma a_0 (1 - e_0^2)\}} \cos \phi,$$

and thence

$$dt = \frac{a_0^{\frac{3}{2}}(1-e_0^2)^{\frac{3}{2}}\,dv_0}{\sqrt{(\sigma)}\cos\phi\,\{\sec\lambda + e_0\cos(v_0-\varpi_0)\}^2}$$

$$= \frac{a_1^{\frac{3}{2}}(1-e_1^2)^{\frac{3}{2}}\,dv_0}{\sqrt{(\sigma)}\sqrt{(1+\gamma^2)}\,[\sqrt{\{1+\gamma^2\sin^2(v_0-\varpi_0)\}}+e_0\cos(v_0-\varpi_0)]^2},$$

or as it may be written

$$\frac{dv_0}{[\sqrt{\{1+\gamma^2\sin^2(v_0-\varpi_0)\}}+e_0\cos(v_0-\varpi_0)]^2} = \sqrt{\left(\frac{\sigma}{a_1^3}\right)}(1-e_1^2)^{-\frac{3}{2}}(1+\gamma^2)^{\frac{1}{2}}\,dt,$$

$$= n(1-e_1^2)^{-\frac{3}{2}}(1+\gamma^2)^{\frac{1}{2}}\,dt.$$

But it is easy to see that if the mean longitude $nt+\epsilon_0$ is expanded in terms of v_0, the relation between these quantities must be of the form $nt+\epsilon_0 = v_0 +$ periodic terms. It follows that in the preceding equation the non-periodic part of the function which multiplies dv_0 (the expansion being in multiple cosines of v_0) must be equal to $(1-e_1^2)^{-\frac{3}{2}}(1+\gamma^2)^{\frac{1}{2}}$. Hence, putting for e_1 its value, we find that the non-periodic part of

$$\frac{1}{[\sqrt{\{1+\gamma^2\sin^2(v_0-\theta)\}}+e_0\cos(v_0-\varpi_0)]^2},$$

expanded in multiple cosines of v_0 is

$$\left[1-e_0^2\left\{1-\frac{\gamma^2}{1+\gamma^2}\sin^2(\varpi_0-\theta)\right\}\right]^{-\frac{3}{2}}(1+\gamma^2)^{\frac{1}{2}},$$

a theorem which might, it is probable, be verified without much difficulty.

2, *Stone Buildings, October*, 1856.

181.

ON SIR W. R. HAMILTON'S METHOD FOR THE PROBLEM OF THREE OR MORE BODIES.

[From the *Quarterly Mathematical Journal*, vol. II. (1858), pp. 66—73.]

THE problem of three or more bodies is considered by Sir W. R. Hamilton in his two well-known memoirs on a general method in Dynamics, *Phil. Trans.* 1834 and 1835, and the differential equations for the relative motion, with respect to the central body, of all the other bodies are obtained in a form containing a single disturbing function only. Several methods of integration are given or indicated, among others, one which is in fact the method of the variation of the elements as applied to the particular form of the equations of motion. But the investigation shows (and Sir W. R. Hamilton notices this as a defect in his theory, as compared with the ordinary theory of the variation of the elements), that in the method in question, the elements are not osculating elements, i.e. that the positions only, and not the velocities of the bodies, can be calculated as if the elements remained constant during an element of time. The peculiar advantage of the method is of course the having a single disturbing function only, and this seems so important, that if I may venture to express an opinion, I cannot but think that the method will ultimately be employed for the purposes of Physical Astronomy. But, however this may be, it has appeared to me that it may be useful to present the method in a separate and distinct form, disengaged from the general theory as an illustration of which it was given by the author; and this is what I propose now to do.

Consider a central body M, and two other bodies M_1, M_2, and let the coordinates of M referred to a fixed origin be x, y, z, and the coordinates of M_1, M_2 referred to the body M as origin be x_1, y_1, z_1 and x_2, y_2, z_2 respectively. Then the coordinates of M_1, M_2 referred to the fixed origin, are $x+x_1$, $y+y_1$, $z+z_1$ and $x+x_2$, $y+y_2$, $z+z_2$

respectively, and if as usual T denotes the Vis-viva or half sum of each mass into the square of its velocity, and U denote the force function, then we have

$$\begin{aligned}T = \ & \tfrac{1}{2}M\ (x'^2 + y'^2 + z'^2),\\ & + \tfrac{1}{2}M_1\ \{(x' + x_1')^2 + (y' + y_1')^2 + (z' + z_1')^2\},\\ & + \tfrac{1}{2}M_2\ \{(x' + x_2')^2 + (y' + y_2')^2 + (z' + z_2')^2\},\end{aligned}$$

$$\begin{aligned}U = \ & \frac{MM_1}{\sqrt{(x_1^2 + y_1^2 + z_1^2)}}\\ & + \frac{MM_2}{\sqrt{(x_2^2 + y_2^2 + z_2^2)}}\\ & + \frac{M_1M_2}{\sqrt{(x_1 - x_2)^2 + (y_1 - y_2)^2 + (z_1 - z_2)^2}},\end{aligned}$$

and the equations of motion are as usual

$$\frac{d}{dt}\frac{dT}{dx'} - \frac{dT}{dx} = \frac{dU}{dx},$$

&c.

If we assume that the centre of gravity of the bodies is at rest, then we have

$$Mx' + M_1(x' + x_1') + M_2(x' + x_2') = 0, \text{ \&c.},$$

and consequently

$$x' = -\frac{M_1x_1' + M_2x_2'}{M + M_1 + M_2},\quad y' = -\frac{M_1y_1' + M_2y_2'}{M + M_1 + M_2},\quad z' = -\frac{M_1z_1' + M_2z_2'}{M + M_1 + M_2}.$$

Now the value of T is

$$\begin{aligned}T = \ & \tfrac{1}{2}(M + M_1 + M_2)(x'^2 + y'^2 + z'^2)\\ & + x'(M_1x_1' + M_2x_2') + y'(M_1y_1' + M_2y_2') + z'(M_1z_1' + M_2z_2')\\ & + \tfrac{1}{2}M_1(x_1'^2 + y_1'^2 + z_1'^2)\\ & + \tfrac{1}{2}M_2(x_2'^2 + y_2'^2 + z_2'^2),\end{aligned}$$

or, putting for x', y', z' their values,

$$\begin{aligned}T = \ & \tfrac{1}{2}M_1(x_1'^2 + y_1'^2 + z_1'^2)\\ & + \tfrac{1}{2}M_2(x_2'^2 + y_2'^2 + z_2'^2)\\ & - \tfrac{1}{2}\frac{1}{M + M_1 + M_2}\{(M_1x_1' + M_2x_2')^2 + (M_1y_1' + M_2y_2')^2 + (M_1z_1' + M_2z_2')^2\},\end{aligned}$$

and with this new value of T the equations of motion still are

$$\frac{d}{dt}\frac{dT}{dx_1'} - \frac{dT}{dx_1} = \frac{dU}{dx_1},\ \text{\&c.}$$

Suppose now that the differential coefficients of T, with respect to x_1', y_1', z_1'; x_2', y_2', z_2', are respectively P_1, Q_1, R_1; P_2, Q_2, R_2, i.e. write

$$\frac{dT}{dx_1'} = P_1, \text{ \&c.},$$

and imagine T expressed as a function of P_1, Q_1, R_1; P_2, Q_2, R_2, and when this is done put $H = T - U$ (so that H stands for a function of P_1, Q_1, R_1; P_2, Q_2, R_2; x_1, y_1, z_1; x_2, y_2, z_2), then the equations of motion in Sir W. R. Hamilton's form are

$$\frac{dx_1}{dt} = \frac{dH}{dP_1}, \quad \frac{dP_1}{dt} = -\frac{dH}{dx_1}, \quad \text{\&c.}$$

Now from the last given value of T

$$P_1 = M_1 x_1' - \frac{M_1}{M + M_1 + M_2}(M_1 x_1' + M_2 x_2'),$$

$$P_2 = M_2 x_2' - \frac{M_2}{M + M_1 + M_2}(M_1 x_1' + M_2 x_2'),$$

and thence

$$P_1 + P_2 = \frac{M}{M + M_1 + M_2}(M_1 x_1' + M_2 x_2'),$$

and consequently

$$M_1 x_1' = P_1 + \frac{M_1}{M}(P_1 + P_2),$$

$$M_2 x_2' = P_2 + \frac{M_2}{M}(P_1 + P_2),$$

and we have

$$\begin{aligned} T = {} & \frac{1}{2M_1}\left[\left\{P_1 + \frac{M_1}{M}(P_1 + P_2)\right\}^2 + \left\{Q_1 + \frac{M_1}{M}(Q_1 + Q_2)\right\}^2 + \left\{R_1 + \frac{M_1}{M}(R_1 + R_2)\right\}^2\right] \\ & + \frac{1}{2M_2}\left[\left\{P_2 + \frac{M_2}{M}(P_1 + P_2)\right\}^2 + \left\{Q_2 + \frac{M_2}{M}(Q_1 + Q_2)\right\}^2 + \left\{R_2 + \frac{M_2}{M}(R_1 + R_2)\right\}^2\right] \\ & - \frac{1}{2M^2}(M + M_1 + M_2)\left[(P_1 + P_2)^2 + (Q_1 + Q_2)^2 + (R_1 + R_2)^2\right], \end{aligned}$$

or, reducing,

$$\begin{aligned} T = {} & \left(\frac{1}{2M_1} + \frac{1}{2M}\right)(P_1^2 + Q_1^2 + R_1^2) \\ & + \left(\frac{1}{2M_2} + \frac{1}{2M}\right)(P_2^2 + Q_2^2 + R_2^2) \\ & + \frac{1}{M}(P_1 P_2 + Q_1 Q_2 + R_1 R_2), \end{aligned}$$

and consequently

$$\begin{aligned} H = &\left(\frac{1}{2M_1}+\frac{1}{2M}\right)(P_1^2+Q_1^2+R_1^2) \\ &+\left(\frac{1}{2M_2}+\frac{1}{2M}\right)(P_2^2+Q_2^2+R_2^2) \\ &+\frac{1}{M}(P_1P_2+Q_1Q_2+R_1R_2) \\ &-\frac{MM_1}{\sqrt{(x_1^2+y_1^2+z_1^2)}} \\ &-\frac{MM_2}{\sqrt{(x_2^2+y_2^2+z_2^2)}} \\ &-\frac{M_1M_2}{\sqrt{\{(x_1-x_2)^2+(y_1-y_2)^2+(z_1-z_2)^2\}}}, \end{aligned}$$

and H having this value, the equations of motion are as before mentioned

$$\frac{dx_1}{dt}=\frac{dH}{dP_1},\quad \frac{dP_1}{dt}=-\frac{dH}{dx_1},\ \&c.$$

Instead of H write $H+\Upsilon$ where

$$\begin{aligned} H = &\left(\frac{1}{2M_1}+\frac{1}{2M}\right)(P_1^2+Q_1^2+R_1^2) \\ &+\left(\frac{1}{2M_2}+\frac{1}{2M}\right)(P_2^2+Q_2^2+R_2^2) \\ &-\frac{MM_1}{\sqrt{(x_1^2+y_1^2+z_1^2)}} \\ &-\frac{MM_2}{\sqrt{(x_2^2+y_2^2+z_2^2)}}, \end{aligned}$$

and

$$\begin{aligned} \Upsilon = &\frac{1}{M}(P_1P_2+Q_1Q_2+R_1R_2) \\ &-\frac{M_1M_2}{\sqrt{\{(x_1-x_2)^2+(y_1-y_2)^2+(z_1-z_2)^2\}}}, \end{aligned}$$

and the function Υ is to be treated as a disturbing function. The equations of motion for the body M_1 become

$$\begin{aligned} \frac{dx_1}{dt}&=\frac{M+M_1}{MM_1}P_1+\frac{d\Upsilon}{dP_1}, & \frac{dP_1}{dt}&=-\frac{MM_1x_1}{(x_1^2+y_1^2+z_1^2)^{\frac{3}{2}}}-\frac{d\Upsilon}{dx_1}, \\ \frac{dy_1}{dt}&=\frac{M+M_1}{MM_1}Q_1+\frac{d\Upsilon}{dQ_1}, & \frac{dQ_1}{dt}&=-\frac{MM_1y_1}{(x_1^2+y_1^2+z_1^2)^{\frac{3}{2}}}-\frac{d\Upsilon}{dy_1}, \\ \frac{dz_1}{dt}&=\frac{M+M_1}{MM_1}R_1+\frac{d\Upsilon}{dR_1}, & \frac{dR_1}{dt}&=-\frac{MM_1z_1}{(x_1^2+y_1^2+z_1^2)^{\frac{3}{2}}}-\frac{d\Upsilon}{dz_1}, \end{aligned}$$

and there is of course a precisely similar system of equations of motion for the body M_2.

If we neglect Υ, the left-hand equations show that P_1, Q_1, R_1 denote the velocities or differential coefficients $\frac{dx_1}{dt}, \frac{dy_1}{dt}, \frac{dz_1}{dt}$ multiplied by the constant factor $\frac{MM_1}{M+M_1}$, and substituting these values in the right-hand equations, we obtain the ordinary equations for the elliptic motion of the body M_1; and similarly for the body M_2. We may, if we please, complete the solution by the method of the variation of the arbitrary constants. Suppose for this purpose that $a_1, b_1, c_1, e_1, f_1, g_1$ are the elements for the elliptic motion of the body M_1, then treating these elements as variable we must have

$$\frac{dx_1}{da_1}\frac{da_1}{dt} + \frac{dx_1}{db_1}\frac{db_1}{dt} \ldots + \frac{dx_1}{dg_1}\frac{dg_1}{dt} = \frac{d\Upsilon}{dP_1}, \text{ \&c.,}$$

$$\frac{dP_1}{da_1}\frac{da_1}{dt} + \frac{dP_1}{db_1}\frac{db_1}{dt} \ldots + \frac{dP_1}{dg_1}\frac{dg_1}{dt} = -\frac{d\Upsilon}{dx_1}, \text{ \&c.,}$$

and it appears from these equations that as already noticed the disturbed values of the velocities are not (as they are in the ordinary theory) identical with the undisturbed values.

The disturbing function Υ may be considered as a function of the elements of the two orbits and of the time, and it is easy to obtain, as in the ordinary theory, the values of the differential coefficients $\frac{da_1}{dt}$, &c. in the form

$$\frac{da_1}{dt} = (a_1, b_1)\frac{d\Upsilon}{db_1} + (a_1, c_1)\frac{d\Upsilon}{dc_1} \ldots + (a_1, g_1)\frac{d\Upsilon}{dg_1},$$

where

$$(a_1, b_1) = \frac{\delta(a_1, b_1)}{\delta(x_1, P_1)} + \frac{\delta(a_1, b_1)}{\delta(y_1, Q_1)} + \frac{\delta(a_1, b_1)}{\delta(z_1, R_1)},$$

if for shortness

$$\frac{\delta(a_1, b_1)}{\delta(x_1, P_1)} = \frac{da_1}{dx_1}\frac{db_1}{dP_1} - \frac{da_1}{dP_1}\frac{db_1}{dx_1}.$$

It will be remembered that in the ordinary theory, if Ω denote Lagrange's disturbing function ($\Omega = -R$ if R is the disturbing function of the Mécanique Céleste) the corresponding formulæ are

$$\frac{da}{dt} = (a, b)\frac{d\Omega}{db} + (a, c)\frac{d\Omega}{dc} \ldots + (a, g)\frac{d\Omega}{dg},$$

where

$$(a, b) = \frac{\delta(a, b)}{\delta(x', x)} + \frac{\delta(a, b)}{\delta(y', y)} + \frac{\delta(a, b)}{\delta(z', z)},$$

if for shortness

$$\frac{\delta(a, b)}{\delta(x', x)} = \frac{da}{dx'}\frac{db}{dx} - \frac{da}{dx}\frac{db}{dx'},$$

or, what is the same thing, where

$$(a, b) = -\frac{\delta(a, b)}{\delta(x, x')} - \frac{\delta(a, b)}{\delta(y, y')} - \frac{\delta(a, c)}{\delta(z, z')},$$

and

$$\frac{\delta(a, b)}{\delta(x, x')} = \frac{da}{dx}\frac{db}{dx'} - \frac{da}{dx'}\frac{db}{dx}.$$

Now the values of the coefficients (a_1, b_1), &c. depend merely on the form of the expressions for a_1, b_1, &c. in terms of P_1, Q_1, R_1, x_1, y_1, z_1 and t; hence comparing the two systems of formulæ and observing P_1, Q_1, R_1 (which in the formulæ for the present theory correspond with x_1', y_1', z_1' in the other system of formulæ) are respectively equal to x_1', y_1', z_1', each of them multiplied by the constant factor $\frac{MM_1}{M+M_1}$, it is easy to see that the formulæ for the variations of any given system of elements in the present theory are at once deduced from the formulæ for the variations of the same system of elements in the ordinary theory by writing $-\Upsilon$ in the place of Ω and multiplying the values of the variations by the constant factor $\frac{MM_1}{M+M_1}$.

Take then as elements Jacobi's canonical system ([1]), viz. if we put

a_1, the semiaxis major,

e_1, the eccentricity,

ϖ_1, the longitude in orbit of pericentre,

ϵ_1, the mean longitude in orbit at epoch,

θ_1, the longitude of node,

ϕ_1, the inclination,

and

n_1, the mean motion $\left\{= \sqrt{\left(\frac{M+M_1}{a_1^3}\right)}\right\}$,

then the canonical elements are

$$\begin{aligned}
\mathfrak{A}_1 &= -\tfrac{1}{2}n_1^2 a_1^2, \\
\mathfrak{B}_1 &= n_1 a_1 \sqrt{(1-e_1^2)}, \\
\mathfrak{C}_1 &= n_1 a_1 \sqrt{(1-e_1^2)} \cos\phi_1, \\
\mathfrak{F}_1 &= \frac{1}{n_1}(\epsilon_1 - \varpi_1), \\
\mathfrak{G}_1 &= \varpi_1 - \theta_1, \\
\mathfrak{H}_1 &= \theta_1,
\end{aligned}$$

[1] I have for uniformity adopted Jacobi's canonical system, see his paper "Neues Theorem der analytischen Mechanik," *Crelle*, t. xxx. pp. 117—120 (1846); but it is proper to remark that Sir W. R. Hamilton, in his Memoirs above referred to, employs a slightly different but equally elegant system of canonical elements, and that the discovery of such a system belongs to Sir W. R. Hamilton, and is part of his general theory.

(the signs of the two elements $\mathfrak{A}_1$, $\mathfrak{F}_1$ have been changed, but this makes no difference in the formulæ) then the equations for the variations of the elements are

$$\frac{d\mathfrak{A}_1}{dt} = -\frac{M+M_1}{MM_1}\frac{d\Upsilon}{d\mathfrak{F}_1},$$

$$\frac{d\mathfrak{B}_1}{dt} = -\frac{M+M_1}{MM_1}\frac{d\Upsilon}{d\mathfrak{G}_1},$$

$$\frac{d\mathfrak{C}_1}{dt} = -\frac{M+M_1}{MM_1}\frac{d\Upsilon}{d\mathfrak{H}_1},$$

$$\frac{d\mathfrak{F}_1}{dt} = +\frac{M+M_1}{MM_1}\frac{d\Upsilon}{d\mathfrak{A}_1},$$

$$\frac{d\mathfrak{G}_1}{dt} = +\frac{M+M_1}{MM_1}\frac{d\Upsilon}{d\mathfrak{B}_1},$$

$$\frac{d\mathfrak{H}_1}{dt} = +\frac{M+M_1}{MM_1}\frac{d\Upsilon}{d\mathfrak{C}_1},$$

and it is easy thence to deduce the formulæ for the variations of any system of elements which it may be thought proper to make use of, for instance the system a_1, e_1, ϖ_1, ϵ_1, θ_1, ϕ_1.

It will be recollected that in the preceding system of formulæ the value of the disturbing function Υ is

$$\Upsilon = \frac{1}{M}(P_1P_2 + Q_1Q_2 + R_1R_2) - \frac{M_1M_2}{\sqrt{\{(x_1-x_2)^2 + (y_1-y_2)^2 + (z_1-z_2)^2\}}},$$

and that as a first approximation P_1, Q_1, R_1 are respectively equal to the velocities x_1', y_1', z_1', each multiplied by the constant factor $\dfrac{MM_1}{M+M_1}$, and P_2, Q_2, R_2 are respectively equal to the velocities x_2', y_2', z_2', each multiplied by $\dfrac{MM_2}{M+M_2}$.

2, *Stone Buildings*, 18*th Oct.*, 1856.

182.

ON LAGRANGE'S SOLUTION OF THE PROBLEM OF TWO FIXED CENTRES.

[From the *Quarterly Mathematical Journal*, vol. II. (1858), pp. 76—83.]

THE following variation of Lagrange's Solution of the Problem of Two Fixed Centres([1]), is, I think, interesting, as showing more distinctly the connection between the differential equations and the integrals. The problem referred to is as follows: viz. to determine the motion of a particle acted upon by forces tending to two fixed centres, such that r, q being the distances of the particle from the two centres respectively, and α, β, γ being constants, the forces are $\frac{\alpha}{r^2} + 2\gamma r$ and $\frac{\beta}{q^2} + 2\gamma q$.

Take the first centre as origin and the line joining the two centres as axis of x; and let h be the distance between the two centres, then writing for symmetry

$$x = x_1 = x_2 + h,$$

(so that x_1 is the coordinate corresponding to the first centre as origin, and x_2 the coordinate corresponding to the second centre as origin) the distances are given by the equations

$$r^2 = x_1^2 + y^2 + z^2, \qquad q^2 = x_2^2 + y^2 + z^2,$$

and the equations of motion are

$$\frac{d^2x}{dt^2} = -\frac{\alpha x_1}{r^3} - \frac{\beta x_2}{q^3} - 2\gamma(x_1 + x_2),$$

$$\frac{d^2y}{dt^2} = -\frac{\alpha y}{r^3} - \frac{\beta y}{q^3} - 4\gamma y,$$

$$\frac{d^2z}{dt^2} = -\frac{\alpha z}{r^3} - \frac{\gamma z}{q^3} - 4\gamma z,$$

[1] Lagrange's Solution was first published in the *Anciens Mémoires de Turin*, t. IV., [1766—69], and is reproduced in the *Mécanique Analytique*.

and we obtain at once the integral of Vis-viva, viz. multiplying the three equations by $\frac{dx}{dt}, \frac{dy}{dt}, \frac{dz}{dt}$, adding and integrating $\left(\text{observing that } \frac{dx}{dt} = \frac{dx_1}{dt} = \frac{dx_2}{dt}\right)$, we have

$$\tfrac{1}{2}\left\{\left(\frac{dx}{dt}\right)^2 + \left(\frac{dy}{dt}\right)^2 + \left(\frac{dz}{dt}\right)^2\right\} = \frac{\alpha}{r} + \frac{\beta}{q} - \gamma(r^2 + q^2) + 2H, \qquad (1, a)$$

and with equal facility, the equation of areas round the line joining the two centres, viz. multiplying the second and third equations by $-z$, y, adding and integrating, we have

$$y\frac{dz}{dt} - z\frac{dy}{dt} = B. \qquad (2, a)$$

So far Lagrange: to obtain a third integral I form the equation

$$\begin{aligned}
&\left[-2(y^2+z^2)\frac{dx}{dt} + (x_1+x_2)\left(y\frac{dy}{dt} + z\frac{dz}{dt}\right)\right] \times \left\{\frac{d^2x}{dt^2} + \frac{\alpha x_1}{r^3} + \frac{\beta x_2}{q^3} + 2\gamma(x_1+x_2)\right\}\\
&+\left[(x_1+x_2)\,y\frac{dx}{dt} - 2x_1x_2\frac{dy}{dt}\right] \times \left\{\frac{d^2y}{dt^2} + \frac{\alpha y}{r^3} + \frac{\beta y}{q^3} + 4\gamma y\right\}\\
&+\left[(x_1+x_2)\,z\frac{dx}{dt} - 2x_1x_2\frac{dz}{dt}\right] \times \left\{\frac{d^2z}{dt^2} + \frac{\alpha z}{r^3} + \frac{\beta z}{q^3} + 4\gamma z\right\} = 0.
\end{aligned}$$

The terms independent of the forces are

$$\begin{aligned}
&-2(y^2+z^2)\frac{dx}{dt}\frac{d^2x}{dt^2} + (x_1+x_2)\left(y\frac{dy}{dt} + z\frac{dz}{dt}\right)\frac{d^2x}{dt^2}\\
&+(x_1+x_2)\frac{dx}{dt}\left(y\frac{d^2y}{dt^2} + z\frac{d^2z}{dt^2}\right) - 2x_1x_2\left(\frac{dy}{dt}\frac{d^2y}{dt^2} + \frac{dz}{dt}\frac{d^2z}{dt^2}\right),
\end{aligned}$$

which are equal to

$$\frac{d}{dt}\left[-(y^2+z^2)\left(\frac{dx}{dt}\right)^2 + (x_1+x_2)\frac{dx}{dt}\left(y\frac{dy}{dt} + z\frac{dz}{dt}\right) - x_1x_2\left\{\left(\frac{dx}{dt}\right)^2 + \left(\frac{dy}{dt}\right)^2\right\}\right],$$

and the terms depending on the forces are readily reduced to the form

$$\frac{d}{dt}\left\{-\frac{h\alpha x_1}{r} + \frac{h\beta x_2}{q} + h^2\gamma(y^2+z^2)\right\},$$

in fact, considering first the terms multiplied by α, these are

$$\begin{aligned}
&\frac{x_1}{r^3}\left\{-2(y^2+z^2)\frac{dx}{dt} + (x_1+x_2)\left(y\frac{dy}{dt} + z\frac{dz}{dt}\right)\right\}\\
&+\frac{1}{r^3}\left\{(x_1+x_2)(y^2+z^2)\frac{dx}{dt} - 2x_1x_2\left(y\frac{dy}{dt} + z\frac{dz}{dt}\right)\right\},
\end{aligned}$$

which is equal to

$$\frac{1}{r^3}\left\{(x_2-x_1)(y^2+z^2)\frac{dx}{dt}+x_1(x_1-x_2)\left(y\frac{dy}{dt}+z\frac{dz}{dt}\right)\right\}$$

$$=\frac{1}{r^3}\left\{-h(y^2+z^2)\frac{dx}{dt}+hx_1\left(y\frac{dy}{dt}+z\frac{dz}{dt}\right)\right\}$$

$$=\frac{h}{r^3}\left\{-(x_1{}^2+y^2+z^2)\frac{dx_1}{dt}+x_1\left(x\frac{dx_1}{dt}+y\frac{dy}{dt}+z\frac{dz}{dt}\right)\right\}$$

$$=\frac{h}{r^3}\left\{-r^2\frac{dx_1}{dt}+rx_1\frac{dr}{dt}\right\}$$

$$=-h\frac{d}{dt}\frac{x_1}{r},$$

and similarly the term multiplied by β is

$$=\quad h\frac{d}{dt}\frac{x_2}{q},$$

lastly, the term multiplied by γ is

$$-4(x_1+x_2)(y^2+z^2)\frac{dx}{dt}+2(x_1+x_2)^2\left(y\frac{dy}{dt}+z\frac{dz}{dt}\right)$$

$$+4(x_1+x_2)(y^2+z^2)\frac{dx}{dt}-\quad 8x_1x_2\quad\left(y\frac{dy}{dt}+z\frac{dz}{dt}\right)$$

$$=\quad 2(x_1-x_2)^2\left(y\frac{dy}{dt}+z\frac{dz}{dt}\right)$$

$$=\quad 2h^2\left(y\frac{dy}{dt}+z\frac{dz}{dt}\right)$$

$$=\quad \frac{d}{dt}h^2(y^2+z^2).$$

The preceding combination of the differential equations gives therefore an equation integrable per se, and effecting the integration we have

$$-(y^2+z^2)\left(\frac{dx}{dt}\right)^2+(x_1+x_2)\frac{dx}{dt}\left(y\frac{dy}{dt}+z\frac{dz}{dt}\right)-x_1x_2\left\{\left(\frac{dy}{dt}\right)^2+\left(\frac{dz}{dt}\right)^2\right\}$$

$$-\frac{h\alpha x_1}{r}+\frac{h\beta x_2}{q}+h^2\gamma(y^2+z^2)=K, \qquad (3, a)$$

which is the third integral equation. It may be convenient to mention here (what appears by the comparison of the formulæ obtained in the sequel with the corresponding formulæ of Lagrange) that the value of Lagrange's constant of integration C is

$$C=K-2Hh^2-B^2+\tfrac{1}{4}\gamma h^4.$$

Making use of the ordinary transformation

$$(y^2+z^2)\left\{\left(\frac{dy}{dt}\right)^2+\left(\frac{dz}{dt}\right)^2\right\}=\left(y\frac{dy}{dt}+z\frac{dz}{dt}\right)^2+\left(y\frac{dz}{dt}-z\frac{dy}{dt}\right)^2,$$

the integral equations may be written under the forms

$$\tfrac{1}{2}\left(\frac{dx}{dt}\right)^2+\frac{1}{2(y^2+z^2)}\left(y\frac{dy}{dt}+z\frac{dz}{dt}\right)^2=\frac{\alpha}{r}+\frac{\beta}{q}-\gamma(r^2+q^2)+2H-\frac{B^2}{2(y^2+z^2)}, \qquad (1,\ b)$$

$$y\frac{dz}{dt}-z\frac{dy}{dt}=B, \qquad (2,\ b)$$

$$-(y^2+z^2)\left(\frac{dx}{dt}\right)^2+(x_1+x_2)\left(y\frac{dy}{dt}+z\frac{dz}{dt}\right)\frac{dx}{dt}-\frac{x_1x_2}{y^2+z^2}\left(y\frac{dy}{dt}+z\frac{dz}{dt}\right)^2$$

$$-\frac{h\alpha x_1}{r}+\frac{h\beta x_2}{q}+h^2\gamma(y^2+z^2)=K+\frac{B^2x_1x_2}{y^2+z^2}, \qquad (3,\ b)$$

and observing that y^2+z^2, x_1, x_2 are in fact functions of r, q, it is clear that the determination of r, q in terms of t depends upon the first and third equations alone. Moreover the form of the equations shows that we can at once eliminate dt and thus obtain a differential equation between r, q alone. It would be difficult to discover *à priori*, before actually obtaining the differential equation in question, that it would be possible to effect the separation of the variables, but we know that this can be done by taking instead of r, q the new variables $u=r+q$, $s=r-q$. In order to complete the solution the first step is to introduce the variables r, q into the first and third equations: for this purpose we have

$$x_1=\frac{h^2+r^2-q^2}{2h}, \quad -x_2=\frac{h^2-r^2+q^2}{2h}, \quad x_1+x_2=\frac{r^2-q^2}{h},$$

$$y^2+z^2=\frac{\nabla}{4h^2},$$

if for shortness

$$\nabla=2r^2q^2+2h^2r^2+2h^2q^2-h^4-r^4-q^4,$$

and consequently

$$\frac{dx}{dt}=\frac{1}{h}\left(r\frac{dr}{dt}-q\frac{dq}{dt}\right),$$

$$y\frac{dy}{dt}+z\frac{dz}{dt}=\frac{1}{2h^2}\left\{(h^2-r^2+q^2)\,r\frac{dr}{dt}+(h^2+r^2-q^2)\,q\frac{dq}{dt}\right\}.$$

Substituting these values in the two equations, we find for the first equation

$$\nabla(rdr-qdq)^2+\{(h^2-r^2+q^2)\,rdr+(h^2+r^2-q^2)\,qdq\}^2$$

$$=2h^2\left\{\frac{\alpha\nabla}{r}+\frac{\beta\nabla}{q}-\gamma(r^2+q^2)\nabla+2H\nabla-2h^2B^2\right\}dt^2, \qquad (1,\ c)$$

and for the second equation

$$\begin{aligned}
&-\nabla^2 (rdr - qdq)^2 \\
&+ 2\nabla (rdr - qdq)(r^2 - q^2)\{(h^2 - r^2 + q^2)\, rdr + (h^2 + r^2 - q^2)\, qdq\} \\
&+ \{h^4 - (r^2 - q^2)^2\}\{(h^2 - r^2 + q^2)\, rdr + (h^2 + r^2 - q^2)\, qdq\}^2 \\
&= h^4 \left[\frac{2\alpha\nabla}{r}(h^2 + r^2 - q^2) + \frac{2\beta\nabla}{q}(h^2 - r^2 + q^2) - \gamma\nabla^2 + 4K\nabla - 4B^2\{h^4 - (r^2 - q^2)^2\}\right] dt^2. \quad (2, c)
\end{aligned}$$

The first equation is easily reduced to

$$4h^2\{r^2q^2(dr^2 + dq^2) + (h^2 - r^2 - q^2)\, rq\, dr\, dq\} = 2h^2\left\{\frac{\alpha\nabla}{r} + \frac{\beta\nabla}{q} - \gamma(r^2 + q^2)\nabla + 2H\nabla - 2h^2B^2\right\} dt^2,$$

the second equation gives

$$h^4\{(h^2 - r^2 + q^2)\, rdr + (h^2 + r^2 - q^2)\, qdq\}^2 - h^4\{(h^2 - r^2 - 3q^2)\, rdr - (h^2 - 3r^2 + q^2)\, qdq\}^2$$

$$= h^4\left[\frac{2\alpha\nabla}{r}(h^2 + r^2 - q^2) + \frac{2\beta\nabla}{q}(h^2 - r^2 + q^2) - \gamma\nabla^2 + 4K\nabla - 4B^2\{h^4 - (r^2 - q^2)^2\}\right] dt^2,$$

and the function on the left-hand side is

$$8h^4(h^2 - r^2 - q^2)\, q^2r^2(dr^2 + dq^2) + 4h^4\{(h^2 - r^2 - q^2)^2 + 4q^2r^2\}\, rq\, dr\, dq.$$

Hence putting for a moment

$$M = \frac{\alpha\nabla}{r} + \frac{\beta\nabla}{q} - \gamma\nabla(r^2 + q^2) + 2H\nabla - 2h^2B^2,$$

$$N = \frac{\alpha\nabla}{r}(h^2 + r^2 - q^2) + \frac{\beta\nabla}{q}(h^2 - r^2 + q^2) - \tfrac{1}{2}\gamma\nabla^2 + 2K\nabla - 2B^2\{h^4 - (r^2 - q^2)^2\},$$

we have

$$2r^2q^2(dr^2 + dq^2) + (h^2 - r^2 - q^2)\, 2rq\, dr\, dq = Mdt^2,$$

$$2(h^2 - r^2 - q^2)\, 2r^2q^2(dr^2 + dq^2) + \{(h^2 - r^2 - q^2)^2 + 4q^2r^2\}\, 2rq\, dr\, dq = Ndt^2,$$

and thence recollecting that

$$-(h^2 - r^2 - q^2)^2 + 4q^2r^2 = \nabla,$$

we find

$$\nabla\,.\, 2rq\, dr\, dq = \{(h^2 - r^2 - q^2)\, 2M - N\}\, dt^2,$$

$$\nabla\,.\, 2r^2q^2(dr^2 + dq^2) = [-\{(h^2 - r^2 - q^2)^2 + 4q^2r^2\}\, M + (h^2 - r^2 - q^2)\, N]\, dt^2,$$

and substituting for M, N their values, the functions on the right-hand side contain ∇ as a factor, and dividing by ∇, we obtain

$$\begin{aligned}
2rq\, dr\, dq = \Big[&\frac{\alpha}{r}(3r^2 + q^2 - h^2) + \frac{\beta}{q}(r^2 + 3q^2 - h^2) \\
&- \tfrac{1}{2}\gamma(3r^4 + 3q^4 + 10q^2r^2 - 2h^2r^2 - 2h^2q^2 - h^4) \\
&+ 4H(q^2 + r^2 - h^2) + 2K - 2B^2\Big]\, dt^2, \quad (1, d)
\end{aligned}$$

$$2r^2q^2(dr^2+dq^2) = [2\alpha r(r^2+3q^2-h^2)+2\beta q(3r^2+q^2-h^2)$$
$$-\tfrac{1}{2}\gamma\{r^6+q^6+15q^2r^2(r^2+q^2)-h^2(r^4+q^4+6r^2q^2)-h^4(r^2+q^2)+h^6\}$$
$$+2H\{r^4+q^4+6r^2q^2-2h^2(r^2+q^2)+h^4\}$$
$$+2K(r^2+q^2-h^2)-2(r^2+q^2)B^2]\,dt^2, \qquad (2,\ d)$$

and by comparing the first of these formulæ with the corresponding formulæ of Lagrange, we find, as already observed, that the relation between the constant K and Lagrange's constant C is $K = C+2Hh^2+B^2-\frac{1}{4}\gamma h^4$. And substituting this value of K, the two equations become identical with those of Lagrange[1].

The equation $y\dfrac{dz}{dt}-z\dfrac{dy}{dt}=B$, (putting $y=\sqrt{(y^2+z^2)}\cos\phi$, $z=\sqrt{(y^2+z^2)}\sin\phi$) gives at once $(y^2+z^2)\,d\phi = Bdt$, and substituting for y^2+z^2 its value $=\dfrac{\nabla}{4h^2}$, we find

$$d\phi = \frac{4h^2B}{4q^2r^2-(h^2-r^2-q^2)^2}\,dt, \qquad (3,\ d)$$

which is the third of Lagrange's equations.

To complete the solution, the combination of the first and second equations gives

$$r^2q^2(dr\pm dq)^2 = [\alpha\ \{(r\pm q)^3-h^2(r\pm q)\}\pm\beta\{(r\pm q)^3-h^2(r\pm q)\}$$
$$-\tfrac{1}{4}\gamma\ \{(r\pm q)^6-h^2(r\pm q)^4-h^4(r\pm q)^2+h^6\}$$
$$+H\ \{(r\pm q)^4-2h^2(r\pm q)^2+h^4\}$$
$$+2K\{(r\pm q)^2-h^2\}-2B^2(r\pm q)^2]\,dt^2,$$

and thence putting $r+q=s$, $r-q=u$ and writing for shortness

$$S=(\alpha+\beta)(s^3-s^2h)$$
$$-\tfrac{1}{4}\gamma(s^6-h^2s^4-h^4s^2+h^6)$$
$$+H(s^4-2h^2s^2+h^4)$$
$$+2K(s^2-h^2)-2B^2s^2,$$

and

$$U=(\alpha-\beta)(u^3-u^2h)$$
$$-\tfrac{1}{4}\gamma(u^6-h^2u^4-h^4s^2+h^6)$$
$$+H(u^4-2h^2u^2-h^4)$$
$$+2K(u^2-h^2)-2B^2u^2,$$

[1] The formulæ referred to are the formulæ (b), (c), *Méc. Anal.* t. II. page 112 of the second edition and page 97 of the third edition, but there is an inaccuracy in the formulæ (c), B^2 ought to be changed into B^2h^2; the error is continued in the subsequent formulæ and moreover the constant term $-Ch^2$ is omitted on the right-hand side of the formulæ (e) and in the subsequent formulæ, i.e. in the functions of s, u, the term $-B^2$ should be $-B^2h^2-Ch^2$.

we have

$$\tfrac{1}{16}(s^2-u^2)^2\,ds^2 = S\,dt^2, \qquad (1,\ e)$$

$$\tfrac{1}{16}(s^2-u^2)^2\,du^2 = U\,dt^2, \qquad (2,\ e)$$

$$d\phi = -\frac{4h^2B}{(s^2-h^2)(u^2-h^2)}\,dt, \qquad (3,\ e)$$

and thence finally

$$\frac{ds}{\sqrt{(S)}} = \frac{du}{\sqrt{(U)}}, \qquad (1,\ f)$$

$$dt = \tfrac{1}{4}\left\{\frac{s^2ds}{\sqrt{(S)}} - \frac{u^2du}{\sqrt{(U)}}\right\}, \qquad (2,\ f)$$

$$d\phi = \frac{Bh^2ds}{(s^2-h^2)\sqrt{(S)}} - \frac{Bu^2du}{(u^2-h^2)\sqrt{(U)}}, \qquad (3,\ f)$$

so that the problem is reduced to quadratures, the functions to be integrated involving the square roots of two rational and integral functions of the sixth degree.

2, *Stone Buildings*, 10*th Nov.*, 1856.

183.

NOTE ON CERTAIN SYSTEMS OF CIRCLES.

[From the *Quarterly Mathematical Journal*, vol. II. (1858), pp. 83—88.]

It will be convenient to remark at the outset that two concentric circles, the radii of which are in the ratio of $1 : i$ (i being as usual the imaginary unit), are orthotomic([1]), and that the most convenient quasi representation of a circle, the centre of which is real and the radius a pure imaginary quantity, is by means of the concentric orthotomic circle. This being premised consider a circle and a point C. The points of contact of the tangents through C to the circle may be termed the taction points; the points where the chord through C perpendicular to the line joining C with the centre meets the circle, may be termed the section points. It is clear that, for an exterior point, the taction points are real and the section points imaginary, while, for an interior point, the section points are real and the taction points imaginary. A circle having C for its centre and passing through the taction points (in fact the orthotomic circle having C for its centre) is said to be the taction circle. A circle having C for its centre and passing through the section points is said to be the section circle. Of course for an exterior point the taction circle is real and the section circle imaginary; while for an interior point the taction circle is imaginary and the section circle is real. It is proper also to remark that the taction circle and the section circle are concentric orthotomic circles.

Passing now to the case of two systems of orthotomic circles, let MM', NN' be lines at right angles to each other intersecting in R, and let M, M' be real or pure imaginary points on the line MM', equidistant from R. Imagine a system of circles,

[1] Two concentric circles are, it is well known, conics having a double contact at infinity, and it appears at first sight difficult to reconcile with this, the idea of two particular concentric circles being orthotomic. The explanation is that any two lines through a circular point at infinity may be considered as being at right angles to each other, and therefore any line through a circular point at infinity may be considered as being at right angles to itself. The two concentric circles in question have, in fact, at each circular point at infinity a common tangent, but this common tangent must be considered as being at right angles to itself. The paradox disappears entirely upon a homographic deformation of the figure; two lines KL, KM are then defined to be at right angles when joining K with the fixed points I, J, the four lines KL, KM, KI, KJ are a harmonic pencil; but when K coincides with I, then KI is indeterminate and may be taken to be the fourth harmonic of the pencil, i.e. any two lines IL, IM through the point I may be considered as being at right angles.

each of them having its centre on the line NN' and passing through the points M, M' (so that MM' is the radical axis of these circles). There are always on the line NN' two pure imaginary or real points N, N' equidistant from R, such that the circles, each of them having its centre on MM' and passing through the points N, N' (NN' being therefore the radical axis of these circles), are orthotomic to the first-mentioned system of circles, [the four points M, M', N, N' being thus as I have since termed them, two pairs of antipoints]. Moreover if R be made the centre of a circle passing through M, M', then the concentric orthotomic circle passes through N, N'; this is in fact only a particular case of the general property.

Suppose now that M, M' being given as the points of intersection of two circles having their centres on NN', it is required to find a circle having for its centre a given point C on NN' and passing through the points M, M'. In the case of M, M' being real, the required circle is obviously given and is always real. But if M, M' are imaginary; then if about any point of MM' as centre a circle be described orthotomic to one of the circles, it will be orthotomic to the other circle, and will meet NN' in the real points N, N'. Now if ρ be the radius of the required circle (i.e. of the circle having C for its centre and passing through the points M, M'), then $\rho^2 = (RC)^2 + (RM)^2 = (RC)^2 - (RN)^2$. Hence if $RC > RN$ or if C lies outside the space NN', ρ^2 is positive or the required circle is real, and the radius is at once constructed from the preceding expression

$$\rho^2 = RC^2 - RN^2.$$

But if $RC < RN$ or C lies within the space NN', then the required circle is imaginary, but the concentric orthotomic circle is at once constructed from the formula

$$\rho'^2 = RN^2 - RC^2.$$

Suppose now the point C is a centre of similitude of the two circles. The circle having C for its centre and passing through the points M, M' is a taction circle of all the taction circles of the two circles, it may be termed the tactaction circle. The concentric orthotomic of the circle having C for its centre and passing through the points M, M' is a section circle of all the taction circles of the two circles, it may be termed the sectaction circle. Consider first the case where the circles intersect in a pair of real points; here the two centres of similitude are on opposite sides of R; the tactaction circles are both real, the sectaction circles both imaginary. Secondly, the case where the two circles are wholly exterior each to the other, the two centres of similitude lie on the same side of R, viz. the centre of inverse similitude between R and N, the centre of direct similitude beyond N. Hence the tactaction circle corresponding to the centre of direct similitude and the sectaction circle corresponding to the centre of inverse similitude are real, the other tactaction circle and sectaction circle are imaginary. Thirdly, the case where one of the circles is wholly interior to the other; here the two centres of similitude are still on the same side of R, but the centre of direct similitude lies between R and N, and the centre of inverse similitude lies beyond N. Hence the sectaction circle corresponding to the centre of direct similitude and the tactaction circle corresponding to the centre of inverse similitude are real, the other sectaction circle and tactaction circle are imaginary.

To obtain a distinct idea of the methods made use of in Gaultier's "Mémoire sur les moyens généraux de construire graphiquement un cercle déterminé par trois conditions," (*Jour. École Polyt.* t. IX. [1813], p. 124), and in Steiner's "Geometrische Betrachtungen," *Crelle*, t. I. [1826], p. 161; it should be remarked that both of these geometers, confining as they do their attention to real circles, do not consider the section circle of an exterior point, or the taction circle of an interior point. The taction circle of an exterior point, or the section circle of an interior point, is Gaultier's "Cercle radical," and Steiner's "Potenzkreis," and Steiner also speaks of the radius of this circle as the "Potenz" of its centre in relation to the given circle. The nature of the Cercle radical, or Potenzkreis, (i.e. whether it is a taction circle or a section circle) is of course determined as soon as it is known whether the centre is an exterior or an interior point, and Gaultier distinguishes the two cases as the "radical réciproque" and the "radical simple," and in like manner Steiner speaks of the Potenz as being "äuszerlich" or "innerlich." Again, for two circles and for a given centre of similitude Gaultier and Steiner employ the tactaction circle or the sectaction circle, whichever of them is real, Gaultier without giving any distinctive appellation to the circle in question, Steiner calling it the Potenzkreis of the two circles, and in particular the "äuszere Potenzkreis" or the "innere Potenzkreis," according as it has for centre the centre of direct similitude or the centre of inverse similitude.

The preceding properties of circles are of course at once extended to conics passing each of them through the same two points; it is I think worth while to notice what the analogue is of a pair of concentric orthotomic circles. If the fixed points are I, J and if the point corresponding to the centre is K, then the conics are of course conics touching the lines KI, KJ in the points I, J, and, one of the conics being given, the other is to be determined. It is easily seen that if an arbitrary line through I meets the conics in P, P' and the line KJ in M, then the points I, M, P, P' are a harmonic range, and this condition gives the construction of the second conic; it of course follows that an arbitrary line through J meets the conics in points Q, Q' and the line KI in a point N such that the points J, N, Q, Q' are also a harmonic range. The two conics in question may be termed "inscribed harmonics" each of the other.

Addition. The equation of the tactaction circle, corresponding to the centre of direct (or inverse) similitude, of two given circles, may be found as follows:

Let the equations of the given circles be

$$(x-\alpha)^2+(y-\beta)^2=c^2,$$
$$(x-\alpha')^2+(y-\beta')^2=c'^2,$$

then the coordinates of the centre of direct similitude are

$$\frac{\alpha c'-\alpha' c}{c'-c'},\quad \frac{\beta c'-\beta' c}{c'-c},$$

which are therefore the coordinates of the centre of the tactaction circle; and the equation of this circle is of the form

$$\lambda\,[(x-\alpha)^2+(y-\beta)^2-c^2]+(1-\lambda)\,[(x-\alpha')^2+(y-\beta')^2-c'^2]=0,$$

or, expanding and reducing,

$$(x^2+y^2)-2[\alpha\lambda+\alpha'(1-\lambda)]x-2[\beta\lambda+\beta'(1-\lambda)]y+\lambda(\alpha^2+\beta^2-c^2)+(1-\lambda)(\alpha'^2+\beta'^2-c'^2)=0.$$

We must therefore have

$$\alpha\lambda+\alpha'(1-\lambda)=\frac{\alpha c'-\alpha' c}{c'-c},$$

$$\beta\lambda+\beta'(1-\lambda)=\frac{\beta c'-\beta' c}{c'-c},$$

which are consistent with each other and give

$$\lambda=\frac{c'}{c'-c},\quad 1-\lambda=\frac{-c}{c'-c};$$

we have then

$$\lambda(\alpha^2+\beta^2-c^2)+(1-\lambda)(\alpha'^2+\beta'^2-c'^2)=\frac{1}{c'-c}[c'(\alpha^2+\beta^2)-c(\alpha'^2+\beta'^2)+cc'(c'-c)];$$

and the equation of the tactaction circle is

$$x^2+y^2-2\frac{\alpha c'-\alpha' c}{c'-c}x-2\frac{\beta c'-\beta' c}{c'-c}y=\frac{1}{c'-c}[c'(\alpha^2+\beta^2)-c(\alpha'^2+\beta'^2)+cc'(c'-c)],$$

which may also be written

$$\left(x-\frac{\alpha c'-\alpha' c}{c'-c}\right)^2+\left(y-\frac{\beta c'-\beta' c}{c'-c}\right)^2=\frac{cc'}{(c'-c)^2}[(\alpha'-\alpha)^2+(\beta'-\beta)^2-(c'-c)^2].$$

We have thus the equation of the tactaction circle corresponding to the centre of direct similitude, and that of the tactaction circle corresponding to the centre of inverse similitude is at once obtained from it by changing the sign of one of the two radii c, c'.

Consider any three circles and combining them in pairs, by what has preceded the equations of the tactaction circles corresponding to the centres of direct similitude will be

$$\begin{aligned}&(c''-c')(x^2+y^2)-2(\alpha'c''-\alpha''c')x-2(\beta'c''-\beta''c')y\\&\qquad+c''(\alpha'^2+\beta'^2)-c'(\alpha''^2+\beta''^2)+c'c''(c''-c')=0,\\&(c-c'')(x^2+y^2)-2(\alpha''c-\alpha c'')x-2(\beta''c-\beta c'')y\\&\qquad+c(\alpha''^2+\beta''^2)-c''(\alpha^2+\beta^2)+c''c(c-c'')=0,\\&(c'-c)(x^2+y^2)-2(\alpha c'-\alpha' c)x-2(\beta c'-\beta' c)y\\&\qquad+c'(\alpha^2+\beta^2)-c(\alpha'^2+\beta'^2)+cc'(c'-c)=0,\end{aligned}$$

and representing these equations by $U=0$, $U'=0$, $U''=0$, we have identically

$$cU+c'U'+c''U''=0,$$

hence the three tactaction circles pass through the same two points, or what is the same thing, have a common radical axis.

184.

A THEOREM RELATING TO SURFACES OF THE SECOND ORDER.

[From the *Quarterly Mathematical Journal*, vol. II. (1858), pp. 140—142.]

GIVEN a surface of the second order

$$(a, b, c, d, f, g, h, l, m, n)(x, y, z, w)^2 = 0,$$

and a fixed plane

$$\alpha x + \beta y + \gamma z + \delta w = 0,$$

imagine a variable plane

$$\xi x + \eta y + \zeta z + \omega w = 0,$$

subjected to the condition that it always touches a surface of the second order, or what is the same thing such that the parameters ξ, η, ζ, ω satisfy a condition

$$(\mathrm{a}, \mathrm{b}, \mathrm{c}, \mathrm{d}, \mathrm{f}, \mathrm{g}, \mathrm{h}, \mathrm{l}, \mathrm{m}, \mathrm{n})(\xi, \eta, \zeta, \omega)^2 = 0.$$

The given surface of the second order, and the variable plane meet in a conic, and the fixed plane and the variable plane meet in a line, it is required to find the locus of the pole of the line with respect to the conic.

The pole in question is the point in which the variable plane is intersected by the polar of the line with respect to the surface of the second order: this polar is the line joining the pole of the fixed plane with respect to the surface of the second order, and the pole of the variable plane with respect to the surface of the second order. Let α_1, β_1, γ_1, δ_1, be given linear functions of α, β, γ, δ, and ξ, η, ζ, ω, be given linear functions of ξ, η, ζ, ω, viz., if

$$(\mathfrak{A}, \mathfrak{B}, \mathfrak{C}, \mathfrak{D}, \mathfrak{F}, \mathfrak{G}, \mathfrak{H}, \mathfrak{L}, \mathfrak{M}, \mathfrak{N}),$$

are the inverse system to $(a, b, c, d, f, g, h, l, m)$, then let

$$\begin{aligned}
\alpha_1 &= \mathfrak{A}\alpha + \mathfrak{H}\beta + \mathfrak{G}\gamma + \mathfrak{L}\delta,\\
\beta_1 &= \mathfrak{H}\alpha + \mathfrak{B}\beta + \mathfrak{F}\gamma + \mathfrak{M}\delta,\\
\gamma_1 &= \mathfrak{G}\alpha + \mathfrak{F}\beta + \mathfrak{C}\gamma + \mathfrak{N}\delta,\\
\delta_1 &= \mathfrak{L}\alpha + \mathfrak{M}\beta + \mathfrak{N}\gamma + \mathfrak{D}\delta,
\end{aligned}$$

and in like manner,

$$\begin{aligned}
\xi_1 &= \mathfrak{A}\xi + \mathfrak{H}\eta + \mathfrak{G}\zeta + \mathfrak{L}\omega,\\
\eta_1 &= \mathfrak{H}\xi + \mathfrak{B}\eta + \mathfrak{F}\zeta + \mathfrak{M}\omega,\\
\zeta_1 &= \mathfrak{G}\xi + \mathfrak{F}\eta + \mathfrak{C}\zeta + \mathfrak{N}\omega,\\
\omega_1 &= \mathfrak{L}\xi + \mathfrak{M}\eta + \mathfrak{N}\zeta + \mathfrak{D}\omega,
\end{aligned}$$

then the coordinates of the pole of the fixed plane are as

$$\alpha_1 : \beta_1 : \gamma_1 : \delta_1,$$

and the coordinates of the pole of the variable plane are as

$$\xi_1 : \eta_1 : \zeta_1 : \delta_1,$$

whence the equations of the polar are

$$\left\|\begin{matrix} x, & y, & z, & w \\ \alpha_1, & \beta_1, & \gamma_1, & \delta_1 \\ \xi_1, & \eta_1, & \zeta_1, & \omega_1 \end{matrix}\right\| = 0,$$

a system of equations which may be thus represented

$$\begin{aligned}
\xi_1 &= \mathrm{K}\lambda x + \mu\alpha_1,\\
\eta_1 &= \mathrm{K}\lambda y + \mu\beta_1,\\
\zeta_1 &= \mathrm{K}\lambda z + \mu\gamma_1,\\
\omega_1 &= \mathrm{K}\lambda w + \mu\delta_1,
\end{aligned}$$

where K is the discriminant of the system

$$(a, b, c, d, e, f, g, h, l, m, n).$$

Write

$$\begin{aligned}
\mathrm{x} &= ax + hy + gz + lw,\\
\mathrm{y} &= hx + by + fz + mw,\\
\mathrm{z} &= gx + fy + cz + nw,\\
\mathrm{w} &= lx + my + nz + dw,
\end{aligned}$$

the last preceding system of equations may be written

$$\begin{aligned}\xi &= \lambda \mathrm{x} + \mu\alpha,\\ \eta &= \lambda \mathrm{y} + \mu\beta,\\ \zeta &= \lambda \mathrm{z} + \mu\gamma,\\ \omega &= \lambda \mathrm{w} + \mu\delta,\end{aligned}$$

equations in which λ, μ are indeterminate, and where x, y, z, w may be considered as current coordinates, and this system represents the polar above referred to. Combining the equations in question with the equation

$$\xi x + \eta y + \zeta z + \omega w = 0,$$

of the variable plane, we have

$$\lambda (x\mathrm{x} + y\mathrm{y} + z\mathrm{z} + w\mathrm{w}) + \mu (\alpha x + \beta y + \gamma z + \delta w) = 0,$$

i.e.

$$\lambda (a,\ldots)(x,\ y,\ z,\ w)^2 + \mu (\alpha x + \beta y + \gamma z + \delta w) = 0,$$

or what is the same thing

$$\lambda : \mu = \alpha x + \beta y + \gamma z + \delta w : (a,\ldots)(x,\ y,\ z,\ w)^2,$$

and substituting these values in the expressions for ξ, η, ζ, ω we have ξ, η, ζ, ω in terms of the coordinates x, y, z, w of the pole above referred to, i.e., if for shortness,

$$\begin{aligned}U &= (a,\ b,\ c,\ d,\ f,\ g,\ h,\ l,\ m,\ n)(x,\ y,\ z,\ w)^2,\\ P &= \alpha x + \beta y + \gamma z + \delta w,\end{aligned}$$

then

$$\begin{aligned}\xi &= \tfrac{1}{2} P d_x U - \alpha U,\\ \eta &= \tfrac{1}{2} P d_y U - \beta U,\\ \zeta &= \tfrac{1}{2} P d_z U - \gamma U,\\ \omega &= \tfrac{1}{2} P d_w U - \delta U,\end{aligned}$$

and combining with these equations the equation

$$(a,\ldots)(\xi,\ \eta,\ \zeta,\ \omega)^2 = 0,$$

we have

$$(a\ldots)(\tfrac{1}{2} P d_x U - \alpha U,\ \tfrac{1}{2} P d_y U - \beta U,\ \tfrac{1}{2} P d_z U - \gamma U,\ \tfrac{1}{2} P d_w U - \delta U)^2 = 0,$$

for the required locus of the pole of the line of intersection of the variable plane and the fixed plane, with the conic of intersection of the given surface of the second order and the variable plane. The locus in question is a surface of the fourth order; and it may be remarked that this surface touches the given surface of the second order along the conic of intersection with the fixed plane.

7th April, 1857.

185.

NOTE ON THE 'CIRCULAR RELATION' OF PROF. MÖBIUS.

[From the *Quarterly Mathematical Journal*, vol. II. (1858), p. 162.]

THEOREM.

Given the points

$$A,\ B,\ C;\ P,$$

and the points

$$A',\ B',\ C';$$

describe the circles α, β, γ, ω as follows: viz.

α through $(B,\ C,\ P)$,
β „ $(C,\ A,\ P)$,
γ „ $(A,\ B,\ P)$,
ω „ $(A,\ B,\ C)$,

and the circles α', β', γ', ω' as follows: viz.,

α' through $(B',\ C')$ cutting ω' at the angle at which α cuts ω,
β' „ $(C',\ A')$ „ ω' „ β „ ω,
γ' „ $(A',\ B')$ „ ω' „ γ „ ω, and
ω' „ $(A',\ B',\ C')$

then will α', β', γ' meet in a point P', i.e. we shall have the points A', B', C'; P'

such that the circles α', β', γ', ω' pass

$$\begin{array}{lll} \alpha' & \text{through} & (B',\ C',\ P'), \\ \beta' & ,, & (C',\ A',\ P'), \\ \gamma' & ,, & (A',\ B',\ P'), \\ \omega' & ,, & (A',\ B',\ C'). \end{array}$$

We may construct in this manner two figures, such that to three points of the first figure there correspond in the second figure three points which may be taken at pleasure, but these once selected to every other point of the first figure there will correspond in the second figure a perfectly determinate point. And the two figures will be such that whenever in the first figure four or more points lie in a circle, then in the second figure the corresponding points will also lie in a circle. The relation in question is due to Prof. Möbius, who has termed it the *Kreis-verwandschaft* (circular relation) of two plane figures. See his paper *Crelle*, t. LII. [1856], pp. 218—228, extracted from the Berichte of the *Royal Saxon Society of Sciences* of the 5th Feb. 1853, [and *Werke*, t. II. pp. 207—217].

186.

ON THE DETERMINATION OF THE VALUE OF A CERTAIN DETERMINANT.

[From the *Quarterly Mathematical Journal*, vol. II. (1858), pp. 163—166.]

CONSIDERING the determinant

$$\begin{vmatrix} \theta, & 1, & . & . & . & \dots \\ n, & \theta, & 2 & & & \\ . & n-1, & \theta, & 3 & & \\ & & n-2, & \theta, & 4 & \\ \vdots & & & & & \end{vmatrix}$$

let the successive diagonal minors be U_0, U_1, $U_2,\dots U_x \dots$, it is easy to find

$$\begin{aligned} U_0 &= 1, \\ U_1 &= \theta, \\ U_2 &= (\theta^2-1)-(n-1), \\ U_3 &= \theta\,(\theta^2-4)-3\,(n-2)\,\theta, \\ U_4 &= (\theta^2-1)\,(\theta^2-9)-6\,(n-3)\,(\theta^2-1)+3\,(n-3)\,(n-1), \end{aligned}$$

which in fact suggests the law, viz.

$$\begin{aligned} U_x = {} & (\theta+x-1)\,(\theta+x-3)\,(\theta+x-5)\dots(\theta-x+5)\,(\theta-x+3)\,(\theta-x+1) \\ & -\frac{x\,(x-1)}{2}\,(n-x+1)\,(\theta+x-3)\,(\theta+x-5)\dots(\theta-x+5)\,(\theta-x+3) \\ & +\frac{x\,(x-1)\,(x-2)\,(x-3)}{2\,.\,4}\,(n-x+1)\,(n-x+3)\,(\theta+x-5)\dots(\theta-x+5) \\ & - \&\text{c.} \\ & \vdots \\ & +(-)^s\,\frac{x\,(x-1)\dots(x-2s+1)}{2\,.\,4\dots2s}\,(n-x+1)\,(n-x+3)\dots(n-x+2s-1) \\ & \qquad (\theta+x-2s-1)\,(\theta+x-2s-3)\dots(\theta-x+2s+1) \\ & \vdots \end{aligned}$$

to $s=\frac{1}{2}x$ or $\frac{1}{2}(x-1)$, as x is even or odd.

And of course if x denote the number of lines or columns of the determinant, then U_x is the value of the determinant. This theorem, or a particular case of it, is due to Prof. Sylvester: I have not been able to find an easier demonstration than the following one, which, it must be admitted, is somewhat complicated. I observe that U_x satisfies the equation

$$U_x - \theta U_{x-1} + (x-1)(n-x+2)\, U_{x-2} = 0.$$

Hence writing $x-1$ and $x-2$ for x, we have the system

$$\begin{aligned}
&U_x - \theta U_{x-1} + (x-1)(n-x+2)\, U_{x-2} &&= 0,\\
&\qquad U_{x-1} - \qquad \theta U_{x-2} + (x-2)(n-x+3)\, U_{x-3} &&= 0,\\
&\qquad\qquad U_{x-2} - \qquad \theta U_{x-3} + (x-3)(n-x+4)\, U_{x-4} &&= 0,
\end{aligned}$$

or, eliminating U_{x-1} and U_{x-3},

$$\begin{aligned}
U_x + \{(x-1)(n-x+2) + (x-2)(n-x+3) - \theta^2\}\, U_{x-2}&\\
+ (x-2)(x-3)(n-x+3)(n-x+4)\, U_{x-4} &= 0.
\end{aligned}$$

Suppose, for shortness,

$$(\theta+x-1)(\theta+x-3)(\theta+x-5)\ldots(\theta-x+5)(\theta-x+3)(\theta-x+1) = H_x,$$

and assume

$$U_x = A_{x,0}\, H_x - A_{x,1}\, H_{x-2} \ldots + (-)^s A_{x,s}\, H_{x-2s} \ldots,$$

where $A_{x,s}$ is independent of θ, then

$$\begin{aligned}
&U_x \quad \text{contains the term } (-)^s A_{x,s} \quad H_{x-2s},\\
&U_{x-2} \text{ contains the term } (-)^s A_{x-2,s}\, H_{x-2s-2},
\end{aligned}$$

which is to be multiplied by

$$(x-1)(n-x+2) + (x-2)(n-x+3) - \theta^2.$$

This multiplier may be written under the form

$$\begin{aligned}
&(x-1)(n-x+2) + (x-2)(n-x+3) - (x-2s-1)^2 - \{\theta^2 - (x-2s-1)^2\}\\
&\quad = M_{x,s} - \{\theta^2 - (x-2s-1)^2\},
\end{aligned}$$

if, for shortness,

$$M_{x,s} = (x-1)(n-x+2) + (x-2)(n-x+3) - (x-2s-1)^2.$$

Now

$$M_{x,s} - \{\theta^2 - (x-2s-1)^2\}$$

multiplied into

$$(-)^s A_{x-2,s}\, H_{x-2s-2}$$

gives rise to the terms

$$(-)^s M_{x,s} A_{x-2,s} H_{x-2s-2} - (-)^s A_{x-2,s} H_{x-2s},$$

(since $\{\theta^2 - (x-2s-1)^2\} H_{x-2s-2} = H_{x-2s}$), or, what is the same thing,

$$-(-)^s M_{x,s-1} A_{x-2,s-1} H_{x-2s} - (-)^s A_{x-2,s} H_{x-2s}$$
$$= -(-)^s \{M_{x,s-1} A_{x-2,s-1} + A_{x-2,s}\} H_{x-2s},$$

and moreover

$$U_{x-4} \text{ contains the term } (-)^s A_{x-4,s} H_{x-2s-4},$$

or, what is the same thing, $(-)^s A_{x-4,s-2} H_{x-2s}$.

Hence we must have

$$A_{x,s} - (A_{x-2,s} + M_{x,s-1} A_{x-2,s-1}) + (x-2)(x-3)(n-x+3)(n-x+4) A_{x-4,s-2} = 0,$$

where

$$M_{x,s-1} = (x-1)(n-x+2) + (x-2)(n-x+3) - (x-2s+1)^2.$$

This may be satisfied by assuming

$$A_{x,s} \quad = B_{x,s}(n-x+1)(n-x+3)\ldots(n-x+2s-1);$$

for then

$$A_{x-2,s} \quad = B_{x-2,s} \quad (n-x+3)\ldots(n-x+2s-1)(n-x+2s+1),$$
$$A_{x-2,s-1} = B_{x-2,s-1}(n-x+3)\ldots(n-x+2s-1),$$
$$(n-x+3)(n-x+4) A_{x-4,x-2} = B_{x-4,s-2}(n-x+4)(n-x+3)\ldots(n-x+2s-1),$$

and consequently

$$\begin{aligned} &B_{x,s} \quad (n-x+1) \\ -&B_{x-2,s}(n-x+2s+1) \\ -&B_{x-2,s-1} M_{s,x-1} \\ +&B_{x-4,s-2}(x-2)(x-3)(n-x+4) = 0; \end{aligned}$$

and if this equation be satisfied independently of n, we must have

$$B_{x,s} - \qquad B_{x-2,s} - \qquad (2x-3) B_{x-2,s-1} + \quad (x-2)(x-3) B_{x-4,s-2} = 0,$$
$$B_{x,s} - (2s+1) B_{x-2,s} - \{5x - 8 - (x-2s+1)^2\} B_{x-2,s-1} + 4(x-2)(x-3) B_{x-4,s-2} = 0,$$

and these are both satisfied by

$$B_{x,s} = \frac{x \,.\, x-1 \ldots x-2s+1}{2^s.1.2.3\ldots s},$$

in fact, substituting this value and omitting the factor

$$\frac{(x-2)(x-3)\ldots(x-2s+1)}{2^s.1.2.3\ldots s},$$

the first equation becomes

$$x(x-1) - \qquad (x-2s)(x-2s-1) - \qquad (2x-3)\,2s + \ 4s(s-1) = 0,$$

and the second equation becomes

$$x(x-1) - (2s+1)(x-2s)(x-2s-1) - \{5x-8-(x-2s+1)^2\}\,2s + 16s(s-1) = 0,$$

which are each of them an identical equation, the first being

$$\begin{aligned} x^2 - \qquad & x \\ - x^2 + (4s+1)\,& x - 2s(2s+1) \\ - & 4sx + 6s \\ & + 4s(s-1) = 0, \end{aligned}$$

and the second being

$$\begin{aligned} x^2 - \qquad & x \\ -(2s+1)\,\{x^2 - (4s+1)\,& x + \ 2s(2s+1)\} \\ + 2s\,\{x^2 - (4s+3)\,& x + (2s-1)^2 + 8\} \\ & + 16s(s-1) = 0, \end{aligned}$$

as may be easily verified.

Hence writing for $B_{x,s}$ its value and recapitulating, the equation

$$U_x + \{(x-1)(n-x+2) + (x-2)(n-x+3) - \theta^2\}\,U_{x-2}$$
$$+ (x-2)(x-3)(n-x+3)(n-x+4)\,U_{x-4} = 0$$

is satisfied by

$$U_x = A_{x,0}\,H_x - A_{x,1}\,H_{x-2} \ldots + (-)^s A_{x,s}\,H_{x-2s} \ldots \text{to } s = \tfrac{1}{2}x \text{ or } \tfrac{1}{2}(x-1), \text{ as } x \text{ is even or odd,}$$

where

$$H_x = (\theta + x - 1)(\theta + x - 3)(\theta + x - 5) \ldots (\theta - x + 5)(\theta - x + 3)(\theta - x + 1),$$

$$A_{x,s} = \frac{x(x-1)\ldots(x-2s+1)}{2^s.1.2.3\ldots s}(n-x+1)(n-x+3)\ldots(n-x+2s-1),$$

and since for $x = 0, 1, 2, 3$ the values of the expression U_x coincide with those of the first four diagonal minors, the expression gives in general the value of the diagonal minor, or when x denotes the number of lines or columns of the determinant, then the value of the determinant.

2, *Stone Buildings*, 1st *April*, 1857.

187.

ON THE SUMS OF CERTAIN SERIES ARISING FROM THE EQUATION $x = u + tfx$.

[From the *Quarterly Mathematical Journal*, vol. II. (1858), pp. 167—171.]

LAGRANGE has given the following formula for the sum of the inverse n^{th} powers of the roots of the equation $x = u + tfx$,

$$\Sigma(z^{-n}) = u^{-n} + (-nu^{-n-1}fu)\frac{t}{1} + (-nu^{-n-1}f^2u)'\frac{t^2}{1.2} + \&c. \qquad (1)$$

where n is a positive integer and the series on the second side of the equation is to be continued as long as the exponent of u remains negative (*Théorie des Équations Numériques*, p. 225). Applying this to the equation $x = 1 + tx^s$, we have

$$\Sigma(z^{-n}) = 1^{-n} - \frac{n}{1}t \,.\, 1^{-n+s-1} + \frac{n(n-2s+1)}{1.2}t^2 \,.\, 1^{-n+2s-2} \ldots$$

$$+ (-)^q \frac{n(n-qs+q-1)\ldots(n-qs+1)}{1.2\ldots q}t^q \,.\, 1^{-n+qs-q} - \&c. \qquad (2)$$

to be continued while the exponent of 1 remains negative.

Let $n = \mu s + \rho$, ρ being not greater than $s-1$, the series may always be continued up to $q = \mu$, and no further. In fact writing the above value for n and putting $q = \mu + \theta$, the general term is

$$(-)^{\mu+\theta}\frac{t^{\mu+\theta}}{1.2\ldots(\mu+\theta)}(\mu s + \rho)(\rho - \theta s + \mu + \theta - 1)\ldots(\rho - \theta s + 1)\, 1^{-(\rho - \theta s + \mu + \theta)}.$$

Now if $\rho + \mu - \theta(s-1)$ is negative or zero, the term is to be rejected on account of the index of 1 not being negative, and if this quantity be positive, then since

$\rho - \theta s + 1$ is necessarily negative for any value of θ greater than zero, the factorial $(\rho - \theta s + \mu + \theta - 1)\ldots(\rho - \theta s + 1)$ begins with a positive and ends with a negative factor, and since the successive factors diminish by unity, one of them is necessarily equal to zero, or the term vanishes; hence the series is always to be continued up to $q = \mu$.

Hence

$$\Sigma\,(z^{-\mu s - \rho}) = 1 - \frac{\mu s + \rho}{1}\,t + \frac{(\mu s + \rho)\,\{(\mu - 2)\,s + \rho + 1\}}{1\,.\,2}\,t^2 \ldots$$

$$+ (-)^q \frac{(\mu s + \rho)\,\{(\mu - q)\,s + \rho + q - 1\}\ldots\{(\mu - q)\,s + \rho + 1\}}{1\,.\,2 \ldots q}\,t^q$$

$$- \&\text{c.} \qquad (3)$$

continued to $q = \mu$.

By taking the terms in a reverse order, it is easy to derive

$$(-)^\mu t^{-\mu}\,\Sigma\,(z^{-\mu s - \rho}) = (\mu s + \rho) \left\{ \begin{array}{l} \dfrac{(\mu + \rho - 1)\ldots(\mu + 1)}{2\,.\,3 \ldots \rho} - \dfrac{(\mu + \rho + s - 2)\ldots \mu}{2\,.\,3 \ldots \rho + s}\,t^{-1} \\[2ex] + (-)^q \dfrac{(\mu + \rho + qs - q - 1)\ldots(\mu + 1 - q)}{2\,.\,3 \ldots (\rho + q)\,s}\,t^{-q} \\[2ex] - \&\text{c.} \end{array} \right. \qquad (4)$$

continued to $q = \mu$.

Suppose in particular $s = 2$, and $t = -\dfrac{\alpha + 1}{\alpha^2}$, so that the equation in x becomes $\dfrac{x - 1}{x^2} = -\dfrac{\alpha + 1}{\alpha^2}$, whence $x = -\alpha$ or $x = \dfrac{\alpha}{\alpha + 1}$, or substituting in (2), we find

$$\frac{(\alpha + 1)^n}{\alpha^n} + \frac{(-)^n}{\alpha^n} = 1 + \frac{n}{1}\,\frac{\alpha + 1}{\alpha^2} + \frac{n\,(n - 3)}{2}\left(\frac{\alpha + 1}{\alpha^2}\right)^2 + \&\text{c.} \qquad (5)$$

continued to the term involving $\left(\dfrac{\alpha + 1}{\alpha^2}\right)^{\frac{1}{2}n}$ or $\left(\dfrac{\alpha + 1}{\alpha^2}\right)^{\frac{1}{2}(n-1)}$

Put $\alpha = -\dfrac{a + b}{a}$; and therefore

$$\alpha + 1 = -\frac{b}{a}, \quad \frac{\alpha + 1}{\alpha} = \frac{b}{a + b}, \quad \frac{\alpha + 1}{\alpha^2} = \frac{ab}{(a + b)^2};$$

we obtain

$$\frac{a^n + b^n}{(a + b)^n} = 1 - \frac{n}{1}\,\frac{ab}{(a + b)^2} + \frac{n\,(n - 3)}{1\,.\,2}\,\frac{a^2 b^2}{(a + b)^4} - \&\text{c.} \qquad (6),$$

or

$$\frac{(a + b)^n - a^n - b^n}{nab\,(a + b)} = (a + b)^{n-3} - \frac{n - 3}{2}\,(a + b)^{n-5}\,ab$$

$$+ \frac{(n - 4)\,(n - 5)}{2\,.\,3}\,(a + b)^{n-7}\,a^2 b^2 - \&\text{c.} \qquad (7),$$

to be continued as long as the exponent of $(a + b)$ on the second side is negative.

This formula, which is easily deducible from that for the expansion of $\cos n\theta$ in powers of $\cos\theta$, is employed by M. Stern, *Crelle*, t. XX. [1840], in proving the following theorem: If

$$S = 1 - \frac{n-3}{2} + \frac{(n-4)(n-5)}{2.3} - \&c. \tag{8}$$

continued to the first term that vanishes, then according as n is of the form $6k+3$, $6k \pm 1$, $6k$ or $6k \pm 2$,

$$S = \frac{3}{n},\quad S = 0,\quad S = -\frac{1}{n},\quad S = \frac{2}{n}, \tag{9}$$

which is in fact immediately deduced from it by writing $b = \omega a$, ω being one of the impossible cube roots of unity. Substituting the above values of x in the equation (4),

$$(1+\alpha)^{p+1} - (1+\alpha)^{-p} = (2p+1)\,\alpha\left\{1 + \frac{(p+1)p}{2.3}\frac{\alpha^2}{\alpha+1} + \frac{(p+2)(p+1)p(p-1)}{2.3.4.5}\frac{\alpha^4}{(\alpha+1)^2} + \dots\right\}, \tag{10}$$

$$(1+\alpha)^{p} + (1+\alpha)^{-p} = 2p\left\{\frac{1}{p} + \frac{p}{2}\frac{\alpha^2}{\alpha+1} + \frac{(p+1)p(p-1)}{2.3.4}\frac{\alpha^4}{(\alpha+1)^2} + \dots\right\}, \tag{11}$$

whence

$$(1+\alpha)^{p+1} + (1+\alpha)^{p} = (2p+1)\,\alpha\left\{1 + \frac{(p+1)p}{2.3}\frac{\alpha^2}{\alpha+1} + \dots\right\}$$
$$+\,2p\left\{\frac{1}{p} + \frac{p}{2}\frac{\alpha^2}{\alpha+1} + \dots\right\}, = U \text{ suppose}, \tag{12}$$

i.e

$$\Delta(-)^p(1+\alpha)^p = (-)^{p+1}U \text{ or } (1+\alpha)^p = (-)^p\,\Sigma\,(-)^{p+1}U,$$

where Δ and Σ refer to the variable p. The summation is readily effected by means of the formulæ

$$\Sigma\,(-)^{p+1}(2p+1)(p+s+1)\dots(p-s) = (-)^p(p+s+1)\dots(p-s-1),$$

$$\Sigma\,(-)^{p+1}(p+s)\dots(p-s)\,2p = (-)^p(p+s)\dots(p-s-1),$$

and we thence find

$$(1+\alpha)^p = \left\{1 + \frac{p(p-1)}{1.2}\frac{\alpha^2}{1+\alpha} + \frac{(p+1)p(p-1)(p-2)}{1.2.3.4}\frac{\alpha^4}{(1+\alpha)^2} + \dots\right\}$$
$$+\,\alpha\left\{\frac{p}{1} + \frac{(p+1)p(p-1)}{1.2.3}\frac{\alpha^2}{1+\alpha} + \dots\right\}, \tag{13}$$

a formula of Euler's (*Pet. Trans.* 1811) demonstrated likewise by M. Catalan (*Liouville*, t. IX. [1844], pp. 161—174) by induction. It may be expressed also in the slightly different form

$$(1+\alpha)^p = \left\{1 + \frac{(p+1)p}{1.2}\frac{\alpha^2}{1+\alpha} + \frac{(p+2)(p+1)p(p-1)}{1.2.3.4}\frac{\alpha^4}{(1+\alpha)^2} + \dots\right\}$$

$$+\frac{\alpha}{1+\alpha}\left\{\frac{p}{1} + \frac{(p+1)p(p-1)}{1.2.3}\frac{\alpha^2}{1+\alpha} + \dots\right\}. \tag{14}$$

The two series (13), (14) are each of them supposed to contain $p+1$ terms, p being an integer; but since the terms after these all of them vanish, the series may be continued indefinitely. Suppose the two sides expanded in powers of p, the coefficients will be separately equal, and thus the identity of the two sides will be independent of the particular values of p, or the equations (13), (14), and similarly, (10), (11), (12) are true for any values of p whatever. It is to be observed that the series for negative values of p do not differ essentially from those for the corresponding positive values; as may be seen immediately by writing $-p$ for p, and $\dfrac{-\alpha}{1+\alpha}$ for α.

Suppose next $s=3$, or that the equation in x is $x = 1 + tx^3$; to rationalise the roots of this, assume $t = \dfrac{4(\beta^2-1)^2}{(\beta^2+3)^3}$, then values of x are

$$x = \frac{\beta^2+3}{2(\beta+1)},\quad x = -\frac{\beta^2+3}{2(\beta-1)},\quad x = \frac{\beta^2+3}{\beta^2-1},$$

and hence

$$\frac{2^n\{(\beta+1)^n + (-)^n(\beta-1)^n\} + (\beta^2-1)^n}{(\beta^2+3)^r} =$$

$$1 - \frac{n}{1}t + \frac{n(n-5)}{1.2}t^2 - \frac{n(n-7)(n-8)}{1.2.3}t^3 \dots + (-)^r\frac{n(n-2r-1)\dots(n-3r+1)}{1.2\dots r}t^r + \&c. \tag{15}$$

where $t = \dfrac{4(\beta^2-1)^2}{(\beta^2+3)^3}$, and the series is to be continued up to the term involving $t^{\frac{1}{3}n}$, $t^{\frac{1}{3}(n-1)}$ or $t^{\frac{1}{3}(n-2)}$.

Again, from the formula (4) we deduce the three following forms,

$$(-)^\mu\frac{2^3\{(\beta+1)^{3\mu} + (-)^\mu(\beta-1)^{3\mu}\} + (\beta^2-1)^{2\mu}}{2^{2\mu}(\beta^2-1)^{2\mu}} =$$

$$3\mu\left\{\frac{1}{\mu} - \frac{(\mu+1)\mu}{2.3}t^{-1} + \frac{(\mu+3)(\mu+2)(\mu+1)\mu(\mu-1)}{2.3.4.5.6}t^{-2} + \dots(-)^q\frac{(\mu+2q-1)\dots(\mu-q+1)}{2.3\dots3q}t^{-q}\dots\right\}, \tag{16}$$

$$(-)^{\mu}\,\frac{2^{3\mu+1}\{(\beta+1)^{3\mu+1}-(-)^{\mu}(\beta-1)^{3\mu+1}\}+(\beta^2-1)^{3\mu+1}}{2^{2\mu}(\beta^2-1)^{2\mu}(\beta^2+3)}=$$

$$(3\mu+1)\left\{1-\frac{(\mu+2)(\mu+1)\mu}{2.3.4}t^{-1}\ldots+(-)^{q}\frac{(\mu+2q)\ldots(\mu-q+1)}{2.3\ldots3q+1}t^{-q}\ldots\right\},\quad(17)$$

$$(-)^{\mu}\,\frac{2^{3\mu+2}\{(+1)^{3\mu+2}+(-)^{\mu}(\beta-1)^{3\mu+2}\}+(\beta^2-1)^{3\mu+2}}{2^{2\mu}(\beta^2-1)^{2}(\beta^2+3)^{2}}=$$

$$(3\mu+2)\left\{\frac{\mu+1}{2}-\frac{(\mu+3)(\mu+2)(\mu+1)\mu}{2.3.4.5}t^{-1}\ldots+(-)^{q}\frac{(\mu+2q+1)\ldots(\mu-q+1)}{2.3\ldots(3q+2)}t^{-q}\ldots\right\},\quad(18)$$

all of them continued up to $q=\mu$.

2, *Stone Buildings*, 1st *April*, 1857.

188.

ON THE SIMULTANEOUS TRANSFORMATION OF TWO HOMOGENEOUS FUNCTIONS OF THE SECOND ORDER.

[From the *Quarterly Mathematical Journal*, vol. II. (1858), pp. 192—195.]

In a former paper with this title, *Cambridge and Dublin Math. Journal*, t. IV. [1849], pp. 47—50 [74], I gave (founded on the methods of Jacobi and Prof. Boole) a simple solution of the problem, but the solution may I think be presented in an improved form as follows, where as before I consider for greater convenience the case of three variables only.

Suppose that by the linear transformation[1]

$$(x, y, z) = \left(\begin{array}{ccc} \alpha, & \beta, & \gamma \\ \alpha', & \beta', & \gamma' \\ \alpha'', & \beta'', & \gamma'' \end{array}\right)\!\!\between (x_1, y_1, z_1),$$

we have identically

$$(a, b, c, f, g, h \between x, y, z)^2 = (a_1, b_1, c_1, f_1, g_1, h_1 \between x_1, y_1, z_1)^2,$$

$$(A, B, C, F, G, H \between x, y, z)^2 = (A_1, B_1, C_1, F_1, G_1, H_1 \between x_1, y_1, z_1)^2;$$

and write also

$$(\xi_1, \eta_1, \zeta_1) = \left(\begin{array}{ccc} \alpha, & \alpha', & \alpha'' \\ \beta, & \beta', & \beta'' \\ \gamma, & \gamma', & \gamma'' \end{array}\right)\!\!\between \xi, \eta, \zeta).$$

[1] I represent in this manner the system of equations

$$x = \alpha x_1 + \beta y_1 + \gamma z_1, \text{ \&c.}$$

and so in all like cases.

Comparing these with the relations between (x, y, z) and (x_1, y_1, z_1), we see that

$$(\xi, \eta, \zeta \between x, y, z) = (\xi_1, \eta_1, \zeta_1 \between x_1, y_1, z_1),$$

and multiplying the first of the relations between two quadrics by an indeterminate quantity λ, and adding it to the second, we have

$$(\lambda a + A, \ldots \between x, y, z)^2 = (\lambda a_1 + A_1, \ldots \between x_1, y_1, z_1)^2.$$

We have thus a linear function and a quadric transformed into functions of the same form by means of the linear substitutions, and any invariant of the system will remain unaltered to a factor près, such factor being a power of the determinant of substitution. The invariants are, 1° the discriminant of the quadric; 2° the reciprocant, considered not as a contravariant of the quadric, but as an invariant of the system. And if we write

$$K = \text{Disc.}\ (\lambda a + A, \ldots \between x, y, z)^2,$$

$$(\mathfrak{A}, \mathfrak{B}, \mathfrak{C}, \mathfrak{F}, \mathfrak{G}, \mathfrak{H} \between \xi, \eta, \zeta)^2 = \text{Recip.}\ (\lambda a + A, \ldots \between x, y, z)^2,$$

then K_1, &c. being the analogous expressions for the transformed functions, and the determinant of substitution being represented by Π, we have

$$K_1 = \Pi^2 K,$$

$$(\mathfrak{A}_1, \ldots \between \xi_1, \eta_1, \zeta_1)^2 = \Pi^2 (\mathfrak{A}, \ldots \between \xi, \eta, \zeta)^2,$$

and substituting for ξ_1, η_1, ζ_1 their values in terms of ξ, η, ζ, the last equation breaks up into six equations, and we have

$$K_1 = \Pi^2 K,$$

$$(\mathfrak{A}_1, \ldots \between \alpha, \alpha', \alpha'')^2 = \Pi^2 \mathfrak{A},$$

$$\vdots$$

$$(\mathfrak{A}_1, \ldots \between \beta, \beta', \beta'')(\gamma, \gamma', \gamma'') = \Pi^2 \mathfrak{F},$$

$$\vdots$$

which is the system obtained in a somewhat different manner in my former paper. Putting $f_1 = g_1 = h_1 = F_1 = G_1 = H_1 = 0$, and writing also (which is no additional loss of generality) $a_1 = b_1 = c_1 = 1$, the formulæ become

$$(a, b, c, f, g, h \between x, y, z)^2 = (1, 1, 1 \between x_1^2, y_1^2, z_1^2),$$

$$(A, B, C, F, G, H \between x, y, z)^2 = (A_1, B_1, C_1 \between x_1^2, y_1^2, z_1^2),$$

viz. there are two given quadrics which are to be by the same linear substitution transformed, one of them into the form $x_1^2 + y_1^2 + z_1^2$ and the other into the form $A_1 x_1^2 + B_1 y_1^2 + C_1 z_1^2$, where A_1, B_1, C_1 have to be determined. The solution is contained in the following system of formulæ, viz.

$$(A_1 + \lambda)(B_1 + \lambda)(C_1 + \lambda) = \Pi^2\ \text{Disc.}\ (\lambda a + A, \ldots),$$

which gives A_1, B_1, C_1 as the roots of a cubic equation, and gives also

$$1 = \Pi^2 \text{ Disc. } (a, \ldots) = \Pi^2\kappa, \text{ or } \Pi^2 = \frac{1}{\kappa} \text{ suppose,}$$

and we have then, writing for shortness, $(*\between X, Y, Z)$ for

$$((B_1+\lambda)(C_1+\lambda),\ (C_1+\lambda)(A_1+\lambda),\ (A_1+\lambda)(B_1+\lambda)\between X, Y, Z),$$

$$(*\between \alpha^2,\ \alpha'^2,\ \alpha''^2) = \frac{1}{\kappa}\mathfrak{A},$$

$$(*\between \beta^2,\ \beta'^2,\ \beta''^2) = \frac{1}{\kappa}\mathfrak{B},$$

$$(*\between \gamma^2,\ \gamma'^2,\ \gamma''^2) = \frac{1}{\kappa}\mathfrak{C},$$

$$(*\between \beta\gamma,\ \beta'\gamma',\ \beta''\gamma'') = \frac{1}{\kappa}\mathfrak{F},$$

$$(*\between \gamma\alpha,\ \gamma'\alpha',\ \gamma''\alpha'') = \frac{1}{\kappa}\mathfrak{G},$$

$$(*\between \alpha\beta,\ \alpha'\beta',\ \alpha''\beta'') = \frac{1}{\kappa}\mathfrak{H},$$

where $(\mathfrak{A}, \mathfrak{B}, \mathfrak{C}, \mathfrak{F}, \mathfrak{G}, \mathfrak{H})$ are the coefficients of the reciprocant of $(\lambda a + A, \ldots\between x, y, z)^2$. Writing $\lambda = -A_1$, $-B_1$, or $-C_1$ the quadric functions on the left-hand side become mere monomials, and we have the actual values of the squares and products α^2, $\beta\gamma$, &c. of the coefficients of the linear substitutions: thus α^2, β^2, γ^2, $\beta\gamma$, $\gamma\alpha$, $\alpha\beta$ are respectively equal to $\mathfrak{A}_0$, $\mathfrak{B}_0$, $\mathfrak{C}_0$, $\mathfrak{F}_0$, $\mathfrak{G}_0$, $\mathfrak{H}_0$ each into the common factor

$$\frac{1}{\kappa}(B_1 - A_1)(C_1 - A_1),$$

the suffix denoting that we are to write in the expressions for $\mathfrak{A}, \mathfrak{B}, \mathfrak{C}, \mathfrak{F}, \mathfrak{G}, \mathfrak{H}$ the value $-A_1$ for λ; and similarly for the sets $(\alpha', \beta', \gamma')$ and $(\alpha'', \beta'', \gamma'')$.

2, *Stone Buildings*, 27*th March*, 1857.

189.

NOTE ON A FORMULA IN FINITE DIFFERENCES.

[From the *Quarterly Mathematical Journal*, vol. II. (1858), pp. 198—201.]

IN Jacobi's Memoir "De usu legitimo formulæ summatoriæ Maclaurinianæ," *Crelle*, t. XII. [1834], pp. 263—273 (1834), expressions are given for the sums of the odd powers of the natural numbers 1, 2, 3...x in terms of the quantity.

$$u = x(x+1),$$

viz. putting for shortness

$$Sx^r = 1^r + 2^r + \ldots + x^r,$$

the expressions in question are

$$\begin{aligned}
Sx^3 &= \tfrac{1}{4}u^2,\\
Sx^5 &= \tfrac{1}{6}u^2\,(u - \tfrac{1}{2}),\\
Sx^7 &= \tfrac{1}{8}u^2\,(u^2 - \tfrac{4}{3}u + \tfrac{2}{3}),\\
Sx^9 &= \tfrac{1}{10}u^2\,(u^3 - \tfrac{5}{2}u^2 + 3u - \tfrac{3}{2}),\\
Sx^{11} &= \tfrac{1}{12}u^2\,(u^4 - 4u^3 + \tfrac{17}{2}u^2 - 10u + 5),\\
Sx^{13} &= \tfrac{1}{14}u^2\,(u^5 - \tfrac{35}{6}u^4 + \tfrac{287}{15}u^3 - \tfrac{118}{3}u^2 + \tfrac{691}{15}u - \tfrac{691}{30}),\\
&\&c.,
\end{aligned}$$

which, especially as regards the lower powers, are more simple than the ordinary expressions in terms of x.

The expressions are continued by means of a recurring formula, viz. if

$$Sx^{2p-3} = \frac{1}{2p-2}\{u^{p-1} - a_1u^{p-2}\ldots + (-)^{p-1}\,a_{p-3}u^2\},$$

$$Sx^{2p-1} = \frac{1}{2p}\{u^p - b_1u^{p-2}\ldots + (-)^p\,b_{p-3}u^2\},$$

then

$$\begin{aligned}
2p(2p-1)\,a_1 &= (2p-2)(2p-3)\,b_1 - p\,(p-1),\\
2p(2p-1)\,a_2 &= (2p-4)(2p-5)\,b_2 - (p-1)(p-2)\,b_1,\\
2p(2p-1)\,a_3 &= (2p-6)(2p-7)\,b_3 - (p-2)(p-3)\,b_2,\\
&\vdots\\
2p(2p-1)\,a_{p-3} &= 5\,.\,6\;b_{p-3} - 3\,.\,4\;b_{p-4},\\
0 &= 3\,.\,4\;b_{p-2} - 2\,.\,3\;b_{p-3},
\end{aligned}$$

by means of which the coefficients b can be determined when the coefficients a are known.

Jacobi remarks also that the expressions for the sums of the even powers may be obtained from those for the odd powers by means of the formula

$$Sx^{2p} = \frac{1}{2p+1}\,\partial_x Sx^{2p+1},$$

which shows that any such sum will be of the form $(2x+1)\,u$ into a rational and integral function of u: thus in particular

$$Sx^2 = \tfrac{1}{6}(2x+1)\,u.$$

To show *à priori* that Sx^{2p+1} can be expressed as a rational and integral function of u, it may be remarked that $Sx^{2p+1} = \phi_1 x$ where $\phi_1 x$ denotes the summatory integral $\Sigma(x+1)^{2p+1}$, taken so as to vanish for $x=0$: $\phi_1 x$ is a rational and integral function of x of the degree $2p+2$, and which, as is well known, contains x^2 as a factor. Suppose that y is any positive or negative integer less than x, we have

$$\phi_1 x - \phi_1 y = (y+1)^{2p+1} + (y+2)^{2p+1} \ldots + x^{2p+1},$$

and in particular putting $y = -1-x$,

$$\phi_1 x - \phi_1(-1-x) = (-x)^{2p+1} + (1-x)^{2p+1} \ldots + x^{2p+1}, = 0,$$

since the terms destroy each other in pairs; we have therefore $\phi_1 x = \phi_1(-1-x)$. Now $u = x^2 + x$, or writing this equation under the form $x^2 = -x + u$, we see that any rational and integral function of x may be reduced to the form $Px + Q$, where P and Q are rational and integral functions of u. Write therefore $\phi_1 x = Px + Q$: the substitution of $-1-x$ in the place of x leaves u unaltered, and the equation $\phi_1 x = \phi_1(-1-x)$ thus shows that $P = 0$; we have therefore $\phi_1 x = Q$, a rational and integral function of u. Moreover $\phi_1 x$ as containing the factor x^2, must clearly contain the factor u^2, and the expressions for Sx^{2p+1} are thus shown to be of the form given by Jacobi.

We may obtain a finite expression for Sx^n in terms of the differences of 0^n as follows: we have

$$Sx^n = 1^n + 2^n \ldots + x^n = \{(1+\Delta) + (1+\Delta)^2 \ldots + (1+\Delta)^x\}\,0^n = \frac{1+\Delta}{\Delta}\{(1+\Delta)^x - 1\}0^n,$$

and putting $(1+\Delta)^x = e^{x\log(1+\Delta)}$ and observing that the term independent of x vanishes, and that the terms containing powers higher than x^{n+1} also vanish, we have

$$Sx^n = S_k\left\{\frac{1+\Delta}{\Delta}\log^k(1+\Delta)\right\}0^n \, . \, \frac{x^k}{\Pi k},$$

where the summation with respect to k, extends from $k=1$ to $k=n+1$, or what is the same thing (since the term corresponding to $k=1$ in fact vanishes) from $k=2$ to $k=n+1$.

The equation $x^2 = -x+u$ gives

$$x^k = P_k x + Q_k,$$

and it is easy to see that writing for shortness

$$M_k = 1 + \frac{k-3}{1}u + \frac{k-4\,.\,k-5}{1\,.\,2}u^2 + \frac{k-5\,.\,k-6\,.\,k-7}{1\,.\,2\,.\,3}u^3 + \ldots,$$

where the series is to be continued to the term $u^{\frac{1}{2}(k-2)}$ or $u^{\frac{1}{2}(k-3)}$) according as k is even or odd, we have

$$P_k = (-)^{k+1}M_{k+1}, \quad Q_k = (-)^k u M_k,$$

we have consequently

$$Sx^n = \quad xS_k\left\{\frac{1+\Delta}{\Delta}\log^k(1+\Delta)\right\}0^n \, . \, \frac{(-)^{k+1}M_{k+1}}{\Pi k}$$

$$+ S_k\left\{\frac{1+\Delta}{\Delta}\log^k(1+\Delta)\right\}0^n \, . \, \frac{(-)^k u M_k}{\Pi k}.$$

If n is odd, $=2p+1$, then (by what precedes) the first term vanishes, or we have

$$S_k\left\{\frac{1+\Delta}{\Delta}\log^k(1+\Delta)\right\}0^{2p+1} \, . \, \frac{(-)^{k+1}M_{k+1}}{\Pi k} = 0, \ (k=1 \text{ to } k=2p+2),$$

and the formula becomes

$$Sx^{2p+1} = S_k\left\{\frac{1+\Delta}{\Delta}\log^k(1+\Delta)\right\}0^{2p+1} \, . \, \frac{(-)^k u M_k}{\Pi k}, \qquad (k=1 \text{ to } k=2p+2),$$

which it may be noticed puts in evidence the factor u but not the factor u^2.

If n is even, $=2p$, then (by what precedes) the coefficient of x is to the constant term in the ratio 2 : 1, or we have

$$S_k\left\{\frac{1+\Delta}{\Delta}\log^k(1+\Delta)\right\}0^{2p} \, . \, \frac{(-)^{k+1}(M_{k+1}-2uM_k)}{\Pi k} = 0, \ (k=1 \text{ to } k=2p+1),$$

and the formula becomes

$$Sx^{2p} = (2x+1)\,S_k\left\{\frac{1+\Delta}{\Delta}\log^k(1+\Delta)\right\}0^{2p} \, . \, \frac{(-)^k u M_k}{\Pi k}, \ (k=1 \text{ to } k=2p+1).$$

The values of the functions M are as follows:

$$M_1 = 0,$$
$$M_2 = 1,$$
$$M_3 = 1,$$
$$M_4 = 1 + u,$$
$$M_5 = 1 + 2u,$$
$$M_6 = 1 + 3u + u^2,$$
$$M_7 = 1 + 4u + 3u^2,$$
&c.

As a simple example of the formulæ, we have

$$Sx^3 = \left\{\frac{1+\Delta}{\Delta}\log^2(1+\Delta)\right\} 0^3 . \tfrac{1}{2}u$$
$$+ \left\{\frac{1+\Delta}{\Delta}\log^3(1+\Delta)\right\} 0^3 . -\tfrac{1}{6}u$$
$$+ \left\{\frac{1+\Delta}{\Delta}\log^4(1+\Delta)\right\} 0^3 . \tfrac{1}{24}(u+u^2),$$

and the coefficients are

$$(\Delta - \tfrac{1}{12}\Delta^3)\, 0^3 = 1 - \tfrac{1}{12}6 = \tfrac{1}{2},$$
$$(\Delta^2 - \tfrac{1}{2}\Delta^3)\, 0^3 = 6 - \tfrac{1}{2}6 = 3,$$
$$\Delta^3 \; 0^3 = \qquad 6,$$

and therefore

$$Sx^3 = \tfrac{1}{4}u - \tfrac{1}{2}u + \tfrac{1}{4}(u+u^2) = \tfrac{1}{4}u^2,$$

which is right; the example shows however that the calculation for the higher powers would be effected more readily by means of Jacobi's recurring formula.

2, *Stone Buildings*, 27*th Oct.*, 1857.

190.

ON THE SYSTEM OF CONICS WHICH PASS THROUGH THE SAME FOUR POINTS.

[From the *Quarterly Mathematical Journal*, vol. II. (1858), pp. 206—207.]

I CONSIDER the system of conics passing through the same four points; these points may be real or imaginary, but it is assumed that there is a real system of conics, this will in fact be the case if two conics of the system are real. The four points are therefore given as the points of intersection of two real conics, and it will be proper to assume in the first instance that the conics intersect in four separate and distinct points, none of them at infinity. The four points may be all real, or two real and two imaginary, or all imaginary.

First, if the points are all real, we have here two cases, viz. each of the points may lie outside of the triangle formed by the other three, or as this may be expressed, the points may form a convex quadrangle; or else one of the points may be inside the triangle formed by the other three, or as this may be expressed, the points may form a triangle and interior point. In each case the pairs of lines joining the points, two and two together, will be conics (degenerate hyperbolas) forming part of the system of conics. Consider the two cases separately.

A

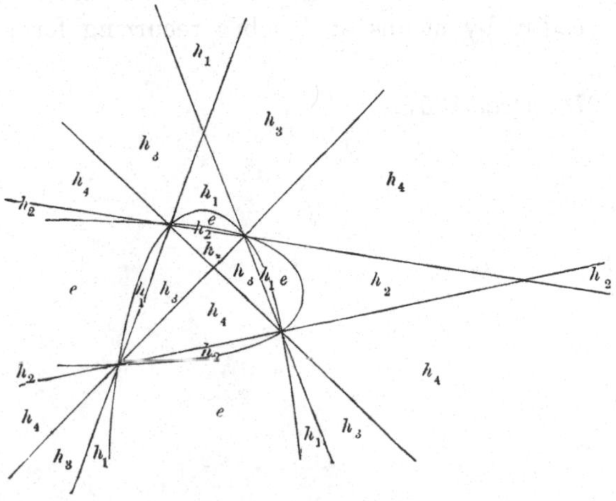

Fig. A. Four real points forming a convex quadrangle. The system contains two parabolas, and the pairs of lines and the parabolas divide the plane of the figure into five distinct regions, one of which contains only ellipses, and the other four contain each of them hyperbolas.

Fig. A′. Four real points forming a triangle and interior point. The system does not contain any parabolas, the three pairs of lines divide the plane of the figure into three distinct regions, each of which contains only hyperbolas.

A′

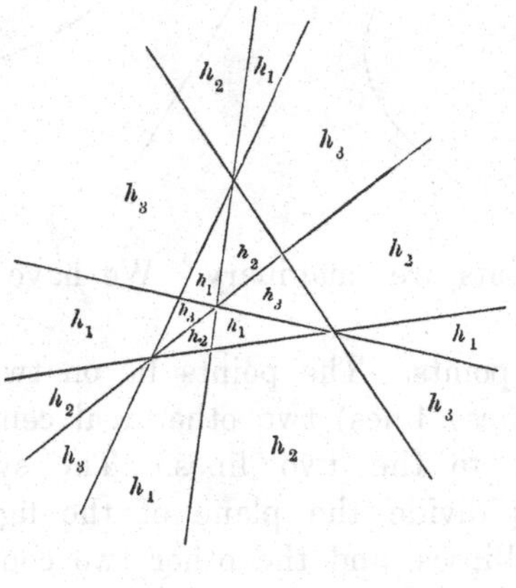

Next, if the points are two of them real and two of them imaginary. The line joining the two imaginary points will be real and this line may meet the line joining the two real points, in a point outside the two real points, or included between them, i.e. the real centre of the quadrangle may lie outside the real points, or may be included between them; I consider the two cases separately.

Fig. B. Two real and two imaginary points, the real centre of the quadrangle lying outside the real points. The system contains two parabolas, and these with the line joining the two real points and the line joining the two imaginary points divide the plane of the figure into three regions, one of which contains ellipses and the other two contain each of them hyperbolas.

B

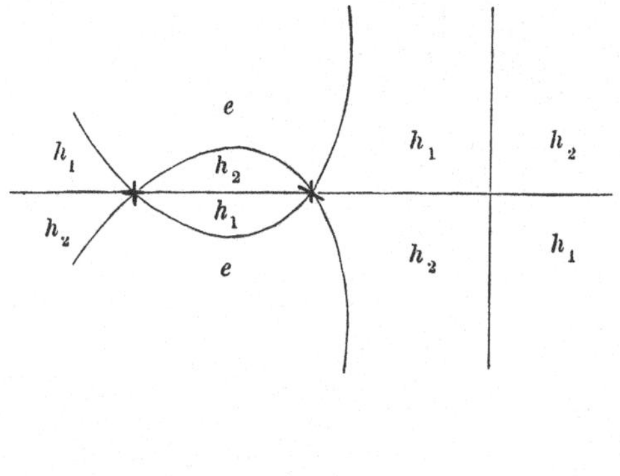

Fig. B′. Two real and two imaginary points, the real centre of the quadrangle lying between the real points. There are no parabolas, and the system contains only hyperbolas.

B′

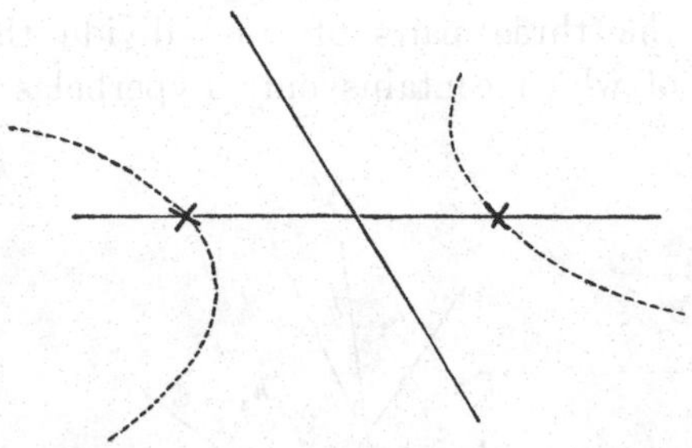

Lastly, when the four points are imaginary. We have here only a single case.

Fig. C. Four imaginary points. The points lie on two real lines, there are (besides the point of intersection of these lines) two other real centres of the quadrangle, which lie harmonically with respect to the two lines. The system contains two parabolas and these and the two lines divide the plane of the figure into four regions, two of which contain each of them ellipses, and the other two contain each of them hyperbolas.

C

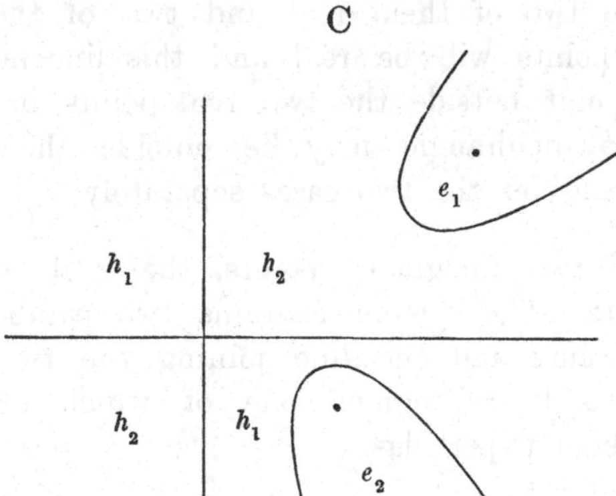

191.

NOTE ON THE EXPANSION OF THE TRUE ANOMALY.

[From the *Quarterly Mathematical Journal*, vol. II. (1858), pp. 229—232.]

IF the true anomaly and the mean anomaly are respectively denoted by u, m, and if e be the eccentricity, then as usual $u - e \sin u = m$; and if we write

$$\lambda = \frac{1 - \sqrt{(1 - e^2)}}{e}$$

and take c to denote the base of the hyperbolic system of logarithms, we have

$$u = m + 2\Sigma_1^\infty A_r \frac{\sin rm}{r},$$

and

$$A_r = \lambda^r c^{-\frac{1}{2}re(\lambda - \lambda^{-1})} + \lambda^{-r} c^{\frac{1}{2}re(\lambda - \lambda^{-1})},$$

where, after expanding the exponentials, the negative powers of λ are to be rejected and the term independent of λ is to be multiplied by $\frac{1}{2}$ (see *Camb. Math. Journal*, t. I. [1839] p. 228 and t. III. [1843] p. 165, [4]).

It is easily seen that e^r is the lowest power of e which enters into the value of A_r and the question arises to find the numerical coefficient of the term in question; this is readily obtained from the formula; in fact considering first a term of the form

$$\lambda^{-r} e^s (\lambda - \lambda^{-1})^s,$$

since λ is itself of the order e, when the negative powers of λ are rejected this is at least of the order e^s and it is consequently to be neglected if $s > r$. But if $s < r$ all the powers of λ are negative and the term is to be rejected. The only case to be considered is therefore that of $s=r$, in which case there is a term containing e^r. We thus obtain from $\lambda^{-r}c^{\frac{1}{2}re(\lambda-\lambda^{-1})}$ the term

$$\tfrac{1}{2}\frac{r^r e^r}{2^r.1.2.3...r}.$$

In the next place a term of the form $\lambda^r e^s(\lambda-\lambda^{-1})^s$ is at least of the order e^s if $s>r$, or the terms to be considered are those for which $s=$ or $<r$. But in such term the only part of the order e^r is

$$(-)^s\lambda^{r-s}e^s,$$

or, since neglecting higher powers of e we have $\lambda=\frac{1}{2}e$, this is

$$(-)^s\,2^{-r+s}\,e^r,$$

and the set of terms arising from

$$\lambda^r\,c^{-\frac{1}{2}re(\lambda-\lambda^{-1})},$$

is

$$\frac{e^r}{2^r}\left\{1+\frac{r}{1}+\frac{r^2}{1.2}...+\frac{r^{r-1}}{1.2...(r-1)}+\tfrac{1}{2}\frac{r^r}{1.2...r}\right\},$$

the last term being divided by 2 because arising from a term independent of λ. Hence the first term of A_r is

$$\frac{e^r}{2^r}\left\{1+\frac{r}{1}+\frac{r^2}{1.2}...+\frac{r^r}{1.2...r}\right\},$$

a result which it may be remarked is contained in the general formula given in Hansen's Memoir "Entwickelung des Products u. s. w.," *Leipzig Trans.*, t. II. p. 277 (1853).

The preceding expression is

$$=\frac{e^r c^r}{2^r}\frac{1}{\Gamma(r+1)}\int_r^\infty x^r c^{-x}\,dx,$$

and to find its value when r is large, we have

$$\int_r^\infty x^r c^{-x}\,dx=\int_0^\infty (y+r)^r e^{-y-r}\,dy=r^r c^{-r}\int_0^\infty\left(1+\frac{y}{r}\right)^r e^{-y}\,dy$$

$$=r^r c^{-r}\int_0^\infty c^{-y+r\log\left(1+\frac{y}{r}\right)}\,dy$$

$$=r^r c^{-r}\int_0^\infty c^{-\frac{y^2}{2r}+\frac{y^3}{3r^2}-\&\text{c.}}\,dy$$

$$=r^r c^{-r}\int_0^\infty\left(1+\frac{y^3}{3r^2}+...\right)e^{-\frac{y^2}{2r}}\,dy$$

$$=r^r c^{-r}\sqrt{2r}\int_0^\infty\left(1+\frac{2\sqrt{2}}{3\sqrt{r}}z^3+...\right)e^{-z^2}\,dz,$$

or neglecting all the terms except the first, this is

$$= r^r c^{-r} \sqrt{2r} \int_0^\infty e^{-z^2} dz$$

$$= \sqrt{2\pi r}\, r^r c^{-r}.$$

Hence multiplying by $\frac{1}{2^r} e^r c^r \frac{1}{\Gamma(r+1)}$ and observing that when r is large, we have, by a well-known formula,

$$\Gamma(r+1) = \sqrt{2\pi r}\, r^r c^{-r},$$

we obtain finally the result that when r is large the first term of A_r is approximately

$$= \left(\frac{ec}{2}\right)^r.$$

I take the opportunity of mentioning the following somewhat singular theorem, which seems to belong to a more general theory: viz. if $u - e \sin u = m$, then we have

$$\log(1 - e \cos u) = \frac{1}{\alpha} \log(1 - \alpha e \cos \phi),$$

where

$$\phi - \frac{1}{\alpha} \tan \phi = m,$$

provided that the negative powers of α are rejected, and α is then put equal to unity.

To show this, we have by Lagrange's theorem, observing that

$$\frac{d}{dm} F(1 - e \cos m) = e \sin m F'(1 - e \cos m),$$

$$F(1 - e \cos u) = F(1 - e \cos m) + \frac{e^2}{1} \sin^2 m F'(1 - e \cos m)$$

$$+ \frac{e^3}{1.2} \frac{d}{dm} \sin^3 m F'(1 - e \cos m) + \&c.,$$

and the coefficient of e^r in $F(1 - e \cos u)$ is

$$\frac{(-)^r}{1.2\ldots(r-1)} \left\{ \frac{1}{r} F_r \cos^r m + \frac{r-1}{1} F_{r-1} \cos^{r-2} m \sin^2 m \right.$$

$$\left. - \frac{(r-1)(r-2)}{1.2} F_{r-2} \frac{d}{dm} (\cos^{r-3} m \sin^3 m) + \&c. \right\},$$

where $F_r = F^r(1)$.

Hence in particular when $Fx = \log x$, $F_r = (-)^{r-1}\, 1 . 2 \ldots (r-1)$ and thence the coefficient of e^r in $\log(1 - e\cos u)$ is

$$-\left\{\frac{1}{r}\cos^r m - \frac{1}{1}\cos^{r-2} m \sin^2 m - \frac{1}{1 . 2}\frac{d}{dm}(\cos^{r-3} m \sin^2 m) - \&c.\right\},$$

continued as long as the exponent of $\cos m$ is not negative. Now in the expansion of $\frac{1}{\alpha}\log(1 - \alpha e \cos\phi)$, where $\phi - \frac{1}{\alpha}\tan\phi = m$, the coefficient of e^r is $-\frac{1}{r}\alpha^{r-1}\cos^r\phi$, and this (by Lagrange's theorem) is equal to

$$-\frac{1}{r}\alpha^{r-1}\left\{\cos^r m - \frac{1}{1 . \alpha}\, r\cos^{r-1} m \sin m \tan m - \frac{1}{1 . 2 . \alpha^2}\frac{d}{dm}(r\cos^{r-1} m \sin m \tan^2 m) - \&c.\right\}$$

$$= -\left\{\frac{1}{r}\alpha^{r-1}\cos^r m - \frac{1}{1}\alpha^{r-2}\cos^{r-2} m \sin^2 m - \frac{1}{1 . 2}\alpha^{r-3}\cos^{r-3} m \sin^3 m - \&c.\right\},$$

where the series is continued indefinitely; but if we reject the negative powers of α and then put α equal to unity this is precisely equal to the former expression for the coefficient of e^r, and the formula is thus shown to be true.

2, *Stone Buildings, W.C.*, 17*th Nov.*, 1857.

192.

ON THE AREA OF THE CONIC SECTION REPRESENTED BY THE GENERAL TRILINEAR EQUATION OF THE SECOND DEGREE.

[From the *Quarterly Mathematical Journal*, vol. II. (1858), pp. 248—253.]

[THE original title was "Direct Investigation of the Question discussed in the Foregoing Paper," viz. a Paper with the present title by N. M. Ferrers (now Dr Ferrers), pp. 247—248. The area S of the conic section represented by the general equation $(A, B, C, A', B', C' \between x, y, z)^2 = 1$, where the coordinates are connected by the equation $x + y + z = 1$, was by considerations founded on the form of the function found to be

$$S = \frac{2\pi (AA'^2 + BB'^2 + CC'^2 - ABC - 2A'B'C')\,\Delta}{\{A'^2 - BC + B'^2 - CA + C'^2 - AB + 2(B'C' - AA') + 2(C'A' - BB') + 2(A'B' - CC')\}^{\frac{3}{2}}},$$

where Δ is the area of the fundamental triangle: and it was remarked that a similar method might be applied to determine the area of the conic section when it is defined by the distances of its several tangents from three given points.]

The position of a point P being determined as in the foregoing paper, let α, β, γ denote in like manner the coordinates of a point O, we have

$$\alpha + \beta + \gamma = 1,$$

and consequently if ξ, η, ζ are the relative coordinates $x - \alpha$, $y - \beta$, $z - \gamma$, we have

$$\xi + \eta + \zeta = 0.$$

The expression for the distance of the two points O, P is readily obtained in terms of the relative coordinates, viz. calling this distance r, we have

$$r^2 = L\xi^2 + M\eta^2 + N\zeta^2,$$

where, if l, m, n are the sides of the triangle ABC, we have

$$\begin{aligned} L &= \tfrac{1}{2}(m^2 + n^2 + l^2\ ),\\ M &= \tfrac{1}{2}(n^2\ + l^2\ - m^2),\\ N &= \tfrac{1}{2}(l^2\ + m^2 - n^2\ );\end{aligned}$$

and it is to be remarked that these values give

$$MN + NL + LM = \tfrac{1}{4}(2m^2n^2 + 2n^2l^2 + 2l^2m^2 - l^4 - m^4 - n^4),\ = 4\Delta^2,$$

if Δ denote the area of the triangle ABC.

Consider now a conic

$$(a,\ b,\ c,\ f,\ g,\ h \between x,\ y,\ z)^2,$$

and suppose as usual that $\mathfrak{A}$, $\mathfrak{B}$, $\mathfrak{C}$, $\mathfrak{F}$, $\mathfrak{G}$, $\mathfrak{H}$ are the inverse coefficients and that K is the discriminant, suppose also for shortness

$$P = (\mathfrak{A},\ \mathfrak{B},\ \mathfrak{C},\ \mathfrak{F},\ \mathfrak{G},\ \mathfrak{H} \between 1,\ 1,\ 1)^2.$$

The coordinates of the centre being α, β, γ, we have

$$\alpha = \frac{1}{P}(\mathfrak{A},\ \mathfrak{H},\ \mathfrak{G} \between 1,\ 1,\ 1),$$

$$\beta = \frac{1}{P}(\mathfrak{H},\ \mathfrak{B},\ \mathfrak{F} \between 1,\ 1,\ 1),$$

$$\gamma = \frac{1}{P}(\mathfrak{G},\ \mathfrak{F},\ \mathfrak{C} \between 1,\ 1,\ 1),$$

and writing as before ξ, η, ζ for $x-\alpha$, $y-\beta$, $z-\gamma$, so that ξ, η, ζ are the coordinates of a point P of the conic, in relation to the centre, we have x, y, z respectively equal to $\xi+\alpha$, $\eta+\beta$, $\zeta+\gamma$, and the equation of the conic gives

$$(a,\ \ldots \between \xi+\alpha,\ \eta+\beta,\ \zeta+\gamma)^2 = 0,$$

which may be written

$$\begin{aligned} &(a,\ \ldots \between \xi,\ \eta,\ \zeta)^2\\ &+ 2(a,\ \ldots \between \alpha,\ \beta,\ \gamma)(\xi,\ \eta,\ \zeta)\\ &+\ (a,\ \ldots \between \alpha,\ \beta,\ \gamma)^2 = 0.\end{aligned}$$

Now observing the equations

$$(a,\ h,\ g\between\alpha,\ \beta,\ \gamma) = \frac{K}{P},$$

$$(h,\ b,\ f\between\alpha,\ \beta,\ \gamma) = \frac{K}{P},$$

$$(g,\ f,\ c\between\alpha,\ \beta,\ \gamma) = \frac{K}{P},$$

we have

$$(a,\ \ldots\between\alpha,\ \beta,\ \gamma)\ (\xi,\ \eta,\ \zeta) = \frac{K}{P}(\xi+\eta+\zeta) = 0,$$

$$(a,\ \ldots\between\alpha,\ \beta,\ \gamma)^2 \qquad = \frac{K}{P}(\alpha+\beta+\gamma) = \frac{K}{P};$$

and the equation of the conic gives therefore

$$(a,\ \ldots\between\xi,\ \eta,\ \zeta)^2 + \frac{K}{P} = 0,$$

and we have as before

$$\xi + \eta + \zeta = 0.$$

To find the axes we have only to make

$$r^2, = L\xi^2 + M\eta^2 + N\zeta^2,$$

a maximum or minimum, ξ, η, ζ varying subject to the preceding two conditions; this gives

$$(a,\ h,\ g\between\xi,\ \eta,\ \zeta) + \lambda L\xi\ + \mu = 0,$$

$$(h,\ b,\ f\between\xi,\ \eta,\ \zeta) + \lambda M\eta + \mu = 0,$$

$$(g,\ f,\ c\between\xi,\ \eta,\ \zeta) + \lambda N\zeta + \mu = 0,$$

and multiplying by ξ, η, ζ, adding and reducing, we have

$$-\frac{K}{P} + \lambda r^2 = 0,$$

which gives

$$\lambda = \frac{K}{Pr^2}.$$

Substituting this value, and joining to the resulting three equations the equation

$$\xi + \eta + \zeta = 0,$$

we may eliminate ξ, η, ζ, μ, and the result is

$$\left|\begin{array}{cccc} a+\dfrac{KL}{Pr^2}, & h\;, & g\;, & 1 \\ h\;, & b+\dfrac{KM}{Pr^2}, & f\;, & 1 \\ g\;, & f\;, & c+\dfrac{KN}{Pr^2}, & 1 \\ 1\;, & 1\;, & 1 & \end{array}\right| = 0,$$

which may also be written

$$(\mathfrak{A}', \mathfrak{B}', \mathfrak{C}', \mathfrak{F}', \mathfrak{G}', \mathfrak{H}' \between 1, 1, 1)^2 = 0,$$

where $(\mathfrak{A}', \ldots)$ are what $(\mathfrak{A}, \ldots)$ become when a, b, c are changed into

$$a+\frac{KL}{Pr^2},\quad b+\frac{KM}{Pr^2},\quad c+\frac{KN}{Pr^2};$$

we in fact have

$$\begin{aligned} \mathfrak{A}' &= \mathfrak{A} + \frac{K}{Pr^2}(bN + cM) + \frac{K^2}{P^2r^4}MN, \\ &\;\vdots \\ \mathfrak{F}' &= \mathfrak{F} - \frac{K}{Pr^2}Lf, \\ &\;\vdots \end{aligned}$$

and (observing the value of P) the result consequently is

$$P + \frac{K}{Pr^2}\{(b+c-2f)L + (c+a-2g)M + (a+b-2h)N\} + \frac{K^2}{P^2r^4}(MN + NL + LM) = 0,$$

which may also be written

$$P^3r^4 + PKr^2\{(b+c-2f)L + (c+a-2g)M + (a+b-2h)N\} + 4\Delta^2K^2 = 0.$$

Hence if r_1, r_2 are the two semiaxes, we have

$$r_1^2r_2^2 = \frac{4\Delta^2K^2}{P^3},$$

and the area is $\pi r_1 r_2$ which is equal to

$$\frac{2\pi K\Delta}{\sqrt{(P^3)}},$$

which agrees with Mr Ferrers' result.

The formula $r^2 = L\xi^2 + M\eta^2 + N\zeta^2$ which is assumed in the preceding investigation may be proved as follows:

Writing a, b, c (instead of l, m, n) for the sides of the fundamental triangle and A, B, C for the angles, the equation in question is

$$r^2 = bc \cos A\ \xi^2 + ca \cos B\ \eta^2 + ab \cos C\ \zeta^2.$$

Now writing α, β, γ for the inclinations of the line r to the sides of the triangle, we have

$$A = \beta - \gamma,$$

$$B = \gamma - \alpha,$$

$$C = \pi + \alpha - \beta.$$

Moreover taking for a moment λ, μ, ν to denote the perpendiculars from the angles on the opposite sides, we have

$$\lambda = c \sin B = b \sin C,$$

$$\mu = a \sin C = c \sin A,$$

$$\nu = b \sin A = a \sin B,$$

and

$$\xi = \frac{r \sin \alpha}{\lambda}, \quad \eta = \frac{r \sin \beta}{\mu}, \quad \zeta = \frac{r \sin \gamma}{\nu};$$

the values of ξ^2, η^2, ζ^2 consequently are

$$\frac{r^2 \sin^2 \alpha}{bc \sin B \sin C}, \quad \frac{r^2 \sin^2 \beta}{ca \sin C \sin A}, \quad \frac{r^2 \sin^2 \gamma}{ab \sin A \sin B},$$

and the equation to be proved becomes

$$1 = \frac{\cos A \sin^2 \alpha}{\sin B \sin C} + \frac{\cos B \sin^2 \beta}{\sin C \sin A} + \frac{\cos C \sin^2 \gamma}{\sin A \sin B},$$

or, what is the same thing,

$$\sin A \sin B \sin C = \sin A \cos A \sin^2 \alpha + \sin B \cos B \sin^2 \beta + \sin C \cos C \sin^2 \gamma,$$

or again

$$4 \sin A \sin B \sin C = \sin 2A\,(1 - \cos 2\alpha) + \sin 2B\,(1 - \cos 2\beta) + \sin 2C\,(1 - \cos 2\gamma),$$

or putting for A, B, C their values in terms of α, β, γ this is

$$\begin{aligned} -4 \sin(\beta - \gamma) \sin(\gamma - \alpha) \sin(\alpha - \beta) = {} & \sin(2\beta - 2\gamma)(1 - \cos 2\alpha) \\ & + \sin(2\gamma - 2\alpha)(1 - \cos 2\beta) \\ & + \sin(2\alpha - 2\beta)(1 - \cos 2\gamma), \end{aligned}$$

which is an identical equation; it is most readily proved by writing x, y, z for $\tan\alpha$, $\tan\beta$, $\tan\gamma$; the equation thus becomes

$$\frac{-4}{(1+x^2)(1+y^2)(1+z^2)}(y-z)(z-x)(x-y)$$
$$=\Sigma\frac{1}{(1+y^2)(1+z^2)}\{2y(1-z^2)-2z(1-y^2)\}\frac{2x^2}{1+x^2},$$

or multiplying out

$$-(y-z)(z-x)(x-y)=\Sigma\,(y-z)(1+yz)\,x^2=\Sigma x^2(y-z)+xyz\,\Sigma x\,(y-z),$$

that is

$$-(y-z)(z-x)(x-y)=\Sigma x^2(y-z) \qquad =x^2(y-z)+y^2(z-x)+z^2(x-y),$$

which is an identity.

[A different investigation of the formula $r^2=L\xi^2+M\eta^2+N\zeta^2$, by Dr Ferrers, was appended to the original Paper.]

193.

ON RODRIGUES' METHOD FOR THE ATTRACTION OF ELLIPSOIDS.

[From the *Quarterly Mathematical Journal*, vol. II. (1858), pp. 333—337.]

THE following is in substance the method given in the "Mémoire sur l'attraction des Spheroïdes," par M. Rodrigues, *Corresp. sur l'École Polyt.*, t. III. pp. 361—385 (1815). It will be seen that the method is very similar to that given two years before by Gauss, see my paper "On Gauss' Method for the Attraction of Ellipsoids," *Journal*, t. I. pp. 162—166 [164]: the solution in fact depends upon the geometrical theorem therein quoted, viz. if M be any point, P a point of a closed surface, PQ the normal (lying outside the surface) at the point P, dS the element of the surface at that point, and if MQ denotes the angle MPQ and $\overline{MP}$ the distance of the points M and P, then, theorem, the integral

$$\iint \frac{dS \cos MQ}{\overline{MP}^2}$$

has for its value

$$0,\ -2\pi \text{ or } -4\pi$$

according as M is exterior to, upon, or interior to the surface.

Suppose that M is the attracted point and taking A, B, C for the semiaxes of the surface of the attracting ellipsoid, or, if we please, for any semiaxes of an arbitrary ellipsoidal surface confocal with the surface of the attracting ellipsoid, let P be a point on the surface of the interior similar ellipsoid whose semiaxes are rA, rB, rC. The coordinates of M are taken to be a, b, c, and those of P are taken to be x, y, z, and the value of the potential is

$$V = \int \frac{dm}{\overline{MP}},$$

where dm is the element of mass.

We may write

$$x = rA\xi,$$

$$y = rB\eta,$$

$$z = rC\zeta,$$

and then ξ, η, ζ will be the coordinates of a point P' on the surface of a sphere, radius unity, corresponding in a definite manner to the point P on the surface of the internal similar ellipsoid. And if $d\sigma$ be the element of the spherical surface, then we have

$$dm = ABCr^2 dr\, d\sigma,$$

and therefore

$$\frac{V}{ABC} = \int \frac{r^2 dr\, d\sigma}{MP},$$

where, in order to obtain the value of the potential V for the ellipsoid whose semi-axes are A, B, C, the integrations must be extended over the spherical surface and from $r = 0$ to $r = 1$.

Suppose that dS is the element of the internal similar surface at P, and let p be the perpendicular from the centre upon the tangent plane at P, we have

$$dS = \frac{r^3 ABC}{p} d\sigma.$$

Let P_0 be the point on the ellipsoid (A, B, C) similarly situated to the point P on the ellipsoid (rA, rB, rC); the coordinates of P_0 are $A\xi$, $B\eta$, $C\zeta$; and if p_0 be the perpendicular from the centre upon the tangent plane at P_0, then $p = rp_0$, and the preceding equation becomes

$$dS = \frac{r^2 ABC}{p_0} d\sigma.$$

Imagine now an ellipsoidal surface confocal with the surface (A, B, C) and having for its semiaxes

$$A + \delta A, \quad B + \delta B, \quad C + \delta C,$$

and let P_0' be the point on this surface which *corresponds* with the point P_0 on the surface (A, B, C); that is, let P_0' be the point whose coordinates are

$$(A + \delta A)\,\xi, \quad (B + \delta B)\,\eta, \quad (C + \delta C)\,\zeta;$$

and let P' be in like manner the point whose coordinates are

$$r(A + \delta A)\,\xi, \quad r(B + \delta B)\,\eta, \quad r(C + \delta C)\,\zeta;$$

the points P, P' will be in like manner *corresponding* points on the surface (rA, rB, rC) and on the confocal surface $\{r(A+\delta A), r(B+\delta B), r(C+\delta C)\}$; and if the normal distance at the point P_0 of the first two surfaces is δN, then the normal distance at the point P of the second two surfaces will be $r\delta N$. The decrement of $\overline{MP}$ will be equal to the normal distance $r\delta N$ of the two surfaces at the point P multiplied into the cosine of the angle MQ, and we have, by a property of confocal surfaces,

$$A\delta A = B\delta B = C\delta C = p_0 \delta N = \text{(suppose) } \tfrac{1}{2}\delta\theta,$$

we have therefore

$$d\overline{MP} = -\frac{\tfrac{1}{2}r\delta\theta}{p_0}\cos MQ.$$

Hence from the equation

$$\frac{V}{ABC} = \int \frac{r^2 dr\, d\sigma}{\overline{MP}}.$$

We deduce

$$\delta\frac{V}{ABC} = \int r^2 dr\, d\sigma \frac{\tfrac{1}{2}r\delta\theta}{P_0}\frac{\cos MQ}{\overline{MP}^2}.$$

But we have

$$\frac{r^2 d\sigma}{p_0} = \frac{dS}{ABC};$$

and the equation thus becomes

$$\delta\frac{V}{ABC} = \frac{\tfrac{1}{2}\delta\theta}{ABC}\int r dr \frac{dS\cos MQ}{\overline{MP}^2}.$$

It may be proper to remark here by way of recapitulation that the course of the investigation has been as follows: viz. that, with a view to obtaining the potential V of an attracting ellipsoid, we have found the increment of $\frac{V}{ABC}$ in passing from the ellipsoidal surface (A, B, C) to the ellipsoidal surface $(A+\delta A, B+\delta B, C+\delta C)$, each of them *confocal* with the surface of the attracting ellipsoid; and that for finding such increment we have had to consider the two surfaces (rA, rB, rC) and $\{r(A+\delta A)$ $r(B+\delta B)$, $r(C+\delta C)\}$ confocal to each other and respectively *similar* to the first-mentioned two surfaces.

Resuming the formula just obtained, the integral with respect to dS is taken over the entire surface of the internal similar ellipsoid (rA, rB, rC), and if the attracted point M is external to the ellipsoid (A, B, C) it will be external to the interior similar ellipsoid (rA, rB, rC): hence in this case the double integral vanishes for all values of r, or we have

$$\delta\frac{V}{ABC} = 0;$$

that is the function $V \div ABC$, which represents the ratio of the potential to the mass, is not altered in passing from the ellipsoid (A, B, C) to the confocal ellipsoid

$$(A+\delta A,\quad B+\delta B,\quad C+\delta C),$$

or, what is the same thing, the potentials (and therefore the attractions) of confocal ellipsoids upon the same external point are proportional to their masses; this is in fact Maclaurin's theorem for the attraction of ellipsoids upon an external point.

But if the attracted point M is interior to the ellipsoid (A, B, C), then writing

$$\frac{a^2}{A^2}+\frac{b^2}{B^2}+\frac{c^2}{C^2}=r'^2,$$

where r' is less than unity, the double integral is $=0$ from $r=0$ to $r=r'$ and is $=-4\pi$ from $r=r'$ to $r=1$, and we have

$$\begin{aligned}\delta\frac{V}{ABC} &= -\frac{2\pi\delta\theta}{ABC}\int_{r'}^{1} r\,dr\\ &= -\frac{\pi\delta\theta}{ABC}(1-r'^2)\\ &= \frac{\pi\delta\theta}{ABC}\left(\frac{a^2}{A^2}+\frac{b^2}{B^2}+\frac{c^2}{C^2}-1\right);\end{aligned}$$

that is, the right-hand side of the equation is the increment (or taken with its sign reversed so as to be positive, it is the decrement) of the function $V \div ABC$ in passing from the ellipsoid (A, B, C) to the confocal ellipsoid

$$(A+\delta A,\quad B+\delta B,\quad C+\delta C),$$

where

$$\tfrac{1}{2}\delta\theta = A\delta A = B\delta B = C\delta C.$$

The preceding formula gives at once the potential for an interior point; in fact taking α, β, γ for the semiaxes of the ellipsoid and writing

$$A^2=\alpha^2+\theta,\quad B^2=\beta^2+\theta,\quad C^2=\gamma^2+\theta,$$

and using the ordinary symbol d instead of δ, we have

$$\frac{d}{d\theta}\frac{V}{\sqrt{\{(\alpha^2+\theta)(\beta^2+\theta)(\gamma^2+\theta)\}}}=\frac{\pi}{\sqrt{\{(\alpha^2+\theta)(\beta^2+\theta)(\gamma^2+\theta)\}}}\left\{\frac{a^2}{\alpha^2+\theta}+\frac{b^2}{\beta^2+\theta}+\frac{c^2}{\gamma^2+\theta}-1\right\},$$

and integrating from $\theta=0$ to $\theta=\infty$, we have

$$-\frac{V}{\alpha\beta\gamma}=\pi\int_0^\infty\frac{d\theta}{\sqrt{\{(\alpha^2+\theta)(\beta^2+\theta)(\gamma^2+\theta)\}}}\left\{\frac{a^2}{\alpha^2+\theta}+\frac{b^2}{\beta^2+\theta}+\frac{c^2}{\gamma^2+\theta}-1\right\},$$

where V is now the potential for the ellipsoid whose semiaxes are α, β, γ; and we have therefore

$$V = -\pi\alpha\beta\gamma \int_0^\infty \frac{d\theta}{\sqrt{\{(\alpha^2+\theta)(\beta^2+\theta)(\gamma^2+\theta)\}}} \left\{\frac{a^2}{\alpha^2+\theta} + \frac{b^2}{\beta^2+\theta} + \frac{c^2}{\gamma^2+\theta} - 1\right\}.$$

To find the potential for an external point it is only necessary to remark that by the theorem above demonstrated $V \div \alpha\beta\gamma$ is equal to the corresponding function for the confocal ellipsoid through the attracted point, that is for the ellipsoid whose semiaxes are $\sqrt{(\alpha^2+\theta_1)}$, $\sqrt{(\beta^2+\theta_1)}$, $\sqrt{(\gamma^2+\theta_1)}$, where θ_1 is a positive quantity such that

$$\frac{a^2}{\alpha^2+\theta_1} + \frac{b^2}{\beta^2+\theta_1} + \frac{c^2}{\gamma^2+\theta_1} = 1,$$

hence in the value of $V \div \alpha\beta\gamma$ we have only to write the above values in the place of α, β, γ; and if we then write $\theta - \theta_1$ in the place of θ the limits will be ∞, θ_1, and the expression for the potential is

$$V = -\pi\alpha\beta\gamma \int_{\theta_1}^\infty \frac{d\theta}{\sqrt{\{(\alpha^2+\theta)(\beta^2+\theta)(\gamma^2+\theta)\}}} \left\{\frac{a^2}{\alpha^2+\theta} + \frac{b^2}{\beta^2+\theta} + \frac{c^2}{\gamma^2+\theta} - 1\right\};$$

this completes the investigation.

194.

NOTE ON THE THEORY OF ATTRACTION.

[From the *Quarterly Mathematical Journal*, vol. II. (1858), pp. 338—339.]

IMAGINE a closed surface, the equation of which contains the two parameters m, h. Call this the surface (m, h), and suppose also that for shortness the shell of uniform density included between the surfaces (m, h), (n, h) is called the shell (m, n, h). Suppose now that the surface is such:

1°. That the infinitesimal shell $(m, m+dm, h)$ exerts no attraction upon an internal point.

2°. That the equipotential surfaces of the shell in question for external points are the surfaces (m, k), where k is arbitrary.

Then, first, the attraction of the shell on a point of the equipotential surface (m, k) is proportional to the normal thickness at that point of the shell $(m, m+\delta m, k)$; or (more precisely) taking the density of the attracting shell as unity, the attraction is $=4\pi\times$ mass of shell $(m, m+dm, h)$ into normal thickness of shell $(m, m+\delta m, k)$ divided by mass of the last-mentioned shell.

In fact the shell $(m, m+\delta m, k)$ exerts no attraction on an internal point, consequently if over the surface (m, k) we distribute the mass of the original shell $(m, m+dm, h)$ in such manner that the density at any point is proportional to the normal thickness of the shell $(m, m+\delta m, k)$ the distribution will be such that the attraction on an internal point may vanish; but in order that this may be the case, the density must be equal to $\frac{1}{4\pi}$ into the attraction upon that point of the shell $(m, m+\delta m, k)$. Hence the attraction is proportional to the normal thickness, and if the whole mass distributed over the surface (m, k) is precisely equal to the mass of the shell $(m, m+dm, h)$, then the density at any point must be equal to the mass into normal thickness divided by mass of $(m, m+\delta m, k)$, and attraction $=4\pi$ into density, $=4\pi\times$ mass of shell $(m, m+dm, h)$ into normal thickness of shell $(m, m+\delta m, k)$ divided by mass of the last-mentioned shell.

And, secondly, the attractions of the solids bounded by the two surfaces (n, h), (n, h_1) respectively upon the same exterior point are proportional to their masses.

For the solid (n, h) may be divided into shells $(m, m+dm, h)$ and for this shell the equipotential surface is (m, k) and the attraction of the shell varies as mass of $(m, m+dm, h)$ into normal thickness of the shell $(m, m+\delta m, k)$. But in like manner the solid (n, h_1) may be divided into shells $(m, m+dm, h_1)$ and the attraction of the shell varies as mass of $(m, m+dm, h_1)$ into normal thickness of the shell $(m, m+\delta m, k)$ and the attractions are in each case in the direction normal to the shell (m, k), and therefore in the same direction; that is, the attraction of the shell $(m, m+dm, h)$ is in the same direction as that of the shell $(m, m+dm, h_1)$ and the two attractions are proportional to the masses. Hence integrating from $m=0$ (if for this value the included space is zero) to $m=n$, the attractions of the solids (n, h), (n, h_1) are composed of elements proportional and parallel, the elements of the attraction of (n, h) to the elements of the attraction of (n, h_1); and consequently the total attractions are in the same direction and proportional to the masses.

Thirdly, the attractions of the two surfaces (m, h), (n, h) upon the same interior point are equal.

A surface having the properties in question is of course the ellipsoidal surface

$$\frac{x^2}{m(a^2+h)}+\frac{y^2}{m(b^2+h)}+\frac{z^2}{m(c^2+h)}=1,$$

where if m varies (h being constant) the several surfaces are similar to each other, but if h varies (m being constant) the several surfaces are confocal to each other: for it is in fact well known that the infinitesimal shell bounded by similar ellipsoidal surfaces has the properties assumed with respect to the shell $(m, m+dm, h)$. The first theorem in effect reduces the problem of the determination of an ellipsoid upon an exterior point to a single integration, and constitutes the foundation of Poisson's method for the attraction of ellipsoids. The second theorem (Maclaurin's theorem for the attraction of ellipsoids on the same external point) shows that the attraction of an ellipsoid upon an external point can be found by means of the attraction of the confocal ellipsoid through the attracted point; and by the third theorem the attraction of an ellipsoid upon an interior point is equal to that of the similar ellipsoid through the attracted point; hence the second and third theorems reduce the determination of the attraction of an ellipsoid upon an external or internal point to that of an ellipsoid upon a point on the surface.

2, *Stone Buildings, W.C., 7th April*, 1858.

195.

REPORT ON THE RECENT PROGRESS OF THEORETICAL DYNAMICS.

[From the *Report of the British Association for the Advancement of Science*, 1857, pp. 1—42.]

THE object of the *Mécanique Analytique* of Lagrange is described by the author in the "Avertissement" to the first edition as follows:—"On a déjà plusieurs traités de mécanique, mais le plan de celui-ci est entièrement neuf. Je me suis proposé de réduire la théorie de cette science et l'art de résoudre tous les problêmes qui s'y rapportent à des formules générales dont le simple développement donne toutes les équations nécessaires pour la solution de chaque problême." And the intention is carried out; the principle of virtual velocities furnishes the general formulæ for the solution of statical problems, and d'Alembert's principle then leads to the general formulæ for the solution of dynamical problems. The general theory of statics would seem to admit of less ulterior development; but as regards dynamics, the formulæ of the first edition of the *Mécanique Analytique* have been the foundation of a series of profound and interesting researches constituting the science of analytical dynamics. The present report is designed to give, so far as I am able, a survey of these researches; there will be found at the end a list, in chronological order, of the works and memoirs referred to, and I shall in the course of the report preserve as far as possible the like chronological order. It is proper to remark that I confine myself to the general theories of dynamics. There are various *special problems* of great generality, and susceptible of the most varied and extensive developments, such for instance as the problem of the motion of a single particle (which includes as particular cases the problem of central forces, that of two fixed centres, and that of the motion of a conical pendulum, either with or without regard to the motion of the earth round its axis), the problem of three bodies, and the problem of the rotation of a solid body about a fixed point. But a detailed account of the researches of geometers in relation to these special problems would properly form the subject of a separate report, and it is not my intention to enter upon them otherwise than incidentally, so far as it may appear desirable to do so. One problem, however, included in the first of the above-mentioned special problems, I shall have frequent occasion to allude to: I mean the problem of the variation of the elements of a planet's orbit, which has a close historical connexion with the general theories which form the subject of this report. The so-called ideal coordinates of Hansen, and the principles of his method of integration in the planetary and lunar theories, have a bearing on the general subject, and might have been considered in the present report; but on the whole I have considered it better not to do so.

1. Lagrange, *Mécanique Analytique*, 1788.—The equations of motion are obtained, as before mentioned, by means of the principle of virtual velocities and d'Alembert's principle. In their original forms they involve the coordinates x, y, z of the different particles m or dm of the system, quantities which in general are not independent. But Lagrange introduces, in place of the coordinates x, y, z of the different particles, any variables or (using the term in a general sense) coordinates $\xi, \psi, \phi, \ldots$ whatever, determining the position of the system at the time t: these may be taken to be independent, and then if $\xi', \psi', \phi', \ldots$ denote as usual the differential coefficients of $\xi, \psi, \phi, \ldots$ with respect to the time, the equations of motion assume the form

$$\frac{d}{dt}\frac{dT}{d\xi'} - \frac{dT}{d\xi} + \Xi = 0;$$
$$\vdots$$

or when $\Xi, \Psi, \Phi, \ldots$ are the partial differential coefficients with respect to $\xi, \psi, \phi, \ldots$ of one and the same function V, then the form

$$\frac{d}{dt}\frac{dT}{d\xi} - \frac{dT}{d\xi'} + \frac{dV}{d\xi} = 0.$$
$$\vdots$$

In these equations, T, or the *vis viva* function, is the *vis viva* of the system, or sum of all the elements each into the half square of its velocity, expressed by means of the coordinates $\xi, \psi, \phi, \ldots$; and (when such function exists) V, or the force function([1]), is a function depending on the impressed forces and expressed in like manner by means of the coordinates $\xi, \psi, \phi, \ldots$; the two functions T and V are given functions, by means of which the equations of motion for the particular problem in hand are completely expressed. In any dynamical problem whatever, the *vis viva* function T is a given function of the coordinates $\xi, \psi, \phi, \ldots$, of their differential coefficients $\xi', \psi', \phi', \ldots$ and of the time t; and it is of the second order in regard to the differential coefficients $\xi', \psi', \phi', \ldots$; and (when such function exists) the force function V is a given function of the coordinates $\xi, \psi, \phi, \ldots$ and of the time t. This is the most general form of the functions T, V, as they occur in dynamical problems, but in an extensive class of such problems the forms are less general, viz. T and V are each of them independent of the time, and T is a homogeneous function of the second order in regard to the differential coefficients $\xi', \psi', \phi', \ldots$; the equations of motion

[1] The sign attributed to V is that of the *Mécanique Analytique*, but it would be better to write $V = -U$, and to call U (instead of V) the force function.

have in this case an integral $T + V = h$, which is the equation of *vis viva*, and the problems are distinguished as those in which the principle of *vis viva* holds good. It is to be noticed also that in this case since t does not enter into the differential equations, the integral equations will contain t in the form $t + c$, that is, in connexion with an arbitrary constant c attached to it by addition.

2. The above-mentioned form is *par excellence* the Lagrangian form of the equations of motion, and the one which has given rise to almost all the ulterior developments of the theory; but it is proper just to refer to the form in which the equations are in the first instance obtained, and which may be called the unreduced form, viz. the equations for the motion of a particle whose rectangular coordinates are x, y, z, are

$$m\frac{d^2x}{dt^2} = X + \lambda\frac{dL}{\delta x} + \mu\frac{dM}{\delta x} + \ldots$$
$$\vdots$$

where $L = 0$, $M = 0$, ... are the equations of condition connecting the coordinates of the different points of the system, and λ, μ, ... are indeterminate multipliers.

3. The idea of a force function seems to have originated in the problems of physical astronomy. Lagrange, in a memoir "On the Secular Equation of the Moon," crowned by the French Academy of Sciences in the year 1774, expressed the attractive forces, decomposed in the directions of the axes of coordinates, by the partial differential coefficients of one and the same function with respect to these coordinates. And it was in these problems natural to distinguish the forces into principal and disturbing forces, and thence to separate the force function into two parts, a principal force function and a disturbing function. The problems of physical astronomy led also to the idea of the variation of the arbitrary constants of a mechanical problem. For as a fact of observation the planets move in ellipses the elements of which are slowly varying; the motion in a fixed ellipse was accounted for by the principal force, the attraction of the sun; the effect of the disturbing force is to produce a continual variation of the elements of such elliptic orbit. Euler, in a memoir published in 1749 in the *Memoirs of the Academy of Berlin* for that year, obtained differential equations of the first order for two of the elements, viz. the inclination and the longitude of the node, by making the arbitrary constants which express these elements in the fixed orbit to vary: this seems to be the first attempt at the method of the variation of the arbitrary constants. Euler afterwards treated the subject in a more complete manner, and the method is also made use of by Lagrange in his "Memoir on the Perturbations of the Planets" in the Berlin Memoirs for 1781, 1782, 1783, and by Laplace in the *Mécanique Céleste*, t. I. 1799. The method in its original form seeks for the expressions of the variations of the elements in terms of the differential coefficients of the disturbing function *with respect to the coordinates*. As regards one element, the longitude of the epoch, such expression (at least in a finite form) was first obtained by Poisson in his memoir of 1808, to be spoken of presently; but I am not able to refer to any place where such expressions in their best form are even now to be found; the question seems to have been unduly passed over in consequence of the new form immediately afterwards assumed by the method. It was very early

observed that the variation of one of the elements, viz. the mean distance, was expressible in a remarkable form by means of the differential coefficients of the disturbing function *taken with respect to the time t, in so far as it entered into the function through the coordinates of the disturbed planet.* I am not able to say at what time, or whether by Euler, Lagrange, or Laplace, it was observed that such differential coefficient with respect to the time was equivalent to the differential coefficient of the disturbing function with respect to one of the elements. But however this may be, the notion of the representation of the variations of the elements by means of the differential coefficients of the disturbing function *with respect to the elements* had presented itself *à posteriori*, and was made use of in an irregular manner prior to the year 1800, and therefore some eight years at any rate before the establishment by Lagrange of the general theory to which these forms belong.

4. Poisson's memoir of the 20th of June, 1808, "On the Secular Inequalities of the Mean Motion of the Planets," was presented by him to the Academy at the age of twenty-seven years. It contains, as already remarked, an expression in finite terms for the variation of the longitude of the epoch. But the memoir is to be considered rather as an application of known methods to an important problem of physical astronomy, than as a completion or extension of the theory of the variation of the planetary elements. The formulæ made use of are those involving the differential coefficients of the disturbing function with respect to the *coordinates;* and there is nothing which can be considered an anticipation of Lagrange's idea of the investigation, *à priori*, of expressions involving the differential coefficients with respect to the elements. But, as well for its own sake as historically, the memoir is a very important one. Lagrange, in his memoir of the 17th of August, 1808, speaks of it as having recalled his attention to a subject with which he had previously occupied himself, but which he had quite lost sight of; and Arago records that, on the death of Lagrange, a copy in his own handwriting of Poisson's memoir was found among his papers; and the memoir is referred to in, and was probably the occasion of, Laplace's memoir also of the 17th of August, 1808.

5. With respect to Laplace's memoir of the 17th of August, 1808, it will be sufficient to quote a sentence from the introduction to Lagrange's memoir:—"Ayant montré à M. Laplace mes formules et mon analyse, il me montra de son côté en même temps des formules analogues qui donnent les variations des élémens elliptiques par les différences partielles d'une même fonction, relatives à ces élémens. J'ignore comment il y est parvenu; mais je présume qu'il les a trouvées par une combinaison adroite des formules qu'il avait données dans la *Mécanique Céleste.*" This is, in fact, the character of Laplace's analysis for the demonstration of the formulæ.

6. In Lagrange's memoir of the 17th of August, 1808, "On the Theory of the Variations of the Elements of the Planets, and in particular on the Variations of the Major Axes of their Orbits," the question treated of appears from the title. The author obtains formulæ for the variations of the elements of the orbit of a planet in terms of the differential coefficients of the disturbing function with respect to the elements; but the method is a general one, quite independent of the particular form

of the integrals, and the memoir may be considered as the foundation of the general theory. The equations of motion are considered under the form,

$$\frac{d^2x}{dt^2} - \frac{1+m}{r^3}\,x = \frac{d\Omega}{dx},$$

$$\frac{d^2y}{dt^2} - \frac{1+m}{r^3}\,y = \frac{d\Omega}{dy},$$

$$\frac{d^2z}{dt^2} - \frac{1+m}{r^3}\,z = \frac{d\Omega}{dz},$$

and it is assumed that the terms in Ω being neglected, the problem is completely solved, viz., that the three coordinates, x, y, z, and their differential coefficients, x', y', z', are each of them given as functions of t, and of the constants of integration a, b, c, f, g, h; the disturbing function Ω is consequently also given as a function of t, and of the arbitrary constants. The velocities are assumed to be the same as in the undisturbed orbit. This gives the conditions

$$\delta x = 0, \qquad \delta y = 0, \qquad \delta z = 0;$$

and then the equations of motion give

$$\delta\frac{dx}{dt} = \frac{d\Omega}{dx}, \quad \delta\frac{dy}{dt} = \frac{d\Omega}{dy}, \quad \delta\frac{dz}{dt} = \frac{d\Omega}{dz},$$

equations in which δx, &c. denote the variations of x, &c., arising from the variations of the arbitrary constants, viz., $\delta x = \frac{dx}{da}\,\delta a + \frac{dx}{db}\,\delta b +$, &c. The differential coefficients $\frac{d\Omega}{\delta x}$, &c., can of course be expressed by means of $\frac{d\Omega}{da}$, &c.; and, by a simple combination of the several equations, Lagrange deduces expressions for $\frac{d\Omega}{da}$, &c., in terms of $\frac{da}{dt}$, &c.; viz.

$$\frac{d\Omega}{da} = (a,\ b)\frac{db}{dt} + (a,\ c)\frac{dc}{dt} + (a,\ f)\frac{df}{dt} + (a,\ g)\frac{dg}{dt} + (a,\ h)\frac{dh}{dt},$$

$$\vdots$$

where [1]

$$(a,\ b) = \frac{\partial(x,\ x')}{\partial(a,\ b)} + \frac{\partial(y,\ y')}{\partial(a,\ b)} + \frac{\partial(z,\ z')}{\partial(a,\ b)},$$

in which, for shortness,

$$\frac{\partial(x,\ x')}{\partial(a,\ b)} \text{ stands for } \frac{dx}{da}\frac{dx'}{db} - \frac{dx'}{da}\frac{dx}{db}.$$

[1] These are substantially the formulæ of Lagrange; but I have introduced here and elsewhere the very convenient abbreviation, due, I think, to Prof. Donkin, of the symbols $\frac{\partial(x,\ x')}{\partial(a,\ b)}$.

The form of the expressions shows at once that $(a, b) = -(b, a)$, so that the number of the symbols (a, b) is in fact fifteen.

Lagrange proceeds to show, that the differential coefficient with respect to t of the expression represented by the symbol (a, b) vanishes identically; and it follows, that the coefficients (a, b) are *functions of the elements only, without the time t.*

The general formulæ are applied to the problem in hand; and, in consequence of the vanishing of several of the coefficients (a, b), it is easy in the particular problem to pass from the expressions for $\frac{d\Omega}{da}$, &c. in terms of $\frac{da}{dt}$, &c. to those for $\frac{da}{dt}$, &c. in terms of $\frac{d\Omega}{da}$, &c. The author thus obtains an elegant system of formulæ for the variations of the elements of a planet's orbit, in terms of the differential coefficients of the disturbing function with respect to the elements; but it is not for the present purpose necessary to consider the form of the system, or the astronomical consequences deduced by means of it.

7. Lagrange's memoir of the 13th of March, 1809, "On the General Theory of the Variation of the Arbitrary Constants in all the Problems of Mechanics."—The method of the preceding memoir is here applied to the general problem; the equations of motion are considered under the form

$$\frac{d}{dt}\frac{dT}{dr'} - \frac{dT}{dr} + \frac{dV}{dr} = \frac{d\Omega}{dr},$$
$$\vdots$$

where T and V are of the degree of generality considered in the *Mécanique Analytique*, viz., T is a function of $r, s \ldots r', s', \ldots$ homogeneous of the second order as regards the differential coefficients $r', s', \ldots$, and V is a function of $r, s, \ldots$ only; or, rather, the equations are considered in a form obtained from the above, by writing $T - V = R$, viz., in the form

$$\frac{d}{dt}\frac{dR}{dr'} - \frac{dR}{dr} = \frac{d\Omega}{dr},$$

and, as in the preceding memoir, expressions are investigated for the differential coefficients $\frac{d\Omega}{dt}$, &c., in terms of $\frac{db}{dt}$, &c.: these are, as before, of the form

$$\frac{d\Omega}{da} = (a, b)\frac{db}{dt} +, \text{ \&c.}$$

where (a, b), &c., are in the body of the memoir obtained under a somewhat complicated form, and this complicates also the demonstration which is there given of the theorem that (a, b), &c. *are functions of the elements only, without the time t*; but in the Addition (published as part of the memoir, and without a separate date) and in the Supplement the investigation is simplified, and the true form of the functions (a, b) obtained viz., writing $\frac{dT}{dr} = \rho, \ldots$ then

$$(a, b) = \frac{\partial (r, \rho)}{\partial (a, b)} + \frac{\partial (s, \sigma)}{\partial (a, b)} + \ldots$$

if, for shortness,

$$\frac{\partial (r, \rho)}{\partial (a, b)} = \frac{dr}{da}\frac{d\rho}{db} - \frac{d\rho}{da}\frac{dr}{db}, \text{ \&c.}$$

The representation of $\frac{dT}{dr'}$, $\frac{dT}{ds'}$, &c. by single letters *is* made by Lagrange in the Addition, No. 26 (Lagrange writes $\frac{dT}{dr'} = T'$, $\frac{dT}{ds'} = T''$, &c.), but quite incidentally in that number only, for the sake of the formula just stated: I have noticed this, as the step is an important one.

8. It is proper to remark that, in order to prove that the expressions (a, b), &c. are independent of the time, Lagrange, instead of considering the differential coefficients of each of these functions separately, establishes a general equation (see Nos. 25, 34, 35 of the Addition, and also the Supplement)

$$\frac{d}{dt}\left(\Delta r \delta \frac{dR}{dr'} - \delta r' \Delta \frac{dR}{dr'} + \ldots\right) = 0,$$

where, if Δa, $\Delta b, \ldots$ denote any arbitrary increments whatever of the constants of integration a, $b \ldots$ then Δr, &c. are the corresponding increments of the coordinates r, &c.; this is, in fact, a grouping together of several distinct equations by means of arbitrary multipliers, and it is extremely elegant as a method of demonstration, and has been employed as well by Lagrange, here and elsewhere, as by others who have written on the subject; but I think the meaning of the formulæ is best seen when the component equations of the group are separately exhibited, and in the citation of formulæ I have therefore usually followed this course. Lagrange gives also an equation which is in fact a condensed form of the preceding expression for $\frac{d\Omega}{da}$, but which it is proper to mention, viz.:

$$\frac{d\Omega}{da}\, dt = \frac{dr}{da}\, \delta \frac{dR}{dr'} + \ldots - \delta r \frac{d}{da}\frac{dR}{dr'} - \ldots.$$

In fact, in the formula $\delta \frac{dR}{dr}$ stands for $\left(\frac{d}{da}\frac{dR}{dr'}\frac{da}{dt} + \frac{d}{db}\frac{dR}{dr}\frac{db}{dt} + \ldots\right) dt$, and δr for $\left(\frac{dr}{da}\frac{da}{dt} + \frac{dr}{dt}\frac{db}{dt} + \right) dt$; and, on substituting these values, the identity of the two expressions is seen without difficulty.

9. Lagrange remarks, that in the case where the condition of *vis viva* holds good, then if a be the constant of *vis viva* $(T + V = a)$, and c the constant attached by addition to the time, then $\frac{da}{dt} = \frac{d\Omega}{dc}$, which, he observes, is an equation remarkable as

well from its simplicity and generality as because it can be obtained *à priori*, independently of the variations of the other arbitrary constants: this is obviously the generalisation of the expression for the variation of the mean distance of a planet.

10. The consideration of Lagrange's function (a, b) originated, as appears from what has preceded, in the theory of the variation of the elements; but it is to be noticed, that the function (a, b) is altogether independent of the disturbing function, and the fundamental theorem that (a, b) is a function of the elements only, without the time, is a property of the undisturbed equations of motion. The like remark applies to Poisson's function (a, b), in the memoir next spoken of.

11. Poisson's memoir of the 16th of October, 1809.—The formulæ of this memoir are, so to speak, the reciprocals of those of Lagrange. The relations between the differential coefficients $\frac{d\Omega}{da}$, &c. of the disturbing function and the variations $\frac{da}{dt}$, &c. of the elements, depend with Lagrange, upon expressions for the coordinates and their differential coefficients in terms of the time and the elements; with Poisson, on expressions for the elements in terms of the time, and of the coordinates and their differential coefficients. The distinction is far more important than would at first sight appear, and the theory of Poisson gives rise to developments which seem to have nothing corresponding to them in the theory of Lagrange. The reason is as follows: when the system of differential equations is completely integrated, it is of course the same thing whether we have the integral equations in the form made use of by Lagrange, or in that by Poisson, the two systems are precisely equivalent the one to the other; but when the equations are not completely integrated, suppose, for instance, we have an expression for one of the coordinates in terms of the time and the elements, it is impossible to judge whether this is or is not one of the integral equations; the differential equations are not satisfied by means of this equation alone, but only by this equation with the assistance of the other integral equations. On the other hand, when we have an expression for one of the constants of integration in terms of the time and of the coordinates and their differential coefficients, it is possible, by mere substitution in the differential equations, and without the knowledge of any other integral equations, to see that the differential equations are satisfied, and that the assumed expression is, in fact, one of the system of integral equations. An expression of the form just referred to, viz., $c = \phi(t, x, y, \dots x', y' \dots)$, where the right-hand side does not contain any of the arbitrary constants, may, with great propriety, be termed an "integral," as distinguished from an integral equation, in which the constants and variables may enter in any conceivable manner; it is convenient also to speak of such equation simply as the integral c. [These locutions were introduced by Jacobi.]

12. Returning now to the consideration of Poisson's memoir, the equations of motion are considered under the same form as by Lagrange, viz., putting $T - V = R$, under the form

$$\frac{d}{dt}\frac{dR}{d\phi'} - \frac{dR}{d\phi} = \frac{d\Omega}{d\phi}:$$

$$\vdots$$

but Poisson writes

$$\frac{dR}{d\phi} = s, \dots$$

thus, in effect, introducing a new set of variables, $s, \dots$ equal in number to the coordinates $\phi, \dots$, but he does not complete the transformation of the differential equations by the introduction therein of the new variables $s, \dots$ in the place of the differential coefficients $\phi', \dots$; this very important transformation was only effected a considerable time afterwards by Sir W. R. Hamilton. Poisson then assumes that the undisturbed equations are integrated in the form above adverted to, viz., that the several elements $a, b, \dots$ are given as functions of the time t, and of the coordinates ϕ, &c. and their differential coefficients ϕ', &c. or, what is the form ultimately assumed, as functions of the time t, of the coordinates $\phi, \dots$, and of the new variables s, &c.; and he then forms the functions

$$(a, b) = \frac{\partial (a, b)}{\partial (s, \phi)} + \dots$$

where

$$\frac{\partial (a, b)}{\partial (s, \phi)} = \frac{da}{ds}\frac{db}{d\phi} - \frac{db}{ds}\frac{da}{d\phi},$$

(the notation is the abbreviated one before referred to), and he proves by differentiation that the differential coefficient of (a, b) with respect to the time vanishes: that is, that (a, b) which, by its definition is given as a function of t and of the coordinates $\phi, \dots$, and of the new variables $s, \dots$, is really a constant. Upon which Poisson remarks—"On conçoit que la constante...sera en général une fonction de a et b et des constantes arbitraires contenues dans les autres intégrales des équations du mouvement; quelquefois il pourra arriver que sa valeur ne renferme ni la constante a ni la constante b; dans d'autres cas elle ne contiendra aucune constante arbitraire, et se réduira à une constante déterminée; mais afin &c."

13. The importance of the remark seems to have been overlooked until the attention of geometers was called to it by Jacobi; it has since been developed by Bertrand and Bour.

It is clear from the definition that $(a, b) = -(b, a)$. It may be as well to remark that the denominator of the functional symbol is (s, ϕ) and not (ϕ, s), which would reverse the sign. [It may be noticed that throughout the Report, I speak of the Lagrange's Coefficients (a, b), and Poisson's Coefficients (a, b), distinguishing them in this manner, and not by any difference of notation.]

14. The equations for the variations of the elements are without difficulty shown to be

$$\frac{da}{dt} = (a, b)\frac{d\Omega}{db} + \dots,$$

$$\vdots$$

which have the advantage over those of Lagrange of giving directly $\frac{da}{dt}$, &c. in terms of $\frac{d\Omega}{da}$, &c., instead of these expressions having to be determined from the value of $\frac{d\Omega}{da}$, &c. in terms of $\frac{da}{dt}$, &c.

15. Poisson applies his formulæ to the case of a body acted upon by a central force varying as any function of the distance, and also to the case of a solid body revolving round a fixed point. There is, as Poisson remarks, a complete similarity between the formulæ for these apparently very different problems, but this arises from the analogy which exists between the arbitrary constants chosen in the memoir for the two problems. The formulæ obtained form a very simple and elegant system, and one which, although not actually of the canonical form (the meaning of the term will be presently explained), might by a slight change be reduced to that form.

16. I may notice here a problem suggested by Poisson in a report to the Institute in the year 1830, on a manuscript work by Ostrogradsky on Celestial Mechanics, viz., in the case of a body acted upon by a central force, the effect of a disturbing function, *which is a function only of the distance from the centre*, is merely to alter the amount of the central force; and the expressions for the variations of the elements should therefore, in the case in question, admit of exact integration; the report is to be found in *Crelle*, t. VII. [1831], pp. 97—101.

17. The two memoirs of Lagrange and Poisson, which have been considered, establish the general theory of the variation of the arbitrary constants, and there is not, I think, very much added to them by Lagrange's memoir of 1810, the second edition of the *Mécanique Analytique*, 1811, or Poisson's memoir of 1816. The memoir by Maurice, in 1844, belongs to this part of the subject, and as its title imports, it is in fact a development of the theories of Lagrange and Poisson.

18. There is, however, one important point which requires to be adverted to. Lagrange, in the memoir of 1810, and the second edition of the *Mécanique Analytique*, remarks, that for a particular system of arbitrary constants, viz., if $\alpha, \ldots$ denote the initial values of the coordinates $\xi, \ldots$ and $\lambda, \ldots$ denote the initial values of $\frac{dT}{d\xi'}, \ldots$ then the equations for the variations of the elements take the very simple form

$$\frac{d\alpha}{dt} = -\frac{d\Omega}{d\lambda}, \ldots, \quad \frac{d\lambda}{dt} = \frac{d\Omega}{d\alpha}, \ldots;$$

this is, in fact, the original idea and simplest example of a system of canonical elements; viz. of a system composed of pairs of elements, α, λ, the variations of which are given in the form just mentioned.

19. The "Avertissement" to the second edition of the *Mécanique Analytique*, contains the remark, that it is not necessary that the disturbing function Ω should actually exist; $\frac{d\Omega}{dx}, \frac{d\Omega}{dy}, \frac{d\Omega}{dz}$ may be considered as mere conventional symbols standing for forces X, Y, Z, not the differential coefficients of one and the same function, and then $\frac{d\Omega}{da}$ will be a conventional symbol standing for $\frac{d\Omega}{dx}\frac{dx}{da} + \frac{d\Omega}{dy}\frac{dy}{da} + \frac{d\Omega}{dz}\frac{dz}{da}$, and similarly for $\frac{d\Omega}{db}$, &c.; and this being so, all the formulæ will subsist as in the case of an actually existing disturbing function.

20. Cauchy, in a note in the *Bulletin de la Société Philomatique* for 1819 (reproduced in the "Mémoire sur l'Intégration des Equations aux Dérivées Partielles du Premier Ordre," *Exer. d'Anal. et de Physique Math.*, t. II. pp. 238—272 (1841)), showed that the integration of a partial differential equation of the first order could be reduced to that of a single system of ordinary differential equations. A particular case of this general theorem was afterwards obtained by Jacobi in the course of his investigations (founded on those of Sir W. R. Hamilton) on the equations of dynamics, and he was thence led to a slightly different form of the general theorem previously established by Cauchy, viz., Cauchy's method gives the *general*, Jacobi's the *complete* integral, of the partial differential equation. The investigations of the geometers who have written on the theory of dynamics are based upon those of Sir W. R. Hamilton and Jacobi, and it is therefore unnecessary, in the present report, to advert more particularly to Cauchy's very important discovery.

21. I come now to Sir W. R. Hamilton's memoirs of 1834 and 1835, which are the commencement of a second period in the history of the subject. The title of the first memoir shows the object which the author proposed to himself, viz., the discovery of a function by means of which the integral equations can be all of them actually represented. The method given for the determination of this function, or rather of each of the several functions which answer the purpose, presupposes the knowledge of the integral equations; it is therefore not a *method of integration*, but a theory of the representation of the integral equations assumed to be known. I venture to dissent from what appears to have been Jacobi's opinion, that the author missed the true application of his discovery; it seems to me, that Jacobi's investigations were rather a theory collateral to, and historically arising out of the Hamiltonian theory, than the course of development which was of necessity to be given to such theory. But the new form obtained in Sir W. R. Hamilton's memoirs for the equations of motion, is a result of not less importance than that which was the professed object of the memoirs.

22. Hamilton's principal function V.—The formulæ are given for the case of any number of free particles, but, for simplicity, I take the case of a single particle. The equations of motion are taken to be

$$m\frac{d^2x}{dt^2}=\frac{dU}{dx},$$

$$m\frac{d^2y}{dt^2}=\frac{dU}{dy},$$

$$m\frac{d^2y}{dt^2}=\frac{dU}{dz};$$

so that the *vis viva* function is

$$T=\tfrac{1}{2}m\,(x'^2+y'^2+z'^2),$$

and the force function, taken with Lagrange's sign, would be $-U$. It is assumed that the condition of *vis viva* holds, that is, that U is a function of x, y, z only. The initial values of the coordinates are denoted by a, b, c, and those of the velocities by a', b', c'. The equation of *vis viva* is $T = U + H$, and this gives rise to an equation $T_0 = U_0 + H$ of the same form for the initial values of the coordinates. The author then writes

$$V = \int_0^t 2T dt,$$

an equation, the form of which implies that T is expressed as a function of the time and of the constants of integration a, b, c, a', b', c'. The method of the calculus of variations leads to the equation

$$\delta V = m(x'\delta x + y'\delta y + z'\delta z) - m(a'\delta a + b'\delta b + c'\delta c) + t\delta H,$$

to understand which, it should be remarked that the coordinates x, y, z, and the velocities x', y', z', being functions of t and of a, b, c, a', b', c', then V is, in the first instance, given as a function of these quantities. But x, y, z being functions of a, b, c, a', b', c', t, we may conversely consider a', b', c' as functions of x, y, z, a, b, c, t, and thus V becomes a function of x, y, z, a, b, c, t. In like manner H is a function of x, y, z, a, b, c, t, and, eliminating t, we have V a function of x, y, z, a, b, c, H, which is the form in which in the last equation V is considered to be expressed. The equation then gives

$$\frac{dV}{dx} = mx', \quad \frac{dV}{dy} = my', \quad \frac{dV}{dz} = mz',$$

$$\frac{dV}{da} = -ma', \quad \frac{dV}{db} = -mb', \quad \frac{dV}{dc} = -mc',$$

$$\frac{dV}{dH} = t;$$

and, considering V as a known function of x, y, z, a, b, c, H, the elimination of H gives a set of equations which are in fact the integral equations of the problem, viz., the first three equations and the last equation give equations containing x, y, z, x', y', z', t and a, b, c, that is, the intermediate integrals; the second three equations and the last equation, give equations containing $x, y, z, t, a, b, c, a', b', c'$, that is, the final integrals.

The function V satisfies the two partial differential equations

$$\frac{1}{2m}\left\{\left(\frac{dV}{dx}\right)^2 + \left(\frac{dV}{dy}\right)^2 + \left(\frac{dV}{dz}\right)^2\right\} = U + H,$$

$$\frac{1}{2m}\left\{\left(\frac{dV}{da}\right)^2 + \left(\frac{dV}{db}\right)^2 + \left(\frac{dV}{dc}\right)^2\right\} = U_0 + H;$$

which, if they could be integrated, would give V as a function of x, y, z, a, b, c, H, and thus determine the motion of the system.

23. Hamilton's principal function S.—This is connected with the function V by the equation

$$V = tH + S;$$

or, what is the same thing, the new principal function S is defined by the equation

$$S = \int_0^t (T + U)\, dt\,;$$

but S is considered (not like V as a function of x, y, z, a, b, c, H, but) as a function of x, y, z, a, b, c, t. The expression for the variation of S is

$$\delta S = -H\delta t + m\,(x'\delta x + y'\delta y + z'\delta z) - m\,(a'\delta a + b'\delta b + c'\delta c)$$

which is equivalent to the system

$$\frac{dS}{dx} = mx', \quad \frac{dS}{dy} = my', \quad \frac{dS}{dz} = mz',$$

$$\frac{dS}{da} = -ma', \quad \frac{dS}{db} = -mb', \quad \frac{dS}{dc} = -mc',$$

$$\frac{dS}{dt} = -H\,;$$

the first three and the second three of which give, respectively, the intermediate and the final integrals; the last equation leads only to the expression of the supernumerary constant H in terms of the initial coordinates a, b, c, and it may be omitted from the system.

The function S satisfies the partial differential equations

$$\frac{dS}{dt} + \frac{1}{2m}\left\{\left(\frac{dS}{dx}\right)^2 + \left(\frac{dS}{dy}\right)^2 + \left(\frac{dS}{dz}\right)^2\right\} = U,$$

$$\frac{dS}{dt} + \frac{1}{2m}\left\{\left(\frac{dS}{da}\right)^2 + \left(\frac{dS}{db}\right)^2 + \left(\frac{dS}{dc}\right)^2\right\} = U_0\,;$$

which, if they could be integrated, would give S as a function of x, y, z, a, b, c, t, and thus determine the motion of the system.

24. Hamilton's form of the equations of motion.—This is in fact the form obtained by carrying out the idea of introducing into the differential equations, in the place of the differential coefficients of the coordinates, the derived functions (with respect to these differential coefficients) of the *vis viva* function T. Taking η to denote any one of the series of coordinates, then the original system may be denoted by

$$\frac{d}{dt}\frac{dT}{d\eta'} - \frac{dT}{d\eta} = \frac{dU}{d\eta},$$
$$\vdots$$

(U is the force function taken with a contrary sign to that of Lagrange), and writing in like manner ϖ to denote any one of the new variables connected with the coordinates η by the equations

$$\frac{dT}{d\eta'} = \varpi,$$
$$\vdots$$

then T, in its original form, is a function of $\eta, \ldots \eta', \ldots$, homogeneous of the second order as regards the differential coefficients $\eta', \ldots$; and, consequently, these being linear functions (without constant terms) of the new variables ϖ, the *vis viva* function T can be expressed as a function of $\eta, \ldots \varpi, \ldots$, homogeneous of the second order as regards the variables $\varpi, \ldots$. And when T has been thus expressed, the equations of motion take the form

$$\frac{d\eta}{dt} = \frac{dH}{d\varpi}, \quad \frac{d\varpi}{dt} = -\frac{dT}{d\eta} + \frac{dU}{d\eta},$$
$$\vdots \qquad \vdots$$

which is the required transformation. The force function U is independent of the differential coefficients $\eta', \ldots$ and, consequently, of the variables $\varpi, \ldots$, hence, writing $H = T - U$, the equations take the form

$$\frac{d\eta}{dt} = \frac{dH}{d\varpi}, \quad \frac{d\varpi}{dt} = -\frac{dH}{d\eta},\text{[1]}$$
$$\vdots \qquad \vdots$$

which correspond to the condensed form obtained by writing $T - V = R$ in Lagrange's equations. It is hardly necessary to remark that H is to be considered as a given function of $\eta, \ldots \varpi, \ldots$, viz., it is what $T - U$ becomes when the differential coefficients $\eta', \ldots$ are replaced by their values in terms of the new variables $\varpi, \ldots$.

25. I have, for greater simplicity, explained the theory of the functions V and S in reference to a very special form of the equations of motion; but the theory is, in fact, applicable to any form whatever of these equations; and, as regards the function V, is in the first memoir examined in detail with reference to Lagrange's general form of the equations of motion. The function S is considered at the end of the memoir, in reference only to the special form. The new form of the equations of motion is first established in the second memoir, and the theory of the functions V and S is there considered in reference to this form. The author considers also another function Q, which, when the matter is looked at from a somewhat more general point of view, is not really distinct from the function S.

26. The first memoir contains applications of the method to the problem of two bodies, and the problem of three or more bodies, and researches in reference to the approximate integration of the equations of motion by the separation of the function V into two parts, one of them depending on the principal forces, the other on the disturbing forces. The method, or one of the methods, given for this purpose, involves the consideration of the variation of the arbitrary constants, but it is not easy to single out any precise results, or explain their relation to the results of Lagrange and Poisson. The like remark applies to the investigations contained in Nos. 7 to 12 of the second memoir, but it is important to consider the theory described in the heading

[1] I find it stated in a note to M. Houel's "Thèse sur l'intégration des équations différentielles de la Mécanique," Paris, 1855, that this form of the equations of motion had been previously employed in an *unpublished* memoir by Cauchy, written in 1831. [Cauchy "Extrait du Mémoire présenté à l'Academie de Turin le 11 Oct. 1831" published in lithograph under the date Turin, 1832, with an Addition dated 6 Mar. 1833.]

of No. 13, as "giving formulæ for the variation of elements more analogous to those already known." The function H is considered as consisting of two parts, one of them being treated as a disturbing function; the equations of motion assume therefore the form

$$\frac{d\eta}{dt}=\frac{dH}{d\varpi}+\frac{d\Upsilon}{d\varpi},\quad \frac{d\varpi}{dt}=-\frac{dH}{d\eta}-\frac{d\Upsilon}{d\eta},$$

(I have written H, Υ instead of the author's H_1, H_2). The terms involving Υ are in the first instance neglected, and it is assumed that the integrals of the resulting equations are presented in the form adopted by Poisson, viz., the constants of integration a, b, &c. are considered as given in terms of t, and of the two sets of variables $\eta, \ldots$ and $\varpi, \ldots$; the integrals are then extended to the complete equations by the method of the variation of the elements. The resulting expressions are the same in form as those of Poisson, viz.:

$$\frac{da}{dt}=(a,\ b)\frac{d\Upsilon}{db}+\ldots,$$

where

$$(a,\ b)=\frac{\partial(a,\ b)}{\partial(\eta,\ \varpi)}+\ldots,$$

if, for shortness,

$$\frac{\partial(a,\ b)}{\partial(\eta,\ \varpi)}=\frac{da}{d\eta}\frac{db}{d\varpi}-\frac{db}{d\eta}\frac{da}{d\varpi},$$

and conversely the values of $\frac{d\Upsilon}{da}$, &c. in terms of $\frac{da}{dt}$, &c. might have been exhibited in a form such as that of Lagrange. The expressions (a, b), considered as functional symbols, have the same meanings as in the theories of Poisson and Lagrange; and, as in these theories, the differential coefficient of (a, b) with respect to the time, vanishes, or (a, b) is a function of the elements only.

27. It is to be observed that the disturbing function Υ is not necessarily in the same problem identical with the disturbing function Ω of Lagrange and Poisson (indeed, in any problem, the separation of the forces into principal forces and disturbing forces is an arbitrary one). Sir W. R. Hamilton, in the second memoir, gives a very beautiful application of his theory to the problem of three or more bodies, which has the peculiar advantage of making the motion of all the bodies depend upon one and the same disturbing function[1]. This disturbing function contains (as in the last-mentioned general formulæ) both sets of variables, and the consequence is that, as the author remarks, the varying elements employed by him are essentially different from those made use of in the theories of Lagrange and Poisson; the velocities cannot, in his theory, be obtained by differentiating the coordinates as if the elements were

[1] Lagrange has given formulæ for the determination of the motion of three or more bodies referred to their common centre of gravity by means of one and the same disturbing function. In Sir W. R. Hamilton's theory there is one central body to which all the others are referred. The method of Sir W. R. Hamilton is made use of in M. Houel's "Thèse d'Astronomie: Application de la Méthode de M. Hamilton au Calcul des Perturbations de Jupiter."—Paris, 1855.

constant. The investigation applies to the case where the attracting force is any function whatever of the distance, and the six elements ultimately adopted form a canonical system.

28. The precise relation of Sir W. R. Hamilton's form of the equations of motion to that of Lagrange, is best seen by considering Lagrange's equations, not as a system of differential equations of the second order between the coordinates and the time t, but as a system of twice as many differential equations of the first order between the coordinates, their differential coefficients treated as a new system of variables, and the time. It will be convenient to write $-U$, instead of Lagrange's force-function V, and (to conform to the usage of later writers who have treated the subject in the most general manner) to represent the coordinates by $q, \ldots$, their differential coefficients by $q', \ldots$, and the new variables which enter into the Hamiltonian form by $p, \ldots$; then the Lagrangian system will be

$$\frac{dq}{dt} = q', \quad \frac{d}{dt}\frac{dT}{dq'} - \frac{dT}{dq} = \frac{dU}{dq};$$
$$\vdots \qquad \vdots$$

or putting $T + U = Z$ (this is the same as Lagrange's substitution, $T - V = R$), the system becomes

$$\frac{dq}{dt} = q', \quad \frac{d}{dt}\frac{dZ}{dq'} = \frac{dZ}{dq},$$
$$\vdots \qquad \vdots$$

while the Hamiltonian system is

$$\frac{dq}{dt} = \frac{dT}{dp}, \quad \frac{dp}{dt} = -\frac{dT}{dq} + \frac{dU}{dq};$$
$$\vdots \qquad \vdots$$

or putting as before $T - U = H$, the system is

$$\frac{dq}{dt} = \frac{dH}{dp}, \quad \frac{dp}{dt} = -\frac{dH}{dq};$$
$$\vdots \qquad \vdots$$

where, in the Lagrangian systems, T and U, and consequently Z, are given functions of a certain form of $t, q, \ldots q', \ldots$, and in like manner, in the Hamiltonian system, T and U, and consequently H, are given functions of a certain form of $t, q, \ldots p \ldots$. The generalisation has since been made (it is not easy to say precisely when first made) of considering Z as standing for any function whatever of $t, q, \ldots q', \ldots$, and in like manner of considering H as standing for any function whatever of $t, q, \ldots p, \ldots$. It is to be noticed that in Sir W. R. Hamilton's memoir, the demonstration which is given of the transformation from Lagrange's equations to the new form depends essentially on the special form of the function T as a homogeneous function of the second order in regard to the differential coefficients of the coordinates; indeed the transformation itself, as regards the actual value of the new function T (= T expressed in terms of the new variables), which enters into the transformed equations, depends essentially upon the special form just referred to of the function T, although, as will be seen in the sequel, there is a like transformation applying to the most general form of the function T.

29. In the greater part of what has preceded, and especially in the above-mentioned substitutions $T+U=Z$ and $T-U=H$, it is of course assumed that the force function U exists; when there is no force function these substitutions cannot be made, but the forms corresponding to the untransformed forms in T and U are as follows, viz. the Lagrangian form is

$$\frac{dq}{dt}=q', \quad \frac{d}{dt}\frac{dT}{dq'}-\frac{dT}{dq}=Q,$$
$$\vdots \qquad \vdots$$

and the Hamiltonian form is

$$\frac{dq}{dt}=\frac{dT}{dp}, \quad \frac{dp}{dt}=-\frac{dT}{dq}+Q\,;$$
$$\vdots \qquad \vdots$$

that is, the only difference is, that the functions Q, instead of being the differential coefficients with respect to the variables $q\ldots$ of one and the same force function U, are so many separate and distinct functions of the variables $q, \ldots$, or more generally of the variables $q, \ldots p, \ldots$ of both sets.

30. Jacobi's letter of 1836.—This is a short note containing a mere statement of two results. The first is as follows, viz. the equations for the motion of a point *in plano* being taken to be

$$\frac{d^2x}{dt^2}=\frac{dU}{dx}, \quad \frac{d^2y}{dt^2}=\frac{dU}{dy},$$

where U is a function x, y without t; one integral is the equation of *vis viva* $\frac{1}{2}(x'^2+y'^2)=U+h$. Assume that another integral is $a=F(x, y, x', y')$, then x', y' will in general be functions of x, y, a, h, and considering them as thus expressed, it is stated that not only $x'dx+y'dy$ will be an exact differential, but its differential coefficients with respect to a, h will be so likewise, and the remaining integrals are

$$b=\int\left(\frac{dx'}{da}dx+\frac{dy'}{da}dy\right),$$

$$t+T=\int\left(\frac{dx'}{dh}dx+\frac{dy'}{dh}dy\right),$$

a theorem, the relation of which to the general subject will presently appear.

The second result does not relate to the general subject, but I give it in a note for its own sake[1].

31. Poisson's memoir of 1837.—This contains investigations suggested by Sir W. R. Hamilton's memoir, and relating to the aid to be derived from a system of given integral equations (equal in number to the coordinates) in the determination of the principal function V. The equations $\frac{dV}{dx} = mx'$, &c. give $dV = m(x'dx + y'dy + z'dz)$, or in the case of a system of points, $dV = \Sigma m(x'dx + y'dy + z'dz)$. If the points, instead of being free, are connected together by any equations of condition, then, by means of these equations, the coordinates x, y, z of the different points and their differential coefficients, x', y', z', can be expressed as functions of a certain number of independent variables ϕ, ψ, θ, &c., and of their differential coefficients ϕ', ψ', θ', &c.; dV then takes the form $dV = Xd\phi + Yd\psi + Zd\theta + \ldots$, where X, Y, Z are functions of $\phi, \psi, \ldots$ $\phi', \psi', \ldots$. Imagine now a system of integrals (one of them the equation of *vis viva*) equal in number to the independent variables $\phi, \psi, \theta \ldots$; then, by the aid of these equations, $\phi', \psi', \theta' \ldots$, and, consequently, $X, Y, Z, \ldots$ can be expressed as functions (of the constants of integration and) of the variables $\phi, \psi, \theta, \ldots$. Hence, attending only to the variables, $dV = Xd\phi + Yd\psi + Zd\theta + \ldots$ is a differential expression involving only the variables $\phi, \psi, \theta \ldots$; but, as Poisson remarks, this expression is not in general a complete differential. In the cases in which it is so, V can of course be obtained directly by integrating the differential expression, viz. the function so obtained is in value, but not in form, Sir W. R. Hamilton's principal function V; for, with him, V is a function of the coordinates, and of a particular set of the constants of integration, viz. the constant of *vis viva* h, and the initial values of the coordinates. Poisson adds the very important remark, that V being determined by his process as above, then h being the constant of *vis viva*, and the constants of the other given integral equations being e, f, &c., the remaining integrals of the problem are[2]

$$\frac{dV}{dh} = t + \tau, \quad \frac{dV}{de} = l, \quad \frac{dV}{df} = m, \ldots$$

[1] Jacobi imagines a point *without mass* revolving round the sun and disturbed by a planet moving in a circular orbit, which is taken for the plane of x, y; the coordinates of the point are x, y, z, those of the planet $a'\cos n't$, $a'\sin n't$, m' is the mass of the planet, M the mass of the sun; then we have accurately

$$\tfrac{1}{2}\left\{\left(\frac{dx}{dt}\right)^2 + \left(\frac{dy}{dt}\right)^2 + \left(\frac{dz}{dt}\right)^2\right\} - n'\left(x\frac{dy}{dt} - y\frac{dx}{dt}\right) =$$

$$\frac{M}{(x^2+y^2+z^2)}^{\frac{1}{2}} + m'\left\{\frac{1}{(x^2+y^2+z^2-2a'(x\cos n't + y\sin n't) + a'^2)^{\frac{1}{2}}} - \frac{x\cos n't + y\sin n't}{a'^2}\right\} + \text{const.}$$

which Jacobi suggests might be found useful in the lunar theory. The point being without mass, means only that it is considered as not disturbing the circular motion of the planet; the problem is properly a case of the problem of two centres, viz. one centre is fixed, and the other one revolves round it in a circle with a uniform velocity.

[2] Poisson writes $\frac{dV}{dh} = -t + \epsilon$; there seems to be a mistake as to the sign of h running through the memoir. Correcting this, and putting $-\tau$ for ϵ, we have the formula $\frac{dV}{dh} = t + \tau$ given in the text.

where τ, l, m, ... are new arbitrary constants. But, as before remarked, the expression for dV is not always a complete differential. Poisson accordingly inquires into and determines (but not in a precise form) the conditions which must be satisfied, in order that the expression in question may be a complete differential. He gives, as an example, the case of the motion of a body in space under the action of a central force; and, secondly, the case considered in Jacobi's letter of 1836, which he refers to, viz., here $dV = x'dx + y'dy$, and when the two integral equations are one of them, the equations of *vis viva* $\frac{1}{2}(x'^2 + y'^2) = U + h$, and the other of them any integral equation $a = F(x, y, x', y')$ whatever (subject only to the restriction that a is not a function of x, y, $x'^2 + y'^2$, the necessity of which is obvious) the condition is satisfied *per se*, and, consequently, $x'dx + y'dy$ is a complete differential, and its integral gives (in value, although as before remarked not in form) the principal function V; and such value of V gives the two integral equations obtained in Jacobi's letter.

32. Jacobi's note of the 29th of November, 1836, "On the Calculus of Variations, and the Theory of Differential Equations."—The greater part of this note relates to the differential equations which occur in the calculus of variations, including, indeed, the differential equations of dynamics, but which belong to a different field of investigation. The latter part of the note relates more immediately to the differential equations of dynamics. The author remarks, that, in any dynamical problem of the motion of a single particle for which the principle of *vis viva* holds good, if, besides the integral of *vis viva*, there is given any other integral, the problem is reducible to the integration of an ordinary differential equation of two variables, and that it is always possible to integrate this equation, or at least *discover by a precise and general rule the factor which renders it integrable.* This would seem to refer to Jacobi's researches on the theory of the ultimate multiplier, but the author goes on to refer to a preceding communication to the Academy of Paris (the before-mentioned letter of 1836), which does not belong (or, at least, does not obviously belong) to this theory. He speaks also of a class of dynamical problems, viz. that of the motion of a system of bodies which mutually attract each other, and which may besides be acted upon by forces in parallel lines, or directed to fixed centres, or even to centres the motion of which is given; and, he remarks, in the solution of such a problem, the system of differential equations being in the first instance of the order $2n$ (that is, being a system admitting of $2n$ arbitrary constants), then if one integral is known, it is possible by a proper choice of the quantities selected for variables to reduce the system to the order $2n-2$. If another integral is known, the equation may in like manner be reduced to a system of the order $2n-4$, and so on until there are no more equations to be integrated; and thus the operations to be effected depend only upon quadratures. All this seems to refer to researches of Jacobi, which, so far as I am aware, have not hitherto been published. The results correspond with those recently obtained by Bour, *post*, Nos. 66 and 67.

33. Jacobi's memoir of 1837.—Jacobi refers to the memoirs of Sir W. R. Hamilton, and he reproduces, in a slightly different form, the investigation of the fundamental property of the principal function S. The case considered is that of a system of n

particles, the coordinates of which are connected together by any number of equations; but it will be sufficient here to attend to the case of a single free particle. The equations of motion are assumed to be

$$m\frac{d^2x}{dt^2}=\frac{dU}{dx},\quad m\frac{d^2y}{dt^2}=\frac{dU}{dy},\quad m\frac{d^2z}{dt^2}=\frac{dU}{dz};$$

but U is considered as being a function of x, y, z and of the time t, that is, it is assumed that the condition of *vis viva* is not of necessity satisfied. The definition of the function S is

$$S+\int_0^t\left[U=\tfrac{1}{2}m\,(x'^2+y'^2+z'^2)\right]dt,$$

which, when the equation of *vis viva* is satisfied, that is, when

$$T=\tfrac{1}{2}m\,(x'^2+y'^2+z'^2)=U+h,$$

agrees with Sir W. R. Hamilton's definition $S=2\int_0^t Udt+ht$. The function S is considered as being, by means of the integral equations assumed as known, expressed as a function of t, of the coordinates x, y, z, and of their initial values a, b, c. And then it is shown that S satisfies the equations

$$\frac{dS}{dx}=\ mx',\quad \frac{dS}{dy}=\ my',\quad \frac{dS}{dz}=\ mz',$$

$$\frac{dS}{da}=-\,ma',\quad \frac{dS}{db}=-\,mb',\quad \frac{dS}{dc}=-\,mc';$$

so that the intermediate and final integrals are expressed by means of the principal function S.

34. But Jacobi proceeds, "the definition assumes the integration of the differential equations of the problem. The results, therefore, are only interesting in so far as they have reduced the system of integral equations into a remarkable form. We may, however, define the function S in a quite different and *very much more general* manner." And then, attending only to the case of a system of free particles, he gives a definition, which, in the case of a single particle, is as follows:

Jacobi's principal function S.—The equations of motion being as before

$$\frac{d^2x}{dt^2}=m\frac{dU}{dx},\quad \frac{d^2y}{dt^2}=m\frac{dU}{dy},\quad \frac{d^2z}{dt^2}=m\frac{dU}{dz},$$

(where U is in general a function of x, y, z and t), then S is defined to be a *complete* solution of the partial differential equation

$$\frac{dS}{dt}+\frac{1}{2m}\left\{\left(\frac{dS}{dx}\right)^2+\left(\frac{dS}{dy}\right)^2+\left(\frac{dS}{dz}\right)^2\right\}=U.$$

A complete solution, it will be recollected, means a solution containing as many arbitrary constants as there are independent variables in the partial differential equation;

in the present case, therefore, four arbitrary constants. But one of these constants may be taken to be a constant attached to the function S by mere addition, and which disappears from the differential coefficients, and it is only necessary to attend to the other three arbitrary constants. S is consequently a function of t, x, y, z, and of the arbitrary constants α, β, γ, satisfying the partial differential equation. And this being so, it is shown that the integrals of the problem are

$$\frac{dS}{dx} = mx', \quad \frac{dS}{dy} = my', \quad \frac{dS}{dz} = mz',$$

$$\frac{dS}{d\alpha} = \lambda, \quad \frac{dS}{d\beta} = \mu, \quad \frac{dS}{d\gamma} = \nu\,;$$

where λ, μ, ν are any other arbitrary constants, viz., the first three equations give the intermediate integrals, and the last three equations give the final integrals of the problem.

Jacobi proceeds to give an analogous definition of the principal function V as follows:

35. Jacobi's principal function V.—First, when the condition of *vis viva* is satisfied. Here V is a complete solution of the partial differential equation

$$\frac{1}{2m}\left\{\left(\frac{dV}{dx}\right)^2 + \left(\frac{dV}{dy}\right)^2 + \left(\frac{dV}{dz}\right)^2\right\} = U + h,$$

where h is the constant of *vis viva*. The partial differential equation contains only three independent variables; and since as before one of the constants of the complete solution may be taken to be a constant attached to V by mere addition, and which disappears from the differential coefficients, we may consider V as a function of t, x, y, z, and of the two constants of integration α and β. But V will of course also contain the constant h, which enters into the partial differential equation. The integrals of the problem are then shown to be

$$\frac{dV}{dx} = mx', \quad \frac{dV}{dy} = my', \quad \frac{dV}{dz} = mz',$$

$$\frac{dV}{dh} = t + \tau, \quad \frac{dV}{d\alpha} = \lambda, \quad \frac{dV}{d\beta} = \mu,$$

where τ, λ, μ are new arbitrary constants.

36. Jacobi's principal function V.—Secondly, when the equation of *vis viva* is not satisfied. Here U contains the time t, and we have no such equation as $T = U + h$, but along with the coordinates x, y, z there is introduced a new variable H, and V is defined to be a complete integral of the partial differential equation

$$\frac{1}{2m}\left\{\left(\frac{dV}{dx}\right)^2 + \left(\frac{dV}{dy}\right)^2 + \left(\frac{dV}{dz}\right)^2\right\} = U + H\,;$$

where, in the expression for U, it is assumed that t is replaced by $\frac{dV}{dH}$. There are, consequently, four independent variables, and a complete solution must contain, exclusively of the constant attached to V by mere addition, and which disappears from the differential coefficients, three arbitrary constants α, β, γ. The integral equations are shown to be

$$\frac{dV}{dx} = mx', \quad \frac{dV}{dy} = my', \quad \frac{dV}{dz} = mz',$$

$$\frac{dV}{d\alpha} = \lambda, \quad \frac{dV}{d\beta} = \mu, \quad \frac{dV}{d\gamma} = \nu,$$

$$\frac{dV}{dH} = t;$$

where λ, μ, ν are arbitrary constants, viz., eliminating H from the first three equations by the assistance of the last equation, we have the intermediate integrals; and eliminating H from the second three equations by the assistance of the last equation, we have the final integrals. The substitution of the above values $\frac{dV}{dx}$, &c. in the partial differential equation gives $T = U + H$, that is, $H (= T - U)$ is that function which, when the condition of *vis viva* is satisfied, becomes equal to h, the constant of *vis viva*.

Jacobi's extension of the theory to the case where the condition of *vis viva* is not satisfied, appears to have attracted very little attention; it is indeed true, as will be noticed in the sequel, that this general case can be reduced to the particular one in which the condition of *vis viva* is satisfied, but there is not it would seem any advantage in making this reduction; the formulæ for the general case are at least quite as elegant as those for the particular case.

37. Jacobi, after considering some particular dynamical applications, proceeds to apply the theory developed in the first part of the memoir to the general subject of partial differential equations; the differential equations of a dynamical problem lead to a partial differential equation, a complete solution of which gives the integral equations. Conversely, the integral equations give the complete solution of the partial differential equation, and applying similar considerations to any partial differential equation of the first order whatever, it is shown (what, but for Cauchy's memoir of 1819, which Jacobi was not acquainted with (1), would have been a new theorem) that the solution of the partial differential equation depends on the integration of a single system of differential equations. The remainder of the memoir is devoted to the discussion of this theory and of the integration of the Pfaffian system of ordinary differential

1 Jacobi refers to Lagrange's "Leçons sur la Théorie des Fonctions," and to a memoir by Pfaff in the Berlin Transactions for 1814, as containing, so far as he was aware, everything essential which was known in reference to the integration of partial differential equations of the first order; he refers also to his own memoir "Ueber die Pfaffsche Methode u. s. w." *Crelle*, t. II. pp. 347—358 (1827), as presenting the method in a more symmetrical and compendious form, but without adding to it anything essentially new.

equations, a system which is also treated of in Jacobi's memoir of 1844, "Theoria Novi Multiplicatoris &c." I take the opportunity of referring here to a short note by Brioschi, "Intorno ad una Proprietà delle Equazioni alle Derivate Parziali del Primo Ordine," *Tortolini*, t. VI. pp. 426—429 (1855), where the theory of the integration of a partial differential equation of the first order is presented under a singularly elegant form.

38. Jacobi's note of 1837, "On the Integration of the Differential Equations of Dynamics."—Jacobi remarks that it is possible to derive from Lagrange's form of the equations of motion an important profit for the integration of these equations, and he refers to his communication of the 29th of November 1839 to the Academy of Berlin, and to his former note to the Academy of Paris. He proceeds to say, that whenever the condition of *vis viva* holds good, he had found that it was possible in the integration of the equations of motion to follow a course such that each of the given integrals successively lowers by two unities the order of the system; and that the like theorem holds good when the condition of *vis viva* is not satisfied, that is, when the force function involves the time (this seems to be a restatement, in a more general form, of the theorems referred to in the note of the 29th of November 1836 to the Academy of Berlin); and he mentions that he had been, by his researches on the theory of numbers, led away from composing an extended memoir on the subject. The note then passes on to other subjects, and it concludes with two theorems, which are given without demonstration as extracts from the intended work he had before spoken of. These theorems are in effect as follows:

I. Let

$$m\frac{d^2x}{dt^2}=\frac{dU}{dx},\quad m\frac{d^2y}{dt^2}=\frac{dU}{dy},\quad m\frac{d^2z}{dt^2}=\frac{dU}{dz},\ \&\text{c.}$$

be the $3n$ differential equations of the motion of a free system, and

$$\tfrac{1}{2}\Sigma m\,(x'^2+y'^2+z'^2)\,dt=U+h,$$

the equation of *vis viva*.

Let V be a complete solution of the partial differential equation

$$\tfrac{1}{2}\Sigma\frac{1}{m}\left\{\left(\frac{dV}{dx}\right)^2+\left(\frac{dV}{dy}\right)^2+\left(\frac{dV}{dz}\right)^2\right\}=U+h,$$

that is, a solution containing, besides the constants attached to V by mere addition, $3n-1$ constants $\alpha\,(\alpha_1,\ \alpha_2,\dots\alpha_{3n-1})$, then first the integral equations are

$$\frac{dV}{d\alpha}=\beta,\dots\quad \frac{dV}{dh}=t+\tau;$$

where $\beta\,(\beta_1,\ \beta_2,\dots\beta_{3n-1})$ and τ are new arbitrary constants: this is in fact the theorem already quoted from Jacobi's memoir of 1837, and it is in the present place referred to as an easy generalisation of Sir W. R. Hamilton's formulæ. But Jacobi proceeds (and this is given as entirely new) that the disturbed equations being

$$m\frac{d^2x}{dt^2}=\frac{dU}{dx}+\frac{d\Omega}{dx},\quad m\frac{d^2y}{dt^2}=\frac{dU}{dy}+\frac{d\Omega}{dy},\quad m\frac{d^2z}{dt^2}=\frac{dU}{dz}+\frac{d\Omega}{dz},$$

then the equations for the variations of the above system of arbitrary constants are

$$\frac{d\alpha}{dt}=\frac{d\Omega}{d\beta},\ldots\quad \frac{dh}{dt}=\frac{d\Omega}{d\tau},$$

$$\frac{d\beta}{dt}=-\frac{d\Omega}{d\alpha},\ldots\quad \frac{d\tau}{dt}=-\frac{d\Omega}{dh};$$

so that the constants form (I think the term is here first introduced) a canonical system.

Jacobi observes that, in the theory of elliptic motion, certain elements which he mentions form a system of canonical elements, and he remarks that, since one complete solution of a partial differential equation gives all the others, the theorem leads to the solution of another interesting problem, viz. "Given one system of canonical elements, to find all the other systems." This is effected by means of the second theorem, which is as follows:

II. Given the systems of differential equations between the variables $a\,(a_1,\ a_2\ldots a_m)$ and $b\,(b_1,\ b_2\ldots b_m)$

$$\frac{da}{dt}=-\frac{dH}{db},\ldots\quad \frac{db}{dt}=\frac{dH}{da},\ldots$$

where H is *any function* of the variables $a,\ldots$ and $b,\ldots$; let $\alpha\,(\alpha_1,\ \alpha_2,\ldots \alpha_m)$ and $\beta\,(\beta_1,\ \beta_2,\ldots \beta_m)$ be two new systems of variables connected with the preceding ones by the equations

$$\frac{d\psi}{d\alpha}=\beta,\ldots\quad \frac{d\psi}{da}=-b,\ldots$$

where ψ is a function of $\alpha,\ldots b,\ldots$ without t or the other variables, then expressing H as a function of t and the new variables $\alpha,\ldots$ and $\beta,\ldots$, these last variables are connected together by equations of the like form with the original system, viz.:

$$\frac{d\alpha}{dt}=-\frac{dH}{d\beta},\ldots\quad \frac{d\beta}{dt}=-\frac{dH}{d\alpha},\ldots$$

Jacobi concludes with the remark, that other theorems no less general may be deduced by putting $\psi+\lambda\psi_1+\mu\psi_2+\ldots$ instead of ψ, and eliminating the multipliers $\lambda,\ \mu,\ldots$ by means of the equations $\psi_1=0,\ \psi_2=0,\ldots$, and that the demonstrations of the theorems are obtained without difficulty.

39. Jacobi's note of the 21st of November, 1838.—Jacobi refers to a memoir by Encke in the Berlin *Ephemeris* for 1837, "über die speciellen Störungen," where expressions are given for the partial differential coefficients of the values in the theory

of elliptic motion of the coordinates x, y, z and the velocities x', y', z' with respect to the elements; and he remarks, that if Encke's elements are replaced by a system of elements α, β, γ, α', β', γ' which he mentions, connected with those of Encke by equations of a simple form, then considering first x, y, z, x', y', z' as given functions of t and the elements, and afterwards the elements α, β, γ, α', β', γ' as given functions of t and x, y, z, x', y', z', there exists the remarkable theorem that the thirty-six partial differential coefficients $\frac{d\alpha}{dx}$, $\frac{d\beta}{dx}$, &c., and the thirty-six partial differential co-coefficients $\frac{dx}{d\alpha}$, $\frac{dx}{d\beta}$, &c., are equal to each other, or differ only in their sign, viz.

$$\frac{dx}{d\alpha}=-\frac{d\alpha'}{dx'},\quad \frac{dx}{d\alpha'}=\frac{d\alpha}{dx'},\quad \frac{dx'}{d\alpha}=\frac{d\alpha'}{dx},\quad \frac{\partial x'}{\partial \alpha'}=-\frac{d\alpha}{dx};$$

thirty-six equations in all, viz. the pair α, α' of corresponding elements may be replaced by the pair β, β' or γ, γ': and then in each of the twelve equations y, y' or z, z' may be written instead of x, x'. The like applies to a system of constants which are the initial values of any system whatever of coordinates $p,\ldots$, and the initial values of the differential coefficients $q'=\frac{dT}{dp}$, &c. of the force function T with respect to $p,\ldots$; *and for every system of elements which possess the property first mentioned, the formulæ for the variations assume the simplest possible form, inasmuch as the variations of each element is equal to a single partial differential coefficient of the disturbing function with the coefficient* $+1$ *or* -1, *as is known to be the case with the last-mentioned system of elements;* in other words, if $a,\ldots$ and $b,\ldots$ be a system of elements corresponding to each other in pairs, such that

$$\frac{dp}{da}=-\frac{db}{dq},\quad \frac{dp}{db}=\frac{da}{dq},\quad \frac{dq}{da}=\frac{db}{dp},\quad \frac{dq}{db}=-\frac{da}{dp}$$
$$\vdots$$

(where a, b may be replaced by any other corresponding pair of elements, and p, q by any other corresponding pair of variables), then the elements $a,\ldots$ and $b,\ldots$ form a canonical system.

40. Jacobi's note of 1840 in the *Comptes Rendus*, calls attention to the theorem contained in the passage quoted above from Poisson's memoir of 1808, a theorem which Jacobi characterizes as "la plus profonde découverte de M. Poisson," and as the theorem "le plus important de la Mécanique et de cette partie du calcul intégral qui s'attache à l'intégration d'un système d'équations différentielles ordinaires"; and he proceeds, "le théorème dont il est question énoncé convenablement est le suivant—un nombre quelconque de points matériels étant tirés par des forces et soumis à des conditions telles que le principe des forces vives ait lieu, si l'on connaît outre que l'intégrale fournie par ce principe deux autres intégrales, on en peut déduire une troisième d'une manière directe et sans même employer des quadratures. En poursuivant le même procédé on pourra trouver une quatrième, une cinquième intégrale, et en général on parviendra à cette manière à déduire des deux intégrales données toutes

les intégrales, ou ce qui revient au même l'intégration complète du problême. *Dans des cas particuliers on retombera sur une combinaison des intégrales déjà trouvées avant qu'on soit parvenu à toutes les intégrales du problême, mais alors les deux intégrales données jouissent des propriétés particulières dont on peut tirer un autre profit pour l'intégration des équations dynamiques proposées.* C'est ce qu'on verra dans un ouvrage auquel je travaille depuis plusieurs années et dont peut-être je pourrai bientôt faire commencer l'impression."

41. Liouville's addition to Jacobi's letter of 1840.—This contains the demonstration of a theorem similar to that given in Jacobi's letter of 1836, and Poisson's memoir of 1837, but somewhat more general; the system considered is a system of four differential equations of the first order:

$$\frac{dx}{dt} = \lambda \frac{dU}{dx'}, \quad \frac{dx'}{dt} = -\lambda \frac{dU}{dx}, \quad \frac{dy}{dt} = \lambda \frac{dU}{dy'}, \quad \frac{dy'}{dt} = -\lambda \frac{dU}{dy},$$

where U is a function of x, y, x', y', and λ is a function of x, y, x', y' and t. One integral is $U = a$, and if there be another integral $V = b$ where V is a function of x, y, x', y' only, then x', y' being by means of these two integrals expressed as a function of x, y, a, b, it is shown that $x'dx + y'dy$ is an exact differential, and putting $\int (x'dx + y'dy) = \theta$, then that $\frac{d\theta}{db} = \beta$ is a new integral of the given equations; and in the case where λ is a function of t only, the remaining integral is $\frac{d\theta}{da} = \int \lambda dt + \alpha$.

42. Binet's memoir of 1841 contains an exposition of the theory of the variation of the arbitrary constants as applied to the general system of equations

$$\frac{d}{dt}\frac{dF}{dx'} = \frac{dF}{dx'} + \lambda \frac{dL}{dx} + \mu \frac{dM}{dx} + \dots,$$
$$\vdots$$

where F is any function of t, and of the coordinates $x, y, z \dots$ of the different points of the system, and of their differential coefficients x', y', z', &c., and $L = 0$, $M = 0$, &c. are any equations of equation between the coordinates $x, y, z, \dots$ of the different points of the system; these equations may contain t, but they must not contain the differential coefficients $x', y', z', \dots$ The form is a more general one than that considered by Lagrange and Poisson. The memoir contains an elegant investigation of the variations of the elements of the orbit of a body acted upon by a central force, the expressions for the variations being obtained in a canonical form; and there is also a discussion of the problem suggested in Poisson's report of 1830 on the manuscript work of Ostrogradsky.

43. Jacobi's note of 1842, in the *Comptes Rendus*, announces the general principle (being a particular case of the theorem of the ultimate multiplier) stated and demonstrated in the memoir next referred to, and gives also the rule for the formation of the multiplier in the case to which the general principle applies.

44. Jacobi's memoir of 1842, "De Motu Puncti singularis": the author remarks, that the greater the difficulties in the general integration of the equations of dynamics, the greater the care which should be bestowed on the examination of the dynamical problems in which the integration can be reduced to quadratures; and the object of the memoir is stated to be the examination of the simplest case of all, viz. the problems relating to the motion of a single point. The first section, entitled, "De Extensione quadam Principii Virium vivarum," contains a remark which, though obvious enough, is of considerable importance: the forces X, Y, Z which act upon a particle, may be such that $Xdx + Ydy + Zdz$ is not an exact differential, so that if the particle were free, there would be no force function, and the equations of motion would not be expressible in the standard form. But if the point move on a surface or a curve, then in the former case $Xdx + Ydy + Zdz$ will be reducible to the form $Pdp + Qdq$, which will be an exact differential if a single condition (instead of the three conditions which are required in the case of a free particle) be satisfied, and in the latter case it will be reducible to the form Pdp, which is, *per se*, an exact differential. In the case of a surface, the requisite transformation is given by the Hamiltonian form of the equations of motion, which Jacobi demonstrates for the case in hand; and then in the third section, with a view to its application to the particular case, he enumerates the general proposition "quæ pro novo principio mechanico haberi potest," which is as follows:

"Consider the motion of a system of material points subjected to any conditions, and let the forces acting on the several points in the direction of the axes be functions of the coordinates alone: if the determination of the orbits of the several points is reduced to the integration of a single differential equation of the first order between two variables, for this equation there may be found, by a general rule, a multiplier which will render it integrable by quadratures only."

And for the particular case the theorem is thus stated:

"Given three differential equations of the first order between the four quantities q_1, q_2, p_1, p_2,

$$dq_1 : dq_2 : dp_1 : dp_2 = \frac{dT}{dp_1} : \frac{dT}{dp_2} : -\frac{dT}{dq_1} + Q_1 : -\frac{dT}{dq_2} + Q_2,$$

in which Q_1, Q_2 are functions of q_1, q_2 only; suppose that there are known two integrals α, β, and that by the aid of these p_1, p_2, $\frac{dT}{dp_1}$, $\frac{dT}{dp_2}$ are expressed by means of the quantities q_1, q_2 and the arbitrary constants α, β; there then remains to be integrated an equation of the first order, $\frac{dT}{dp_1}dq_2 - \frac{dT}{dp_2}dq_1 = 0$ between the quantities q_1, q_2, by which is determined the orbit of the point on the given surface: I say that the left-hand side of the equation multiplied by the factor

$$\frac{dp_1}{d\alpha}\frac{dp_2}{d\beta} - \frac{dp_2}{d\alpha}\frac{dp_1}{d\beta},$$

will be a complete differential, or will be integrable by quadratures alone," and the demonstration of the theorem is given. The remainder of the memoir, sections 4 to 7, is occupied by a very interesting discussion of various important special problems.

45. There is an important memoir by Jacobi, which, as it relates to a special problem, I will merely refer to, viz. the memoir "Sur l'Elimination des Nœuds dans le problême des trois Corps," *Crelle*, t. XXVII. pp. 115—131 (1843). The solution is made to depend upon six differential equations, all of them of the first order except one, which is of the second order, and upon a quadrature.

46. Jacobi's memoir of 1844, "Theoria Novi Multiplicatoris &c."—This is an elaborate memoir establishing the definition and developing the properties of the "multiplier" of a system of ordinary differential equations, or of a linear partial differential equation of the first order, with applications to various systems of differential equations, and in particular to the differential equations of dynamics. The definition of the multiplier is as follows, viz. the multiplier of the system of differential equations

$$dx : dy : dz : dw \ldots = X : Y : Z : W \ldots$$

or of the linear partial differential equation of the first order

$$X\frac{df}{dx} + Y\frac{df}{dy} Z\frac{df}{dz} + W\frac{df}{dw} + \ldots = 0$$

is a function M, such that

$$\frac{dMX}{dx} + \frac{dMY}{dy} + \frac{dMZ}{dz} + \frac{dMW}{dw} + \ldots = 0.$$

One of the properties of the multiplier is that contained in the theorem of the ultimate multiplier, viz. that when all the integrals (except one) of the system of partial differential equations are known, and the system is thereby reduced to a single differential equation between two variables, then the multiplier (in the ordinary sense of the word) of this last equation is $M\nabla$, where M is the multiplier of the system, and ∇ is a given derivative of the known integrals, so that the multiplier of the system being known, the integration of the last differential equation is reduced to a mere quadrature. To explain the theorem more particularly, suppose that the system of given integrals, that is, all the integrals (except one) of the system are represented by $p = \alpha$, $q = \beta$, ..., and let u, v be any two functions whatever of the variables, so that $p, q, \ldots u, v$ are in number equal to the system $x, y, z, w, \ldots$ then if

$$X\frac{du}{dx} + Y\frac{du}{dy} + Z\frac{du}{dz} + W\frac{du}{dw} + \ldots = U,$$

$$X\frac{dv}{dx} + Y\frac{dv}{dy} + Z\frac{dv}{dz} + W\frac{dv}{dw} + \ldots = V,$$

the last differential equation takes the form

$$Udv - Vdu = 0,$$

where it is assumed that U and V are, by the assistance of the given integrals, expressed as functions of u, v and the constants of integration. The multiplier of the last-mentioned equation is $M\nabla$, where M is the multiplier of the system, and ∇ may be expressed in either of the two forms

$$\nabla = \frac{\partial (x, y, z, w, \ldots)}{\partial (\alpha, \beta, \ldots u, v)}$$

and

$$\nabla = \left\{\frac{\partial (p, q, \ldots u, v)}{\partial (x, y, z, w, \ldots)}\right\}^{-1};$$

where the symbols on the right-hand sides represent functional determinants; in the first form it is assumed that $x, y, z, w, \ldots$ are expressed as functions of $\alpha, \beta, \ldots u, v$, and in the second form that $p, q, \ldots u, v$ are expressed as functions of $x, y, z, w, \ldots$, but that ultimately $p, q, \ldots$ are replaced by their values in terms of the constants and u, v; the first of the two forms, from its not involving this transformation backwards, appears the more convenient.

47. I have thought it worth while to quote the theorem in its general form, but we may take for u, v any two of the original variables, and if, to fix the ideas, it is assumed that there are in all the four variables x, y, z, w, then the theorem will be stated more simply as follows:—given the system of differential equations

$$dx : dy : dz : dw = X : Y : Z : W,$$

and suppose that two of the integrals are $p = \alpha$, $q = \beta$, the last equation to be integrated will be

$$Wdz - Zd W = 0,$$

where, by the assistance of the given integrals, W, Z are expressed as functions of z, w. And the multiplier of this equation is $M\nabla$, where M is the multiplier of the system, and ∇, attending only to the first of the two forms, is given by the equation

$$\nabla = \frac{\partial (x, y)}{\partial (\alpha, \beta)},$$

which supposes that x, y are expressed as functions of α, β, z, w.

48. Jacobi applies the theorem of the ultimate multiplier to the differential equations of dynamics, considered first in the unreduced Lagrangian form, where the coordinates are connected by any given system of equations of condition; secondly, in the reduced or ordinary Lagrangian form; and, thirdly, in the Hamiltonian form. The multiplier can be found for the first two forms, and the expressions obtained are simple and elegant; but, as regards the third form, there is a further simplification: the multiplier M of the system is equal to unity, and the multiplier of the last equation is therefore equal to ∇. The two cases are to be distinguished in which t does not, or does enter into the equations of motion; in the latter case the theorem furnished by the principle of the ultimate multiplier is the same as for the general

case of a system the multiplier of which is known, viz., the theorem is, given all the integrals except one, the remaining integral can be found by quadratures only. But in the former case, which is the ordinary one, including all the problems in which the condition of *vis viva* is satisfied, there is a further consequence deduced. In fact, the time t may be separated from the other variables, and the system of differential equations reduced to a system not involving the time, and containing a number of equations less by unity than the original system, and the theorem of the ultimate multiplier applies to this new system. But when the integrals of the new system have been obtained, the system may be completed by the addition of a single differential equation involving the time, and which is integrable by quadratures; the theorem consequently is, given all the integrals except two, these given integrals being independent of the time, the remaining integrals can be found by quadratures only. This is, in fact, the "Principium generale mechanicum" of the memoir of 1842.

The last of the published writings of Jacobi on the subject of dynamics is the "Auszug zweier Schreiben des Professors Jacobi an Herrn Director Hansen," *Crelle*, t. XLII. pp. 12—31 (1851): these relate chiefly to Hansen's theory of ideal coordinates.

49. The very interesting investigations contained in several memoirs by Liouville (*Liouville*, t. XI. XII. and XIV., and the additions to the "Connaissance des Temps" for 1849 and 1850) in relation to the cases in which the equations of motion of a particle or system of particles admit of integration, are based upon Jacobi's theory of the S function, that is, of the function which is the complete solution of a certain partial differential equation of the first order; the equation is given, in the first instance, in rectangular coordinates, and the author transforms it by means of elliptic coordinates or otherwise, and he then inquires in what cases, that is, for what forms of the force function, the equation is one which admits of solution. A more particular account of these memoirs does not come within the plan of the present report.

50. Desboves' memoir of 1848 contains a demonstration of the two theorems given in Jacobi's note of 1837, in the *Comptes Rendus;* and, as the title imports, there is an application of the theory to the problem of the planetary perturbations; the author refers to the above-mentioned memoirs of Liouville as containing a solution of the partial differential equation on which the problem depends, and also to a memoir of his own relating to the problem of two centres, where the solution is also given; and from this he deduces the solution just referred to, and which is employed in the present memoir. Jacobi's theorem gives at once the formulæ for the variation of the arbitrary constants contained in the solution. The material thing is to determine the signification of these constants, which can of course be done by a comparison of the formulæ with the known formulæ of elliptic motion; the author is thus led to a system of canonical elements similar to, but not identical with, those obtained by Jacobi.

51. Serret's two notes of 1848 in the *Comptes Rendus*.—These relate to the theory of Jacobi's S function, that is, of the function considered as the complete solution of a given partial differential equation of the first order. In the first of the two notes,

which relates to a single particle, the author gives a demonstration founded on a particular choice of variables, viz., those which determine orthotomic surfaces to the curve described by the moving point. The process seems somewhat artificial.

52. Sturm's note of 1848, in the *Comptes Rendus*, relates to the theory of Jacobi's S function, that is, of this function considered as the complete solution of a given partial differential equation of the first order. The force function is considered as involving the time t, which, however, is no more than had been previously done by Jacobi.

53. Ostrogradsky's note of 1848.—This contains an important step in the theory of the forms of the equations of motion, viz. it is shown how, in the case where the force function contains the time, the equations of motion may be transformed from the form of Lagrange to that of Sir W. R. Hamilton. If, as before, the force function (taken with the contrary sign to that of Lagrange) is represented by U, then putting, as before, $T+U=Z$ (the author writes V instead of Z), in the case under consideration Z will contain not only terms of the second order and terms of the order zero in the differential coefficients of the coordinates $q, \ldots$, but also terms of the first order, that is, Z will be of the form $Z=Z_2+Z_1+Z_0$, and putting $H=Z_2-Z_0$, this new function H being expressed as a function of the coordinates $q, \ldots$ and of the new variables $p, \ldots$, then the equations of motion take the Hamiltonian form, viz.—

$$\frac{dq}{dt}=\frac{dH}{dp}, \quad \frac{dp}{dt}=-\frac{dH}{dq};$$
$$\vdots$$

in the theory of the transformation, as originally given by Sir W. R. Hamilton, $Z_2=T$, $Z_1=0$, $Z_0=U$, and, consequently, $H=Z_2-Z_0=T-U$ as before.

[Ostrogradsky's memoir of 1850.—This among other important researches contains, and that *in the most general form*, the transformation of the equations of motion from the Lagrangian to the Hamiltonian form, and indeed the transformation of the general isoperimetric system (that is the system arising from any problem in the calculus of variations) to the Hamiltonian form.]

54. Brassinne's memoir of 1851.—The author reproduces for the Lagrangian equations of motion $\frac{d}{dt}\frac{dZ}{d\xi'}-\frac{dZ}{d\xi}=0$, &c. the demonstration of the theorem

$$\frac{d}{dt}\left(\delta\frac{dZ}{d\xi'}\Delta\xi-\Delta\frac{dZ}{d\xi'}\delta\xi+\ldots\right)=0;$$

and he shows that a similar theorem exists with regard to the system

$$-\frac{d^2}{dt^2}\frac{dZ}{d\xi''}+\frac{d}{dt}\frac{dz}{d\xi'}-\frac{dZ}{d\xi}=0;$$
$$\vdots$$

and with respect to the corresponding system of the mth order. The system in question, which is, in fact, the general form of the system of equations arising from a problem in the calculus of variations, had previously been treated of by Jacobi, but the theorem is probably new. In conclusion, the author shows in a very elegant manner the interdependence of the theorem relating to Lagrange's coefficients (a, b), and of the corresponding theorem for the coefficients of Poisson.

55. Bertrand's memoir of 1851, "On the Integrals common to several Mechanical Problems," is one of great importance, but it is not very easy to explain its relation to other investigations. The author remarks that, given the integral of a mechanical problem, it is in general a question admitting of determinate solution to find the expression for the forces; in other words, to determine the problem which has given rise to the integral; at least, this is the case when it is assumed that the forces are functions of the coordinates, without the time or the velocities; and he points out how the solution of the question is to be obtained. But, in certain cases, the method fails, that is, it leads to expressions which are not sufficient for the determination of the forces; these are the only cases in which the given integral can belong to several different problems; and the method shows the conditions necessary in order that these cases may present themselves. It is to be remarked that the given integral must be understood to be one of an absolutely definite form, such for instance as the equations of the conservation of the motion of the centre of gravity or of areas, but not such as the equation of *vis viva*, which is a property common indeed to a variety of mechanical problems, but which involves the forces, and is therefore not the *same* equation for different problems. The author studies in particular the case where the system consists of a single particle; he shows, that when the motion is in a plane, the integrals capable of belonging to two or more different problems are two in number, each of them involving as a particular case the equation of areas. When the point moves on a surface, he arrives at the remarkable theorem—"In order that the equations of motion of a point moving on a surface may have an integral independent of the time, and common to two or more problems, it is necessary that the surface should be a surface of revolution, or one which is developable upon a surface of revolution." When the condition is satisfied, he gives the form of the integral, and the general expression of the forces in the problems for which such integral exists. He examines, lastly, the general case of a point moving freely in space. The number of integrals common to several problems is here infinite. After giving a general form which comprehends them all, the author shows how to obtain as many particular forms as may be desired: it is, in fact, only necessary to resolve any problem relative to motion in a plane, and to effect a certain simple transformation on the integrals; one thus obtains a new equation which is the integral of an infinite number of different problems relating to the motion in space.

As an instance of the analytical forms on which these remarkable results depend, I quote the following, which is one of the most simple:—"If an integral of the equations of motion of a point in a plane belongs to two different problems, it is of the form

$$\alpha = F(\phi', x, y, t),$$

where ϕ' is the derivative with respect to t, of a function of x, y, which equated to zero gives the equation of a system of right lines."

56. Bertrand's memoir of 1851, "On the Integration of the Differential Equations of Dynamics."—The author refers to Jacobi's note of 1840, in relation to Poisson's theorem; and after remarking that there are very few problems of which two integrals are not known, and which therefore might not be solved by the method if it never failed; he observes that unfortunately there are (as was known) cases of exception, and that, as his memoir shows, these cases are far more numerous than those to which the method applies; thus for example the equation of *vis viva*, combined with any other integral whatever, leads to an illusory result. The theorem of Poisson may lead to an illusory result in two ways; either the resulting integral may be an identity $0 = 0$, or it may be an integral contained in the integrals already known, and which consequently does not help the solution. It appears by the memoir that the two cases are substantially the same, and that it is sufficient to study the case in which the two integrals lead to the identity $0 = 0$. Suppose that one integral is given, the author shows that there always exist integrals which, combined with the given integral, lead to an illusory result, and he shows how the integrals which, combined with the given integral, leads to such illusory result, are to be obtained.

For instance, in the case of a body moving round a fixed centre [and attracted to it by a force which is a function of the distance], there are here two known integrals; first, the equation of *vis viva* (but this, as already remarked, combined with any other integral whatever, leads to an illusory result); secondly, the equation of areas. The question arises, what are the integrals which, combined with the equation of areas, lead to an illusory result? The integrals in question are, in fact, the other two integrals of the problem; so that the inquiry into the integrals which give an illusory result, leads here to the completion of the solution.

The like happens in two other cases which are considered, viz. first, the problem of motion *in plano* when the force function is a homogeneous function of the coordinates of the degree -2; secondly, the problem of two fixed centres. Indeed the case is the same for all problems whatever, where the coordinates of the points of the system can be expressed by means of two independent variables.

The next problem considered is that where two bodies attract each other, and are attracted to a fixed centre. Suppose, first, the motion is *in plano*, then as in the former case *all* the integrals will be found by seeking for the integrals which, combined with the equation of areas, give an illusory result. When the motion is in space, the principle of areas furnishes three integrals (the equation of *vis viva* is contained in these three equations); the integrals which, combined with the integrals in question, give illusory results, are eight in number, and, to complete the solution, there must be added to these one other integral, which alone does not put the method in default. The problem of three bodies is then shown to be reducible to the last-mentioned problem; and the same consequences therefore hold good with respect to the problem of three bodies, viz., there are eight integrals which, combined with the integrals furnished by the principle of areas, give illusory results. To complete the solution it would be necessary to add to these a ninth integral, which alone would not put the method into default.

57. The author remarks that it appears by the preceding enumeration that the method of integration, based on the theorem of Poisson, is far from having all the

importance attributed to it by Jacobi. The cases of exception are numerous; they constitute, in certain cases, the complete solutions of the problems, and embrace in other cases eleven integrals out of twelve. But it would be a misapprehension of his meaning to suppose that, according to him, the cases in which Poisson's theorem is usefully applicable ought to be considered as exceptions. The expression would not be correct even for the problems which are completely resolved in seeking for the integrals which put the method into default; there exists for these problems, it is true, a system of integrals which give illusory results; but these integrals, combined in a suitable manner, might furnish others to which the theorem could be usefully applied.

The author remarks, that, in seeking the cases of exception to Poisson's theorem, there is obtained a new method of integration, which may lead to useful results; and, after referring to Jacobi's memoir on the elimination of the nodes in the problem of three bodies, he remarks that, by his own new method, the problem is reduced to the integration of six equations, all of them of the first order; so that he effectuates one more integration than had been done by Jacobi [this is incorrect]; and he refers to a future memoir, (not, I believe, yet published) [? the Memoir of 1857] for the further development of his solution.

58. To give an idea of the analytical investigations, the equations of motion are considered under the Hamiltonian form

$$\frac{dq}{dt}=\frac{dH}{dp}, \quad \frac{dp}{dt}=-\frac{dH}{dq},$$

where H is any function whatever of $q, \ldots p, \ldots$ without t, and then a given integral being

$$\alpha=\phi(q, \ldots p, \ldots),$$

the question is shown to resolve itself into the determination of an integral $\beta=\psi(q, \ldots p, \ldots)$ such that identically $(\alpha, \beta)=0$ or else $(\alpha, \beta)=1$, where (α, β) represents, as before, Poisson's symbol, viz.

$$(\alpha, \beta)=\frac{\partial(\alpha, \beta)}{\partial(q, p)}+\ldots,$$

if for shortness

$$\frac{\partial(\alpha, \beta)}{\partial(q, p)}=\frac{d\alpha}{dq}\frac{d\beta}{dp}-\frac{d\alpha}{dp}\frac{d\beta}{dq}.$$

The partial differential equations $(\alpha, \beta)=0$ or $(\alpha, \beta)=1$, satisfied by certain integrals β, are in certain cases, as Bertrand remarks, a precious method of integration leading to the classification of the integrals of a problem, so as to facilitate their ulterior determination: it is in fact by means of them that the several results before referred to are obtained in the memoir.

59. Bertrand's note of 1852 in the *Comptes Rendus.*—This contains the demonstration of a theorem analogous to Poisson's theorem $(\alpha, \beta)=\text{const.}$, but the function on the left-hand side is a function involving four of the arbitrary constants and binary combinations of pairs of corresponding variables, instead of two arbitrary constants and the series of pairs of corresponding variables.

60. Bertrand's notes, VI. and VII., to the third edition of the *Mécanique Analytique*, 1853, contain a concise and elegant exposition of various theorems which have been considered in the present report. The latter of the two notes relates to the above-mentioned theorem of Poisson, and places the theorem in a very clear light, in fact, establishing its connexion with the theory of canonical integrals. Bertrand in fact shows, that, given any integral α of the differential equations (in the last-mentioned form, the whole number of equations being $2k$), then the solution may be completed by joining to the integral α a system of integrals $\beta_1, \beta_2 \ldots \beta_{2k-1}$, which, combined with the integral α, give to Poisson's equation an identical form, viz. which are such that

$$(\alpha, \beta_1) = 1, (\alpha, \beta_2) = 0, \ldots (\alpha, \beta_{2k-1}) = 0.$$

This, he remarks, shows, that, given any integral α, the solution of the problem *may* be completed by integrals $\beta_1, \beta_2 \ldots \beta_{2k-1}$, which, combined with α, give all of them an identical form to the theorem of Poisson. But it is not to be supposed that all the integrals of the problem are in the same case. In fact, the most general integral is $\eta = \varpi(\alpha, \beta_1, \beta_2 \ldots \beta_{2k-1})$, and it is at once seen that $(\alpha, \eta) = (\alpha, \beta_1)\dfrac{d\eta}{d\beta_1} = \dfrac{d\eta}{d\beta_1}$, consequently the expression (α, η) will not be identically constant unless $\dfrac{d\eta}{d\beta_1}$ is so: but the integrals, in number infinite, which result from the combination of α, with $\beta_2, \beta_3 \ldots \beta_{2k-1}$ combined with the integral α, give identical results. Only the integrals which contain β_1 lead to results which are not identical. The integrals α and β_1, connected together in the above special manner, are termed by the author *conjugate integrals.*

61. Brioschi's two notes of 1853.—The memoir "Sulla Variazione &c." contains reflections and developments in relation to Bertrand's method of integration and to canonical systems of integrals, but I do not perceive that any new results are obtained.

The note, "Intorno ad un Teorema di Meccanica," contains a demonstration of the theorem in Bertrand's note of 1852 in the *Comptes Rendus*, and an extension of the theorem to the case of a combination of any even number of the arbitrary constants; the value of the symbol is shown by the theory of determinants to be a function of the Poissonian coefficients (α, β), and as these are constants, the value of the symbol considered is also constant.

62. Liouville's note of the 29th of June 1853 [1], contains the enunciation of a theorem which completes the investigations contained in Poisson's memoir of 1837. The equations considered are the Hamiltonian equations in their most general form, viz., H is any function whatever of t and the other variables: it is assumed that

[1] The date is that of the communication of the note to the Bureau of Longitudes, but the note is only published in Liouville's *Journal* in the May Number for 1855, which is subsequent to the date of the second part of Professor Donkin's memoir in the *Philosophical Transactions*, which contains the theorem in question. I have not had the opportunity of seeing a thesis by M. Adrien Lafon, Paris, 1854, where Liouville's theorem is quoted and demonstrated.

half of the integrals are known, and that the given integrals are such that for any two of them α, β, Poisson's coefficient (α, β) is equal to zero; this being so, the expression $pdq + \ldots - Hdt$, where, by means of the known integrals, the variables $p, \ldots$ are expressed in terms of $q, \ldots t$, is a complete differential in respect to $q, \ldots t$, viz. it will be the differential of Sir W. R. Hamilton's principal function V, which is thus determined by means of the known integrals, and the remaining integrals are then given at once by the general theory.

63. Professor Donkin's memoir of 1854 and 1855, Part I. (sections 1, 2, 3, articles 1 to 48).—The author refers to the researches of Lagrange, Poisson, Sir W. R. Hamilton, and Jacobi, and he remarks that his own investigations do not pretend to make any important step in advance. The investigations contained in section 1, articles 1 to 14, establish by an inverse process (that is, one setting out from the integral equations) the chief conclusions of the theories of Sir W. R. Hamilton and Jacobi, and in particular those relating to the canonical system of elements as given by Jacobi's theory. The theorem (3), article 1, which is a very general property of functional determinants, is referred to as probably new. The most important results of this portion of the memoir are recapitulated in section 4, in the form of seven theorems there given without demonstration; some of these will be presently again referred to. Articles 17 and 18 contain, I believe, the only demonstration which has been given of the equivalence of the generalised Lagrangian and Hamiltonian systems. The transformation is as follows: the generalised Lagrangian system is

$$\frac{d}{dt}\frac{dZ}{dq'} = \frac{dZ}{dq},$$
$$\vdots$$

where Z is any function of t and of $q, \ldots q', \ldots$; writing herein $\frac{dZ}{dq'} = p, \ldots$, and also $H = -Z + q'p + \ldots$, where, on the right-hand side, $q', \ldots$ are expressed in terms of $t, q, \ldots p, \ldots$, so that H is a function of $t, q, \ldots p, \ldots$; then the theorems in the preceding articles show that

$$\frac{dq}{dt} = \frac{dH}{dp}, \quad \frac{dp}{dt} = -\frac{dH}{dq},$$
$$\vdots \qquad \vdots$$

which is the generalised Hamiltonian system.

In section 2, articles 21 and 22, there is an elegant demonstration, by means of the Hamiltonian equations, of the theorem in relation to Poisson's coefficients (a, b), viz., that these coefficients are functions of the elements only. And there are contained various developments as to the consequences of this theorem; and as to systems of canonical, or, as the author calls them, *normal* elements. The latter part of the section and section 3, relate principally to the special problems of the motion of a body under the action of a central force, and of the motion of rotation of a solid body.

64. Part II. (sections 4, 5, 6 and 7, articles 49—93, appendices).—Section 4 contains the seven theorems before referred to. Although not given as new theorems, yet, to a considerable extent, and in form and point of view, they are new theorems.

Theorem 1 is a theorem standing apart from the others, and which is used in the demonstration of the transformation from the Lagrangian to the Hamiltonian system. It is as follows: viz., if X be a function of the n variables $x, \ldots$, and if $y, \ldots$ be n other variables connected with these by the n equations

$$\frac{dX}{dx} - y, \ldots$$

then will the values of $x, \ldots$, expressed by means of these equations in terms of $y, \ldots$, be of the form

$$x = \frac{dY}{dy}, \ldots$$

and if p be any other quantity explicitly contained in X, then also

$$\frac{dX}{dp} + \frac{dY}{dp} = 0,$$

the differentiation with respect to p being in each case performed only so far as p appears explicitly in the function.

The value of Y is given by the equation

$$Y = -X + xy + \ldots$$

where, on the right-hand side, $x, \ldots$ are expressed in terms of $y, \ldots$

Theorems 2, 3 and 4, and a supplemental theorem in article 50, relate to the deduction of the generalised Hamiltonian system of differential equations from the integral equations assumed to be known. In fact (writing $V,\ q, \ldots\ p, \ldots\ b, \ldots\ a, \ldots$, instead of the author's $X,\ x_1, \ldots x_n,\ y_1, \ldots y_n,\ a_1, \ldots a_n,\ b_1, \ldots b_n$), it is assumed that V is a given function of t, of the n variables $q, \ldots$, and of the n constants $b, \ldots$, and that the n variables $p, \ldots$, and the n constants $a, \ldots$, are determined by the conditions

$$\frac{dV}{dq} = p, \ldots \quad (1)$$

$$\frac{dV}{db} = a, \ldots \quad (2)$$

so that in fact by virtue of these $2n$ equations the $2n$ variables $X,\ q, \ldots\ p, \ldots$ may be considered as functions of t, and the $2n$ constants $b, \ldots\ a, \ldots$ (hypothesis 1), or conversely, the $2n$ constants $b, \ldots\ a, \ldots$, may be considered as functions of t and of the $2n$ variables $q, \ldots\ p, \ldots$ (hypothesis 2).

Theorem 2 is as follows: viz., if from the $2n$ equations (1, 2) and their total differential coefficients with respect to t, the $2n$ constants be eliminated, there will result the following $2n$ simultaneous differential equations of the first order, viz.:

$$\frac{dq}{dt} = \frac{dH}{dp}, \ldots$$

$$\frac{dp}{dt} = -\frac{dH}{dq}, \ldots$$

H is here a function of $q, \ldots\ p, \ldots$ (which will in general also contain t explicitly), given by the equation

$$H = -\frac{dV}{dt},$$

where, on the right-hand side, the differential coefficient $\frac{dV}{dt}$ is taken with respect to t, in so far as t appears explicitly in the original expression for V in terms of $q, \ldots\ b, \ldots$ and t, and after the differentiation, $b, \ldots$, are to be expressed in terms of the variables and t, by means of the equations (1).

Theorem 3 is, that there exists the following relations, viz.:

$$\frac{dq}{db} = -\frac{da}{bp}, \quad \frac{dq}{da} = \frac{db}{dp}, \ldots$$

$$\frac{dp}{db} = \frac{da}{dq}, \quad \frac{dp}{da} = -\frac{db}{dq}, \ldots$$

where (p, q) are *any* corresponding pair out of the systems $p, \ldots$ and $q, \ldots$, and (b, a) are any corresponding pair out of the systems $b, \ldots$ and $a, \ldots$, so that the total number of equations is $4n^2$: in each of the equations the left-hand side refers to hypothesis 1, and the right-hand side to hypothesis 2.

To these theorems should be added the supplemental theorem contained in article 50, viz., that there subsists also the system of equations

$$\frac{db}{dt} = \frac{dH}{da}, \ldots$$

$$\frac{da}{dt} = -\frac{dH}{db}, \ldots$$

where the left-hand sides refer to hypothesis 2, while the right-hand sides refer to hypothesis 1; as before $H = -\frac{dV}{dt}$, but here H is differentially expressed, being what the H of theorem 3 becomes when the variables are expressed according to hypothesis 1.

In theorem 4 the author's symbol (p, q) has a signification such as Poisson's (a, b), and if we write as before

$$(a, b) = \frac{\partial (a, b)}{\partial (p, q)} + \ldots,$$

where

$$\frac{\partial (a, b)}{\partial (p\ q)} = \frac{da}{dp}\frac{db}{dq} - \frac{db}{dp}\frac{da}{dq}$$

(this refers of course to hypothesis 2), the theorem is, that the following equations subsist identically, viz., b, a being corresponding constants out of the two series $b, \ldots$ and $a, \ldots$, then

$$(b, a) = -(a, b) = 1,$$

but that for any other pairs b, a, or for any pairs whatever b, b or a, a, the corresponding symbol is $= 0$: in fact, that the constants $b, \ldots$ and $a, \ldots$ form a canonical system of elements.

Theorem 5 is a theorem including theorem 4, and relating to any two functions u, v either of the two $2n$ constants or else of the $2n$ variables, and which may besides contain t explicitly; it establishes, in fact, a relation between Poisson's coefficient (u, v) and the corresponding coefficient of Lagrange.

Theorem 6 is as follows: viz., if $q, \ldots\ p, \ldots$ are any $2n$ variables concerning which no supposition is made, except that they are connected by the n equations

$$b = \phi (q, \ldots\ p, \ldots),$$

which equations are only subject to the condition of being sufficient for the determination of $p, \ldots$ in terms of $q, \ldots$ and $a, \ldots$, and they may contain explicitly any other quantities, for example, a variable t. Then, in order that the $\tfrac{1}{2}n(n-1)$ equations

$$\frac{dp_i}{dq_j} = \frac{dp_j}{dq_i}$$

may subsist identically, it is only necessary that each of the $\tfrac{1}{2}n(n-1)$ equations $(b_i, b_j) = O$ may be satisfied identically.

Theorem 7 is, in fact, the theorem previously established in its general form in Liouville's note of the 29th of June, 1853, viz., if, of the system of $2n$ differential equations

$$\frac{dq}{dt} = \frac{dH}{dp}, \quad \frac{dp}{dt} = -\frac{dH}{dq},$$

there be given n integrals involving the n arbitrary constants $b, \ldots$, so that each of these constants can be expressed as a function of the variables $q, \ldots, p, \ldots$ (with or without t); then, *if the* $\tfrac{1}{2}n(n-1)$ *conditions* $(b_i, b_j) = O$ *subsist identically, the remaining* n *integrals can be found as follows:*—By means of the n integrals, let the n variables

$p, \ldots$ be expressed in terms of $x, \ldots$ $b, \ldots$ and t, and let H stand for what H, as originally given, becomes when $q, \ldots$ are thus expressed. Then the values of $p, \ldots$ and $-H$ are the differential coefficients of one and the same function of $p, \ldots$ and t; call this function V, then, since its differential coefficients are all given (by the equations $\frac{dV}{dq} = p, \ldots$ $\frac{dV}{dt} = -H$), V may be found by integration; and it is therefore to be considered as a given function of $p, \ldots$ and t and of the constants $b, \ldots$. The remaining n integrals are given by the n equations

$$\frac{dV}{db} = a, \ldots$$

where the n quantities $a, \ldots$ are new arbitrary constants.

65. Section 5 of the memoir relates to the theory of the variation of the elements considered in relation to the following very general problem: viz., $Q, \ldots$ $P, \ldots$ being any functions whatever of the $2n$ variables $q, \ldots$ $p, \ldots$ and t; it is required to express the integrals of the system $2n$ differential equations

$$\begin{array}{ll} \frac{dq}{dt} = P, & \frac{dp}{dt} = Q, \\ \vdots & \vdots \end{array}$$

in the same form as the integrals (supposed given) of the standard system

$$\begin{array}{ll} \frac{dq}{dt} = \frac{dH}{dp}, & \frac{dp}{dt} = -\frac{dH}{dq}, \\ \vdots & \vdots \end{array}$$

by substituting functions of t for the constant elements of the latter system. And section 6 contains some very general researches on the general problem of the transformation of variables, a problem of which, as the author remarks, the method of the variation of elements is a particular, and not the only useful case. In particular, the author considers what he terms a normal transformation of variables, and he obtains the theorem 8, which includes as a particular case the second of the two theorems in Jacobi's note of 1837, in the *Comptes Rendus*. This theorem is as follows: viz., if the original variables $q, \ldots$ $p, \ldots$ are given by the $2n$ equations

$$\begin{array}{ll} \frac{dq}{dt} = \frac{dH}{dp}, & \frac{dp}{dt} = -\frac{dH}{dq}; \\ \vdots & \end{array}$$

and if the new variables $\eta, \ldots$ $\varpi, \ldots$ are connected with the original variables by the equations

$$\begin{array}{ll} \frac{dK}{dp} = q, & \frac{dK}{d\eta} = \varpi; \\ \vdots & \vdots \end{array}$$

where K is any function of $\eta, \ldots\ p, \ldots$ which may also contain t explicitly, then will the transformed equations be

$$\frac{d\eta}{dt} = \frac{d\Phi}{d\varpi}, \quad \frac{d\varpi}{dt} = -\frac{d\Phi}{d\eta};$$

in which Φ is defined by the equation

$$\Phi = H - \frac{dK}{dt},$$

and is to be expressed in terms of the new variables, the substitution of the new variables in $\frac{dK}{dt}$ being made after the differentiation. In particular, if K does not contain t explicitly, then $\frac{dK}{dt} = 0$ and $\Phi = H$, so that, in this case, the transformation is effected merely by expressing H in terms of the new variables. There is also an important theorem relating to the *transformation of coordinates.* To explain this, it is necessary to go back to the generalised Lagrangian form

$$\frac{d}{dt}\frac{dZ}{dq'} = \frac{dZ}{dq};$$
$$\vdots$$

where the variables $q, \ldots$ correspond to the coordinates of a dynamical problem; if the new variables $\eta, \ldots$ are any given functions whatever of the original variables $q, \ldots$ and of t, this is what may be termed a transformation of coordinates. But the proposed system can be expressed, as shown in the former part of the memoir, in the generalised Hamiltonian form with the variables $q, \ldots$ and the derived variables $p, \ldots$ (the values of which are given by $\frac{dZ}{dq} = p, \ldots$): the problem is to transform the last-mentioned system by introducing, instead of the original coordinates $q, \ldots$, the new coordinates $\eta, \ldots$, and instead of the derived variables $p, \ldots$ the new derived variables $\varpi, \ldots$ defined by the analogous equations $\frac{dZ}{d\eta} = \varpi, \ldots$, in which Z is supposed to be expressed as a function of $\eta, \ldots$ and t. The method of transformation is given by the theorem 9, which states that the transformation is a normal transformation, and that the modulus of transformation (that is, the function corresponding to K in theorem 8) is

$$K = qp + \ldots$$

where $q, \ldots$ are to be expressed in terms of $\eta, \ldots$. The latter part of the same section contains researches relating to the case where the proposed equations are symbolically, but not actually, in the Hamiltonian form, viz., where the function H is considered as containing functions of $q, \ldots\ p, \ldots$ which are exempt from differentiation in forming the differential equations (the author calls this a pseudo-canonical system), and where, in like manner, the transformation of variables is a pseudo-normal transformation; the theorems 10 and 11 relate to this question, which is treated still more generally in Appendix C. The general methods are illustrated by applications to the problem of three bodies and the problem of rotation; the former problem is specially discussed in section 7; but the results obtained (and which, as the author remarks, affords an example of the so-called "elimination of the nodes") do not come within the plan of the present report.

66. Bour's memoir of 1855, "On the Integration of the Differential Equations of Analytical Mechanics."—It has been already seen that the knowledge of half the entire system of the integrals of the differential equations (these known integrals satisfying certain conditions) leads by quadratures only to the knowledge of the remaining integrals; the researches contained in this most interesting and valuable memoir show that this theorem is, in fact, only the last of a series of theorems, here first established, relating to the successive reduction which results from the knowledge of each new integral. Speaking in general terms, it may be stated that the author operates on the linear partial differential equation of the first order, which is satisfied by the integrals of the differential equations; and that he effectuates upon this equation a reduction of two unities in the number of variables for every suitable new integral which is obtained[1]. The author shows also that an equal or greater reduction may sometimes be obtained by means of integrals which appear at first foreign to his method. Before going further, it may be convenient to remark that the author restricts himself to the case in which H is independent of the time, and where, consequently, the condition of *vis viva* is satisfied; it was, however, remarked by Liouville that the analysis, slightly modified, applies to the most general case where H is any function of t and the variables, and it is possible that when the entire memoir is published (it is given in Liouville's *Journal* as an extract), the theory will be exhibited under this more general form.

67. To give an idea of the analytical results, the equations are considered under the form

$$\frac{dp_i}{dt}=\frac{dH}{dq_i},\quad \frac{dq_i}{dt}=-\frac{dH}{dq_i}\ (i=1 \text{ to } i=n)$$

(where, as already remarked, H is independent of t). The integrals admit, therefore, of representation in the canonical form $\alpha, \beta, \alpha_1, \alpha_2, \ldots \alpha_{2n-2}$ where $\alpha\,(=H)$ is the equation of *vis viva*; $\beta\,(=G-t)$ is the integral conjugate to this, and the only integral involving the time; and the remaining integrals α_1 and α_2, α_3 and $\alpha_4 \ldots \alpha_{2n-3}$ and α_{2n-2} are conjugate pairs: we have

$$(\alpha_1, \alpha)\ (=(\alpha_1, H))=0,\quad (\alpha_1, \beta)\ (=(\alpha_1, G))=0,\quad (\alpha_1, \alpha_2)=1,\quad (\alpha_1, \alpha_3)=0, \ldots (\alpha_1, \alpha_{2n-2})=0.$$

The integrals $\alpha_1, \alpha_2, \ldots \alpha_{2n-2}$ verify the linear partial differential equation

$$\sum_{i=1}^{i=n}\left(\frac{dH}{dq_i}\frac{d\zeta}{dp_i}-\frac{dH}{dp_i}\frac{d\zeta}{dq_i}\right)=0 \text{ or, } (H, \zeta)=0 \qquad (1)$$

[1] I have borrowed this and the next sentence from Liouville's report. It would, I think, be more accurate to say, for every suitable new integral after the first one; in the case considered in the memoir, the condition of *vis viva* is satisfied, and there is always one integral, the equation of *vis viva*, which is known; but this alone, and in the general case the first known integral, will not cause a reduction of two unities.

which is also satisfied by $\zeta = H$, and of which the general solution is $\zeta = \phi(H, \alpha_1, \alpha_2 \dots \alpha_{2n-2})$, while, on the contrary, the first member of the equation (1), becomes unity for $\zeta = G$, in other words $(H, G) = 1$. The equation (1) replaces the original differential equations; it is to the equation (1) that the theorems of Poisson and Bertrand may be supposed to be applied, and it is this equation (1) which is studied in the memoir, where it it shown how the order may be diminished when one or more integrals are known.

In the first place, the integral $\alpha = H$ which is known, may be made use of to eliminate one of the variables, suppose p_n; the result is found to be

$$\sum_{i=1}^{l=n-1}\left(\frac{dp_n}{dq_i}\frac{d\zeta}{dp_i} - \frac{dp_n}{dp_i}\frac{d\zeta}{dq_i}\right) + \frac{d\zeta}{dq_n} = 0, \qquad (2)$$

which has the same integrals as the equation (1), except the integral of *vis viva* $\zeta = H$; it is this equation (2) which would have to be integrated if only the integral α were known.

Suppose now there is known a new integral α_1; this gives rise to the partial differential equation

$$\sum_{i=1}^{l=n}\left(\frac{d\alpha_1}{dq_i}\frac{d\zeta}{dp} - \frac{d\alpha_1}{dp_i}\frac{d\zeta}{dq_i}\right) = 0 \text{ or } (\alpha_1, \zeta) = 0, \qquad (4)$$

which is satisfied by $\zeta = H, G, \alpha_1, \alpha_3, \alpha_4 \dots \alpha_{2n-2}$, but not by α_2, which gives $(\alpha_1, \alpha_2) = 0$. The equation (4) is satisfied by $\zeta = H$, and it may be therefore transformed in the same manner as the equation (1) was, viz. p_n may be expressed in terms of the other variables and of α. The author remarks that it will happen, what causes the success of the method, that this operation (the object of which is to get rid of the solution $\zeta = H$) conducts to two different equations, according as $\zeta = G$ or $\zeta =$ any other integral of the equation (4); so that in the second form of the transformed equation the unknown integral $\zeta = G$ is also eliminated. This second form is found to be

$$\sum_{i=1}^{l=n-1}\left(\frac{d\alpha_1}{dq_i}\frac{d\zeta}{dp_i} - \frac{d\alpha_1}{dp_i}\frac{d\zeta}{dq_i}\right) = 0 \text{ or } (\alpha_1, \zeta) = 0, \qquad (5)$$

which is precisely similar to the equation (1) (only the number of variables is diminished by two unities), and is possessed of the same properties. Its integrals are $\alpha_1, \alpha_3, \alpha_4 \dots \alpha_{2n-2}$, which are all of them integrals of the problem, and give $(\alpha_1, \alpha_i) = 0$. And the theorems of Poisson and Bertrand apply equally to this equation; the only difference is, that the number of terms in the expressions (α, β) is less by two unities. A new integral (α_3) leads in like manner to an equation (8) similar to (5), but with the number of variables further diminished by two unities, and so on, until the half series of integrals $\alpha, \alpha_1, \alpha_3 \dots \alpha_{2n-3}$ are known; the conjugate integrals $\beta, \alpha_2, \alpha_4 \dots \alpha_{2n-2}$ are then obtained by quadratures only, in the method explained in the memoir, and which is in fact identical with that given by the theorem of Liouville and Donkin. The memoir contains other results, which have been already alluded to in a general manner; some of these are made use of by the author in his "Mémoire sur le problème des trois corps," *Journal École Polyt.*, t. XXI. pp. 35—58 (1856).

68. Liouville's note of July, 1855, on the occasion of Bour's memoir, mentions that the author of the memoir had recognized that, according to the remark made to him, his formulæ subsist with even increased elegance when H is considered as a function of t and the other variables. But (it is remarked) the general case can be always reduced to the particular one considered in the memoir, provided that the number of equations is augmented by two unities by the introduction of the new variables τ and u, the former of them, τ, equal to $t+$ constant, so that

$$\frac{dt}{d\tau}=1$$

the latter of them, u, defined by the equation

$$\frac{du}{d\tau}=-\frac{dH}{dt}.$$

Suppose in fact that

$$V=H+u,$$

then, since τ and u do not enter into H, which is a function only of t and the variables $q, \ldots p, \ldots$, we have

$$\frac{dV}{du}=1=\frac{dt}{d\tau};$$

and, moreover, the differential coefficients with respect to $t, q, \ldots p, \ldots$ of the functions H and V are equal. The system may be written

$$\frac{dt}{d\tau}=\frac{dV}{du}, \quad \frac{du}{d\tau}=-\frac{dV}{dt},$$

$$\frac{dp}{d\tau}=\frac{dV}{dq}, \quad \frac{dq}{d\tau}=-\frac{dV}{dp},$$

$$\vdots \qquad \vdots$$

which is a system containing two more variables, but in which V is independent of the variable τ, which stands in the place of t. The transformation is an elegant and valuable one, but it is not in anywise to be inferred that there is any advantage in considering the particular case (which is thus shown to be capable of including the general one), rather than the general one itself: such inference does not seem to be intended, and would, I think, be a wrong one.

69. Brioschi's note of 1855 contains an elegant demonstration (founded on the theory of skew determinants) of a property, which appears to be a new one, of the canonical integrals of a dynamical problem, viz. if q, p stand for a corresponding pair of the variables $q, \ldots p, \ldots$ then

$$\Sigma\frac{\partial(\alpha,\ \beta)}{\partial(q,\ p)}=1$$

where the summation refers to all the different pairs of conjugate integrals α, β of the canonical system, the pair q, p in the denominator being the same in each term;

but if the variables in the denominator are a non-corresponding pair out of the two series $q, \ldots$ and $p, \ldots$, or else a pair out of one series only (that is, both q's or both p's), then the expression on the left-hand side is equal to zero. This is in fact a sort of reciprocal theorem to the theorem which defines the canonical system of integrals. There are two or three memoirs of Brioschi in Crelle's *Journal* connected with this note and the note of 1853; but as they relate professedly to skew determinants and not to the equations of dynamics, it is not necessary here to refer to them more particularly.

70. Bertrand's memoir of 1857 forms a sequel to the memoir of 1851, on the integrals common to several problems of mechanics. The author calls to mind that he has shown in the first memoir, that, given an integral of a mechanical problem, and assuming only that the forces are functions of the coordinates, it is possible to determine the problem and find the forces which act upon each point; and (he proceeds) it is important to remark, that the solution leads often to contradictory results,—that, in fact, an equation assumed at hazard is not in general an integral of any problem whatever of the class under consideration: and he thereupon proposes to himself in the present memoir to develope some of the consequences of this remark, and to seek among the most simple forms, the equations which can present themselves as integrals, and the problems to which such integrals belong. The various special results obtained in the memoir are interesting and valuable.

71. In what precedes I have traced as well as I have been able the series of investigations of geometers in relation to the subject of analytical dynamics. The various theorems obtained have been in general stated with sufficient fulness to render them intelligible to mathematicians; the attempt to state them in a uniform notation and systematic order would be out of the province of the present report. The leading steps are,—first, the establishment of the Lagrangian form of the equations of motion; secondly, Lagrange's theory of the variation of the arbitrary constants, a theory perfectly complete in itself; and it would not have been easy to see *à priori* that it would be less fruitful in results than the theory of Poisson; thirdly, Poisson's theory of the variation of the arbitrary constants, and the method of integration thereby afforded; fourthly, Sir W. R. Hamilton's representation of the integral equations by means of a single characteristic function determinable *à posteriori* by means of the integral equations assumed to be known, or by the condition of its simultaneous satisfaction of two partial differential equations; fifthly, Sir W. R. Hamilton's form of the equations of motion; sixthly, Jacobi's reduction of the integration of the differential equations to the problem of finding a complete integral of a single partial differential equation, and the general theory of the connexion of the integration of a system of ordinary differential equations, and of a partial differential equation of the first order, a theory, however, of which Jacobi can only be considered as the second founder; seventhly, the notion (arising from the researches of Lagrange and Poisson) and ulterior development of the theory of a system of canonical integrals.

I remark in conclusion, that the differential equations of dynamics (including in the expression, as I have done throughout the report, the generalized Lagrangian and

Hamiltonian forms) are only one of the classes of differential equations which have occupied the attention of geometers. The greater part of what has been done with respect to the general theory of a system of differential equations is due to Jacobi, and he has also considered in particular, besides the differential equations of dynamics, the Pfaffian system of differential equations (including therein the system of differential equations which arise from any partial differential equation of the first order), and the so-called isoperimetric system of differential equations, that is, the system arising from any problem in the calculus of variations. In a systematic treatise it would be proper to commence with the general theory of a system of differential equations, and as a branch of this general theory, to consider the generalized Hamiltonian system, and in relation thereto to develope the various theorems which have a dynamical application. It would be shown that the generalized Lagrangian form could be transformed into the Hamiltonian form, but the first-mentioned form would, I think, properly be treated as a particular case of the isoperimetric system of differential equations.

List of Memoirs and Works above referred to.

Lagrange. Mécanique Analytique. 1st edition. 1788.

Laplace. Mécanique Céleste, t. I. 1799.

Poisson. Sur les inégalités séculaires des moyens mouvemens des planètes. Read to the Institute 20th June, 1808.—Jour. École Polyt., t. VIII. pp. 1—56. 1808.

Laplace. Mémoire.... Read to the Bureau of Longitudes 17th Aug., 1808.—Forms an Appendix to the 3rd volume of the Mécanique Céleste. 1808.

Lagrange. Mémoire sur la théorie des variations des éléments des planètes et en particulier des variations des grands axes de leurs orbites. Read to the Bureau of Longitudes the 17th Aug. and to the Institute the 22nd Aug., 1808.—Mém. de l'Instit., 1808, pp. 1—72. 1808.

Lagrange. Mémoire sur la théorie générale de la variation des constantes arbitraires dans tous les problêmes de la Mécanique. Read to the Institute 13th March, 1809.—Mém. de l'Instit., 1808, pp. 257—302 (includes an undated addition), and there is a Supplement also without date, pp. 363, 364. 1809.

Poisson. Mémoire sur la variation des constantes arbitraires dans les questions de Mécanique. Read to the Institute 16th Oct., 1809.—Jour. École Polyt., t. VIII. pp. 266—344. 1809.

Lagrange. Seconde mémoire sur la variation des constantes arbitraires dans les problêmes de Mécanique, dans lequel on simplifie l'application des formules générales à ces problêmes. Read to the Institute 19th Feb., 1810.—Mém. de l'Instit., 1809, pp. 343—352. 1810.

Lagrange. Mécanique Analytique. 2nd edition, t. I., 1811; t. II., 1813. 1811.

Poisson. Mémoire sur la variation des constantes arbitraires dans les questions de Mécanique. Read to the Academy, 2nd Sept., 1816.—Bulletins de la Soc. Philom., 1816, p. 109; and also, Mém. de l'Instit., t. I. pp. 1—70. 1816.

Hamilton, Sir W. R. On a general method in Dynamics, by which the study of all free systems of attracting or repelling points is reduced to the search and differentiation of one central relation or characteristic function.—Phil. Trans. for 1834, pp. 247—308. 1834.

Hamilton, Sir W. R. Second Essay on a general method in Dynamics.—Phil. Trans. for 1835, pp. 95—144. 1835.

Poisson. Remarques sur l'intégration des équations différentielles de la dynamique.—Liouville, t. II. pp. 317—337. 1837.

Jacobi. Lettre à l'Académie.—Comptes Rendus, t. III. pp. 59—61. 1836.

Jacobi. Zur Theorie der Variations-rechnung und der Differential-gleichungen. Extract from a letter of the 29th Nov. 1836, to Professor Encke, secretary of the mathematical class of the Academy of Berlin.—Crelle, t. XVII. pp. 61—82. 1836.

Jacobi. Ueber die Reduction der Integration der partiellen Differential-gleichungen erster Ordnung zwischen irgend einer Zahl Variabeln auf die Integration eines einzigen Systemes gewöhnlicher Differential-gleichungen.—Crelle, t. XXVII. pp. 97—162, and translated into French, Liouville, t. III. pp. 60—96, and 161—201. 1837.

Jacobi. Note sur l'intégration des équations différentielles de la dynamique.—Comptes Rendus, t. V. p. 61. 1837.

Jacobi. Neues Theorem der analytischen Mechanik.—Monatsbericht of the Academy of Berlin for 1838 (paper is dated 21st Nov., 1838); and Crelle, t. XXX. pp. 117—120. 1838.

Jacobi. Lettre adressée à M. le Président de l'Académie des Sciences à Paris.—Comptes Rendus, t. XI. p. 529; Liouville, t. V. pp. 350—351 (with an addition by Liouville, pp. 351—355). 1840.

Binet. Mémoire sur la variation des constantes arbitraires dans les formules générales de la dynamique et dans un système d'équations analogues plus étendues.—Jour. École Polyt., t. XVII. pp. 1—94. 1841.

Jacobi. Sur un nouveau principe général de la Mécanique Analytique.—Comptes Rendus, t. XV. pp. 202—205. 1842.

Jacobi. De motu puncti singularis.—Crelle, t. XXIV. pp. 5—27. 1842.

Maurice. Mémoire sur la variation des constantes arbitraires comme l'ont établie dans sa généralité les mémoires de Lagrange et celui de Poisson. Read to the Academy the 3rd June, 1844.—Mém. de l'Institut, t. XIX. pp. 553—638. 1844.

Jacobi. Theoria novi multiplicatoris systemati æquationum differentialium vulgarium applicandi.—Crelle, t. XXVII. pp. 199—268; and t. XXIX. pp. 213—279, and 333—376. 1844.

Desboves. Démonstration de deux théorèmes de M. Jacobi, application au problême des perturbations planétaires. Thesis presented to the Faculty of Sciences the 3rd April, 1848.—Liouville, t. XIII. pp. 397—411. 1848.

Serret. Sur l'intégration des équations différentielles du mouvement d'un point matériel.—Comptes Rendus, t. XXVI. pp. 605—610. 1848.

Serret. Sur l'intégration des équations différentielles de la dynamique.—Comptes Rendus, t. XXVI. pp. 639—643. 1848.

Sturm. Note sur l'intégration des équations générales de la dynamique.—Comptes Rendus, t. XXVI. pp. 658—673. 1848.

Ostrogradsky. Sur les intégrales des équations générales de la dynamique.—Mélanges de l'Acad. de St Pétersbourg, $\frac{6}{18}$th Oct., 1848.

[**Ostrogradsky.** Mémoire sur les équations différentielles relatives au problême des Isopérimètres.—Mém. de l'Acad. de St Pétersb., t. VI. pp. 385—517. 1850.]

Brassinne. Théorème relatif à une classe d'équations différentielles simultanées analogue à un théorème employé par Lagrange dans la théorie des perturbations.—Liouville, t. XVI. pp. 283—288. 1851.

Bertrand. Mémoire sur les intégrales communes à plusieurs problêmes de Mécanique. Presented to the Academy the 12th May, 1851.—Liouville, t. XVII. pp. 121—174. 1851.

Bertrand. Mémoire sur l'intégration des équations différentielles de la dynamique.—Liouville, t. XVII. pp. 393—436. 1852.

Bertrand. Sur un nouveau théorème de Mécanique Analytique.—Comptes Rendus, t. XXXV. pp. 698—699. 1852.

Bertrand's notes VI. and VII. to the third edition of the Mécanique Analytique, t. I. pp. 409—428, viz. note VI.—Sur les équations différentielles des problêmes de la Mécanique et la forme que l'on peut donner à leurs intégrales; and note VII.—Sur un théorème de Poisson. 1853.

Brioschi. Sulla variazione delle costanti arbitrarie nei problemi della Dinamica.—Tortolini, Annali, t. IV. pp. 298—311. 1853.

Brioschi. Intorno ad un teorema di Meccanica.—Tortolini, Annali, t. IV. pp. 395—400. 1853.

Liouville. Note sur l'intégration des équations différentielles de la dynamique, présentée au Bureau des Longitudes le 29 Juin, 1853.—Liouville, t. XX. pp. 137—138. 1853.

Donkin, Prof. On a Class of Differential Equations, including those which occur in Dynamical Problems. Part I.—Phil. Trans. 1854, pp. 71—113. Received Feb. 23rd, read Feb. 23rd, 1854. 1854.

Donkin, Prof. On a Class of Differential Equations, including those which occur in Dynamical Problems. Part II.—Phil. Trans. 1855, pp. 299—358. Received Feb. 17th, read March 22nd, 1855. 1855.

Liouville. Rapport sur un mémoire de M. Bour concernant l'intégration des équations différentielles de la Mécanique Analytique.—Comptes Rendus, t. XL. p. 661, séance du 26 Mars, 1855; Liouville, t. XX. pp. 135--136. 1855.

Bour. Sur l'intégration des équations différentielles de la Mécanique Analytique (extrait d'un mémoire présenté à l'Académie le 5 Mars, 1855).—Liouville, t. XX. pp. 185—200. 1855. [And Mém. Savants Étrang. t. XIV. pp. 35—58, 1856.]

Liouville. Note à l'occasion du mémoire précédent de M. Edmond Bour.—Liouville, t. XX. pp. 201—202. 1855.

Brioschi. Sopra una nuova proprietà degli integrali di un problema di dinamica.—Tortolini, t. VI. pp. 430—432. 1855.

Bertrand. Mémoire sur quelqu'une des formes les plus simples que puissent prendre les intégrales des équations différentielles du mouvement d'un point matériel.—Liouville, t. II. (2e série), pp. 113—140. 1857.

196.

NOTE SUR UN PROBLÈME D'ANALYSE INDÉTERMINÉE.

[From the *Nouvelles Annales de Mathématiques* (Terquem and Gerono), t. XVI. (1857), pp. 161—165.]

EULER a donné dans le Mémoire: *Regula facilis problemata Diophantea per numeros integros expedite solvendi* (*Comment. Arith. Coll.*, t. II., p. 263) la solution que voici de l'équation indéterminée

$$\alpha x^2 + \beta x + \gamma = \zeta y^2 + \eta y + \theta; \tag{1}$$

en supposant que l'on ait la solution $x = a$, $y = b$ de manière que

$$\alpha a^2 + \beta a + \gamma = \zeta b^2 + \eta b + \theta,$$

et en posant

$$s = \sqrt{\alpha r^2 + 1},$$

où r est une quantité quelconque, l'équation sera satisfaite par les valeurs

$$x = sa + \zeta rb + \frac{(s-1)\beta}{2\alpha} + \tfrac{1}{2}r\eta,$$

$$y = \alpha ra + sb + \frac{(s-1)\eta}{2\zeta} + \tfrac{1}{2}r\beta;$$

en effet, on voit sans peine que ces valeurs donnent identiquement

$$\alpha x^2 + \beta x + \gamma - (\zeta y^2 + \eta y + \theta) = \alpha a^2 + \beta a + \gamma - (\zeta b^2 + \eta b + \theta) = 0. \tag{2}$$

En supposant de plus que les coefficients α, β, γ, ζ, η, θ soient des nombres entiers tels que $\alpha\zeta$ soit un entier positif non carré, on peut toujours déterminer le nombre entier r de manière que s soit un nombre entier; cela étant, et en supposant que a, b soient des entiers, il est évident que x, y seront des nombres rationnels. Euler a de plus remarqué que l'on peut toujours faire en sorte que x, y soient des nombres entiers. En effet, si les formules donnent $x = a'$, $y = b'$ des valeurs non entières, en substituant dans les formules au lieu de a, b les valeurs a', b', on obtiendra pour x, y des valeurs entières; cela se vérifie sans peine.

L'équation indéterminée (2) rentre dans celle-ci

$$(a,\ b,\ c,\ f,\ g,\ h\big)\!\big(x',\ y',\ z')^2=(a,\ b,\ c,\ f,\ g,\ h\big)\!\big(x,\ y,\ z)^2\,; \tag{3}$$

en supposant que la forme ternaire

$$(a,\ b,\ c,\ f,\ g,\ h\big)\!\big(x,\ y,\ z)^2$$

se transforme en elle-même au moyen d'une substitution linéaire quelconque, on peut supposer que cette substitution soit telle que l'on ait $z'=z$; cela étant, en écrivant $z'=z=1$ et en mettant de plus $h=0$, l'équation (3) se réduit évidemment à une forme telle que l'équation (2). Or on peut trouver par la méthode générale de M. Hermite la solution convenable de l'équation (3). En supposant, comme à l'ordinaire,

$$\mathfrak{A}=bc-f^2,\ldots \qquad \mathfrak{F}=gh-af,\ldots$$

$$K=abc-af^2-bg^2-ch^2+2fgh,$$

il faut pour cela écrire

$$x'=2\xi-x,\ \ y'=2\eta-y,\ \ z'=2\zeta-z,$$

et

$$\begin{aligned}
ax+hy+gz&=a\xi+h\eta+g\zeta &&-q\mathfrak{C}\eta+q\mathfrak{F}\zeta,\\
hx+bx+fz&=h\xi+b\eta+f\zeta+q\mathfrak{C}\xi &&-q\mathfrak{G}\zeta,\\
gx+fy+cz&=g\xi+f\eta+c\zeta-q\mathfrak{F}\xi+q\mathfrak{G}\eta &&\quad ,
\end{aligned}$$

où q est une quantité arbitraire. En effet, en multipliant ces équations par ξ, η, ζ et en ajoutant, on obtient

$$(a,\ b,\ c,\ f,\ g,\ h\big)\!\big(\xi,\ \eta,\ \zeta\big)\!\big(x,\ y,\ z)=(a,\ b,\ c,\ f,\ g,\ h\big)\!\big(\xi,\ \eta,\ \zeta)^2$$

et, au moyen de cette équation et des valeurs

$$x'=2\xi-x,\ y'=2\eta-y,\ z'=2\zeta-z,$$

on forme tout de suite l'équation (3). De plus, en multipliant les trois équations par $\mathfrak{C}$, $\mathfrak{F}$, $\mathfrak{G}$ et en ajoutant, on obtient $Kz=K\zeta$, c'est-à-dire $z=\zeta$ et de là $z'=z$.

Cela étant, les deux équations donnent, en remplaçant ζ par z,

$$\begin{aligned}
a\xi+(h-q\mathfrak{C})\,\eta&=ax+hy+(g-q\mathfrak{F})\,z,\\
(h+q\mathfrak{C})\,\xi\qquad\quad+b\eta&=hx+by+(f+q\mathfrak{G})\,z,
\end{aligned}$$

et de là, en remarquant que

$$ab-(h-q\mathfrak{C})\,(h+q\mathfrak{C})=\mathfrak{C}+q^2\mathfrak{C}^2=\mathfrak{C}\,(1+q^2\mathfrak{C}),$$

on obtient très-facilement, éliminant successivement η et ξ,

$$(1+q^2\mathfrak{C})\,\xi=(1+qh)\,x+qby+(qf+q^2\mathfrak{G})\,z,$$
$$(1+q^2\mathfrak{C})\,\eta=-qax+(1-qh)\,y+(-qg+q^2\mathfrak{F})\,z$$

et ensuite

$$(1+q^2\mathfrak{C})\,x'=(1+2qh-q^2\mathfrak{G})\,x+2qby+2(qf+q^2\mathfrak{G})\,z,$$
$$(1+q^2\mathfrak{C})\,y'=-2qax+(1-2qh-q^2\mathfrak{C})\,y+2(-qg+q^2\mathfrak{F})\,z,$$

c'est-à-dire, en écrivant $z=z'=1$, les valeurs

$$\left.\begin{aligned}(1+q^2\mathfrak{C})\,x'&=(1+2qh-q^2\mathfrak{C})\,x+2qby+2(qf+q^2\mathfrak{G})\\(1+q^2\mathfrak{C})\,y'&=-2qax+(1-2qh-q^2\mathfrak{C})\,y+2(-qg+q^2\mathfrak{F})\end{aligned}\right\}\quad(4),$$

satisfont identiquement à l'équation

$$(a,\ b,\ c,\ f,\ g,\ h\between x',\ y',\ 1)^2=(a,\ b,\ c,\ f,\ g,\ h\between x,\ y,\ 1)^2\quad(5).$$

En prenant $h=0$, on a

$$\mathfrak{C}=ab,\quad \mathfrak{F}=-af,\quad \mathfrak{G}=-bg,$$

et les formules deviennent

$$\left.\begin{aligned}(1+q^2ab)\,x'&=(1-q^2ab)\,x+2qby+2q\,(f-qbg)\\(1+q^2ab)\,y'&=-2qax+(1-q^2ab)\,y-2q\,(g+qaf)\end{aligned}\right\}\quad(6)$$

valeurs qui satisfont identiquement à l'équation

$$(ax'^2+2gx'+l)+(by'^2+2fy'+m)=(ax^2+2gx+l)+(by^2+2fy+m)\quad(7),$$

[où pour c j'ai mis $l+m$] et en y écrivant

$$\frac{1-q^2ab}{1+q^2ab}=s=\sqrt{1-abr^2},$$

on obtient des formules qui correspondent précisément aux équations données par Euler pour x, y en termes de a, b.

Londres, 10 *Mars*, 1857.

197.

NOTE ON THE THEORY OF LOGARITHMS.

[From the *Philosophical Magazine*, vol. XI. (1856), pp. 275—280.]

An imaginary quantity $x+yi$ may always be expressed in the form

$$x+yi=r(\cos\theta+i\sin\theta)=re^{\theta i},$$

where r is positive, and θ is included between the limits $-\pi$ and $+\pi$. We have in fact

$$r=\sqrt{x^2+y^2};$$

and when x is positive,

$$\theta=\tan^{-1}\frac{y}{x};$$

but when x is negative,

$$\theta=\tan^{-1}\frac{y}{x}\pm\pi;$$

where $\tan^{-1}$ denotes an arc between the limits $-\frac{1}{2}\pi$, $+\frac{1}{2}\pi$, and where the upper or under sign is to be employed according as y is positive or negative. I use for convenience the mark $\equiv$ to denote identity of sign; we may then write

$$\theta=\tan^{-1}\frac{y}{x}+\epsilon\pi,$$

where

$$x\equiv+,\quad \epsilon=0,$$

$$x\equiv-,\quad \epsilon=\pm1\equiv y.$$

It should be remarked that θ has a unique value except in the single case $x\equiv-$, $y=0$, where θ is indeterminately $\pm\pi$. We have, in fact, $\theta=+\pi$ or $\theta=-\pi$

according as x is considered as the limit of $x+yi$, $y\equiv +$, or of $x+yi$, $y\equiv -$. It is natural to write

$$\log(x+yi)=\log r+\theta i,$$

or what is the same thing,

$$\log(x+yi)=\log\sqrt{x^2+y^2}+\left(\tan^{-1}\frac{y}{x}+\epsilon\pi\right)i;$$

and I take this equation as the definition of the logarithm of an imaginary quantity. The question then arises, to find the value of the expression

$$\log(x+yi)+\log(x'+y'i)-\log(x+yi)(x'+y'i).$$

The preceding definition is, in fact, in the case of x positive, that given by M. Cauchy in the *Exercises de Mathématique*, vol. I. [1826]; and he has there shown that x, x', $xx'-yy'$ being all of them positive, the above-mentioned expression reduces itself to zero. The general definition is that given in my *Mémoire sur quelques Formules du Calcul Intégral*, Liouville, vol. XII. [1847], p. 231 [49]; but I was wrong in asserting that the expression always reduced itself to zero. We have, in fact, in general

$$\tan^{-1}\alpha+\tan^{-1}\beta=\tan^{-1}\frac{\alpha+\beta}{1-\alpha\beta},$$

when $1-\alpha\beta$ is positive; but when $1-\alpha\beta$ is negative (which implies that α, β have the same sign), then

$$\tan^{-1}\alpha+\tan^{-1}\beta=\tan^{-1}\frac{\alpha+\beta}{1-\alpha\beta}\pm\pi,$$

where the upper or under sign is to be employed according as α and β are positive or negative; or what is the same thing,

$$\tan^{-1}\alpha+\tan^{-1}\beta=\tan^{-1}\frac{\alpha+\beta}{1-\alpha\beta}+\epsilon\pi,$$

where

$$1-\alpha\beta\equiv +,\quad \epsilon=0,$$
$$1-\alpha\beta\equiv -,\quad \epsilon=\pm 1\equiv\alpha+\beta\equiv\alpha\equiv\beta.$$

This being premised, then writing

$$\log(x+yi)=\log\sqrt{x^2+y^2}+\left(\tan^{-1}\frac{y}{x}+\epsilon\pi\right)i,$$

$$\log(x'+y'i)=\log\sqrt{x'^2+y'^2}+\left(\tan^{-1}\frac{y'}{x'}+\epsilon'\pi\right)i,$$

$$\log(x+yi)(x'+y'i)=\log[(xx'-yy')+(xy'+yx')i]$$

$$=\log\sqrt{x^2+y^2}\sqrt{x'^2+y'^2}+\left(\tan^{-1}\frac{xy'+x'y}{xx'-yy'}+\epsilon''\pi\right)i,$$

$$\tan^{-1}\frac{y}{x}+\tan^{-1}\frac{y'}{x'}=\tan^{-1}\frac{xy'+x'y}{xx'-yy'}+\epsilon'''\pi,$$

we find

$$\log(x+yi)+\log(x'+y'i)-\log(x+yi)(x'+y'i)=(\epsilon+\epsilon'-\epsilon''+\epsilon''')\pi i.$$

Hence, considering the different cases:

I. $x\equiv +,\quad x'\equiv +,\quad xx'-yy'\equiv +,$

$\epsilon\ =0,$

$\epsilon'=0,$

$\epsilon''=0,$

$\epsilon'''=0,$

and therefore

$$\epsilon+\epsilon'-\epsilon''+\epsilon'''=0.$$

II. $x\equiv +,\quad x'\equiv +,\quad xx'-yy'\equiv -,$

$\epsilon\ =0,$

$\epsilon'\ =0,$

$\epsilon''=\pm 1\equiv (xy'+x'y)\equiv\left(\frac{y}{x}+\frac{y'}{x'}\right),$

$\epsilon'''=\pm 1\qquad\equiv\left(\frac{y}{x}+\frac{y'}{x'}\right),$

and therefore

$$\epsilon+\epsilon'-\epsilon''+\epsilon'''=0.$$

III. $x\equiv +,\quad x'\equiv -,\quad xx'-yy'\equiv +,$

$\epsilon\ =0,$

$\epsilon'\ =\pm 1\equiv y'\equiv -\frac{y'}{x'}\equiv -\left(\frac{y}{x}+\frac{y'}{x'}\right),$

$\epsilon''=0,$

$\epsilon'''=\pm 1\qquad\equiv\left(\frac{y}{x}+\frac{y'}{x'}\right),$

and therefore

$$\epsilon+\epsilon'-\epsilon''+\epsilon'''=0.$$

IV. $x\equiv +,\quad x'\equiv -,\quad xx'-yy'\equiv -,$

$\epsilon\ =0,$

$\epsilon'\ =\pm 1\equiv y'\equiv -\frac{y'}{x'},$

$\epsilon''=\pm 1\equiv xy'+x'y\equiv -\left(\frac{y}{x}+\frac{y'}{x'}\right),$

$\epsilon'''=0,$

and therefore

$$\epsilon + \epsilon' - \epsilon'' + \epsilon''' = 0 \qquad \text{if } \frac{y'}{x'} \equiv \left(\frac{y}{x} + \frac{y'}{x'}\right),$$

but

$$\epsilon + \epsilon' - \epsilon'' + \epsilon''' = \pm 2 \equiv \left(\frac{y}{x} + \frac{y'}{x'}\right) \text{ if } \frac{y'}{x'} \equiv -\left(\frac{y}{x} + \frac{y'}{x'}\right).$$

V.

$$x \equiv -, \quad x' \equiv +, \quad xx' - yy' \equiv +,$$

$$\epsilon = \pm 1 \equiv y \equiv -\frac{y}{x} \equiv -\left(\frac{y}{x} + \frac{y'}{x'}\right),$$

$$\epsilon' = 0,$$

$$\epsilon'' = 0,$$

$$\epsilon''' = \pm 1 \qquad \equiv \left(\frac{y}{x} + \frac{y'}{x'}\right),$$

and therefore

$$\epsilon + \epsilon' - \epsilon'' + \epsilon''' = 0.$$

VI.

$$x \equiv -, \quad x' \equiv +, \quad xx' - yy' \equiv -,$$

$$\epsilon = \pm 1 \equiv y \equiv -\frac{y}{x},$$

$$\epsilon' = 0,$$

$$\epsilon'' = \pm 1 \equiv (xy' + x'y) \equiv -\left(\frac{y}{x} + \frac{y'}{x'}\right),$$

$$\epsilon''' = 0,$$

and therefore

$$\epsilon + \epsilon' - \epsilon'' + \epsilon''' = 0 \qquad \text{if } \frac{y}{x} \equiv \left(\frac{y}{x} + \frac{y'}{x'}\right),$$

but

$$\epsilon + \epsilon' - \epsilon'' + \epsilon''' = \pm 2 \equiv \frac{y}{x} + \frac{y'}{x'} \text{ if } \frac{y}{x} \equiv -\left(\frac{y}{x} + \frac{y'}{x'}\right).$$

VII.

$$x \equiv -, \quad x' \equiv -, \quad xx' - yy' \equiv +,$$

$$\epsilon = \pm 1 \equiv y \equiv -\frac{y}{x},$$

$$\epsilon' = \pm 1 \equiv y' \equiv -\frac{y'}{x'},$$

$$\epsilon'' = 0,$$

$$\epsilon''' = 0;$$

and therefore

$$\epsilon + \epsilon' - \epsilon'' + \epsilon''' = 0 \qquad \text{if } \frac{y}{x} \equiv -\frac{y'}{x'},$$

but

$$\epsilon + \epsilon' - \epsilon'' + \epsilon'' = \pm 2 \equiv -\left(\frac{y}{x} + \frac{y'}{x'}\right) \text{ if } \frac{y}{x} \equiv \frac{y'}{x'}.$$

VIII.
$$x \equiv -,\quad x' \equiv -,\quad xx' - yy' \equiv -,$$

$$\epsilon = \pm 1 \equiv y \equiv -\frac{y}{x} \equiv -\left(\frac{y}{x} + \frac{y'}{x'}\right),$$

$$\epsilon' = \pm 1 \equiv y' \equiv -\frac{y'}{x'} \equiv -\left(\frac{y}{x} + \frac{y'}{x'}\right),$$

$$\epsilon'' = \pm 1 \equiv (xy' + x'y) \equiv \left(\frac{y}{x} + \frac{y'}{x'}\right),$$

$$\epsilon''' = \pm 1 \equiv \left(\frac{y}{x} + \frac{y'}{x'}\right),$$

and therefore
$$\epsilon + \epsilon' - \epsilon'' + \epsilon''' = \pm 2 \equiv -\left(\frac{y}{x} + \frac{y'}{x'}\right).$$

Hence writing
$$\log(x + yi) + \log(x' + y'i) - \log(x + yi)(x' + y'i) = E\pi i,$$

we have $E = 0$, except in the following cases, viz.

1. (See IV.)
$$x \equiv +,\ x' = -,\ xx' - yy' \equiv -,\ \frac{y'}{x'} \equiv -\left(\frac{y}{x} + \frac{y'}{x'}\right),$$

where
$$E = \pm 2 \equiv \left(\frac{y}{x} + \frac{y'}{x'}\right).$$

2. (See VI.)
$$x \equiv -,\ x' \equiv +,\ xx' - yy' \equiv -,\ \frac{y}{x} \equiv -\left(\frac{y}{x} + \frac{y'}{x'}\right),$$

where
$$E = \pm 2 \equiv \left(\frac{y}{x} + \frac{y'}{x'}\right).$$

3. (See VII.)
$$x \equiv -,\ x' \equiv -,\ xx' - yy' \equiv +,\ \frac{y}{x} \equiv \frac{y'}{x'},$$

where
$$E = \pm 2 \equiv -\left(\frac{y}{x} + \frac{y'}{x'}\right).$$

4. (See VIII.)
$$x \equiv -,\ x' \equiv -,\quad xx' - yy' \equiv -,$$

where
$$E = \pm 2 \equiv -\left(\frac{y}{x} + \frac{y'}{x'}\right).$$

It thus appears that when the real parts x, x', $xx'-yy'$ are all three of them positive, or any two of them positive and the third negative, E is equal to zero, or the logarithm of the product is equal to the sum of the logarithms of the factors; but that if the real parts are one of them positive and the other two of them negative, then if a certain relation between the real and imaginary parts is satisfied, but not otherwise, the property holds; and if the real parts are all three of them negative, the property does not hold in any case.

The preceding results do not apply to the case where any one of the arguments $x+yi$, $x'+y'i$, $(x+yi)(x'+y'i)$ is real and negative, for no definition applicable to such case has been given of a logarithm. If, however, we assume as a definition that the logarithm of a negative real quantity is equal to the logarithm of the corresponding positive quantity, then in the case, $x\equiv -$, $y=0$, we have

$$\log x+\log(x'+y'i)-\log x(x'+y'i)=\epsilon\pi i,\quad \epsilon=\pm 1\equiv y';$$

an equation which is, in fact, equivalent to

$$\log(x'+y'i)-\log[-(x'+y'i)]=\epsilon\pi i,\quad \epsilon=\pm 1\equiv y';$$

and in the case $xy'+x'y=0$, $xx'-yy'\equiv -$, which implies $y\equiv y'$, then

$$\log(x+yi)+\log(x'+y'i)-\log(x+yi)(x'+y'i)=\pi i,\quad \epsilon=\pm 1\equiv y \text{ or } y';$$

an equation which is in fact equivalent to

$$\log(x+yi)+\log(-x+yi)-\log(x^2+y^2)=\epsilon\pi i,\ \epsilon=\pm 1\equiv y.$$

The case where both of the arguments $x+yi$, $x'+y'i$ are real and negative, i.e. $x\equiv -$, $y=0$, $x'\equiv -$, $y'=0$ gives of course $\log x+\log x'-\log xx'=0$, the logarithms of the negative real quantities x, x' being by the definition the same as the logarithms of the corresponding positive quantities. It should, however, be remarked that the definition, $(x\equiv -)$, $\log x=\log(-x)$ not only gives for $\log x$ a different value from that which would be obtained from the general definition of a logarithm, by considering $\log x$ as the limit of $\log(x+yi$, $y\equiv +$, or of $\log(x+yi)$, $y\equiv -$, but gives also a value, which, for the particular case in question, contradicts the fundamental equation $e^{\log x}=x$. It is therefore, I think, better not to establish any definition for the logarithm of a negative real quantity x, but to say that such logarithm is absolutely indeterminate and indeterminable, except in the case where, from the nature of the question, x is considered as the limit of $x+yi$, y positive, or of $x+yi$, y negative.

2, *Stone Buildings, March* 15, 1856.

198.

NOTE UPON A RESULT OF ELIMINATION.

[From the *Philosophical Magazine*, vol. XI. (1856), pp. 378—379.]

If the quadratic function

$$(a, b, c, f, g, h \between x, y, z)^2$$

break up into factors, then representing one of these factors by $\xi x + \eta y + \zeta z$, and taking any arbitrary quantities α, β, γ, the factor in question, and therefore the quadratic function is reduced to zero by substituting $\beta\zeta - \alpha\eta$, $\gamma\xi - \alpha\zeta$, $\alpha\eta - \beta\zeta$ in the place of x, y, z. Write

$$(a, b, c, f, g, h \between \beta\zeta - \alpha\eta, \gamma\xi - \alpha\zeta, \alpha\eta - \beta\zeta)^2 = (\mathrm{a}, \mathrm{b}, \mathrm{c}, \mathrm{f}, \mathrm{g}, \mathrm{h} \between \alpha, \beta, \gamma)^2;$$

the coefficients of the function on the right hand are

$$\begin{aligned}
\mathrm{a} &= \quad . \quad c\eta^2 + b\zeta^2 - 2f\eta\zeta \quad . \quad . \quad ,\\
\mathrm{b} &= c\xi^2 + \quad . \quad + a\zeta^2 \quad . \quad - 2g\zeta\xi \quad . \quad ,\\
\mathrm{c} &= b\xi^2 + a\eta^2 \quad . \quad . \quad . \quad - 2h\xi\eta,\\
\mathrm{f} &= -f\xi^2 \quad . \quad . \quad - a\eta\zeta + h\zeta\xi + g\xi\eta,\\
\mathrm{g} &= \quad . \quad - g\eta^2 \quad . \quad + h\eta\zeta - b\zeta\xi + f\xi\eta,\\
\mathrm{h} &= \quad . \quad . \quad - h\zeta^2 + g\eta\zeta + f\zeta\xi - c\xi\eta;
\end{aligned}$$

and it is to be remarked that we have identically

$$\begin{aligned}
\mathrm{a}\xi + \mathrm{h}\eta + \mathrm{g}\zeta &= 0,\\
\mathrm{h}\xi + \mathrm{b}\eta + \mathrm{f}\zeta &= 0,\\
\mathrm{g}\xi + \mathrm{f}\eta + \mathrm{c}\zeta &= 0.
\end{aligned}$$

Hence of the six equations, $\mathrm{a}=0$, $\mathrm{b}=0$, $\mathrm{c}=0$, $\mathrm{f}=0$, $\mathrm{g}=0$, $\mathrm{h}=0$, any three (except $\mathrm{a}=0$, $\mathrm{h}=0$, $\mathrm{g}=0$, or $\mathrm{h}=0$, $\mathrm{b}=0$, $\mathrm{f}=0$, or $\mathrm{g}=0$, $\mathrm{f}=0$, $\mathrm{c}=0$) imply the remaining three.

If from the six equations we eliminate ξ^2, η^2, &c., we obtain

$$\square = \begin{vmatrix} \cdot & c, & b, & -2f, & \cdot & \cdot \\ c, & \cdot & a, & \cdot & -2g, & \cdot \\ b, & a, & \cdot & \cdot & \cdot & -2h \\ -f, & \cdot & \cdot & -a, & h, & g \\ \cdot & -g, & \cdot & f, & -b, & f \\ \cdot & \cdot & -h, & g, & c, & -c \end{vmatrix} = 0\,;$$

and the equation $\square = 0$ is therefore the result of the elimination of ξ, η, ζ from any three (other than the excepted combinations) of the six equations. But from what precedes, it appears that the equation $\square = 0$ must be satisfied when the quadratic function breaks up into factors, and consequently $\square$ must contain as a factor the discriminant

$$K = \begin{vmatrix} a, & h, & g \\ h, & b, & f \\ g, & f, & c \end{vmatrix}$$

of the quadratic function. This agrees perfectly with the results obtained long ago by Prof. Sylvester in his paper, "Examples of the Dialytic Method of Elimination as applied to Ternary Systems of Equations," *Camb. Math. Journ.* vol. II. p. 232; but according to the assumption there made, the value of $\square$ would be (to a numerical factor *près*) $= abcK$. The correct value is by actual development shown to be $\square = -2K^2$. It would be interesting to show *à priori* that $\square$ contains K^2 as a factor.

2, *Stone Buildings, March* 28, 1856.

199.

NOTE ON THE THEORY OF ELLIPTIC MOTION.

[From the *Philosophical Magazine*, vol. XI. (1856), pp. 425—428.]

IF, as usual, r, θ denote the radius vector and longitude, and μ the central mass, then the Vis Viva and Force function are respectively

$$T = \tfrac{1}{2}(r'^2 + r^2\theta'^2),$$

$$U = \frac{\mu}{r};$$

and writing

$$\frac{dT}{dr'} = \quad r' = p,$$

$$\frac{dT}{d\theta'} = r^2\theta' = q,$$

we have $r' = p$, $\theta' = \dfrac{q}{r^2}$, and $T = \tfrac{1}{2}\left(p^2 + \dfrac{q^2}{r^2}\right)$, whence, putting $H = T - U$, the value of H is

$$H = \tfrac{1}{2}\left(p^2 + \frac{q^2}{r^2}\right) - \frac{\mu}{r};$$

and by Sir W. R. Hamilton's theory, the equations of motion are

$$\frac{dr}{dt} = \frac{dH}{dp}, \quad \frac{dp}{dt} = -\frac{dH}{dr},$$

$$\frac{d\theta}{dt} = \frac{dH}{dq}, \quad \frac{dq}{dt} = -\frac{dH}{d\theta}:$$

or substituting for H its value, the equations of motion are

$$\frac{dr}{dt} = p,$$

$$\frac{d\theta}{dt} = \frac{q}{r^2},$$

$$\frac{dp}{dt} = \frac{q^2}{r^3} + \frac{\mu}{r^2},$$

$$\frac{dq}{dt} = 0.$$

Putting, as usual, $\mu = n^2a^3$, and introducing the eccentric anomaly u, which is given as a function of t by means of the equation

$$nt + c = u - e \sin u,$$

$\left(\text{so that } \frac{du}{dt} = \frac{n}{1 - e\cos u}\right)$, the integral equations are

$$q = na^2\sqrt{1-e^2},$$

$$p = \frac{nae \sin u}{1 - e\cos u},$$

$$r = a(1 - e\cos u),$$

$$\theta - \varpi = \tan^{-1}\left(\frac{\sqrt{1-e^2}\sin u}{\cos u - e}\right);$$

where the constants of integration a, e, c, ϖ denote as usual the mean distance, the eccentricity, the mean anomaly at epoch, and the longitude of pericentre.

Suppose that q_0, p_0, r_0, θ_0, u_0 correspond to the time t_0 (q is constant, so that $q_0 = q$), and write

$$V = na^2(u - u_0 + e\sin u - e\sin u_0);$$

joining to this the equations

$$r = a(1 - e\cos u), \quad r_0 = a(1 - e\cos u_0),$$

$$\theta - \theta_0 = \tan^{-1}\left(\frac{\sqrt{1-e^2}\sin u}{\cos u - e}\right) - \tan^{-1}\left(\frac{\sqrt{1-e^2}\sin u_0}{\cos u - e}\right),$$

u, u_0, e will be functions of a, r, r_0, θ, θ_0, and consequently (n being throughout considered as a function of a) V will be a function of a, r, r_0, θ, θ_0. The function V so expressed as a function of a, r, r_0, θ, θ_0 is, in fact, the characteristic function of Sir W. R. Hamilton, and according to his theory we ought to have

$$dV = \tfrac{1}{2}n^2a(t - t_0)\,da + p\,dr + q\,d\theta - p_0\,dr_0 - q_0\,d\theta_0.$$

To verify this, I form the equation

$$\begin{aligned}dV = \; & \tfrac{1}{2}na\,(u - u_0 + e\sin u - e\sin u_0)\,da \\ & + na^2\,[(1+e\cos u)\,du - (1+e\cos u_0)\,du_0] \\ & + na^2\,(\sin u - \sin u_0)\,de \\ & + \frac{nae\sin u}{1-e\cos u}\,\{dr - (1-e\cos u)\,da - ae\sin u\;du + a\cos u\;de\} \\ & - \frac{nae\sin u_0}{1-e\cos u_0}\,\{dr_0 - (1-e\cos u_0)\,da - ae\sin u_0\,du_0 + a\cos u_0\,de\} \\ & + na^2\sqrt{1-e^2}\left\{\;d\theta - \frac{1}{\sqrt{1-e^2}\,(1-e\cos u)}\,[(1-e^2)\,du + \sin u\,de]\right. \\ & \qquad\left. - d\theta_0 + \frac{1}{\sqrt{1-e^2}\,(1-e\cos u_0)}\,[(1-e^2)\,du_0 + \sin u_0\,de]\right\};\end{aligned}$$

the coefficient of du on the right-hand side is

$$\begin{aligned}& na^2\,(1+e\cos u) - \frac{na^2e^2\sin^2 u}{1-e\cos u} - \frac{na^2\,(1-e^2)}{1-e\cos u} \\ & = na^2\left(1 + e\cos u - \frac{1-e^2+e^2\sin^2 u}{1-e\cos u}\right),\end{aligned}$$

which vanishes, and similarly the coefficient of du_0 also vanishes: the coefficient of de is the difference of two parts, the first of which is

$$\begin{aligned}& na^2\sin u + \frac{na^2e\sin u\cos u}{1-e\cos u} - \frac{na^2\sin u}{1-e\cos u} \\ & = na^2\sin u\left(1 - \frac{1-e\cos u}{1-e\cos u}\right),\end{aligned}$$

which vanishes, and the second part in like manner also vanishes; the coefficient of da is the difference of two parts, the first of which is

$$\tfrac{1}{2}na\,(u + e\sin u) - nae\sin u = \tfrac{1}{2}na\,(u - e\sin u),$$

and the second is the like function of u_0; the entire coefficient therefore is

$$\tfrac{1}{2}\,na\,(u - u_0 - e\sin u + e\sin u_0).$$

We have therefore

$$\begin{aligned}dV = \; & \tfrac{1}{2}na\,(u - u_0 - e\sin u + e\sin u_0)\,da \\ & + \frac{nae\sin u}{1-e\cos u}\,dr + na^2\sqrt{1-e^2}\,d\theta \\ & - \frac{nae\sin u_0}{1-e\cos u_0}\,dr_0 - na^2\sqrt{1-e^2}\,d\theta_0\,;\end{aligned}$$

or what is the same thing,

$$dV = \tfrac{1}{2}n^2a\,(t - t_0)\,da + p\,dr + q\,d\theta - p_0\,dr_0 - q_0\,d\theta_0,$$

the equation which was to be verified.

2, *Stone Buildings, March* 28, 1856.

200.

ON THE CONES WHICH PASS THROUGH A GIVEN CURVE OF THE THIRD ORDER IN SPACE.

[From the *Philosophical Magazine*, vol. XII. (1856), pp. 20—22.]

THE following investigation is connected with the theory of the cubic

$$(a,\ b,\ c,\ d \between x,\ y)^3,$$

and in particular with a theorem that the determinant formed with the second differential coefficients of the discriminant gives the square of the discriminant.

Consider the coefficients a, b, c, d as linear functions of coordinates, the equations

$$ac - b^2 = 0,\quad bc - ad = 0,\quad bd - c^2 = 0$$

(equivalent, of course, to two equations) belong to a curve of the third order in space, the edge of regression of the developable surface obtained by putting the discriminant equal to zero, or which has for its equation

$$-a^2d^2 + 6abcd - 4ac^3 - 4b^3d + 3b^2c^2 = 0\,;$$

and, moreover, the above forms are the general representations of any curve of the third order, and developable surface of the fourth order. The question arises, to find the equation of the cone of the third order having an arbitrary point for its vertex, and passing through the curve of the third order. This may be done by Joachimsthal's method: let α, β, γ, δ be the values of a, b, c, d at the vertex of the cone, and in the equations of the curve, say in

$$ac - b^2 = 0,\quad bd - c^2 = 0,$$

for a, b, c, d write $ua + v\alpha$, $ub + v\beta$, $uc + v\gamma$, $ud + v\delta$; if from the equations so obtained u, v are eliminated, the resulting equation will be that of the cone of the third order.

The substitutions in question give

$$L\,u^2 + 2M\,uv + N\,v^2 = 0,$$
$$L'u^2 + 2M'uv + N'v^2 = 0,$$

where

$$L = 2\,(ac - b^2),\quad M = a\gamma + c\alpha - 2b\beta,\quad N = 2\,(\alpha\gamma - \beta^2),$$
$$L' = 2\,(bd - c^2),\quad M' = b\delta + d\beta - 2c\gamma,\quad N' = 2\,(\beta\delta - \gamma^2);$$

the result of the elimination is

$$4\,(LN - M^2)\,(L'N' - M'^2) - (LN' + L'N - 2MM')^2 = 0,$$

and we have

$$LN - M^2 = 4\,(b\gamma - c\beta)\,(a\beta - b\alpha) - (c\alpha - a\gamma)^2,$$
$$L'N' - M'^2 = 4\,(c\delta - d\gamma)\,(b\gamma - c\beta) - (b\delta - d\gamma)^2,$$
$$LN' + L'N - 2MM' = 4\,(b\gamma - c\beta)^2 + 4\,(a\beta - b\alpha)\,(c\delta - d\gamma) + 2\,(c\alpha - a\gamma)\,(b\delta - d\beta).$$

Write for shortness

$$b\gamma - c\beta = l,\quad c\alpha - a\gamma = m,\quad a\beta - b\alpha = n,$$
$$a\delta - d\alpha = f,\quad b\delta - d\beta = g,\quad c\delta - d\gamma = h,$$

values which give $lf + mg + nh = 0$. Then forming the expression

$$4\,(4ln - m^2)\,(4hl - g^2) - (4l^2 + 4nh + 2mg)^2,$$

this is equal to [1]

$$-16l\,(l^3 - nfh + lmg - 2lnh + ng^2 + m^2h),$$

and the equation of the cone is

$$l^3 - nfh + lmg - 2lnh + ng^2 + m^2h = 0;$$

or, substituting and expanding,

$$\begin{aligned}
&(\alpha\delta^2 - \gamma^3)\,b^3 - (\alpha^2\delta - \beta^3)\,c^3 - \delta\,(\beta\delta - \gamma^2)\,ab^2 + \alpha\,(\alpha\gamma - \beta^2)\,dc^2\\
&+ \delta\,(\beta\delta - \gamma^2)\,a^2c - \alpha\,(\alpha\gamma - \beta^2)\,d^2b + 2\delta\,(\alpha\gamma - \beta^2)\,ac^2 - 2\alpha\,(\beta\delta - \gamma^2)\,db^2\\
&- \gamma\,(\beta\delta - \gamma^2)\,a^2d + \beta\,(\alpha\gamma - \beta^2)\,ad^2 + 3\gamma\,(\beta\gamma - \alpha\delta)\,b^2c - 3\beta\,(\beta\gamma - \alpha\delta)\,bc^2\\
&+ \delta\,(\beta\gamma - \alpha\delta)\,abc - \alpha\,(\beta\gamma - \alpha\delta)\,dbc + (\alpha\gamma\delta - 3\beta\gamma^2 + 2\beta^2\delta)\,abd\\
&- (\alpha\beta\delta - 3\beta^2\gamma + 2\alpha\gamma^2)\,acd = 0.
\end{aligned}$$

[1] With respect to the occurrence of the factor $l\,(=b\gamma - c\beta)$, it is worth noticing, that, putting $b = k\beta$, $c = k\gamma$, we have identically

$$L\,u^2 + 2M\,uv + N\,v^2 = 2\,[(a\gamma - b\beta)\,u + (\alpha\gamma - \beta^2)\,v]\,(ku + v),$$
$$L'u^2 + 2M'uv + N'v^2 = 2\,[(\beta\delta - c\gamma)\,u + (\beta\delta - \gamma^2)\,v]\,(ku + v),$$

i.e. the two functions will contain a common factor if $b = k\beta$, $c = k\gamma$, or what is the same thing, if $b\gamma - c\beta = 0$. But if the functions contain a common factor, their resultant vanishes, i.e. the resultant will vanish in virtue of the relation $b\gamma - c\beta = 0$, or what is the same thing, $b\gamma - c\beta$ is a factor of the resultant.

Now putting

$$bd-c^2=p,\quad bc-ad=q,\quad ac-b^2=r,$$

and in like manner

$$\beta\delta-\gamma^2=P,\quad \beta\gamma-\alpha\delta=Q,\quad \alpha\gamma-\beta^2=R,$$

and reducing, the final result may be expressed in the form

$$\begin{aligned} &P\{\quad -2\alpha bp+(\gamma a-2\beta b)\,q+(\delta a-\gamma c)\,r\}\\ +&Q\{\quad -\alpha cp+(\gamma b-\beta c)\,q+\delta br\}\\ +&R\{-(\alpha d-\beta c)\,p-(\beta d-2\gamma c)\,q+2\delta cr\}=0,\end{aligned}$$

where a, b, c, d are current coordinates, and p, q, r are quadratic functions of a, b, c, d. The equation is (as it should be) satisfied by the equations ($p=0$, $q=0$, $r=0$) of the given curve; it is also satisfied *per se* when $P=0$, $Q=0$, $R=0$, i.e. when the vertex is a point on the curve; this indicates a change in form of the equation, and in fact the cone is in this case of the second order only. Suppose that the coordinates of the vertex are in this case given by $\frac{\alpha}{\beta}=\frac{\beta}{\gamma}=\frac{\gamma}{\delta}=\frac{1}{\sigma}$ (σ an arbitrary quantity), it may be easily shown that the equation of the cone is

$$p+\sigma q+\sigma^2 r=0;$$

or at full length

$$(bd-c^2)+\sigma(bc-ad)+\sigma^2(ac-b^2)=0;$$

in fact this equation is evidently that of a surface of the second order passing through the curve; and there is no difficulty in showing that it is a cone.

2, *Stone Buildings, May* 1, 1856.

201.

SECOND NOTE ON THE THEORY OF LOGARITHMS.

[From the *Philosophical Magazine*, vol. XII. (1856), pp. 354—360.]

THE theory of logarithms, as developed in my first note (*Phil. Mag.* April 1856), [197] may be exhibited in a clearer light by considering, instead of $\log a + \log b - \log ab = E\pi i$, the equation $\log \frac{b}{a} = \log b - \log a + E\pi i$, a form which more readily enables the accounting *à priori* for the discontinuity in the value of E. Writing then for b, a the complex values $x' + y'i$, $x + yi$, we have

$$\log \frac{x' + y'i}{x + yi} = \log (x' + y'i) - \log (x + yi) + E\pi i\,;$$

where, according to the assumed definition of a logarithm,

$$\log (x + yi) = \log \sqrt{x^2 + y^2} + i\left(\tan^{-1}\frac{y}{x} + \epsilon\pi\right),$$

in which $\log \sqrt{x^2 + y^2}$ is the real logarithm of $\sqrt{x^2 + y^2}$, and $\tan^{-1}\frac{y}{x}$ is an arc between the limits $-\frac{1}{2}\pi$, $+\frac{1}{2}\pi$. The coefficient ϵ is equal to zero when x is positive; but when x is negative, then $\epsilon = +1$ or -1, according as y is positive or negative, i.e. we have

$$x \equiv +, \quad \epsilon = 0,$$

$$x \equiv -, \quad \epsilon = \pm 1 \equiv y\,;$$

and of course the other logarithms in the equation have an analogous signification.

Hence, attending to the equation

$$\tan^{-1}\beta - \tan^{-1}\alpha = \tan^{-1}\frac{\beta-\alpha}{1+\alpha\beta} + \epsilon'''\pi,$$

where, when $1+\alpha\beta$ is positive, ϵ''' is equal to zero; but when $1+\alpha\beta$ is negative, ϵ''' is equal to $+1$ or -1, according as $\beta-\alpha$ is positive or negative: or what is the same thing (α, β being of opposite signs when $1+\alpha\beta$ is negative),

$$1+\alpha\beta \equiv +,\quad \epsilon'''=0,$$
$$1+\alpha\beta \equiv -,\quad \epsilon'''=\pm 1 \equiv \beta-\alpha \equiv \beta \equiv -\alpha,$$

we find

$$E = \epsilon - \epsilon' + \epsilon'' - \epsilon''',$$

where ϵ, ϵ', ϵ'', ϵ''' are defined by the conditions

$$\begin{aligned}
&x \equiv +, && \epsilon = 0,\\
&x \equiv -, && \epsilon = \pm 1 \equiv y,\\
&x' \equiv +, && \epsilon' = 0,\\
&x' \equiv -, && \epsilon' = \pm 1 \equiv y',\\
&xx'+yy' \equiv +, && \epsilon'' = 0,\\
&xx'+yy' \equiv -, && \epsilon'' = \pm 1 \equiv xy'-x'y,\\
&1+\frac{y}{x}\frac{y'}{x'} \equiv +, && \epsilon''' = 0,\\
&1+\frac{y}{x}\frac{y'}{x'} \equiv -, && \epsilon''' = \pm 1 \equiv \frac{y'}{x'}-\frac{y}{x} \equiv \frac{y'}{x'} \equiv -\frac{y}{x}.
\end{aligned}$$

Suppose, to fix the ideas, that x, y are each of them positive, we have $\epsilon=0$; and considering the several cases:

1. $x'\equiv +,\ \ y'\equiv +.$

Here $xx'+yy'\equiv +$, $1+\frac{y}{x}\frac{y'}{x'}\equiv +$; and consequently not only $\epsilon'=0$, but also $\epsilon''=0$, $\epsilon'''=0$, and thence $E=0$.

2. $x'\equiv +,\ \ y'\equiv -.$

Here $\epsilon'=0$. Moreover, xx' being positive, $xx'+yy'$ and $1+\frac{y}{x}\frac{y'}{x'}$ will have the same sign. If they are both positive, then $\epsilon''=0$, $\epsilon'''=0$; but if they are both negative, then

$$\epsilon''=\pm 1 \equiv xy'-x'y \equiv \frac{y'}{x'}-\frac{y}{x}$$

(since $xx'\equiv +$) and $\epsilon'''=\pm 1 \equiv \frac{y'}{x'}-\frac{y}{x}$, i.e. $\epsilon''=\epsilon'''$. Hence in either case we have $E=0$.

3. $x' \equiv -$, $y' \equiv +$.

Here $\epsilon' = \pm 1 \equiv y'$, i.e. $\epsilon' = 1$. Also xx' being negative, $xx' + yy'$ and $1 + \frac{y}{x}\frac{y'}{x'}$ will have opposite signs. Suppose first $xx' + yy'$ is positive, then $\epsilon'' = 0$. And $1 + \frac{y}{x}\frac{y'}{x'}$ being negative, we have $\epsilon''' = \pm 1 \equiv \frac{y'}{x'} - \frac{y}{x} \equiv \frac{y'}{x'}$, i.e. $\epsilon''' = -1$. But if $xx' + yy'$ is negative, then $\epsilon'' = \pm 1 \equiv xy' - x'y \equiv -\left(\frac{y'}{x'} - \frac{y}{x}\right)$ (since xx' is negative) $\equiv -\frac{y'}{x'}$, i.e. $\epsilon'' = +1$. And $1 + \frac{y}{x}\frac{y'}{x'}$ being positive, we have $\epsilon''' = 0$. Hence in each case $E = 0$.

4. $x' \equiv -$, $y' \equiv -$.

Here $\epsilon' = \pm 1 \equiv y'$, i.e. $\epsilon' = -1$. Also xx' and yy' being each negative, $xx' + yy'$ will be negative, and therefore

$$\epsilon'' = \pm 1 \equiv xy' - x'y \equiv -\left(\frac{y'}{x'} - \frac{y}{x}\right);$$

i.e. if $\frac{y'}{x'} > \frac{y}{x}$, then $\epsilon'' = -1$; but if $\frac{y'}{x'} < \frac{y}{x}$, then $\epsilon'' = +1$. And $1 + \frac{y}{x}\frac{y'}{x'}$ being positive, $\epsilon''' = 0$. Hence if $\frac{y'}{x'} > \frac{y}{x}$, then $E = 1 - 1 = 0$; but if $\frac{y'}{x'} < \frac{y}{x}$, then $E = 1 + 1 = 2$.

Consider (x, y) (x', y') as the rectangular coordinates of two points A, A'. In the case which has been considered, the point A has been taken in the positive quadrant; and the preceding discussion shows that we have always $E = 0$, except in the case where the finite line AA' meets the negative portion of the axis of x, in which case we have $E = +2$. The same thing is true generally in whichever quadrant

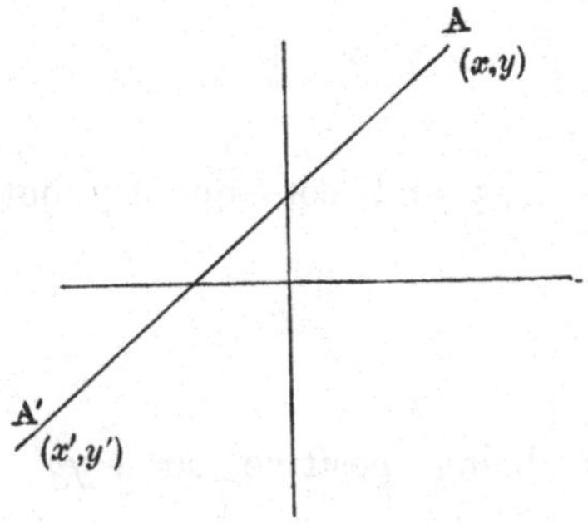

A is situated, i.e. we have always $E = 0$, except in the cases in which the finite line AA' meets the negative portion of the axis of x. But when this happens, then if the line AA', considered as drawn from A to A', passes from above to below the axis of x, we have $E = +2$; but if the line AA', considered as drawn from A to A', passes from below to above the axis of x, then $E = -2$. So that treating the points

A, A' as the geometrical representations of the complex numbers $x+yi$, $x'+y'i$, we have in an exceedingly simple form the precise determination of the discontinuous number $E\,(=0$ or $\pm 2)$ in the formula

$$\log \frac{x'+y'i}{x+yi} = \log (x'+y'i) - \log (x+yi) + E\pi i.$$

Consider in general the definite integral

$$\int_z^{z'} \phi u du,$$

where z', z are complex numbers of the form $x+yi$, $x'+y'i$; and take A, A' as the geometrical representations of these limits, and the variable point P as the geometrical representation of the complex variable u. The value of the definite integral will depend to a certain extent on the series of values which we suppose u successively to assume in passing from z to z', or what is the same thing, on the path of the variable point P from A to A'. For (excluding altogether the case in which the path passes through a point for which ϕu becomes infinite) it is well known that the value of the definite integral is the same for any two paths which do not include between them a point for which ϕu becomes infinite; but when this condition is not satisfied, then the value of the definite integral is not in general the same for the

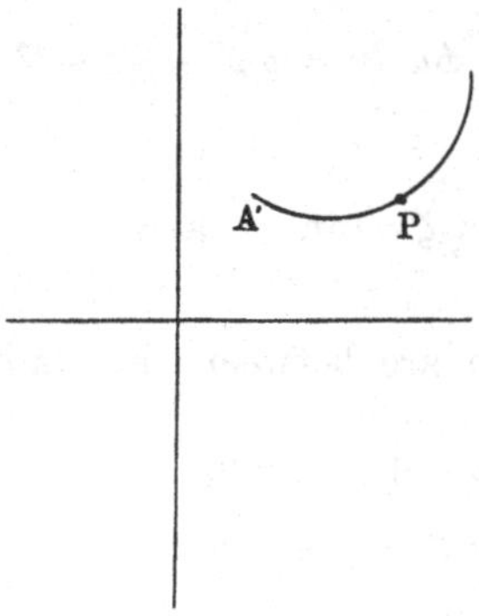

two paths(1). In order therefore to give a precise signification to the notations, we must fix the path of the point P, and it is natural to assume that the path is a right line (of course there are an infinity of paths which give the same value to the definite integral, or as we may call them, paths equivalent to the right line; but the consideration of these would be a needless complication of the definition, and it is better to attend to the single path—the right line). The definition is at once converted into an analytical one; we have only to assume $u = z + r(z'-z)$, and to suppose that the new variable r passes from $r=0$ (which gives $u=z$) *through real*

[1] The theorem is, I believe, due to M. Cauchy. See the memoir of M. Puiseux, *Recherches sur les Fonctions Algébriques*, Liouville, vol. xv. [1850] pp. 365—480, where the subject is elaborately discussed.

values to $r=1$ (which gives $u=z'$), i.e. we have as the equivalent analytical definition of the definite integral between the complex limits z, z' the equation

$$\int_z^{z'} \phi u\, du = (z'-z)\int_0^1 \phi[z+r(z'-z)]\, dr,$$

where the new variable r is real. The only restriction is, that ϕu must not become infinite for any value of u along the path in question, i.e. $\phi[z+r(z'-z)]$ must not become infinite for any real value of r between the limits $r=0$, $r=1$.

Suppose next, the path being defined as above, or in any other manner, that $\phi_, u$ is a function of u such that $\phi_,' u = \phi u$. Then if $\phi_, u$ is continuous along the entire path, we have

$$\int_z^{z'} \phi u\, du = \phi_, z' - \phi_, z;$$

but if $\phi_, u$ is discontinuous at any points of the path, e.g. at the point $u=u_,$, and at no other point, then

$$\int_z^{z'} \phi u\, du = \phi_, z' - \phi_, (u_, + \alpha') + \phi_, (u_, - \alpha) - \phi_, z,$$

where $u_, - \alpha$, $u_, + \alpha'$ are values indefinitely near to $u_,$, the path being from z through $u_1 - \alpha$ to $u_, + \alpha'$ and thence to z'. Or if we represent the break $\phi_, (u_, + \alpha') - \phi_, (u_, + \alpha)$ by the symbol ∇, then we have

$$\int_z^{z'} \phi u\, du = \phi_, z' - \phi_, z - \nabla.$$

Suppose now

$$\zeta = \tan^{-1}\frac{y}{x} + \epsilon\pi,$$

where, as before, $\tan^{-1}\frac{y}{x}$ denotes an arc between the limits $-\frac{1}{2}\pi$, $+\frac{1}{2}\pi$, and

$$x \equiv +,\quad \epsilon = 0,$$
$$x \equiv -,\quad \epsilon = \pm 1 \equiv y;$$

and to fix the ideas, consider ζ as the z-coordinate of a surface, the other two coordinates being x and y. If x be negative and y be indefinitely small and positive, then $\epsilon=+1$, and we have $\zeta=\pi$; but if (x being still negative) y be indefinitely small and negative, then $\epsilon=-1$, and therefore $\zeta=-\pi$, i.e. there is a break or abrupt increment 2π of the coordinate ζ in passing across the negative part of the axis of x from a negative to a positive value of y, or, as we have before called it, from below to above; this is the only discontinuity in the surface, the form of the surface being, in fact, what is intended to be represented in the annexed figure.

Suppose now $z' = x' + y'i$, $z = x + yi$, and consider the definite integral

$$\int_z^{z'} \frac{du}{u},$$

the path being, as before, a right line. We have, by the equivalent analytical definition,

$$\int_z^{z'} \frac{du}{u} = (z' - z) \int_0^1 \frac{dr}{z + r(z' - z)},$$

where the new variable r is real. And in like manner considering the integral $\int_1^{\frac{z'}{z}} \frac{du}{u}$,

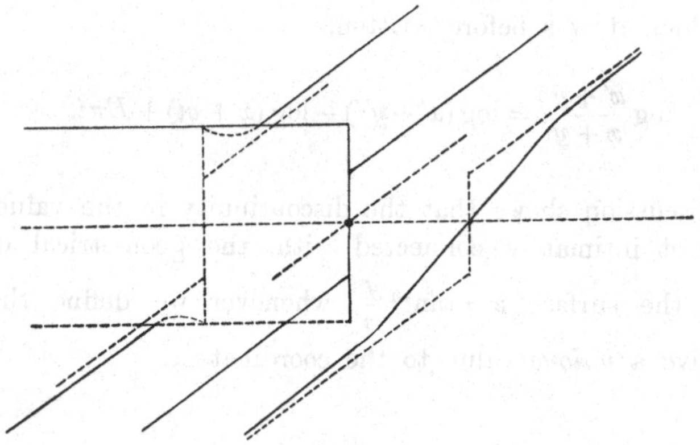

the path being in this case also a right line, we have

$$\int_1^{\frac{z'}{z}} \frac{du}{u} = \left(\frac{z'}{z} - 1\right) \int_0^1 \frac{dr}{1 + r\left(\frac{z'}{z} - 1\right)},$$

where the new variable r is real. The two integrals in r are identically the same, and consequently we have in every case

$$\int_1^{\frac{z'}{z}} \frac{du}{u} = \int_z^{z'} \frac{du}{u}.$$

Now $\log' u = \frac{1}{u}$; and in passing from $u = 1$ to $u = \frac{z'}{z}$, there is no discontinuity in the value of $\log u$,

$$= \log \sqrt{p^2 + q^2} + i\left(\tan^{-1}\frac{q}{p} + \epsilon\pi\right),$$

if for the moment $u = p + qi$; hence the value of the integral on the left-hand side is simply $\log \frac{z'}{z}$. The value of the integral on the right-hand side is in like manner $\log z' - \log z$, in the case in which the finite right line from $u = z$ to $u = z'$ does not meet the negative part of the axis of x; but when this happens, then there is a discontinuity in the value of the logarithm, and the integral on the right-hand side will be $\log z' - \log z - 2\pi i$, or $\log z' - \log z + 2\pi i$, according as the right line considered as drawn from z to z' passes from below to above or from above to below the negative part of the axis of x. We have therefore in every case (E being defined as above) $\log z' - \log z + E\pi i$ for the value of the integral on the right-hand side, and the relation between the two integrals gives, as it ought to do, the equation

$$\log \frac{z'}{z} = \log z' - \log z + E\pi i,$$

or, in the form in which it was before written,

$$\log \frac{x' + y'i}{x + yi} = \log (x' + y'i) - \log (x + yi) + E\pi i.$$

The preceding discussion shows that the discontinuity in the value of E (-0 or ± 2) arises from, or is most intimately connected with, the geometrical discontinuity which necessarily exists in the surface $z = \tan^{-1} \dfrac{y}{x}$, whenever we define the symbol $\tan^{-1}$ in such manner as to give a *unique* value to the coordinate z.

2, *Stone Buildings, Sept.* 19, 1856.

202.

SUPPLEMENTARY REMARKS ON THE PORISM OF THE IN-AND-CIRCUMSCRIBED TRIANGLE.

[From the *Philosophical Magazine*, vol. XIII. (1857), pp. 19—30.]

IN my former papers (see *Phil. Mag.* August and November, 1853, [115, 116]) I established (as part of a more general one) the following theorem, viz. the condition that there may be inscribed in the conic $U=0$ an infinity of triangles circumscribed about the conic $V=0$, is, that if we develope in ascending powers of k the square root of the discriminant of $kU+V$, the coefficient of k^2 in this development must vanish. Thus writing

$$\text{disct.}\,(kV+U) \doteq (K,\ \Theta,\ \Theta',\ K' \between k,\ 1)^3,$$

the condition in question is found to be

$$3\Theta^2 - 4K\Theta' = 0.$$

The following investigations, although relating only to particular cases of the theorem, are, I think, not without interest.

If the equation of the conic containing the angles is

$$U = 2ayz + 2bzx + 2cxy = 0,$$

and the equation of the conic touched by the sides is

$$V = x^2 + y^2 + z^2 - 2yz - 2zx - 2xy = 0,$$

we have

$$\text{disct.}\ (k,\ k,\ k,\ a-k,\ b-k,\ c-k \between x,\ y,\ z)^2 = (K,\ \Theta,\ \Theta',\ K' \between k,\ 1)^3,$$

that is,

$$\begin{aligned} K &= -4, \\ \Theta &= \tfrac{4}{3}(a+b+c), \\ \Theta' &= -\tfrac{1}{3}(a+b+c)^2, \\ K' &= 2abc. \end{aligned}$$

and the equation $3\Theta^2 - 4K\Theta' = 0$ becomes

$$\tfrac{16}{3}(a+b+c)^2 - \tfrac{16}{3}(a+b+c)^2 = 0,$$

which is satisfied identically. This is as it should be; for it is plain that there exists a triangle, viz. the triangle ($x=0$, $y=0$, $z=0$), inscribed in the conic $U=0$, and circumscribed about the conic $V=0$.

Suppose that the equation of the conic containing the angles is

$$y^2 - 4zx = 0,$$

and the equation of the conic touched by the sides is

$$ax^2 + \ by^2 + \ cz^2 = 0,$$

then the tangential equation of the last-mentioned conic is

$$bc\xi^2 + ca\eta^2 + ab\zeta^2 = 0;$$

and if we take for the angles of the triangle $x : y : z = 1 : 2\lambda : \lambda^2$, or $1 : 2\mu : \mu^2$, or $1 : 2\nu : \nu^2$, then the equation of the line joining the angles (μ), (ν) is

$$2\mu\nu x - (\mu+\nu)\,y + z = 0,$$

which will touch the conic $ax^2+by^2+cz^2=0$ if

$$bc \,.\, 4\mu^2\nu^2 + ca\,(\mu+\nu)^2 + ab \,.\, 4 = 0;$$

and it is required to find under what circumstances the equations

$$bc \,.\, 4\mu^2\nu^2 \ + ca\,(\mu+\nu)^2 + ab \,.\, 4 = 0,$$
$$bc \,.\, 4\nu^2\lambda^2 + ca\,(\nu+\lambda)^2 + ab \,.\, 4 = 0,$$
$$bc \,.\, 4\lambda^2\mu^2 + ca\,(\lambda+\mu)^2 + ab \,.\, 4 = 0,$$

become equivalent to two equations only. The condition is of course included in the general formula; and putting

$$\text{disct.}\,(ka,\ kb+1,\ kc,\ 0,\ -2,\ 0 \between x,\ y,\ z^2) = (K,\ \Theta,\ \Theta',\ K' \between k,\ 1)^3,$$

we must have

$$3\Theta^2 - 4K\Theta' = 0.$$

The discriminant in question is

$$k^3abc + k^2ac - k \,.\, 4b - 4 = 0,$$

where $K=1$, $\Theta = \tfrac{1}{3}ac$, $\Theta' = -\tfrac{4}{3}b$, $K' = -4$; the required condition is therefore $ac + 16b^2 = 0$, or say

$$b = -\tfrac{1}{4}i\sqrt{ac}.$$

Substituting this value, the equations become

$$c\mu^2\nu^2 + i\sqrt{ca}\,(\mu+\nu)^2 + a = 0,$$
$$c\nu^2\lambda^2 + i\sqrt{ca}\,(\nu+\lambda)^2 + a = 0,$$
$$c\lambda^2\mu^2 + i\sqrt{ca}\,(\lambda+\mu)^2 + a = 0;$$

the first and second of these are

$$A + 2H\nu + B\nu^2 = 0$$
$$A' + 2H'\nu + B'\nu^2 = 0,$$

where

$$A = (-i\sqrt{a} + \mu^2\sqrt{c})\,i\sqrt{a},\quad H = i\sqrt{ca}\mu,\quad B = (i\sqrt{a} + \mu^2\sqrt{c})\sqrt{c},$$
$$A' = (-i\sqrt{a} + \lambda^2\sqrt{c})\,i\sqrt{a},\quad H' = i\sqrt{ca}\lambda,\quad B' = (i\sqrt{a} + \lambda^2\sqrt{c})\sqrt{c},$$

$$AB' + A'B - 2HH' = 2i\sqrt{ac}\,(a - i\sqrt{ac}\lambda\mu + c\lambda^2\mu^2),$$
$$AB - H^2 = i\sqrt{ac}\,(a - i\sqrt{ac}\mu^2 + c\mu^4),$$
$$A'B' - H'^2 = i\sqrt{ac}\,(a - i\sqrt{ac}\lambda^2 + c\lambda^4);$$

and the result of the elimination therefore is

$$(a - i\sqrt{ac}\lambda^2 + c\lambda^4)(a - i\sqrt{ac}\mu^2 + c\mu^4) - (a - i\sqrt{ac}\lambda\mu + c\lambda^2\mu^2)^2 = 0,$$

viz.

$$2\sqrt{ca}\,(\lambda-\mu)^2\,(c\lambda^2\mu^2 + i\sqrt{ca}\,(\lambda+\mu)^2 + a) = 0;$$

which agrees, as it should do, with the third equation.

To find the condition that it may be possible in the conic

$$x^2 + y^2 + z^2 = 0$$

to inscribe an infinity of triangles, each of them circumscribed about the conic

$$ax^2 + by^2 + cz^2 = 0:$$

let the equations of the sides be

$$l\sqrt{a}x + m\sqrt{b}y + n\sqrt{c}z = 0,$$
$$l'\sqrt{a}x + m'\sqrt{b}y + n'\sqrt{c}z = 0,$$
$$l''\sqrt{a}x + m''\sqrt{b}y + n''\sqrt{c}z = 0;$$

then the conditions of circumscription are

$$l^2 + m^2 + n^2 = 0,$$
$$l'^2 + m'^2 + n'^2 = 0,$$
$$l''^2 + m''^2 + n''^2 = 0;$$

and the conditions of inscription are

$$bc\,(m'n''-m''n')^2+ca\,(n'l''-n''l')^2+ab\,(l'm''-l''m')^2=0,$$
$$bc\,(m''n-mn'')^2+ca\,(n''l-nl'')^2+ab\,(l''m-lm'')^2=0,$$
$$bc\,(mn'-m'n)^2+ca\,(nl'-n'l)^2+ab\,(lm'-l'm)^2=0.$$

Now

$$\begin{aligned}(mn'-m'n)^2 &= (m^2+n^2)(m'^2+n'^2)-(mm'+nn')^2\\ &= l^2l'^2-(mm'+nn')^2\\ &=-(ll'+mm'+nn')(-ll'+mm'+nn')\,;\end{aligned}$$

and making the like change in the analogous expressions, and putting for shortness

$$-bc+ca+ab=\alpha,$$
$$bc-ca+ab=\beta,$$
$$bc+ca-ab=\gamma,$$

the conditions in question become

$$(l'l''+m'm''+n'n'')(\alpha l'l''+\beta m'm''+\gamma n'n'')=0,$$
$$(l''l+m''m+n''n)(\alpha l''l+\beta m''m+\gamma n''n)=0,$$
$$(ll'+mm'+nn')(\alpha ll'+\beta mm'+\gamma nn')=0.$$

The proper solution is that given by the system of equations

$$l^2+m^2+n^2=0,$$
$$l'^2+m'^2+n'^2=0,$$
$$l''^2+m''^2+n''^2=0,$$

$$\alpha l'l''+\beta m'm''+\gamma n'n''=0,$$
$$\alpha l''l+\beta m''m+\gamma n''n=0,$$
$$\alpha ll'+\beta mm'+\gamma nn'=0\,;$$

and by writing $l=\frac{f}{\sqrt{\alpha}}$, $l'=\frac{f'}{\sqrt{\alpha}}$, $l''=\frac{f''}{\sqrt{\alpha}}$, $m=\frac{g}{\sqrt{\beta}}$, &c., $A=\frac{1}{\alpha}$, $B=\frac{1}{\beta}$, $C=\frac{1}{\gamma}$, these equations become

$$Af^2+Bg^2+Ch^2=0,$$
$$Af'^2+Bg'^2+Ch'^2=0,$$
$$Af''^2+Bg''^2+Ch''^2=0,$$

$$f'f''+g'g''+h'h''=0,$$
$$f''f+g''g+h''h=0,$$
$$ff'+gg'+hh'=0.$$

The first of which systems expresses that the points (f, g, h), (f', g', h'), (f'', g'', h'') are points in the conic

$$Ax^2 + By^2 + Cz^2 = 0;$$

and the second condition expresses that each of the points in question is the pole with respect to the conic

$$x^2 + y^2 + z^2 = 0$$

of the line joining the other two points, i.e. that the three points are a system of conjugate points with respect to the last-mentioned conic. The problem is thus reduced to the following one:

To find the condition in order that it may be possible in the conic

$$Ax^2 + By^2 + Cz^2 = 0$$

to inscribe an infinity of triangles such that the angles are a system of conjugate points with respect to the conic

$$x^2 + y^2 + z^2 = 0.$$

Before going further it is proper to remark that if, instead of assuming $\alpha l'l'' + \beta m'm'' + \gamma n'n'' = 0$, we had assumed

$$l'l'' + m'm'' + n'n'' = 0,$$

this, combined with the equations

$$l'^2 + m'^2 + n'^2 = 0, \quad l''^2 + m''^2 + n''^2 = 0,$$

would have given $l' : m' : n' = l'' : m'' : n''$, i.e. two of the angles of the triangle would have been coincident: this obviously does not give rise to any proper solution. Returning now to the system of equations in f, g, h, &c., since the equations give only the ratios $f : g : h$; $f' : g' : h'$; $f'' : g'' : h''$, we may if we please assume

$$\begin{aligned} f^2 + g^2 + h^2 &= 1, \\ f'^2 + g'^2 + h'^2 &= 1, \\ f''^2 + g''^2 + h''^2 &= 1, \end{aligned}$$

which, combined with the second system of equations, gives

$$\begin{aligned} f^2 + f'^2 + f''^2 &= 1, \\ g^2 + g'^2 + g''^2 &= 1, \\ h^2 + h'^2 + h''^2 &= 1. \end{aligned}$$

We have, consequently,

$$\begin{aligned} A + B + C &= A(f^2 + f'^2 + f''^2) + B(g^2 + g'^2 + g''^2) + C(h^2 + h'^2 + h''^2) \\ &= (Af^2 + Bg^2 + Ch^2) + (Af'^2 + Bg'^2 + Ch'^2) + (Af''^2 + Bg''^2 + Ch''^2), \end{aligned}$$

i.e.

$$A + B + C = 0,$$

for the condition that it may be possible in the conic

$$Ax^2 + By^2 + Cz^2 = 0$$

to describe an infinity of triangles the angles of which are conjugate points with respect to the conic $x^2 + y^2 + z^2 = 0$.

The equation of the conic $Ax^2 + By^2 + Cz^2 = 0$ may be written in the form

$$(b^2c^2 - c^2a^2 - a^2b^2 + 2a^2bc)\,x^2 + (c^2a^2 - a^2b^2 - b^2c^2 + 2ab^2c)\,y^2 + (a^2b^2 - b^2c^2 - c^2a^2 + 2abc^2)\,z^2 = 0,$$

which gives the values of A, B, C; or again in the form

$$\begin{aligned} &2(bc + ca + ab)(bcx^2 + cay^2 + abz^2) \\ &- (bc + ca + ab)^2 (\ x^2 + \ y^2 + \ z^2) \\ &+ 4abc\,(ax^2 + by^2 + cz^2) = 0; \end{aligned}$$

where it should be observed that $bcx^2 + cay^2 + abz^2 = 0$ is the equation of the conic which is the polar of $ax^2 + by^2 + cz^2 = 0$ with respect to $x^2 + y^2 + z^2 = 0$. It is very easy from the last form to deduce the equation of the auxiliary conic, when the conics $ax^2 + by^2 + cz^2 = 0$, $x^2 + y^2 + z^2 = 0$ are replaced by conics represented by perfectly general equations.

The condition $A + B + C = 0$ gives, substituting the values of A, B, C,

$$b^2c^2 + c^2a^2 + a^2b^2 - 2abc\,(a + b + c) = 0;$$

or in a more convenient form,

$$(bc + ca + ab)^2 - 4abc\,(a + b + c) = 0,$$

as the condition in order that it may be possible to inscribe in the conic $x^2 + y^2 + z^2 = 0$ an infinity of triangles, the sides of which touch the conic $ax^2 + by^2 + cz^2 = 0$: this agrees perfectly with the general theorem.

It is convenient to add (as a somewhat more general form of the equation $A + B + C = 0$), that the condition in order that it may be possible in the conic $Ax^2 + By^2 + Cz^2 = 0$ to inscribe an infinity of triangles the angles of which are conjugate points with respect to the conic $A_1x^2 + B_1y^2 + C_1z^2 = 0$, is

$$\frac{A}{A_1} + \frac{B}{B_1} + \frac{C}{C_1} = 0.$$

But the problem to find the condition in order that it may be possible in the conic $x^2 + y^2 + z^2 = 0$ to inscribe an infinity of triangles the sides of which touch the conic $ax^2 + by^2 + cz^2 = 0$, may, by the assistance of the geometrical theorem to be presently mentioned, be at once reduced to the problem:

To find the condition in order that it may be possible in the conic $x^2 + y^2 + z^2 = 0$ to inscribe an infinity of triangles the sides of which are conjugate points with respect to a conic

$$A_1x^2 + B_1y^2 + C_1z^2 = 0.$$

The theorem referred to is as follows:

Theorem. If the chord PP' of a conic S envelope a conic σ, the points P, P' are harmonics with respect to a conic T which has with S, σ, a system of common conjugate points.

Take for the equation of S,

$$x^2 + y^2 + z^2 = 0,$$

and for the equation of σ,

$$ax^2 + by^2 + cz^2 = 0;$$

then if (x_1, y_1, z_1), (x_2, y_2, z_2) are the coordinates of the points P, P' respectively, we have

$$x_1^2 + y_1^2 + z_1^2 = 0,$$
$$x_2^2 + y_2^2 + z_2^2 = 0,$$

and the condition in order that the chord may touch the conic σ is

$$bc(y_1z_2 - y_2z_1)^2 + ca(z_1x_2 - z_2x_1)^2 + ab(x_1y_2 - x_2y_1)^2 = 0.$$

But we have

$$\begin{aligned}(y_1z_2 - y_2z_1)^2 &= (y_1^2 + z_1^2)(y_2^2 + z_2^2) - (y_1y_2 + z_1z_2)^2 \\ &= x_1^2x_2^2 - (y_1y_2 + z_1z_2)^2 \\ &= (x_1x_2 + y_1y_2 + z_1z_2)(x_1x_2 - y_1y_2 - z_1z_2),\end{aligned}$$

and making the like change in the analogous quantities, and putting for shortness

$$\begin{aligned}\alpha &= -bc + ca + ab, \\ \beta &= bc - ca + ab, \\ \gamma &= bc + ca - ab,\end{aligned}$$

the condition in question becomes

$$(x_1x_2 + y_1y_2 + z_1z_2)(\alpha x_1x_2 + \beta y_1y_2 + \gamma z_1z_2) = 0.$$

But the equation $x_1x_2 + y_1y_2 + z_1z_2 = 0$ must be rejected, as giving with the equations $x_1^2 + y_1^2 + z_1^2 = 0$, $x_2^2 + y_2^2 + z_2^2 = 0$ the relation $x_1 : y_1 : z_1 = x_2 : y_2 : z_2$; we have therefore

$$\alpha x_1x_2 + \beta y_1y_2 + \gamma z_1z_2 = 0,$$

which implies that the points (x_1, y_1, z_1) and (x_2, y_2, z_2) are harmonics with respect to the conic

$$\alpha x^2 + \beta y^2 + \gamma z^2 = 0,$$

which is a conic having with S, σ, a system of common conjugate points. The equation may also be written

$$(-bc + ca + ab)x^2 + (bc - ca + ab)y^2 + (bc + ca - ab)z^2 = 0;$$

or, as it may also be written,

$$(bc+ca+ab)(x^2+y^2+z^2)-2(bcx^2+caz^2+abx^2);$$

and, as before remarked, $bcx^2+cay^2+abz^2=0$ is the equation of the conic which is the polar of $ax^2+by^2+cz^2=0$ with respect to $x^2+y^2+z^2=0$.

The condition in order that there may be inscribed in the conic $x^2+y^2+z^2=0$ an infinity of triangles the angles of which are conjugate points with respect to the conic $\alpha x^2+\beta y^2+\gamma z^2=0$, is

$$\frac{1}{\alpha}+\frac{1}{\beta}+\frac{1}{\gamma}=0;$$

or writing this equation under the form $\beta\gamma+\gamma\alpha+\alpha\beta=0$, and substituting for α, β, γ their values, we have the equation already found, as the condition in order that it may be possible in the conic $x^2+y^2+z^2=0$ to inscribe an infinity of triangles the sides of which touch the conic $ax^2+by^2+cz^2=0$.

THEOREM. Let

$$ax^2+by^2+cz^2=0$$

be the equation of a spherical conic, and let $(\xi:\eta:\zeta)$, a point on the conic, be the pole of a great circle cutting the conic in two points; the conic intersects upon the great circle an arc given by the equation

$$\cos\delta=\frac{(a+b+c)\sqrt{\xi^2+\eta^2+\zeta^2}}{\sqrt{(a+b+c)^2(\xi^2+\eta^2+\zeta^2)-4(bc\xi^2+ca\eta^2+ab\zeta^2)}};$$

hence if $a+b+c=0$, $\delta=90°$; or there may be inscribed in the conic an infinity of triangles having each of their sides equal to 90°.

It is worth while, in connexion with the subject, and for the sake of a remark to which they give rise, to reproduce in a short compass some results long ago obtained by Jacobi and Richelot. The following are Jacobi's formulæ for the chords of a circle subjected to the condition of touching another circle; viz. if in the figure we put

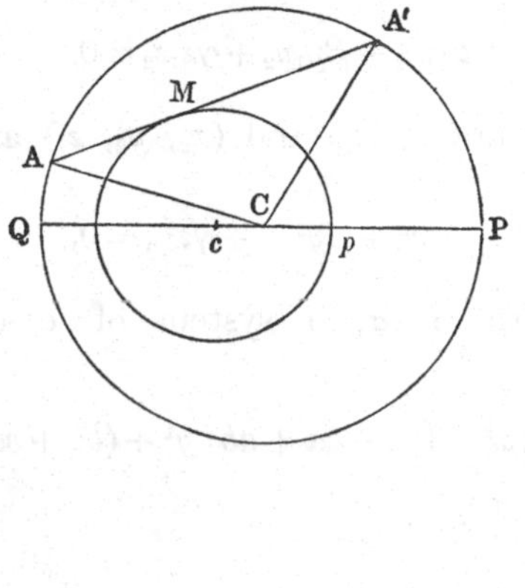

$$CP = R,$$
$$cp = r,$$
$$Cc = a,$$
$$\angle ACP = 2\phi,$$
$$\angle A'CP = 2\phi';$$

then it is clear from geometrical considerations that

$$\frac{d\phi}{MA} = \frac{d\phi'}{MA'}.$$

We have

$$\begin{aligned}\overline{MA}^2 = \overline{cA}^2 - \overline{cM}^2 &= a^2 + R^2 + 2aR\cos 2 - \phi r^2 \\ &= (a+R)^2 - r^2 - 4aR\sin^2\phi \\ &= \{(a+R)^2 - r^2\}(1 - k^2 \sin 2\phi),\end{aligned}$$

or

$$\frac{d\phi}{\sqrt{1-k^2\sin^2\phi}} = \frac{d\phi'}{\sqrt{1-k^2\sin^2\phi'}},$$

where

$$k^2 = \frac{4aR}{(a+R)^2 - r^2},$$

and therefore also

$$k'^2 = \frac{(R-a)^2 - r^2}{(R+a)^2 - r^2}.$$

It will be convenient for comparison with the formulæ of Richelot to write $\angle ACQ = 2\psi$; this gives

$$2\psi = \pi - 2\phi,$$

and the differential equation thus becomes

$$\frac{d\psi}{\sqrt{a^2+R^2-r^2-2aR\cos 2\psi}} = \frac{d\psi'}{\sqrt{a^2+R^2-r^2-2aR\cos 2\psi'}},$$

i.e.

$$\frac{d\psi}{\sqrt{m-n\cos 2\psi}} = \frac{d\psi'}{\sqrt{m-n\cos 2\psi'}};$$

or if

$$\tan\psi = g\tan\theta,$$

and therefore

$$\cos 2\psi = \frac{1-g^2\tan^2\theta}{1+g^2\tan^2\theta},$$

$$m - n\cos 2\psi = \frac{(m-n)\cos^2\theta + (m+n)g^2\sin^2\theta}{\cos^2\theta + g^2\sin^2\theta};$$

or if $g^2 = \dfrac{m-n}{m+n}$, then this is

$$= \frac{m-n}{\cos^2\theta + \dfrac{m-n}{m+n}\sin^2\theta},$$

$$= \frac{m-n}{1-\dfrac{2n}{m+n}\sin^2\theta}$$

and we have also

$$d\psi = \frac{\sqrt{\dfrac{m-n}{m+n}}\,d\theta}{1-\dfrac{2n}{m+n}\sin^2\theta},$$

and thence

$$\frac{d\psi}{\sqrt{m-n\cos 2\psi}} = \frac{d\theta}{\sqrt{m+n}\sqrt{1-\dfrac{2n}{m+n}\sin^2\theta}},$$

that is,

$$\frac{d\psi}{\sqrt{m-n\cos 2\psi}} = \frac{d\theta}{\sqrt{m+n}\sqrt{1-k^2\sin^2\theta}},$$

where $k^2 = \dfrac{2n}{m+n}$ has the same value as before. Hence the relation between θ, θ' is

$$\frac{d\theta}{\sqrt{1-k^2\sin^2\theta}} = \frac{d\theta'}{\sqrt{1-k^2\sin^2\theta'}},$$

which is identical with that between ϕ and ϕ'; and in fact the equation between θ, ϕ is

$$\tan\theta\tan\phi = \frac{1}{k'},$$

which, if $\phi = \text{am}\, u$, gives $\theta = \text{am}\,(K-u)$.

The differential equation contains only a single arbitrary parameter; hence the same differential equation might have been obtained from different values of a, R, the parameters which determine the circle enveloped by the moveable chord. The condition for this of course is $\dfrac{m'}{n'} = \dfrac{m}{n}$, that is

$$\frac{a^2+R^2-r^2}{2aR} = \frac{a'+R^2-r'^2}{2a'R},$$

or as the equation may be written

$$(aa'-R^2)(a-a') - R(a'r^2 - ar'^2) = 0;$$

this implies that the enveloped circles intersect the other circle in the same two points, or that all the circles have a common chord.

Suppose for $\psi = 0$, we have $\psi' = \beta$, then it is easy to see geometrically that

$$\tan^2 \beta = \frac{(R-a)^2 - r^2}{r^2};$$

let the corresponding value of θ' be $\theta' = \alpha$, i.e. suppose that for $\theta = 0$, we have $\theta' = \alpha$, then

$$\tan^2 \alpha = \frac{(R+a)^2 - r^2}{(R-a)^2 - r^2} \cdot \frac{(R-a)^2 - r^2}{r^2},$$

i.e.

$$\tan^2 \alpha = \frac{(R+a)^2 - r^2}{r^2};$$

or, what is the same thing,

$$\sec \alpha = \frac{R}{r} + \frac{a}{r};$$

and α having this value, we have for the finite relation between θ, θ',

$$F\theta' = F\theta + F\alpha.$$

Richelot has shown, by precisely similar reasoning, that for circles of the sphere we have

$$\frac{d\psi}{\sqrt{\cos^2 r - (\cos R \cos a + \sin R \sin a \cos 2\psi)^2}} = \frac{d\psi'}{\sqrt{\cos^2 r - (\cos R \cos a + \sin R \sin a \cos 2\psi')^2}}$$

which is of the form

$$\frac{d\psi}{\sqrt{1 - (\lambda + \mu \cos 2\psi)^2}} = \frac{d\psi'}{\sqrt{1 - (\lambda + \mu \cos 2\psi')^2}},$$

where

$$\lambda = \frac{\cos R \cos a}{\cos r},$$

$$\mu = \frac{\sin R \sin a}{\cos r}.$$

It is very important to remark, that this equation contains the two parameters λ, μ, so that the same equation cannot be obtained with any new values of the parameters a, r; or the formulæ *in plano* for three or more circles do not apply to circles of the sphere: the geometrical reason for this is as follows, viz. in the plane a circle is a conic passing through two fixed points (the circular points at ∞), and consequently any number of circles having a common chord are in fact to be considered as conics each of which passes through the same four points. But circles of the sphere are not spherical conics passing through two fixed points, but are merely spherical conics having a double contact with an imaginary spherical conic (viz. the curve of intersection of the sphere with a sphere radius zero); hence circles of the sphere having a common spherical chord are not spherical conics passing through the same four points. I am not sure whether this remark as to the ground of the distinction between the theory of circles *in plano* and that of circles on the sphere has been explicitly made in any of the treatises on spherical geometry.

To reduce the equation, write

$$\tan\psi = \sqrt{\frac{1-(\lambda+\mu)}{1-(\lambda-\mu)}}\tan\theta;$$

then, after a simple reduction,

$$\frac{d\psi}{\sqrt{1-(\lambda+\mu\cos 2\psi)^2}} = \frac{d\theta}{\sqrt{(1+\mu)^2-\lambda^2}\sqrt{1-\dfrac{4\mu}{(1+\mu)^2-\lambda^2}\sin^2\theta}};$$

or the relation between the two values of θ is

$$\frac{d\theta}{\sqrt{1-k^2\sin^2\theta}} = \frac{d\theta'}{\sqrt{1-k^2\sin^2\theta'}},$$

where

$$k^2 = \frac{4\mu}{(1+\mu)^2-\lambda^2},$$

that is

$$k^2 = \frac{4\dfrac{\tan R}{\tan r}\cdot\dfrac{\sin a}{\cos R\sin r}}{\left(\dfrac{\tan R}{\tan r}+\dfrac{\sin a}{\cos R\sin r}\right)^2-1}.$$

Suppose that for $\psi=0$, we have $\psi'=\beta$, it is easy to see that

$$\tan^2\beta = \frac{\sin^2(R-a)-\sin^2 r}{\cos^2 R\sin^2 r};$$

let the corresponding value of θ' be $\theta'=\alpha$, i.e. suppose that for $\theta=0$, we have $\theta'=\alpha$, then

$$\tan^2\alpha = \frac{1-\dfrac{\cos(R+a)}{\cos r}}{1-\dfrac{\cos(R-a)}{\cos r}}\cdot\frac{\sin^2(R-a)-\sin^2 r}{\cos^2 R\sin^2 r}$$

$$= \frac{\cos r-\cos(R+a)}{\cos r-\cos(R-a)}\cdot\frac{\cos^2 r-\cos^2(R-a)}{\cos^2 R\sin^2 r}$$

$$= \frac{(\cos r-\cos(R+a))(\cos r+\cos(R-a))}{\cos^2 R\sin^2 r}$$

$$= \frac{(\cos r+\sin R\sin a)^2-\cos^2 R\cos^2 a}{\cos^2 R\sin^2 r}$$

$$= \frac{(\cos r\sin R+\sin a)^2-\cos^2 R\sin^2 r}{\cos^2 R\sin^2 r},$$

i.e.

$$\tan^2 \alpha = \left(\frac{\tan R}{\tan r} + \frac{\sin a}{\cos R \sin r}\right)^2 - 1,$$

whence

$$\sec \alpha = \left(\frac{\tan R}{\tan r} + \frac{\sin a}{\cos R \sin r}\right);$$

and α having this value, the finite relation between θ, θ' is

$$F\theta' = F\theta + F\alpha.$$

By comparing with the corresponding formula *in plano*, we arrive at Richelot's conclusion, that the formulæ for the sphere may be deduced from those *in plano* by writing in the place of $\frac{R}{r}$, $\frac{a}{r}$, the functions $\frac{\tan R}{\tan r}$, $\frac{\sin a}{\cos R \sin r}$, respectively.

2, *Stone Buildings, October* 1, 1856.

203.

ON THE THEORY OF THE ANALYTICAL FORMS CALLED TREES.

[*From the Philosophical Magazine*, vol. XIII. (1857), pp. 172—176.]

A SYMBOL such as $A\partial_x + B\partial_y + \ldots$, where A, B, &c. contain the variables x, y, &c. in respect to which the differentiations are to be performed, partakes of the natures of an operand and operator, and may be therefore called an Operandator. Let P, Q, R, ... be any operandators, and let U be a symbol of the same kind, or to fix the ideas, a mere operand; PU denotes the result of the operation P performed on U, and QPU denotes the result of the operation Q performed on PU; and generally in such combinations of symbols, each operation is considered as affecting the operand denoted by means of all the symbols on the right of the operation in question. Now considering the expression QPU, it is easy to see that we may write

$$QPU = (Q \times P)\, U + (QP)\, U,$$

where on the right-hand side $(Q \times P)$ and (QP) signify as follows: viz. $Q \times P$ denotes the mere algebraical product of Q and P, while QP (consistently with the general notation as before explained) denotes the result of the operation Q performed upon P as operand; and the two parts $(Q \times P)\, U$ and $(QP)\, U$ denote respectively the results of the operations $(Q \times P)$ and (QP) performed each of them upon U as operand. It is proper to remark that $(Q \times P)$ and $(P \times Q)$ have precisely the same meaning, and the symbol may be written in either form indifferently. But without a more convenient notation, it would be difficult to find the corresponding expressions for $RQPU$, &c. This, however, can be at once effected by means of the analytical forms called trees (see figs. 1, 2, 3), which contain all the trees which can be formed with one branch, two branches, and three branches respectively.

The inspection of these figures will at once show what is meant by the term in question, and by the terms *root*, *branches* (which may be either main branches, intermediate branches, or free branches), and *knots* (which may be either the root itself, or proper knots, or the extremities of the free branches). To apply this to the question in hand, PU consists of a single term represented by fig. 1 (*bis*); QPU consists, as above, of two terms represented by the two parts of fig. 2 (*bis*), viz. the

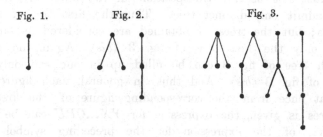

Fig. 1.

Fig. 2.

Fig. 3.

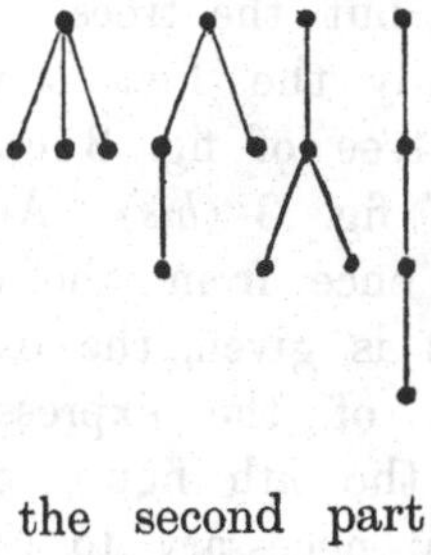

first part represents the term $(Q \times P)\,U$, and the second part represents the term $(QP)\,U$. And it is obvious that fig. 2 (*bis*) is at once formed from the figure 1 (*bis*) by adding on a branch terminated by Q at each of the knots of the single part of fig. 1 (*bis*). In like manner $RQPU$ consists of six terms represented by the six parts of fig. 3 (*bis*), and this figure is at once formed from fig. 2 (*bis*) by adding on a branch terminated by R at each knot of each part of fig. 2 (*bis*). It is

Fig. 1 (*bis*).

Fig. 2 (*bis*).

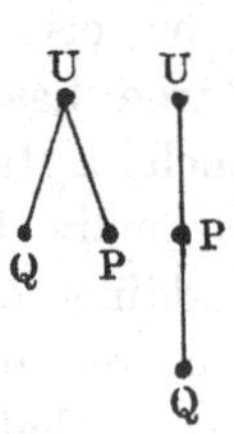

Fig. 3 (*bis*).

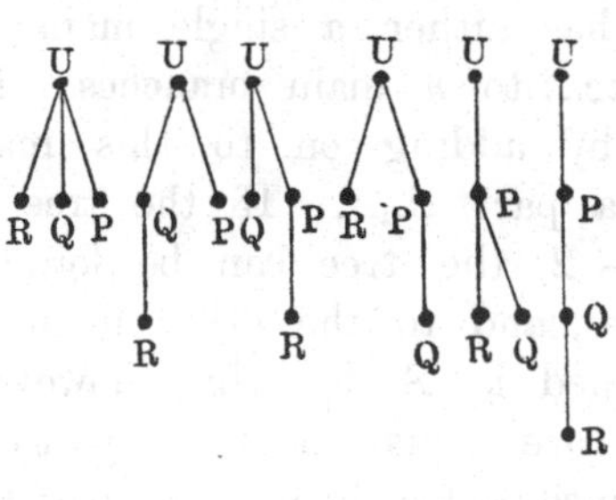

hardly necessary to remark that the first part of fig. 3 (*bis*) denotes what, in the notation first explained, would be denoted by $(R \times Q \times P)\,U$, the second term what would in like manner be denoted by $(RQ \times P)\,U$, and so on, the last part being the term which would be denoted by $\big((RQ)\,P\big)\,U$; viz. R operates upon Q, giving the operandator RQ, which operates upon P, giving the operandator $(RQ)\,P$, which finally operates upon U.

The figures 1 (*bis*), 2 (*bis*), &c. contain the same trees as are contained in the corresponding figures 1, 2, &c.; only, on account of the different modes of filling up, trees are considered as so many distinct trees in a figure of the second set which are considered as one and the same tree in the corresponding figure of the first set. A difference in the number of trees first occurs in the figures 3 and 3 (*bis*), the first of which contains only four, while the latter contains six trees, viz. the first tree, the second, third and fourth trees, the fifth tree and the sixth tree of fig. 3 (*bis*) correspond respectively to the first tree, the second tree, the third tree, and the fourth tree of fig. 3. To derive fig. 3 (*bis*) from fig. 3, we must fill up the trees of fig. 3 with U at the root and R, Q, P at the other knots in every possible manner, subject only to the restriction, that, reckoning up from the extremity of a free branch to the root, there must not be any transposition in the order of the symbols RQP, and taking care to admit only distinct trees. Thus the first tree of fig. 3 might be filled up in six ways; but the trees so obtained are considered as one and the same tree, and we have only the first tree of fig. 3 (*bis*). Again, on account of the restriction, the fourth tree of fig. 3 can be filled up in one way only, and we have thus the sixth tree of fig. 3 (*bis*). And thus, in general, each figure of the second set can be formed at once from the corresponding figure of the first set; or when the first set of figures is given, the expression for $YX...QPU$ can be formed directly without the assistance of the expression for the preceding symbol $X...QPU$; the number of terms for the nth figure of the second set is obviously $1 . 2 . 3 ... n$, and consequently it is only necessary to count the terms in order to ascertain that no admissible mode of filling up has been omitted.

The number of parts in any one of the figures of the first set is much smaller than the number of parts in the corresponding figure of the second set; and the law for the number of parts, i.e. for the number A_n of the trees with n branches, is a very singular one. To obtain this law, we must consider how the trees with n branches can be formed by means of those of a smaller number of branches. A tree with n branches has either a single main branch, or else two main branches, three main branches, &c....to n main branches. If the tree has one main branch, it can only be formed by adding on to this main branch a tree with $(n-1)$ branches, i.e. A_n contains a part A_{n-1}. If the tree has two main branches, then $p+q$ being a partition of $n-2$, the tree can be formed by adding on to one main branch a tree of p branches, and to the other main branch a tree of q branches; the number of trees so obtained is A_pA_q: this, however, assumes that the parts p and q are unequal; if they are equal, it is easy to see that the number of trees is only $\tfrac{1}{2}A_p(A_p+1)$. Hence $p+q$ being any partition of $n-2$, A_n contains the part A_pA_q if p and q are unequal, and the part $\tfrac{1}{2}A_p(A_p+1)$ if p and q are equal. In like manner, considering the trees with three main branches, then if $p+q+r$ is any partition of $n-3$, A_n contains the part $A_pA_qA_r$ if p, q, r are unequal; but if two of these numbers, e.g. p and q, are equal, then the part $\tfrac{1}{2}A_p(A_p+1)A_r$; and if p, q, r are all equal, then the part $\tfrac{1}{6}A_p(A_p+1)(A_p+2)$; and so on, until lastly we have a single tree with n main branches, or A_n contains the part unity. A little consideration will show that the preceding rule for the formation of the number A_n is completely expressed by the equation

$$(1-x)^{-1}(1-x^2)^{-A_1}(1-x^3)^{-A_2}(1-x^4)^{-A_3}... = 1 + A_1x + A_2x^2 + A_3x^3 + A_4x^4 + \&c.,$$

and consequently that we may, by means of this equation, calculate successively for the different values of n the number A_n of the trees with n branches. The calculation may be effected very easily as follows: [the table as originally printed contained at the end of it some errors of calculation which were corrected, *B.A.* Report for 1875, p. 258].

$A_1 =$	1	(1)	1	1	1	1	1	1	1	1	1	$(1-x^2)^{-1}$
			1	1	1	1	1	1	1	1	1	
					1	1	1	1	1	1	1	
							1	1	1	1	1	
									1	1	1	
											1	
$A_2 =$	1	1	(2)	2	3	3	4	4	5	5	5	$(1-x^3)^{-2}$
				2	2	4	4	6	6	8	8	
							3	3	6	6	9	
										4	4	
$A_3 =$	1	1	2	(4)	5	7	11	13	17	23	27	$(1-x^4)^{-4}$
					4	4	8	16	20	28	44	
									10	10	20	
$A_4 =$	1	1	2	4	(9)	11	19	29	47	61	91	$(1-x^5)^{-9}$
						9	9	18	36	81	99	
											45	
$A_5 =$	1	1	2	4	9	(20)	28	47	83	142	235	$(1-x^6)^{-20}$
							20	20	40	80	180	
$A_6 =$	1	1	2	4	9	20	(48)	67	123	222	415	$(1-x^7)^{-48}$
								48	48	96	192	
$A_7 =$	1	1	2	4	9	20	48	(115)	171	318	607	$(1-x^8)^{-115}$
									115	115	230	
$A_8 =$	1	1	2	4	9	20	48	115	(286)	433	837	$(1-x^9)^{-306}$
										286	286	
$A_9 =$	1	1	2	4	9	20	48	115	286	(719)	1123	$(1-x^{10})^{-775}$
											719	
$A_r =$		1	2	4	9	20	48	115	286	719	1842	
for $r =$		1	2	3	4	5	6	7	8	9	10	

I have had occasion, for another purpose, to consider the question of finding the number of trees with a given number of free branches, bifurcations at least. Thus, when the number of free branches is three, the trees of the form in question are

Fig.

those in the annexed figure, and the number is therefore two. It is not difficult to see that we have in this case (B_r being the number of such trees with r free branches),

$$(1-x)^{-1}(1-x^2)^{-B_2}(1-x^3)^{-B_3}(1-x^4)^{-B_4}\ldots = 1 + x + 2B_2x^2 + 2B_3x^3 + 2B_4x^4 + \&c.;$$

and a like process of development gives:

$B_r =$	1	2	5	12	33	90
for $r =$	2	3	4	5	6	7

I may mention, in conclusion, that I was led to the consideration of the foregoing theory of trees by Professor Sylvester's researches on the change of the independent variables in the differential calculus.

2, *Stone Buildings, January* 2, 1856.

204.

ON A PROBLEM IN THE PARTITION OF NUMBERS.

[From the *Philosophical Magazine*, vol. XIII. (1857), pp. 245—248.]

It is required to find the number of partitions into a given number of parts, such that the first part is unity, and that no part is greater than twice the preceding part.

Commencing to form the partitions in question, these are

1	1	1	1	1	1	1	1	1	&c.;
	1	2	1	1	2	2	2	2	
			1	2	1	2	3	4	

and if we were to proceed to the 4-partitions, each 3-partition ending in 1 would give rise to two such partitions; each 3-partition ending in 2 to four such partitions; each 3-partition ending in 3 to six partitions; and each 3-partition ending in 4 to eight such partitions. We form in this manner the Table:

	ending in																
Number of	1	2	3	4	5	6	7	8	9	10	11	12	13	14	15	16	Totals
1-partitions	1																1
2-partitions	1	1															2
3-partitions	2	2	1	1													6
4-partitions	6	6	4	4	2	2	1	1									26
5-partitions	26	26	20	20	14	14	10	10	6	6	4	4	2	2	1	1	166

&c.;

and we are thus led to the series

1,

1, 2,

1, 2, 4, 6,

1, 2, 4, 6, 10, 14, 20, 26,

&c.;

where, considering 0 as the first term of each series, the first differences of any series are the terms twice repeated of the next preceding series: thus the differences of the fourth series are 1, 1, 2, 2, 4, 4, 6, 6. It is moreover clear that the first half of each series is precisely the series which immediately precedes it; we need, in fact, only consider a single infinite series, 1, 2, 4, 6, &c. It is to be remarked, moreover, that in the column of totals, the total of any line is precisely the first number in the next succeeding line.

Consider in general a series A, B, C, D, E, &c., and a series A', B', C', D', E', &c. derived from it as follows:

$$\begin{aligned} A' &= 1A, \\ B' &= 2A, \\ C' &= 2A + B, \\ D' &= 2A + 2B, \\ E' &= 2A + 2B + C, \\ F' &= 2A + 2B + 2C, \\ &\text{\&c.}; \end{aligned}$$

viz. the first differences of the series 0, A', B', C', D', E', &c. are A, A, B, B, C, C, &c. Then multiplying by 1, x, x^2, &c. and adding, we have

$$A' + B'x + C'x^2 + \text{\&c.} = (1 + 2x + 2x^2 + \ldots)(A + Bx^2 + Cx^4 + \text{\&c.})$$

$$= \frac{1+x}{1-x}(A + Bx^2 + Cx^4 + \text{\&c.});$$

and if we form in a similar manner A'', B'', C'', D'', &c. from A', B', C', D', &c. and so on, we have

$$A'' + B''x + C''x^2 + \text{\&c.} = \frac{1+x}{1-x}(A' + B'x^2 + C'x^4 + \text{\&c.})$$

$$= \frac{1+x}{1-x}\,\frac{1+x^2}{1-x^2}(A + Bx^4 + Cx^8 + \text{\&c.}),$$

and so on. Write $A = 1$, and suppose that the process is repeated an indefinite number of times, we have

$$1 + \mathfrak{B}x + \mathfrak{C}x^2 + \mathfrak{D}x^3 + \text{\&c.} = \frac{1+x\,.\,1+x^2\,.\,1+x^4\,.\,\text{\&c.}}{1-x\,.\,1-x^2\,.\,1-x^4\,.\,\text{\&c.}};$$

and the coefficients 1, $\mathfrak{B}$, $\mathfrak{C}$, $\mathfrak{D}$, &c. are precisely those of the infinite series 1, 2, 4, 6, &c. We have more simply

$$1 + \mathfrak{B}x + \mathfrak{C}x^2 + \mathfrak{D}x^3 + \&c. = \frac{1}{(1-x)^2(1-x^2)(1-x^4)(1-x^8)\,\&c.},$$

which gives rise to the following very simple algorithm for the calculation of the coefficients:

<table>
<tr><td>1,</td><td>2,</td><td>3,</td><td>4,</td><td>5,</td><td>6,</td><td>7,</td><td>8,</td><td>9,</td><td>10,</td><td>11,</td><td>12,</td><td>13,</td><td>14,</td><td>15,</td><td>16</td></tr>
<tr><td>0,</td><td>0;</td><td>1,</td><td>2,</td><td>4,</td><td>6,</td><td>9,</td><td>12,</td><td>16,</td><td>20,</td><td>25,</td><td>30,</td><td>36,</td><td>42,</td><td>49,</td><td>56</td></tr>
<tr><td>1,</td><td>2,</td><td>4,</td><td>6,</td><td>9,</td><td>12,</td><td>16,</td><td>20,</td><td>25,</td><td>30,</td><td>36,</td><td>42,</td><td>49,</td><td>56,</td><td>64,</td><td>72</td></tr>
<tr><td>0,</td><td>0,</td><td>0,</td><td>0;</td><td>1,</td><td>2,</td><td>4,</td><td>6,</td><td>10,</td><td>14,</td><td>20,</td><td>26,</td><td>35,</td><td>44,</td><td>56,</td><td>68</td></tr>
<tr><td>1,</td><td>2,</td><td>4,</td><td>6,</td><td>10,</td><td>14,</td><td>20,</td><td>26,</td><td>35,</td><td>44,</td><td>56,</td><td>68,</td><td>84,</td><td>100,</td><td>120,</td><td>140</td></tr>
<tr><td>0,</td><td>0,</td><td>0,</td><td>0,</td><td>0,</td><td>0,</td><td>0,</td><td>0;</td><td>1,</td><td>2,</td><td>4,</td><td>6,</td><td>10,</td><td>14,</td><td>20,</td><td>26</td></tr>
<tr><td>1 |</td><td>2 |</td><td>4,</td><td>6 |</td><td>10,</td><td>14,</td><td>20,</td><td>26 |</td><td>36,</td><td>46,</td><td>60,</td><td>74,</td><td>94,</td><td>114,</td><td>140,</td><td>166|</td></tr>
</table>

&c.

The last line is marked off into periods of (reckoning from the beginning) 1, 2, 4, 8, &c.; and by what has preceded, the series which gives the number of 1-partitions, 2-partitions, 3-partitions, &c. is found by summing to the end of each period and doubling the results; we thus, in fact, obtain (1), 2, 6, 26, 166, 1626, &c.: and the same series is also given by means of the last terms of the several periods.

The preceding expression for $1+\mathfrak{B}x+\mathfrak{C}x^2+$&c. shows that $\mathfrak{B}$, $\mathfrak{C}$, &c. are the number of partitions of 1, 2, 3, 4, 5, 6, &c. respectively into the parts 1, 1′, 2, 4, 8, &c.: and we are thus led to—

THEOREM. The number of x-partitions (first part unity, no part greater than twice the preceding one) is equal to the number of partitions of $2^{x-1}-1$ into the parts 1, 1′, 2, 4,... 2^{x-2}. Or, again, it is equal to twice the sum of the number of partitions of 0, 1, 2,... $2^{x-2}-1$ respectively into the parts 1, 1′, 2, 4,... 2^{x-3} (where the number of partitions of 0 counts for 1).

For example, the partitions of 0, 1, 2, 3, &c. with the parts 1, 1′, 2,... are

(·)

1, 1′,

1 + 1, 1 + 1′, 1′ + 1′, 2,

1 + 1 + 1, 1 + 1 + 1′, 1 + 1′ + 1′, 1′ + 1′ + 1′, 2 + 1, 2 + 1′,

the numbers of which are 1, 2, 4, 6. Hence, by the first part of the theorem, the number of 3-partitions is 6, and by the second part of the theorem, the number of 4-partitions is

$$2\,(1+2+4+6), = 26.$$

2, *Stone Buildings, March* 17, 1857.

205.

NOTE ON THE SUMMATION OF A CERTAIN FACTORIAL EXPRESSION.

[From the *Philosophical Magazine*, vol. XIII. (1857), pp. 419—423.]

MR KIRKMAN some months ago communicated to me a formula for the double summation of a factorial expression, to which formula he had been led by his researches on the partition of polygons. The formula in a slightly altered form is as follows: viz.

$$\Sigma_x \Sigma_y \frac{[x+y+2]^y}{[y+1]^{y+1}} \frac{[x]^y}{[y]^y} \frac{[r+k-x-y]^{k-1-y}}{[k-y]^{k-y}} \frac{[r-1-x]^{k-1-y}}{[k-1-y]^{k-1-y}} = \frac{2k}{r+3} \frac{[r+k+2]^k}{[k+1]^{k+1}} \frac{[r]^k}{[k]^k},$$

the summation extending from $x=0$ to $x=r-1$, and $y=0$ to $y=k-1$. In the particular case when $k=r$, then all the terms of the series except those in which $y=x$ vanish; and putting therefore $k=r$ and $y=x$, and making a slight change in the form of the right-hand side, the formula becomes

$$\Sigma \frac{[2x+2]^x}{[x+1]^{x+1}} \frac{[2r-2x]^{r-1-x}}{[r-x]^{r-x}} = 4 \frac{[2r+1]^{r-2}}{[r-1]^{r-1}},$$

the summation extending from $x=0$ to $x=r-1$.

We have, in the notation of Gauss, $[m]^m = m(m-1)\ldots 2.1 = \Pi m$, and a factorial $[m]^n$ is expressed in terms of the function Π by the formula $[m]^n = \Pi m \div \Pi(m-n)$. Write also

$$\Pi_1(m-\tfrac{1}{2}) = (m-\tfrac{1}{2})(m-\tfrac{3}{2})\ldots\tfrac{3}{2}.\tfrac{1}{2},$$

we have

$$\Pi 2m \quad = 2^{2m} \ \Pi m \Pi_1(m-\tfrac{1}{2}),$$

$$\Pi(2m+1) = 2^{2m+1} \Pi m \Pi_1(m+\tfrac{1}{2});$$

and transforming the factorials by these formulæ, the series becomes

$$\Sigma \frac{\Pi_1(x+\tfrac{1}{2})\,\Pi_1(r-x-\tfrac{1}{2})}{\Pi(x+2)\,\Pi(r-x+1)} = \frac{2r\Pi_1(r+\tfrac{1}{2})}{\Pi(r+3)},$$

the summation, as before, from $x=0$ to $x=r-1$. This may be written

$$\Sigma \frac{\Pi_1(x+\tfrac{1}{2})}{\Pi_1(\tfrac{1}{2})} \cdot \frac{\Pi_1(r-x-\tfrac{1}{2})}{\Pi_1(r-\tfrac{1}{2})} \frac{\Pi 2}{\Pi(x+2)} \frac{\Pi(r+1)}{\Pi(r-x+1)} = \frac{8r(r+\tfrac{1}{2})}{(r+2)(r+3)},$$

the summation from $x=0$ to $x=r-1$. The general term does not vanish for $x=r$ or $x=r+1$, but it vanishes for all greater values of x; hence if we add to the right-hand side the two terms corresponding to $x=r$ and $x=r+1$, the summation may be extended from $x=0$ to $x=r+1$, or what is the same thing, from $x=0$ indefinitely. The two terms in question are

$$\frac{4(r+\tfrac{1}{2})}{r+2} + \frac{-8(r+\tfrac{3}{2})(r+\tfrac{1}{2})}{(r+2)(r+3)}, = \frac{-4r(r+\tfrac{1}{2})}{(r+2)(r+3)},$$

and the resulting equation is

$$\Sigma \frac{\Pi_1(x+\tfrac{1}{2})}{\Pi_1(\tfrac{1}{2})} \frac{\Pi(r-x-\tfrac{1}{2})}{\Pi_1(r-\tfrac{1}{2})} \frac{\Pi 2}{\Pi(x+2)} \frac{\Pi(r+1)}{\Pi(r-x+1)} = \frac{4r(r+\tfrac{1}{2})}{(r+2)(r+3)},$$

the summation from $x=0$ indefinitely; or substituting for the functions Π and Π_1 their values, the formula is

$$1 + \frac{\tfrac{3}{2}.(r+1)}{3.(r-\tfrac{1}{2})} + \frac{\tfrac{3}{2}.\tfrac{5}{2}.(r+1)r}{3.4.(r-\tfrac{1}{2})(r-\tfrac{3}{2})} + \frac{\tfrac{3}{2}.\tfrac{5}{2}.\tfrac{7}{2}.(r+1)r(r-1)}{3.4.5.(r-\tfrac{1}{2})(r-\tfrac{3}{2})(r-\tfrac{5}{2})} + \&c. = \frac{4r(r+\tfrac{1}{2})}{(r+2)(r+3)},$$

which is a formula obviously belonging to the theory of the hypergeometric series

$$F(\alpha, \beta, \gamma, x) = 1 + \frac{\alpha.\beta}{1.\gamma}x + \frac{\alpha.\alpha+1.\beta.\beta+1}{1.2.\gamma.\gamma+1}x^2 + \&c.;$$

but the formula applicable to the case in hand has probably not been given. It may be proved as follows, premising that I disregard all difficulties arising from infinite values of the functions in the definite integrals, convergency, &c. We have

$$\int_0^1 \theta^{\alpha-1}(1-\theta)^{-\alpha-\gamma-1}(1-x\theta)^\beta\,d\theta$$

$$= \frac{\Pi(\alpha-1)\,\Pi(-\alpha-\gamma-1)}{\Pi(-\gamma-1)}\left(1 + \frac{\alpha.\beta}{1.\gamma}x + \frac{\alpha.\alpha+1.\beta.\beta-1}{1.2.\gamma.\gamma-1}x^2 + \ldots\right).$$

Now we have

$$\int_0 dx \int_0 dx\,(1-x\theta)^\beta = \frac{1}{(\beta+1)(\beta+2)\,\theta^2}(1-x\theta)^{\beta+2} - \frac{1}{(\beta+1)(\beta+2)}\frac{1}{\theta^2} + \frac{x}{(\beta+1)\,\theta};$$

and hence multiplying by dx and integrating from $x=0$, and again multiplying by dx and integrating from $x=0$ to $x=1$, we find

$$\int_0^1 d\theta \,.\, \theta^{\alpha-3}(1-\theta)^{-\alpha-\gamma+\beta+1} - \int_0^1 d\theta \,.\, \theta^{\alpha-3}(1-\theta)^{-\alpha-\gamma-1} + (\beta+2)\int_0^1 d\theta \,.\, \theta^{\alpha-2}(1-\theta)^{-\alpha-\gamma-1}$$

$$= \frac{(\beta+1)(\beta+2)\,\Pi(\alpha-1)\,\Pi(-\alpha-\gamma-1)}{\Pi(-\gamma-1)}\,\tfrac{1}{2}S,$$

if for shortness

$$S = 1 + \frac{\alpha\,.\,\beta}{3\,.\,\gamma} + \frac{\alpha\,.\,\alpha+1\,.\,\beta\,.\,\beta-1}{3\,.\,4\,.\,\gamma\,.\,\gamma-1} + \&\text{c.}$$

Substituting for the definite integrals their values,

$$\frac{\Pi(\alpha-3)\,\Pi(-\alpha-\gamma+\beta+1)}{\Pi(-\gamma+\beta-1)} - \frac{\Pi(\alpha-3)\,\Pi(-\alpha-\gamma-1)}{\Pi(-\gamma-3)} + (\beta+2)\frac{\Pi(\alpha-2)\,\Pi(-\alpha-\gamma-1)}{\Pi(-\gamma-2)}$$

$$= \frac{(\beta+1)(\beta+2)\,\Pi(\alpha-1)\,\Pi(-\alpha-\gamma-1)}{\Pi(-\gamma-1)}\,\tfrac{1}{2}S,$$

whence

$$\tfrac{1}{2}(\beta+1)(\beta+2)\,S = \frac{\Pi(\alpha-3)}{\Pi(\alpha-1)}\,\frac{\Pi(-\alpha-\gamma+\beta+1)}{\Pi(-\alpha-\gamma-1)}\,\frac{\Pi(-\gamma-1)}{\Pi(-\gamma+\beta-1)}$$

$$- \frac{\Pi(\alpha-3)}{\Pi(\alpha-1)}\,\frac{\Pi(-\gamma-1)}{\Pi(-\gamma-3)} + (\beta+2)\frac{\Pi(\alpha-2)}{\Pi(\alpha-1)}\,.\,\frac{\Pi(-\gamma-1)}{\Pi(-\gamma-2)}.$$

The second and third terms are

$$-\frac{1}{(\alpha-1)(\alpha-2)}(\gamma+1)(\gamma+2) - (\beta+2)\frac{1}{\alpha-1}(\gamma+1),$$

which are

$$= -\frac{\gamma+1}{(\alpha-1)(\alpha-2)}(\alpha\beta+2\alpha-2\beta+\gamma-2).$$

For the reduction of the first term we have

$$\Pi(-\alpha-\gamma+\beta+1) = [\beta+1-\alpha-\gamma]^{\beta+2}\,\Pi(-\alpha-\gamma-1),$$

$$\Pi(\beta-\gamma-1) = [\beta-\gamma-1]^{\beta}\,\Pi(-\gamma-1);$$

and we thus find

$$\tfrac{1}{2}(\beta+1)(\beta+2)(\alpha-1)(\alpha-2)\,S = \frac{[\beta+1-\alpha-\gamma]^{\beta+2}}{[\beta-\gamma-1]^{\beta}} - (\gamma+1)(\alpha\beta+2\alpha-2\beta+\gamma-2),$$

where, as before,

$$S = 1 + \frac{\alpha\,.\,\beta}{3\,.\,\gamma} + \frac{\alpha\,.\,\alpha+1\,.\,\beta\,.\,\beta-1}{3\,.\,4\,.\,\gamma\,.\,\gamma-1} + \&\text{c.};$$

this is the formula in hypergeometric series required for the present purpose, and it is certainly true when the series is finite.

Write now

$$\alpha = \tfrac{3}{2}, \quad \beta = r + 1, \quad \gamma = r - \tfrac{1}{2};$$

then the first term is $[1]^{r+2} \div [\tfrac{1}{2}]^{r+1}$, which vanishes on account of the numerator, and the second term is $-\tfrac{1}{2}r(r+\tfrac{1}{2})$, and we have consequently

$$-\tfrac{1}{2}r(r+2)(r+3)\,\tfrac{1}{2}\,.\,\tfrac{1}{2}S = -\tfrac{1}{2}r(r+\tfrac{1}{2}),$$

which gives

$$S = \frac{4r(r+\tfrac{1}{2})}{(r+2)(r+3)},$$

S being here the series in r, the sum of which was required, and the particular case of Mr Kirkman's formula is thus verified. It is probable that the general case might be treated in an analogous manner by first grouping together the terms which correspond to a given difference $x \sim y$, and ultimately summing the sums of these partial series; but I have not examined this question.

2, *Stone Buildings, W.C., April* 18, 1857.

206.

NOTE ON A THEOREM RELATING TO THE RECTANGULAR HYPERBOLA.

[From the *Philosophical Magazine*, vol. XIII. (1857), p. 423.]

THE following theorem is given in a slightly different form by Brianchon and Poncelet, *Gergonne*, vol. XI. [1820], p. 205, viz. Any conic whatever which passes through the three angles of a triangle and the point of intersection of the perpendiculars let fall from the angles of the triangle upon the opposite sides is a rectangular hyperbola; and there is an elegant demonstration depending on the properties of the inscribed hexagon. The theorem is, however, a particular case of the following: viz. "Any conic whatever which passes through the four points of intersection of two rectangular hyperbolas is a rectangular hyperbola." And this, again, is a particular case of the following: viz. If there be a conic Ω and a line P, then considering any two conics U, V such that the points of intersection of P, U are harmonics in respect to the points of intersection of P, Ω, and the points of P, V are also harmonics in respect to the points of intersection of P, Ω, then any conic whatever W which passes through the four points of intersection of U, V will have the like property, viz. the points of intersection of P, W will be harmonics in respect of the points of intersection of P, Ω; a theorem which is an immediate consequence of the theorem that three conics which intersect in the same four points are intersected by any line whatever in six points which are in involution.

2, *Stone Buildings, W.C., April* 23, 1857.

207.

ANALYTICAL SOLUTION OF THE PROBLEM OF TACTIONS.

[From the *Philosophical Magazine*, vol. XIII. (1857), pp. 507—509.]

IT is well known that the eight circles, each of which touches three given circles, are determined as follows:—viz. considering any one in particular of the four axes of similitude of the given circles, and the perpendicular let fall from the radical centre (or centre of the orthotomic circle) of the given circles, there are two of the required tangent circles which have their centres upon the perpendicular, and pass through the points of intersection of the orthotomic circle and the axis of similitude, or in other words, the axis of similitude is a common chord (or radical axis) of the orthotomic circle and the two tangent circles. This suggests the choice of the radical centre for the origin of coordinates; and the resulting formulæ then take very simple forms, and the theorem is verified without difficulty.

Take then the centre of the orthotomic circle as the origin of coordinates, and let the radius of this circle be put equal to unity; then if (α, β), (α', β'), (α'', β'') are the coordinates of the centres of the given circles, the equations of these will be

$$x^2 + y^2 + 1 - 2\alpha\, x - 2\beta\, y = 0,$$
$$x^2 + y^2 + 1 - 2\alpha' x - 2\beta' y = 0,$$
$$x^2 + y^2 + 1 - 2\alpha'' x - 2\beta'' y = 0;$$

and the radii of the circles will be $\sqrt{\alpha^2+\beta^2-1}$, $\sqrt{\alpha'^2+\beta'^2-1}$, $\sqrt{\alpha''^2+\beta''^2-1}$. It will be convenient to write

$$\gamma\ = \pm\sqrt{\alpha^2\ + \beta^2\ - 1},$$
$$\gamma'\ = \pm\sqrt{\alpha'^2 + \beta'^2 - 1},$$
$$\gamma'' = \pm\sqrt{\alpha''^2 + \beta''^2 - 1},$$

where the three several signs $\pm$ are fixed once for all in a determinate manner. If, however, all the signs are reversed, the result is the same, so that the system is one of *four* (not of eight) different systems. The coordinates of a centre of similitude of the second and third circles are

$$\frac{\alpha'\gamma''-\alpha''\gamma'}{\gamma''-\gamma'},\ \frac{\beta'\gamma''-\beta''\gamma'}{\gamma''-\gamma'},$$

and forming the corresponding expressions for the coordinates of the centres of similitude of the third and first circles, and of the first and second circles, the three centres of similitude lie on a line which will be an axis of similitude: to find the equation, write

$$\begin{aligned}
A &= \beta\gamma'-\beta'\gamma+\beta'\gamma''-\beta''\gamma'+\beta''\gamma-\beta\gamma'',\\
B &= \gamma\alpha'-\gamma'\alpha+\gamma'\alpha''-\gamma''\alpha'+\gamma''\alpha-\gamma\alpha'',\\
C &= \alpha\beta'-\alpha'\beta+\alpha'\beta''-\alpha''\beta'+\alpha''\beta-\alpha\beta'',\\
\nabla &= \alpha\beta'\gamma''-\alpha\beta''\gamma'+\alpha'\beta''\gamma-\alpha'\beta\gamma''+\alpha''\beta\gamma'-\alpha''\beta'\gamma,
\end{aligned}$$

values which, it will be observed, give

$$\begin{aligned}
A\alpha+B\beta+C\gamma&=\nabla,\\
A\alpha'+B\beta'+C\gamma'&=\nabla,\\
A\alpha''+B\beta''+C\gamma''&=\nabla;
\end{aligned}$$

then the equation of the axis of similitude is found to be

$$Ax+By-\nabla=0;$$

and hence the equation of the perpendicular let fall from the radical centre upon the axis of similitude is

$$Bx-Ay=0.$$

It should therefore be possible to find two circles having their centres on the last-mentioned line and touching the three given circles Take $A\theta$, $B\theta$ as the coordinates of the centre of one of the two circles, and let r be its radius; the conditions of tangency are

$$\begin{aligned}
r&=\sqrt{(A\theta-\alpha)^2+(B\theta-\beta)^2}\pm\gamma,\\
&=\sqrt{(A\theta-\alpha')^2+(B\theta-\beta')^2}\pm\gamma',\\
&=\sqrt{(A\theta-\alpha'')^2+(B\theta-\beta'')^2}\pm\gamma'',
\end{aligned}$$

where the sign $\pm$ has the same value in each expression. We have consequently

$$(r\mp\gamma)^2=(A\theta-\alpha)^2+(B\theta-\beta)^2;$$

or, observing that $A\alpha+B\beta=\nabla-C\gamma$, and reducing,

$$r^2-(A^2+B^2)\theta^2+2\nabla\theta-1=2\gamma(\pm r+C\theta).$$

Forming the two analogous equations, the three equations will be satisfied if only

$$\begin{aligned} r^2 - (A^2 + B^2)\,\theta^2 + 2\nabla\theta - 1 &= 0, \\ \pm\, r + C\theta \qquad\qquad &= 0. \end{aligned}$$

Eliminating r, we have

$$(A^2 + B^2 - C^2)\,\theta^2 - 2\nabla\theta + 1 = 0,$$

which gives for θ the two values

$$(A^2 + B^2 - C^2)\,\theta = \nabla \pm \sqrt{\nabla^2 - (A^2 + B^2 - C^2)}\,;$$

and then r is determined linearly by the equation

$$r = \mp\, C\theta.$$

The equation of the tangent circle is therefore

$$(x - A\theta)^2 + (y - B\theta)^2 = C^2\theta^2\,;$$

or reducing,

$$x^2 + y^2 - 1 - 2\theta\,(Ax + By - \nabla) = 0\,;$$

and recollecting that $Ax + By - \nabla = 0$ is the equation of the axis of similitude, the equation shows that the axis of similitude is a common chord or radical axis of the orthotomic circle and the two tangent circles.

2, *Stone Buildings, W.C., May* 15, 1857.

208.

NOTE ON THE EQUIPOTENTIAL CURVE $\frac{m}{r}+\frac{m'}{r'}=C$.

[From the *Philosophical Magazine*, vol. XIV. (1857), pp. 142—146.]

THE equation $\frac{m}{r}+\frac{m'}{r'}=C$, where m, m', C are constants, and r, r' are the distances of a point P of the locus from two given points M, M' respectively, expresses that the potential of the attracting or repelling masses m, m' has a constant value at all points of the locus. The locus is obviously a surface of revolution, having the line through the points M, M' for its axis; and instead of the surface, we may consider the section by a plane through the axis, or what is the same thing, we may consider r, r' as the distances *in plano* of a point P of the curve from the given points M, M': such curve may be termed the equipotential curve. I propose in the present Note to investigate in a general manner, and without entering into any analytical detail, the general form of the curve corresponding to different values of the quantity C.

It is proper to remark, that the curve is not altered by changing the signs of each or any of the quantities m, m', C (in fact, analytically the distances r, r' are essentially ambiguous in sign), so that we may without loss of generality consider m, m', C as all of them positive. The different branches of the complete analytical or geometrical curve have distinct mechanical significancies; thus, r, r' being positive, $\frac{m}{r}+\frac{m'}{r}=C$ is the curve for which the potential of the attracting masses m, m' is equal to C; but $\frac{m}{r}-\frac{m'}{r'}=C$ is the curve for which the attracting mass m, and the repulsive mass m', have the potential C; but this is a distinction to which I do not attend. I write for homogeneity $\frac{k}{a}$ instead of C, where a is the distance between the points M, M'; the equation thus becomes

$$\frac{m}{r}+\frac{m'}{r'}=\frac{k}{a},$$

where a is a positive distance, m, m', k may be considered as positive abstract numbers. The curve is obviously a curve of the eighth order. When k is large in comparison with m, m', then since r, r' cannot be both of them small in comparison of a (for if one be small, the other will be nearly equal to a), it is clear that one of these distances, for instance r, will be small, and the other, r', nearly equal to a. We in fact have $\left(\text{neglecting in the first instance } \frac{m'}{r'} \text{ in comparison with } \frac{m}{r}\right) \frac{m}{r} = \frac{k}{a}$, or more accurately, $\frac{m}{r} = \frac{k \pm m'}{a}$, i.e. $r = \frac{m}{k \pm m'}$, which shows that a part of the curve consists of two ovals, which are approximately concentric circles, radii $\frac{m'}{k \pm m'}$, a, about the point M as centre. In like manner a part of the curve consists of two ovals, which are approximately concentric circles, radii $\frac{m'}{k \pm m} a$, about the point M' as centre. I denote by A, B, the two ovals about M, viz. A is the exterior, and B the interior oval; and in like manner by A', B' the two ovals about M', viz. A' is the exterior, and B' the interior oval. The distances *inter se* of the ovals A and B, or of the ovals A' and B', are small in comparison with the radii of these ovals respectively; and if, to fix the ideas, m' be greater than M, then the ovals A', B' are greater than the ovals A, B.

It is easy to see that the curve will have a node or double point on the axis if $k = (\sqrt{m'} \pm \sqrt{m})^2$; and we must first consider the case $k = (\sqrt{m'} + \sqrt{m})^2$. The node lies between the points M, M', and its distances from these points are respectively as $\sqrt{m} : \sqrt{m'}$, that is, it is nearest to M. The transition from the original form is very obvious; the exterior ovals A, A' have gradually expanded until they come in contact, and at the instant of doing so the two ovals change themselves into a figure of eight, AA'. The ovals B, B' also expand and change their form, but they preserve the general character of ovals enclosing the points M, M' respectively. The curve consists of a figure of eight AA', and (inside of the two divisions thereof respectively) of the ovals B, B' enclosing the points M, M'. The half of the curve nearest to M' is, as before, preponderant in magnitude.

The next change when k continues to diminish is an obvious one: the figure of eight opens out into an hourglass-shaped oval AA', while the ovals B, B' continue increasing in magnitude and altering their form.

There will be again a node or double point when $k = (\sqrt{m'} - \sqrt{m})^2$; but to explain the transition to this special form, it is necessary to attend more particularly to the change of form in the oval B' as k approaches to the value in question, viz. this oval lengthens out and begins to twist itself round the oval B; and when k' becomes $= (\sqrt{m'} - \sqrt{m})^2$, then the oval B' has completely encircled B, the two extremities of B' meeting together at the double point, which is a point beyond M (i.e. on the other side to M'), such that its distances from M, M' are in the ratio of $\sqrt{m} : \sqrt{m'}$. And at the instant of contact there is, as in the former case, a modification of the form of the portions which come into contact, so that the node is an ordinary double point. The oval B' has, in fact, become what may be termed a re-entrant figure of eight, , the small part of which encloses the oval B which encloses the point M, while the large part encloses the point M'. The curve consists of the exterior oval AA' (which has probably lost wholly or partially its hourglass form, and is more nearly an ordinary oval), of the re-entrant figure of eight, B', and of the enclosed oval B.

As k continues to diminish, the re-entrant figure of eight, B', breaks up into two detached ovals lB', mB', the larger of which, lB', encloses the other one and also the point M'; while the smaller one, mB', does not enclose M', but encloses the oval B which encloses M; the curve consists of the exterior oval AA', the ovals lB' and mB' which have arisen out of the oval B', and the oval B. As k further decreases, the ovals AA' and lB' continually increase in magnitude, and the ovals mB' and B approximate more and more nearly together; and at length, when k becomes $=0$, the ovals AA' and lB' disappear at infinity, while the ovals mB' and B unite themselves into a circle enclosing M, but not enclosing M': the equation of this circle is, in fact, $\frac{m}{r}+\frac{m'}{r'}=0$; or what is the same thing, $r^2=\frac{m^2}{m'^2}r'^2$, and the points M, M' have, in relation to this circle, the well-known relation that each is the image of the other.

The preceding description is, I think, intelligible without the assistance of a series of figures illustrating the different forms of the curve, but there is no difficulty in actually tracing the curve for any particular values of the constant parameters. Thus

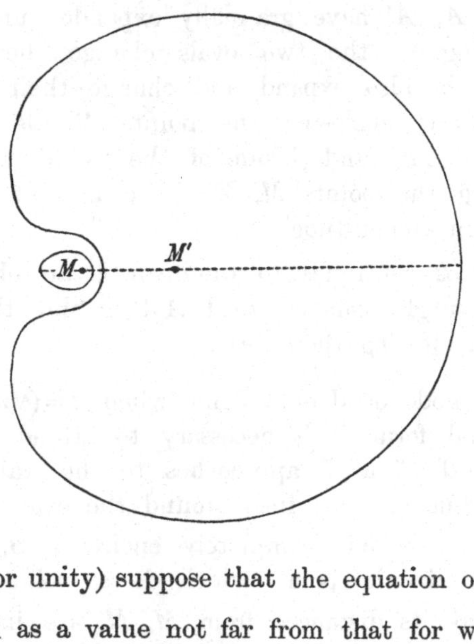

(taking the distance MM' for unity) suppose that the equation of the curve is $\frac{1}{r}+\frac{4}{r'}=1\cdot2$; (the value 1·2 was selected as a value not far from that for which the oval B' becomes a re-entrant figure of eight, though the change of form is so rapid that this value shows only the incipient tendency of the oval B' to take the form in question). The form of the portion of the curve consisting of the two ovals B, B' will be that shown by the figure, which was constructed by points on a double scale with some accuracy.

The case $m=m'$ is an exception, and must be considered separately: the curve is here in all its changes symmetrical about a perpendicular to the axis midway between the two centres M, M'. The curve in the first instance, i.e. when k is greater than $(\sqrt{m}+\sqrt{m})^2, =4m$, consists of the two ovals B, A about M, and the two ovals B', A' about M'. As k decreases to $4m$, the two ovals A, A' gradually increase in magnitude, and at length come together, as before, into a figure of eight, AA'; and as k continues to diminish, the figure of eight opens out into an hourglass form AA', which continues increasing in magnitude, and degenerating into the form of an oval. The interior ovals B, B' approach more and more nearly together, lengthen out in the direction perpendicular to the axis, and present to each other a more and more flattened portion. The second value,

$$k=(\sqrt{m'}-\sqrt{m})^2,$$

which in the general case gives a node, in the present case only arises when $k=0$; and there is not then any node, but the curve degenerates in a similar manner to what happens for $k=0$ in the general case; viz. the oval AA' disappears at infinity, while the ovals B, B' coalesce together (their outer parts disappearing at infinity) into a pair of lines coincident with the perpendicular to the axis midway between the two centres.

2, *Stone Buildings, May* 31, 1857.

209.

A DEMONSTRATION OF SIR W. R. HAMILTON'S THEOREM OF THE ISOCHRONISM OF THE CIRCULAR HODOGRAPH.

[From the *Philosophical Magazine*, vol. XIV. (1857), pp. 427—430.]

IMAGINE a body moving *in plano* under the action of a central force, and let h denote, as usual, the double of the area described in a unit of time; let P be any point of the orbit, then measuring off, on the perpendicular let fall from the centre of force O on the tangent at P to the orbit, a distance OQ equal or proportional to h into the reciprocal of the perpendicular on the tangent, the locus of Q is the hodograph, and the points P, Q are corresponding points of the orbit and hodograph.

It is easy to see that the hodograph is the polar reciprocal of the orbit with respect to a circle having O for its centre, and having its radius equal or proportional to $\sqrt{h}$. And it follows at once that Q is the pole, with respect to this circle, of the tangent at P to the orbit.

In the particular case where the force varies inversely as the square of the distance, the hodograph is a circle. And if we consider two elliptic orbits described about the same centre, under the action of the same central force, and such that the major axes are equal, then (as will be presently seen) the common chord or radical axis of the two hodographs passes through the centre of force.

Imagine an orthotomic circle of the two hodographs (the centre of this circle is of course on the common chord or radical axis of the two hodographs), and consider the arcs intercepted on the two hodographs respectively by the orthotomic circle; then the theorem of the isochronism of the circular hodograph is as follows, viz. the times of hodographic description of the intercepted arcs are equal; in other words, the times of description in the orbits, of the arcs which correspond to the intercepted arcs of the hodographs, are equal. It was remarked by Sir W. R. Hamilton, that the theorem is in fact equivalent to Lambert's theorem, that the time depends only on the chord of the described arc and the sum of the two radius vectors. And this remark suggests a mode of investigation of the theorem. Consider the intercepted arc of one of the hodographs: the tangents to the hodograph at the extremities of this arc are radii of the orthotomic circle; i.e. the corresponding arc of the orbit is the arc cut off by the polar (in respect to the directrix circle by which the hodograph is determined) of the centre of the orthotomic circle; the portion of this polar intercepted by the orbit is the elliptic chord, and this elliptic chord and the sum of the radius vectors at the two extremities of the elliptic chord determine the time of description of the arc; and the values of these quantities, viz. the elliptic chord and the sum of the radius vectors, must be the same in each orbit.

The analytical investigation is not difficult. I take as the equation of the first orbit,

$$r=\frac{a\,(1-e^2)}{1+e\cos(\theta-\varpi)};$$

then the polar of the orbit with respect to a directrix circle $r=c$ is

$$r^2-\frac{c^2e}{a\,(1-e^2)}\,r\cos(\theta-\varpi)-\frac{c^4}{a^2\,(1-e^2)}=0,$$

and putting $c^2=k\sqrt{k}\sqrt{a\,(1-e^2)}$ (where k is a constant quantity, i.e. it is the same in each orbit), the equation becomes

$$r^2-\frac{ek\sqrt{k}}{\sqrt{a\,(1-e^2)}}\,r\cos(\theta-\varpi)-\frac{k^3}{a}=0.$$

But since a is supposed to be the same in each orbit, we may for greater simplicity write $k^3=m^2a$; it will be convenient also to put $e=\sin\kappa$; we have then

$$r=\frac{a\cos^2\kappa}{1+\sin\kappa\cos(\theta-\varpi)}$$

for the equation of the orbit, $r^2=ma\cos\kappa$ for the equation of the directrix circle, and

$$r^2-r\,.\,m\tan\kappa\cos(\theta-\varpi)-m^2=0$$

for the equation of the hodograph.

We have in like manner

$$r=\frac{a\cos^2\kappa'}{1+\sin\kappa'\cos(\theta-\varpi')}$$

for the equation of the second orbit, $r^2=ma\cos\kappa'$ for the equation of the corresponding directrix circle, and

$$r^2-r\,.\,m\tan\kappa'\cos(\theta-\varpi')-m^2=0$$

for that of the hodograph.

The equations of the two hodographs give at once

$$\tan\kappa\cos(\theta-\varpi)-\tan\kappa'\cos(\theta-\varpi')=0$$

for the equation of the common chord or radical axis of the two hodographs,—an equation which shows that, as already noticed, the common chord passes through the origin or centre of force. This equation gives $\theta=\alpha$ if

$$\tan\kappa\cos(\alpha-\varpi)-\tan\kappa'\cos(\alpha-\varpi')=0;$$

i.e. α is a quantity such that the expressions $\tan\kappa\cos(\alpha-\varpi)$ and $\tan\kappa'\cos(\alpha-\varpi')$, which correspond to each other in the two orbits, are equal. We may take R, α as the polar coordinates of the centre of the orthotomic circle (where R is arbitrary); the equation of the polar of this point with respect to the directrix circle $r^2=ma\cos\kappa$, is then at once seen to be

$$r\cos(\theta-\alpha)=\frac{ma\cos\kappa}{R},$$

which is the equation of the line cutting off the arc of the elliptic orbit

$$r=\frac{a\cos^2\kappa}{1+\sin\kappa\sin(\theta-\varpi)}.$$

Writing $\theta-\varpi=\theta-\alpha+(\alpha-\varpi)$, the two equations give

$$\cos(\theta-\alpha)=\frac{A}{r},$$

$$\sin(\theta-\alpha)=\frac{B}{r}+C,$$

if for shortness

$$A=\frac{ma\cos\kappa}{R},$$

$$B=\frac{ma\cos\kappa}{R}\frac{\cos(\alpha-\varpi)}{\sin(\alpha-\varpi)}-\frac{a\cos^2\kappa}{\sin\kappa\sin(\alpha-\varpi)},$$

$$C=\frac{1}{\sin\kappa\sin(\alpha-\varpi)};$$

we have therefore

$$\frac{A^2+B^2}{r^2}+\frac{2BC}{r}+C^2=1,$$

or, what is the same thing,

$$(1-C^2)r^2-2BCr-(A^2+B^2)=0;$$

and thence, if r', r'' are the two values of r,

$$r'+r''=\frac{2BC}{1-C^2},$$

$$r'r''=-\frac{A^2+B^2}{1-C^2}.$$

Let θ', θ'' be the corresponding values of θ, we have

$$\theta' - \theta'' = \theta' - \alpha - (\theta'' - \alpha),$$

and thence

$$\cos(\theta' - \theta'') = \frac{A}{r'} \cdot \frac{A}{r''} + \left(\frac{B}{r'} + C\right)\left(\frac{B}{r''} + C\right), \quad = \frac{A^2 + B^2}{r'r''} + BC\left(\frac{1}{r'} + \frac{1}{r''}\right) + C^2,$$

or adding unity to each side, multiplying by $r'r''$, and on the right-hand side substituting for $r' + r''$, $r'r''$ their values

$$r'r''\left(1 + \cos(\theta' - \theta'')\right) = -\frac{2C^2A^2}{1 - C^2};$$

the square of the chord is $r^2 + r'^2 - 2rr' \cos(\theta' - \theta'')$, or, what is the same thing, $(r' + r'')^2 - 2r'r''\left(1 + \cos(\theta' - \theta'')\right)$; hence to prove the theorem, it is only necessary to show that $r' + r''$ and $r'r''\left(1 + \cos(\theta' - \theta'')\right)$ have the same values in each orbit, that is, that $\frac{2BC}{1 - C^2}$ and $-\frac{2C^2A^2}{1 - C^2}$ have the same values in each orbit. But observing that

$$1 - \sin^2\kappa \sin^2(\alpha - \varpi) = \cos^2\kappa + \sin^2\kappa \cos^2(\alpha - \varpi) = \cos^2\kappa \left\{1 + \tan^2\kappa \cos^2(\alpha - \varpi)\right\},$$

the values of these expressions are respectively

$$-\frac{2a}{R} \frac{\left(m \tan\kappa \cos(\alpha - \varpi) - R\right)}{1 + \tan^2\kappa \cos^2(\alpha - \varpi)},$$

$$\frac{2a^2}{R^2} \frac{m^2}{1 + \tan^2\kappa \cos^2(\alpha - \varpi)},$$

which contain only the quantities m, a, R, $\tan\kappa \cos(\alpha - \varpi)$, which are the same for each orbit, and the theorem is therefore proved, viz. it is made to depend on Lambert's theorem. I may remark, that a geometrical demonstration which does not assume Lambert's theorem is given by Mr Droop in his paper "On the Isochronism of the Circular Hodograph," *Quarterly Mathematical Journal*, vol. I. [1857] pp. 374—378, where the dependence of the theorem on Lambert's theorem is also shown.

By what precedes, the theorem may be stated in a geometrical form as follows:—"Imagine two ellipses having a common focus, and their major axes equal; describe about the focus two directrix circles having their radii proportional to the square roots of the minor axes of the ellipses respectively; the polar reciprocal of each ellipse in respect to its own directrix circle will be a circle (the hodograph), and the common chord or radical axis of the two hodographs will pass through the focus. Consider any point on the common chord, and take the polar with respect to each directrix circle; such polar will cut off an arc of the corresponding ellipse; and then, *theorem*, the elliptic chord, and the sum of the radius vectors through the two extremities of the chord, will be respectively the same for each ellipse."

2, *Stone Buildings, W.C., June* 24, 1857.

210.

ON THE CUBIC TRANSFORMATION OF AN ELLIPTIC FUNCTION.

[From the *Philosophical Magazine*, vol. xv. (1858), pp. 363—364.]

LET

$$z = \frac{(a', b', c', d' \between x, 1)^3}{(a, b, c, d \between x, 1)^3}$$

be any cubic fraction whatever of x, then it is always possible to find quartic functions of z, x respectively, such that

$$\frac{dz}{\sqrt{(\mathrm{a, b, c, d, e} \between z, 1)^4}} = \frac{dx}{\sqrt{(A, B, C, D, E \between x, 1)^4}}.$$

This depends upon the following theorem, viz. putting for shortness,

$$U = (a, b, c, d \between x, y)^3,$$
$$U' = (a', b', c', d' \between x, y)^3,$$

and representing by the notation

$$\text{disct. } (aU' - a'U, \quad bU' - b'U, \quad cU' - c'U, \quad dU' - d'U);$$

or more shortly by

$$\text{disct. } (aU' - a'U, \ldots),$$

the discriminant in regard to the facients (λ, μ) of the cubic function

$$(aU' - a'U, \quad bU' - b'U, \quad cU' - c'U, \quad dU' - d'U \between \lambda, \mu)^3;$$

or what is the same thing, the cubic function

$$(a, b, c, d \between \lambda, \mu)^3 . (a', b', c', d' \between x, y)^3$$
$$-(a', b', c', d' \between \lambda, \mu)^3 . (a, b, c, d \between x, y)^3;$$

and by $J(U, U')$ the functional determinant, or Jacobian, of the two cubics U, U'; the theorem is that the discriminant contains as a factor the square of the Jacobian, or that we have

$$\text{disct. } (aU' - a'U, \ldots) = \{J(U, U')\}^2 . (A, B, C, D, E \between x, y)^4.$$

For assuming this to be the case, then (disregarding a mere numerical factor) we have

$$UdU' - U'dU = J(U,\ U')(ydx - xdy),$$

and the two equations give

$$\frac{UdU' - U'dU}{\sqrt{\text{disct.}(aU' - a'U, \ldots)}} = \frac{ydx - xdy}{\sqrt{(A,\ B,\ C,\ D,\ E \between x,\ y)^4}},$$

whence writing z for $U' \div U$, and putting y equal to unity, we have

$$\frac{dz}{\sqrt{\text{disct.}(az - a', \ldots)}} = \frac{dx}{\sqrt{(A,\ B,\ C,\ D,\ E \between x,\ 1)^4}};$$

where disct. $(az - a', \ldots)$, or at full length,

$$\text{disct.}(az - a',\ bz - b',\ cz - c',\ dz - d'),$$

is a given quartic function of z,

$$= (\mathrm{a},\ \mathrm{b},\ \mathrm{c},\ \mathrm{d},\ \mathrm{e} \between z,\ 1)^4$$

suppose; and this proves the theorem of transformation.

The assumed subsidiary theorem may be thus proved: suppose that the parameter θ is determined so that the cubic

$$U + \theta U'$$

may have a square factor, the cubic may be written

$$(a + \theta a',\ b + \theta b',\ c + \theta c',\ d + \theta d' \between x,\ y)^3,$$

and the requisite condition is

$$\text{disct.}(a + \theta a', \ldots) = 0;$$

there are consequently four roots; and calling these θ_1, θ_2, θ_3, θ_4, we have identically

$$\text{disct.}(a + \theta a', \ldots) = K(\theta - \theta_1)(\theta - \theta_2)(\theta - \theta_3)(\theta - \theta_4),$$

or what is the same thing,

$$\text{disct.}(aU' - a'U, \ldots) = K(U + \theta_1 U')(U + \theta_2 U')(U + \theta_3 U')(U + \theta_4 U').$$

Now any double factor of U or U' (that is the linear factor which enters twice into U or U') is a simple factor of $J(U,\ U')$, and we have $J(U,\ U') = J(U,\ U + \theta U')$, and consequently

$$J(U,\ U') = J(U,\ U + \theta_1 U') = \&\text{c}.;$$

hence the double factors of each of the expressions $U + \theta_1 U'$, $U + \theta_2 U'$, $U + \theta_3 U'$, $U + \theta_4 U'$ are simple factors of $J(U,\ U')$, or what is the same thing, $J(U,\ U')$ is the product of four linear factors, which are respectively double factors of the product

$$(U + \theta_1 U')(U + \theta_2 U')(U + \theta_3 U')(U + \theta_4 U'),$$

or this product contains the factor $\{J(U,\ U')\}^2$, which proves the theorem.

2, *Stone Buildings, W.C., March* 5, 1858.

211.

ON A THEOREM RELATING TO HYPERGEOMETRIC SERIES.

[From the *Philosophical Magazine*, vol. XVI. (1858), pp. 356—357.]

In attempting to verify a formula of Hansen's relating to the development of the disturbing function in the planetary theory, I was led to a theorem in hypergeometric series: viz. writing, as usual,

$$F(\alpha,\ \beta,\ \gamma,\ x) = 1 + \frac{\alpha\,.\,\beta}{1\,.\,\gamma}\,x + \frac{\alpha\,.\,\alpha+1\,.\,\beta\,.\,\beta+1}{1\,.\,2\,.\,\gamma\,.\,\gamma+1}\,x^2 + \ldots$$

then the product

$$F(\alpha,\ \beta,\ \gamma+\tfrac{1}{2},\ x)\,F(\gamma-\alpha,\ \gamma-\beta,\ \gamma+\tfrac{1}{2},\ x)$$

is connected with

$$(1-x)^{-(\gamma-\alpha-\beta)}\,F(2\alpha,\ 2\beta,\ 2\gamma,\ x)$$

by a simple relation; for if the last-mentioned expression is put equal to

$$1 + Bx + Cx^2 + Dx^3 + \ldots$$

then the product in question is equal to

$$1 + \frac{\gamma}{\gamma+\frac{1}{2}}\,Bx + \frac{\gamma\,.\,\gamma+1}{\gamma+\frac{1}{2}\,.\,\gamma+\frac{3}{2}}\,Cx^2 + \frac{\gamma\,.\,\gamma+1\,.\,\gamma+2}{\gamma+\frac{1}{2}\,.\,\gamma+\frac{3}{2}\,.\,\gamma+\frac{5}{2}}\,Dx^3 + \text{\&c.}$$

The form of the identity thus arrived at will be best perceived by considering a particular case. Thus, comparing the coefficients of x^3, we have

$$\frac{\alpha\,.\,\alpha+1\,.\,\alpha+2\,.\,\beta\,.\,\beta+1\,.\,\beta+2}{1\,.\,2\,.\,3\,.\,\gamma+\frac{1}{2}\,.\,\gamma+\frac{3}{2}\,.\,\gamma+\frac{5}{2}}\,.\,1$$

$$+\frac{\alpha\,.\,\alpha+1\,.\,\beta\,.\,\beta+1}{1\,.\,2\,.\,\gamma+\frac{1}{2}\,.\,\gamma+\frac{3}{2}}\,.\,\frac{\gamma-\alpha\,.\,\gamma-\beta}{1\,.\,\gamma+\frac{1}{2}}$$

$$+\frac{\alpha\,.\,\beta}{1\,.\,2}\,.\,\frac{\gamma-\alpha\,.\,\gamma-\alpha+1\,.\,\gamma-\beta\,.\,\gamma-\beta+1}{1\,.\,2\,.\,\gamma+\frac{1}{2}\,.\,\gamma+\frac{3}{2}}$$

$$+1\,.\,\frac{\gamma-\alpha\,.\,\gamma-\alpha+1\,.\,\gamma-\alpha+2\,.\,\gamma-\beta\,.\,\gamma-\beta+1\,.\,\gamma-\beta+2}{1\,.\,2\,.\,3\,.\,\gamma+\frac{1}{2}\,.\,\gamma+\frac{3}{2}\,.\,\gamma+\frac{5}{2}}$$

$$=\left.\begin{array}{l}\dfrac{2\alpha\,.\,2\alpha+1\,.\,2\alpha+2\,.\,2\beta\,.\,2\beta+1\,.\,2\beta+2}{1\,.\,2\,.\,3\,.\,2\gamma\,.\,2\gamma+1\,.\,2\gamma+2}\,.\,1\\[2ex] +\dfrac{2\alpha\,.\,2\alpha+1\,.\,2\beta\,.\,2\beta+1}{1\,.\,2\,.\,2\gamma\,.\,2\gamma+1}\,.\,\dfrac{\gamma-\alpha-\beta}{1}\\[2ex] +\dfrac{2\alpha\,.\,2\beta}{1\,.\,2\gamma}\,.\,\dfrac{\gamma-\alpha-\beta\,.\,\gamma-\alpha-\beta+1}{1\,.\,2}\\[2ex] +1\,.\,\dfrac{\gamma-\alpha-\beta\,.\,\gamma-\alpha-\beta+1\,.\,\gamma-\alpha-\beta+2}{1\,.\,2\,.\,3}\end{array}\right\}\frac{\gamma\,.\,\gamma+1\,.\,\gamma+2}{\gamma+\frac{1}{2}\,.\,\gamma+\frac{3}{2}\,.\,\gamma+\frac{5}{2}}.$$

It may be observed that the function on the right-hand side is, as regards α, a rational and integral function of the degree 3, and as such may be expanded in the form

$$\begin{array}{ll} A\,\alpha\,.\,\alpha+1\,.\,\alpha+2 & \\ +B\,\alpha\,.\,\alpha+1 & \gamma-\alpha \\ +C\alpha\,. & .\,\gamma-\alpha\,.\,\gamma-\alpha+1 \\ +D & .\,\gamma-\alpha\,.\,\gamma-\alpha+1\,.\,\gamma-\alpha+2, \end{array}$$

and that the last coefficient D can be obtained at once by writing $\alpha=0$; this in fact gives

$$D\gamma\,.\,\gamma+1\,.\,\gamma+2=\frac{\gamma-\beta\,.\,\gamma-\beta+1\,.\,\gamma-\beta+2}{1\,.\,2\,.\,3}\,\frac{\gamma\,.\,\gamma+1\,.\,\gamma+2}{\gamma+\frac{1}{2}\,.\,\gamma+\frac{3}{2}\,.\,\gamma+\frac{5}{2}},$$

and thence

$$D=\frac{\gamma-\beta\,.\,\gamma-\beta+1\,.\,\gamma-\beta+2}{1\,.\,2\,.\,3\,.\,\gamma+\frac{1}{2}\,.\,\gamma+\frac{3}{2}\,.\,\gamma+\frac{5}{2}},$$

which agrees with the left-hand side of the equation: and the value of the first coefficient A may be obtained in like manner with a little more difficulty; but I have not succeeded in obtaining a direct proof of the equation. The form of the equation shows that the left-hand side should vanish for $\gamma=-2$, which may be at once verified.

Grassmere, August 25, 1858.

212.

A MEMOIR ON THE PROBLEM OF DISTURBED ELLIPTIC MOTION.

[From the *Memoirs of the Royal Astronomical Society*, vol. XXVII., 1859, pp. 1—29. Read March 9, 1858.]

I VENTURE to take up the problem of disturbed elliptic motion, for the sake of a further elaboration of the analytical theory. The points which present difficulty are the measurement of longitudes in the varying plane of the orbit, and (in the lunar theory) the determination of the position of the orbit by reference to the varying plane of the sun's orbit; it is, in memoirs and works on the lunar and planetary theories, often difficult to discover where or how (or whether at all) account is taken of these variations, and the analytical mode of treatment is for the most part very imperfect. I must except always Hansen's *Fundamenta Nova* [*investigationis orbitæ veræ quam Luna perlustrat*, Gotha 1838] where the points referred to are treated in a perfectly rigorous manner. There is, however, a want of clearness in the form under which his investigations are presented; and the comprehension of them is greatly facilitated by Jacobi's remarks, published under the title "Auszug zweier Schreiben des Prof. Jacobi an Herrn Director Hansen" (*Crelle*, t. XLII. pp. 12—31 (1851)). Jacobi observes that the integration of Hansen's system of differential equations introduces seven arbitrary constants, which, in the expressions for the coordinates referred to fixed axes, reduce themselves to six. The seventh constant, neglecting the disturbing forces, is in fact a constant which determines the position in the orbit of the arbitrary origin from which the longitudes in orbit are reckoned. I have, in my paper "On Hansen's Lunar Theory," *Quarterly Mathematical Journal*, vol. I. pp. 112—125 (1855), [163], termed this origin "the departure-point," and longitudes measured from it "departures." The seventh constant may be taken to be the departure of the node. I reproduce in the present memoir the explanation of what is meant by the departure when the plane of the orbit is variable. If the problem is treated by the method of the variation of the elements, the seventh constant becomes, like the other elements, variable; and we have thus a seventh variable element, the departure of the node. The element just referred to (the departure of the node) forms, with the longitude of the node and the inclination, a group of three elements, which determine the position of the orbit and of the departure-point. The coordinates of the planet are in the first instance taken to be the radius vector, longitude, and latitude; but the before-mentioned three elements being considered as given, the position of the planet depends only on the radius vector and the departure. These may be then expressed in terms of the remaining four elements; as to the choice of these four elements, it is to be remarked that there is one element which only enters through the mean anomaly, and that there is great convenience in representing with Hansen the mean anomaly by a single letter; and that in the various formulæ we may use, in the place of the element implicitly involved in the mean anomaly, the mean anomaly itself, or treat the mean anomaly as an element; the four elements may be taken to be the semi-axis major, the eccentricity, the mean anomaly, and the departure of the pericentre. And joining to these the before-mentioned three elements, we have the system of elements represented in the memoir by a, e, g, ϖ, σ, θ, ϕ. It has been assumed so far that the three elements determine the position of the orbit and departure-point in reference to a fixed plane and origin of longitudes; but we may suppose more generally that, instead of the fixed plane and origin of longitudes, we have a variable plane or orbit of reference and a departure-point in this variable orbit of reference. The quantities which determine the orbit of reference and departure-point are naturally taken to be the departure of the node, longitude of the node, and inclination; these are assumed to be given functions of the time, and they are in the memoir represented by σ', θ', ϕ'. The three elements of the planet's orbit (viz. departure of node, longitude of node, and inclination) in relation to the orbit of reference and departure-point therein, are in the memoir represented by Σ, Θ, Φ, and the system of elements ultimately adopted is therefore a, e, g, ϖ, Σ, Θ, Φ. I obtain formulæ for the variations of these elements under two different modes of expression of the disturbing function: first, when the disturbing function is expressed in terms of the radius vector and departure and of the three elements Σ, Θ, Φ; secondly, when the disturbing function is expressed in terms of the seven elements a, e, g, ϖ, Σ, Θ, Φ. The establishment of the two sets of formulæ just referred to constitutes the chief object of the memoir; but the memoir contains some other investigations and formulæ in relation to the general subject.

The coordinates of the planet are

r, the radius vector,

v, the longitude,

y, the latitude.

The attractive force at distance unity is for convenience represented by n^2a^3, which denotes, therefore, an absolute constant; but the significations of n and a are not yet defined.

The disturbing function, as used by Lagrange, is denoted by Ω, that is $\Omega = -R$, if R be the disturbing function of the *Mécanique Céleste.*

The equations of motion are

$$\frac{d}{dt}\frac{dr}{dt} - r\cos^2 y\left(\frac{dv}{dt}\right)^2 - r\left(\frac{dy}{dt}\right)^2 + \frac{n^2a^3}{r^2} = \frac{d\Omega}{dr},$$

$$\frac{d}{dt}\left(r^2\cos^2 y\,\frac{dv}{dt}\right) = \frac{d\Omega}{dv},$$

$$\frac{d}{dt}\left(r^2\frac{dy}{dt}\right) + r^2\cos y\sin y\left(\frac{dv}{dt}\right)^2 = \frac{d\Omega}{dy},$$

where Ω is regarded as a function of r, v, y, or (as this may be expressed) where $\Omega = \Omega(r, v, y)$.

If we neglect the disturbing forces, the planet moves in an ellipse; and taking a to represent the semi-axis major, the mean motion will be n. The mean anomaly, which I call g, will be a function of the form $nt + c$; but as c only enters through g, it will be convenient to use the mean anomaly g (considered as implicitly involving an arbitrary constant c) in the place of an element, and I write

a, the semi-axis major,

e, the eccentricity,

g, the mean anomaly,

θ, the longitude of node,

ϕ, the inclination,

ϖ, the distance of pericentre from node.

I assume also

f, the true anomaly,

z, the distance of planet from node,

x, the reduced distance from node.

We have then r and f given functions of t and the elements, viz. we may write

$$r = a \text{ elqr } (e, g),$$

$$f = \text{ elta } (e, g),$$

(read elqr. elliptic quotient radius, and elta. elliptic anomaly). These values satisfy $r = \dfrac{a(1-e^2)}{1+e\cos f}$. Moreover z, x, y, are the hypothenuse, base, and perpendicular of a right-angled spherical triangle, the base angle whereof is ϕ; the equations which connect these quantities are therefore

$$\tan x = \tan z \cos\phi,$$

$$\sin y = \sin z \sin\phi,$$

$$\tan y = \sin x \tan\phi,$$

$$\cos z = \cos x \cos y,$$

equivalent, of course, to two equations. The first and second of them give in fact x, y, in terms of z and ϕ.

The value of z is

$$z = \varpi + f,$$

so that x and y are given functions, and the longitude v is given in terms of x and θ by the equation

$$v = x + \theta,$$

and consequently the three coordinates r, v, y, are by the system of equations given in terms of t and the elements.

From the equations which connect z, x, y, ϕ, treating all these quantities as variable we deduce

$$\begin{aligned}
\sec^2 x\, dx &= \cos\phi \sec^2 z\, dz - \tan z \sin\phi\, d\phi,\\
\cos y\, dy &= \sin\phi \cos z\, dz + \sin z \cos\phi\, d\phi,\\
\sec^2 y\, dy &= \tan\phi \cos x\, dx + \sin x \sec\phi\, d\phi,\\
\sin z\, dz &= \cos y \sin x\, dx + \cos x \sin y\, dy,
\end{aligned}$$

equivalent of course to two equations; and the system is easily reduced to the more convenient form

$$\begin{aligned}
dx &= \cos\phi \sec^2 y\, dz - \tan z \cos^2 x \sin\phi\, d\phi,\\
dy &= \sin\phi \cos x\, dz + \cos x \tan z \cos\phi\, d\phi,\\
dx &= \cot\phi \sec x \sec^2 y\, dy - \tan z \operatorname{cosec}\phi\, d\phi,\\
dz &= \cos\phi\, dx + \cos x \sin\phi\, d\phi,
\end{aligned}$$

joining to these equations the

$$\begin{aligned}
dz &= d\varpi + df,\\
dv &= dx + d\theta,
\end{aligned}$$

and considering at present the mere analytical forms, first if $d\phi = 0$, $d\varpi = 0$, we have

$$\begin{aligned}
dx &= \cos\phi \sec^2 y\, dz,\\
dy &= \sin\phi \cos x\, dz,\\
dz &= df,\\
dv &= dx.
\end{aligned}$$

Next, if $dy = 0$, $dv = 0$, we have

$$\begin{aligned}
dx &= -\tan z \operatorname{cosec}\phi\, d\phi,\\
dz &= -\tan z \cot\phi\, d\phi,\\
dz &= \cos\phi\, dx,\\
dx &= -d\theta,\\
dz &+ \cos\phi\, d\theta = 0.
\end{aligned}$$

I remark also that the equation, $\tan y = \sin x \tan\phi$, may be written in the form

$$\cos^2\phi \sec^2 y + \sin^2\phi \cos^2 x = 1.$$

The equations

$$r = a \text{ elqr } (e, g),$$

$$f = \text{ elta } (e, g),$$

treating all the quantities as variable, give

$$dr = \frac{ae \sin f}{\sqrt{1-e^2}} dg + \frac{1-e^2}{1+e\cos f} da - a \cos f\, de,$$

$$df = \frac{(1+e\cos f)^2}{(1-e^2)^{\frac{3}{2}}} dg + \frac{\sin f(2+e\cos f)}{1-e^2} de,$$

to which is to be joined

$$dr = \frac{ae(1-e^2)\sin f}{(1+e\cos f)^2} df + \frac{1-e^2}{1+e\cos f} da + \frac{a(2e - \overline{1+e^2}\cos f)}{(1+e\cos f)^2} de,$$

all which formulæ will be useful.

If we treat the elements as constant, then in the foregoing expressions for dr and df, we must attend only to the part involving dg, and must put this equal to ndt; the values first obtained for dx, dy, dz, dv, correspond to this assumption, and we have

$$\frac{dr}{dt} = \frac{nae \sin f}{\sqrt{1-e^2}},$$

$$\frac{df}{dt} = \frac{na^2\sqrt{1-e^2}}{r^2},$$

$$\frac{dz}{dt} = \frac{na^2\sqrt{1-e^2}}{r^2},$$

$$\frac{dx}{dt} = \cos\phi \sec^2 y \frac{na^2\sqrt{1-e^2}}{r^2},$$

$$\frac{dv}{dt} = \cos\phi \sec^2 y \frac{na^2\sqrt{1-e^2}}{r^2},$$

$$\frac{dy}{dt} = \sin\phi \cos x \frac{na^2\sqrt{1-e^2}}{r^2},$$

and we then deduce

$$\frac{d}{dt}\frac{dr}{dt} = \frac{na^3 e \cos f}{r^2},$$

$$\frac{d}{dt}\left(r^2 \cos^2 y \frac{dv}{dt}\right) = 0,$$

$$\frac{d}{dt}\left(r^2 \frac{dy}{dt}\right) = -\cos^2\phi \sin y \sec^3 y \frac{n^2a^4(1-e^2)}{r^4},$$

$$\frac{d}{dt}\left(\frac{df}{dt}\right) = -\frac{2n^2a^3e\cos f}{r^2},$$

values which satisfy the undisturbed equations.

The disturbed equations may be dealt with in the usual manner by the method of the variation of the elements, and attending only to the variations of the elements we have

$$dr = 0,$$
$$dv = 0,$$
$$dy = 0,$$
$$d\frac{dr}{dt} = \frac{d\Omega}{dr}\,dt,$$
$$d\left(r^2 \cos^2 y\,\frac{dv}{dt}\right) = \frac{d\Omega}{dv}\,dt,$$
$$d\left(r^2\frac{dy}{dt}\right) = \frac{d\Omega}{dy}\,dt,$$

or, what is the same thing,

$$dr = 0,$$
$$dv = 0,$$
$$dy = 0,$$
$$d\frac{nae\sin f}{\sqrt{1-e^2}} = \frac{d\Omega}{dr}\,dt,$$
$$d\,na^2\sqrt{1-e^2}\cos\phi = \frac{d\Omega}{dv}\,dt,$$
$$d\,na^2\sqrt{1-e^2}\sin\phi\cos x = \frac{d\Omega}{dy}\,dt,$$

where as before $\Omega = \Omega(r,\ v,\ y)$.

In virtue of the relations $dv = 0$, $dy = 0$, we have the above-mentioned equations,

$$dx = -\tan z \operatorname{cosec}\phi\,d\phi,$$
$$dz = -\tan z \cot\ \phi\,d\phi,$$
$$dz = \cos\phi\,dx,$$
$$dx = -\,d\theta,$$
$$dr \ + \cos\phi\,d\theta = 0,$$

we have

$$d\sin\phi\cos x = -\sin\phi\sin x\,dx + \cos x\cos\phi\,d\phi,$$
$$= \cos x\cos\phi\sec^2 z\,d\phi,$$
$$= \sec x\cos\phi\sec^2 y\,d\phi\,;$$

and the last two equations for the variations become

$$d\,na^2\sqrt{1-e^2}\cos\phi \qquad - na^2\sqrt{1-e^2}\sin\phi\,d\phi \qquad = \frac{d\Omega}{dv}\,dt,$$
$$d\,na^2\sqrt{1-e^2}\sin\phi\cos x + na^2\sqrt{1-e^2}\sec x\cos\phi\sec^2 y\,d\phi = \frac{d\Omega}{dy}\,dt,$$

and attending to the equations $\cos^2\phi \sec^2 y + \sin^2\phi \cos^2 x = 1$ we deduce at once

$$d\, na^2\sqrt{1-e^2} = \cos\phi \sec^2 y \frac{d\Omega}{dv}dt + \sin\phi \cos x \frac{d\Omega}{dy}dt,$$

$$d\phi = \frac{1}{na^2\sqrt{1-e^2}}\left(-\sin\phi\cos^2 x \frac{d\Omega}{dv}dt + \cos\phi\cos x\frac{d\Omega}{dy}dt\right).$$

Now the position of the planet may be determined by the quantities r, z, θ, ϕ, or we may consider Ω as a function of the last-mentioned quantities. And if on the right-hand side $\Omega = \Omega(r, v, y)$ as before, the formulæ of transformation are

$$\frac{d\Omega}{dr}dr + \frac{d\Omega}{dv}dv + \frac{d\Omega}{dy}dy = \frac{d\Omega}{dr}dr + \frac{d\Omega}{dz}dz + \frac{d\Omega}{d\theta}d\theta + \frac{d\Omega}{d\phi}d\phi$$

where

$$dv = \cos\phi\sec^2 y\, dz - \tan z \cos^2 x \sin\phi\, d\phi + d\theta,$$

$$dy = \sin\phi\cos x\, dz + \tan z\cos x\cos\phi\, d\phi,$$

and we have

$$\frac{d\Omega}{dr} = \frac{d\Omega}{dr},$$

$$\frac{d\Omega}{d\theta} = \frac{d\Omega}{dv},$$

$$\frac{d\Omega}{d\phi} = \tan z\left(-\sin\phi\cos^2 x\frac{d\Omega}{dv} + \cos\phi\cos x\frac{d\Omega}{dy}\right),$$

$$\frac{d\Omega}{dz} = \left(\cos\phi\sec^2 y\frac{d\Omega}{dv} + \sin\phi\cos x\frac{d\Omega}{dy}\right),$$

where on the left-hand side $\Omega = \Omega(r, z, \theta, \phi)$; and these equations give

$$\cot z\frac{d\Omega}{d\phi} = \cot\phi\frac{d\Omega}{dz} - \operatorname{cosec}\phi\frac{d\Omega}{d\theta},$$

an equation which is satisfied by $\Omega = \Omega(r, z, \theta, \phi)$. We have thus

$$dr = 0,$$

$$dv = 0,$$

$$dy = 0,$$

$$d\frac{nae\sin f}{\sqrt{1-e^2}} = \frac{d\Omega}{dr}dt,$$

$$d\, na^2\sqrt{1-e^2} = \frac{d\Omega}{dr}dt,$$

$$d\phi = \frac{\cot z}{na^2\sqrt{1-e^2}}\frac{d\Omega}{d\phi}dt,$$

which may be replaced by

$$dr = 0,$$

$$d\theta = \frac{\operatorname{cosec}\phi}{na^2\sqrt{1-e^2}}\,\frac{d\Omega}{d\phi}\,dt,$$

$$dz = \frac{-\cot\phi}{na^2\sqrt{1-e^2}}\,\frac{d\Omega}{d\phi}\,dt,$$

$$d\,\frac{nae\sin f}{\sqrt{1-e^2}} = \frac{d\Omega}{dr}\,dt,$$

$$dna^2\sqrt{1-e^2} = \frac{d\Omega}{dz}\,dt,$$

$$d\phi = \frac{\cot z}{na^2\sqrt{1-e^2}}\,\frac{d\Omega}{d\phi}\,dt,$$

where as before $\Omega = \Omega\,(r, z, \theta, \phi)$.

I remark that in the case of any central force whatever, we have an element h corresponding to $na^2\sqrt{1-e^2}$ in the elliptic theory, and the system for the variations is

$$dr = 0,$$

$$d\theta = \frac{\operatorname{cosec}\phi}{h}\,\frac{d\Omega}{d\phi}\,dt,$$

$$dz = \frac{-\cot\phi}{h}\,\frac{d\Omega}{d\phi}\,dt,$$

$$d\,\frac{dr}{dt} = \frac{d\Omega}{dr}\,dt,$$

$$dh = \frac{d\Omega}{dz}\,dt,$$

$$d\phi = \frac{\cot z}{h}\,\frac{d\Omega}{d\phi}\,dt,$$

where $\Omega = \Omega\,(r, z, \theta, \phi)$.

Imagine a point in the orbit, which I call the departure-point, the angular distances from this point are termed departures. And I write

þ, the departure of planet,

ϖ, the departure of pericentre,

σ, the departure of node,

so that we have

$$þ = \varpi + f,$$
$$z = þ - \sigma,$$
$$\mho = \varpi - \sigma.$$

I write also

♉, the longitude in orbit of departure-point, or, as it may be termed, the adjustment;

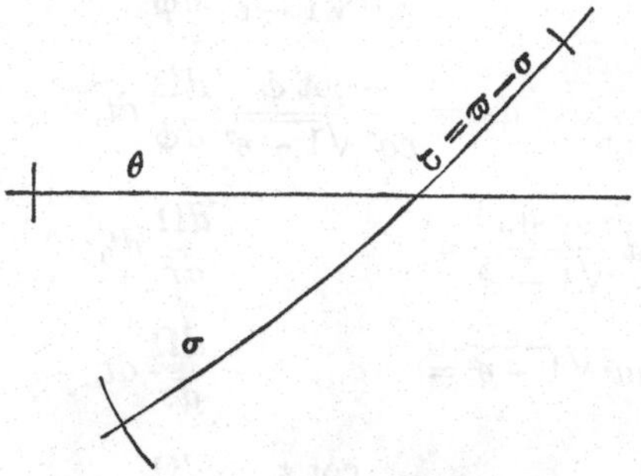

that is

$$♉ = \theta - \sigma.$$

In the undisturbed motion the departure-point is simply a fixed point in the orbit, but when the orbit is variable, the departure-point is taken to be the point of intersection of the orbit with any orthogonal trajectory of the successive positions of the orbit, a definition which is expressed analytically by the equation,

$$d\sigma = \cos\phi \, d\theta.$$

The equation, $z = þ - \sigma$, gives

$$dz = dþ - d\sigma = dþ - \cos\phi \, d\theta,$$

or, what is the same thing,

$$dþ = dz + \cos\phi \, d\theta.$$

But we have $dz + \cos\phi \, d\theta = 0$, and consequently $dþ = 0$, an equation which expresses that the increment of departure, in so far as such increment arises from the variation of the elements, is equal to zero. Or, what is the same thing, the total increment of departure is equal to the infinitesimal angle between two consecutive radius vectors of the planet.

I propose to consider the departure-point as a point which is constantly defined as above, viz., when the orbit is variable, the departure-point is the point of intersection of the orbit with any orthogonal trajectory of the successive positions of the orbit; and as a particular case of the definition, when the orbit is fixed, the departure-point is simply a fixed point on the orbit. The orbit here considered is that of the planet and the position of the planet is determined by the departure and radius vector (the latitude being zero), and this is assumed to be the case whenever the departure is spoken of, and it is such departure which is denoted by the letter þ. But we might consider a departure-point (defined as above), upon any other orbit whatever, and use such departure-point as an origin of longitude (for instance, in the lunar theory we might consider a longitude measured along the variable plane of the sun's orbit from a departure-point, defined as above, in that orbit), and the position of the planet would then be determined by means of the longitude, latitude, and radius vector. The term sidereal longitude is, I think, used in Physical Astronomy rather loosely to denote the longitude in the mean ecliptic from the mean equinox, less the precession; so defined it is not practically different from, and may I think in all cases be replaced by the longitude as measured from a departure-point in the mean ecliptic.

Returning from this digression, the assumed equation, $d\sigma = \cos\phi d\theta$, gives the expression for the variation $d\sigma$ of the departure of the node, and we now have in the place of the former six equations the seven equations

$$dr = 0,$$

$$d\mathfrak{p} = 0,$$

$$d\frac{nae\sin f}{\sqrt{1-e^2}} = \frac{d\Omega}{dr}dt,$$

$$d\,na^2\sqrt{1-e^2} = \frac{d\Omega}{dz}dt,$$

$$d\phi = \frac{\cot z}{na^2\sqrt{1-e^2}}\frac{d\Omega}{d\phi}dt,$$

$$d\sigma = \frac{\cot\phi}{na^2\sqrt{1-e^2}}\frac{d\Omega}{d\phi}dt,$$

$$d\theta = \frac{\operatorname{cosec}\phi}{na^2\sqrt{1-e^2}}\frac{d\Omega}{d\phi}dt,$$

where as before $\Omega = \Omega(r, z, \theta, \phi)$.

But the value of z is $z = \mathfrak{p} - \sigma$, and Ω can be expressed, and that in a single way only, viz. by means of the substitution of $\mathfrak{p} - \sigma$ in the place of z, in the form $\Omega = \Omega(r, \mathfrak{p}, \sigma, \theta, \phi)$, and if on the right-hand side $\Omega = \Omega(r, z, \theta, \phi)$ as before, then we have

$$\frac{d\Omega}{dr} = \frac{d\Omega}{dr},$$

$$\frac{d\Omega}{d\mathfrak{p}} = \frac{d\Omega}{dz},$$

$$-\frac{d\Omega}{d\sigma} = \frac{d\Omega}{dz},$$

$$\frac{d\Omega}{d\phi} = \frac{d\Omega}{d\phi},$$

$$\frac{d\Omega}{d\theta} = \frac{d\Omega}{d\theta},$$

where on the left-hand side $\Omega = \Omega\,(r,\ \mathfrak{p},\ \sigma,\ \theta,\ \phi)$. The function Ω so expressed satisfies, of course, the partial differential equation

$$\frac{d\Omega}{d\mathfrak{p}} + \frac{d\Omega}{d\sigma} = 0$$

(which conversely implies that $\mathfrak{p}$, σ only enters through the function $\mathfrak{p} - \sigma$), and it also satisfies the partial differential equation obtained from the before-mentioned equation

$$\cot z\,\frac{d\Omega}{d\phi} = \cot\phi\,\frac{d\Omega}{dz} - \operatorname{cosec}\phi\,\frac{d\Omega}{d\theta},\quad (\Omega = \Omega\,(r,\ z,\ \theta,\ \phi)),$$

by the introduction of the transformed expressions of the differential coefficients, and which may be written

$$\cot z\,\frac{d\Omega}{d\phi} = -\cot\phi\,\frac{d\Omega}{d\sigma} - \operatorname{cosec}\phi\,\frac{d\Omega}{d\theta},$$

where $\Omega = \Omega\,(r,\ \mathfrak{p},\ \sigma,\ \theta,\ \phi)$.

Using the last-mentioned equation to transform the value of $d\phi$, the expressions for the variations become

$$dr = 0,$$

$$d\mathfrak{p} = 0,$$

$$d\,\frac{nae\sin f}{\sqrt{1-e^2}} = \frac{d\Omega}{dr}\,dt,$$

$$dna^2\sqrt{1-e^2} = \frac{d\Omega}{d\mathfrak{p}}\,dt,$$

$$d\phi = \frac{-\cot\phi}{na^2\sqrt{1-e^2}}\,\frac{d\Omega}{d\sigma}\,dt - \frac{\operatorname{cosec}\phi}{na^2\sqrt{1-e^2}}\,\frac{d\Omega}{d\theta}\,dt,$$

$$d\sigma = \frac{\cot\phi}{na^2\sqrt{1-e^2}}\,\frac{d\Omega}{d\phi}\,dt,$$

$$d\theta = \frac{\operatorname{cosec}\phi}{na^2\sqrt{1-e^2}}\,\frac{d\Omega}{d\phi}\,dt,$$

where $\Omega = \Omega\,(r,\ \mathfrak{p},\ \sigma,\ \theta,\ \phi)$ as before.

I suppose now that the orbit of the planet, instead of being referred to a fixed plane, is referred to a moveable plane or orbit of reference. It is assumed that the longitudes in the orbit of reference are measured from a departure-point defined as above,—that is, from the point in which the orbit of reference is intersected by any orthogonal trajectory of the successive positions of the orbit of reference. And the position in regard to the fixed plane, of the orbit of reference, and of the departure-point in this orbit, are determined by θ', σ', ϕ',—that is, we have for the orbit of reference,

θ', the longitude of node,

σ', the departure of node,

ϕ', the inclination.

The position of the planet's orbit in relation to the moveable orbit of reference is determined in like manner by Θ, Σ, Φ,—that is, we have for the planet's orbit in relation to the orbit of reference,

Θ, the longitude of node,

Σ, the departure of node,

Φ, the inclination.

Hence if, as before, θ, σ, ϕ, belong to the orbit of the planet considered in relation to the fixed plane, $\Sigma - \sigma$, $\Theta - \sigma'$, $\theta - \theta'$, will be the sides of a spherical triangle, the opposite angles of which are ϕ', $180^\circ - \phi$ and Φ.

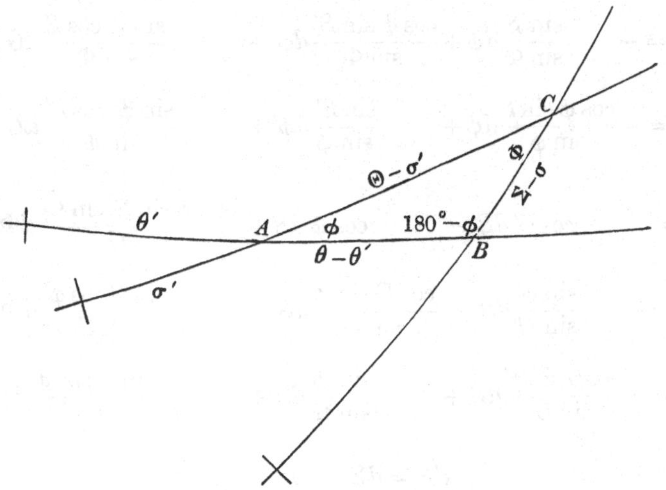

Putting for shortness $S = \Sigma - \sigma$, $S' = \Theta - \sigma'$, $G = \theta - \theta'$, so that these symbols denote

S, the distance of node, along planet's orbit, from fixed plane,

S', the distance of node, along orbit of reference, from fixed plane,

G, the distance in fixed plane of the nodes on fixed plane,

the sides of the spherical triangle are S, S', G, and the opposite angles are ϕ', $180^\circ - \phi$, Φ.

Calling the sides A, B, C, and the opposite angles a, b, c, the general formulæ for a spherical triangle give without difficulty,

$$dC = -\cos b\, dA - \cos a\, dB + \frac{\sin A \sin B \sin c}{\sin C}\, dc,$$

$$da = \frac{\sin b}{\sin C}\, dA + \frac{\cos C \sin a}{\sin C}\, dB + \frac{\sin A \cos b}{\sin C}\, dc,$$

$$db = \frac{\cos C \sin b}{\sin C}\, dA + \frac{\sin a}{\sin C}\, dB + \frac{\sin B \cos a}{\sin C}\, dc,$$

and conversely

$$dc = \cos B\, da + \cos A\, db + \frac{\sin a \sin b \sin C}{\sin C}\, dC,$$

$$dA = \frac{\sin B}{\sin c}\, da - \frac{\cos c \sin A}{\sin c}\, db - \frac{\sin a \cos B}{\sin c}\, dC,$$

$$dB = -\frac{\cos c \sin B}{\sin c}\, da + \frac{\sin A}{\sin c}\, db - \frac{\sin b \cos A}{\sin c}\, dC,$$

which, in the present case, become

$$d\Phi = \cos S\, d\phi - \cos S'\, d\phi' + \frac{\sin \phi \sin \phi' \sin G}{\sin \Phi}\, dG\,,$$

$$dS' = -\frac{\sin S}{\sin \Phi}\, d\phi + \frac{\cos \phi \sin S'}{\sin \Phi}\, d\phi' + \frac{\sin \phi \cos S}{\sin \Phi}\, dG\,,$$

$$dS = -\frac{\cos \phi \sin S}{\sin \phi}\, d\phi + \frac{\sin S'}{\sin \phi}\, d\phi' + \frac{\sin \phi' \cos S'}{\sin \phi}\, dG\,,$$

and

$$dG = \cos \phi'\, dS' - \cos \phi\, dS + \frac{\sin S \sin S' \sin \Phi}{\sin G}\, d\Phi,$$

$$d\phi = -\frac{\sin \phi'}{\sin G}\, dS' + \frac{\cos G \sin \phi}{\sin G}\, dS + \frac{\sin S \cos \phi'}{\sin G}\, d\Phi,$$

$$d\phi' = -\frac{\cos G \sin \phi'}{\sin G}\, dS' + \frac{\sin \phi}{\sin G}\, dS + \frac{\sin S \cos \phi}{\sin G}\, d\phi,$$

and we have also

$$\begin{aligned} dS &= d\Sigma - d\sigma\,, \\ dS' &= d\Theta - d\sigma', \\ dG &= d\theta - d\theta'. \end{aligned}$$

Hence, observing that $\sin S \sin \phi = \sin S' \sin \phi'$, the preceding equations may be written

$$\begin{aligned} d\Phi &= \cos S\, d\phi - \cos S'\, d\phi' + \sin S \sin \phi\, d\theta - \sin S' \sin \phi'\, d\theta' \\ d\Theta &= d\sigma' - \operatorname{cosec} \Phi \sin S\, d\phi + \cot \phi \sin S'\, d\phi' + \operatorname{cosec} \Phi \sin \phi' \cos S\, (d\theta - d\theta'), \\ d\Sigma &= d\sigma - \cot \Phi \sin S\, d\phi + \operatorname{cosec} \phi \sin S'\, d\phi' + \operatorname{cosec} \Phi \sin \phi' \cos S'\, (d\theta - d\theta'), \end{aligned}$$

and

$$d\theta = d\theta' + \cos\phi'(d\Theta - d\sigma') - \cos\phi(d\Sigma - d\sigma) + \sin S\sin\phi\, d\Phi,$$

$$d\phi = -\operatorname{cosec} G\sin\phi'(d\Theta - d\sigma') + \cot G\sin\phi(d\Sigma - d\sigma) + \operatorname{cosec} G\sin S'\cos\phi'\, d\Phi,$$

$$d\phi' = -\cot G\sin\phi'(d\Theta - d\sigma') + \operatorname{cosec} G\sin\phi(d\Sigma - d\sigma) + \operatorname{cosec} G\sin S\cos\phi\, d\Phi,$$

and it is proper to remark, that in obtaining these equations no use has been made of the equations $d\sigma = \cos\phi\, d\theta$, $d\sigma' = \cos\phi'\, d\theta'$.

The term in $d\Theta$ which contains $d\sigma'$, &c. may be written

$$(d\sigma' - \cos\phi'\, d\theta') + \cos\phi'\, d\theta' + \cot\Phi\sin S'\, d\phi' - \operatorname{cosec}\Phi\sin\phi\cos S\, d\theta',$$

which is equal to

$$(d\sigma' - \cos\phi'\, d\theta') + \cot\Phi(\sin S'\, d\phi' - \cos S'\sin\phi'\, d\theta');$$

and the term in $d\Sigma$ which contains $d\sigma$, &c. may be written

$$(d\sigma - \cos\phi\, d\theta) + \cos\phi\, d\theta - \cot\Phi\sin S\, d\phi + \operatorname{cosec}\Phi\sin\phi'\, d\theta,$$

which is equal to

$$(d\sigma - \cos\phi\, d\theta) - \cot\Phi(\sin S\, d\phi - \cos S\sin\phi\, d\theta);$$

reductions which depend on

$$\cos\phi' - \operatorname{cosec}\Phi\sin\phi\cos S = -\cot\Phi\sin\phi'\cos S',$$

$$\cos\phi + \operatorname{cosec}\Phi\sin\phi'\cos S' = -\cot\Phi\sin\phi\cos S,$$

or, what is the same thing,

$$\cos\phi'\sin\Phi - \sin\phi'\cos\Phi\cos S' = \sin\phi\cos S,$$

$$\cos\phi\sin\Phi - \sin\phi\cos\Phi\cos S = -\sin\phi'\cos S',$$

which are relations between the sides and angles of the spherical triangle. And we then have

$$d\Phi = (\cos S\, d\phi + \sin S\sin\phi\, d\theta) - (\cos S'\, d\phi' + \sin S'\sin\phi'\, d\theta'),$$

$$d\Theta = (d\sigma' - \cos\phi' d\theta') - \operatorname{cosec}\Phi(\sin S\, d\phi - \cos S\sin\phi\, d\theta) + \cot\Phi(\sin S'\, d\phi' - \cos S'\sin\phi'\, d\theta'),$$

$$d\Sigma = (d\sigma - \cos\phi\, d\theta) - \cot\Phi(\sin S\, d\phi - \cos S\sin\phi\, d\theta) + \operatorname{cosec}\Phi(\sin S'\, d\phi' - \cos S'\sin\phi'\, d\theta'),$$

expressions which may be simplified by omitting the terms $(d\sigma - \cos\phi'\, d\theta')$ and $(d\sigma - \cos\phi\, d\theta)$.

Next substituting for $d\sigma$, $d\phi$, $d\theta$, their values, we obtain,

$$dr = 0,$$

$$d\mathfrak{p} = 0,$$

$$d\,\frac{nae\sin f}{\sqrt{1-e^2}} = \frac{d\Omega}{dr}\, dt,$$

$$d\, na^2 \sqrt{1-e^2} = \frac{d\Omega}{d\text{þ}}\, dt,$$

$$d\Phi = \frac{1}{na^2\sqrt{1-e^2}}\left(\sin S \frac{d\Omega}{d\phi} - \cos S\left(\cot\phi \frac{d\Omega}{d\sigma} + \operatorname{cosec}\phi \frac{d\Omega}{d\theta}\right)\right) dt - \qquad (\sin S' d\phi' + \sin S' \sin\phi' d\theta'),$$

$$d\Sigma = \frac{\cot\Phi}{na^2\sqrt{1-e^2}}\left(\cos S \frac{d\Omega}{d\phi} + \sin S\left(\cot\phi \frac{d\Omega}{d\sigma} + \operatorname{cosec}\phi \frac{d\Omega}{d\theta}\right)\right) dt + \operatorname{cosec}\Phi(\sin S' d\phi' - \cos S' \sin\phi' d\theta'),$$

$$d\Theta = \frac{\operatorname{cosec}\Phi}{na^2\sqrt{1-e^2}}\left(\cos S \frac{d\Omega}{d\phi} + \sin S\left(\cot\phi \frac{d\Omega}{d\sigma} + \operatorname{cosec}\phi \frac{d\Omega}{d\theta}\right)\right) dt + \quad \cot\Phi(\sin S' d\phi' - \cos S' \sin\phi' d\theta'),$$

where $\Omega = \Omega(r, \text{þ}, \sigma, \theta, \phi)$, as before.

But Ω may be expressed in the form $\Omega = \Omega(r, \text{þ}, \Sigma, \Theta, \Phi, \sigma', \theta', \phi')$, or disregarding σ', θ', ϕ', in the form $\Omega = \Omega(r, \text{þ}, \Sigma, \Theta, \Phi)$, and to effect the transformation of the differential coefficients we must write,

$$d\Phi = \cos S\, d\phi + \sin S \sin\phi\, d\theta,$$
$$d\Theta = \qquad - \operatorname{cosec}\Phi(\sin S\, d\phi - \cos S \sin\phi\, d\theta),$$
$$d\Sigma = (d\sigma - \cos\phi\, d\theta) - \cot\Phi(\sin S\, d\phi - \cos S \sin\phi\, d\theta),$$

or, what is the same thing,

$$d\phi = \cos S\, d\Phi - \sin S \sin\Phi\, d\Theta,$$
$$d\theta = \qquad \operatorname{cosec}\phi(\sin S\, d\Phi + \cos S \sin\Phi\, d\Theta),$$
$$d\sigma = d\Sigma - \cos\Phi\, d\Theta + \cot\phi(\sin S\, d\Phi + \cos S \sin\Phi\, d\Theta),$$

and substituting in

$$\frac{d\Omega}{dr}dr + \frac{d\Omega}{d\text{þ}}d\text{þ} + \frac{d\Omega}{d\sigma}d\sigma + \frac{d\Omega}{d\theta}d\theta + \frac{d\Omega}{d\phi}d\phi$$
$$= \frac{d\Omega}{dr}dr + \frac{d\Omega}{d\text{þ}}d\text{þ} + \frac{d\Omega}{d\Sigma}d\Sigma + \frac{d\Omega}{d\Theta}d\Theta + \frac{d\Omega}{d\Phi}d\Phi,$$

if on the right-hand side $\Omega = \Omega(r, \text{þ}, \sigma, \theta, \phi)$ as before, then we have

$$\frac{d\Omega}{dr} = \frac{d\Omega}{dr},$$

$$\frac{d\Omega}{d\text{þ}} = \frac{d\Omega}{d\text{þ}},$$

$$\frac{d\Omega}{d\Sigma} = \frac{d\Omega}{d\sigma},$$

$$\frac{d\Omega}{d\Theta} = (-\cos\Phi + \cot\phi \cos S \sin\Phi)\frac{d\Omega}{d\sigma} + \operatorname{cosec}\phi \cos S \sin\Phi \frac{d\Omega}{d\theta} - \sin S \sin\Phi \frac{d\Omega}{d\phi},$$

$$\frac{d\Omega}{d\Sigma} = \qquad \cot\phi \sin S \frac{d\Omega}{d\sigma} + \qquad \operatorname{cosec}\phi \sin S \frac{d\Omega}{d\theta} + \qquad \cos S \frac{d\Omega}{d\phi},$$

where on the left-hand side $\Omega = \Omega(r, \text{þ}, \Sigma, \Theta, \Phi)$.

The last three equations give

$$\cos S \frac{d\Omega}{d\phi} + \sin S \left(\cot\phi \frac{d\Omega}{d\sigma} + \operatorname{cosec}\phi \frac{d\Omega}{d\theta} \right) = \frac{d\Omega}{d\Phi},$$

$$-\sin S \frac{d\Omega}{d\phi} + \cos S \left(\cot\phi \frac{d\Omega}{d\sigma} + \operatorname{cosec}\phi \frac{d\Omega}{d\theta} \right) = \cot\Phi \frac{d\Omega}{d\Sigma} + \operatorname{cosec}\Phi \frac{d\Omega}{d\Theta},$$

and the formulæ for the variations become

$$dr \qquad = 0,$$

$$d\mathfrak{p} \qquad = 0,$$

$$d\frac{nae \sin f}{\sqrt{1-e^2}} = \frac{d\Omega}{dr}\,dt,$$

$$dna^2\sqrt{1-e^2} = \frac{d\Omega}{d\mathfrak{p}}\,dt,$$

$$d\Phi \qquad = \frac{-\cot\Phi}{na^2\sqrt{1-e^2}}\frac{d\Omega}{d\Sigma}\,dt - \frac{\operatorname{cosec}\Phi}{na^2\sqrt{1-e^2}}\frac{d\Omega}{d\Theta}\,dt - (\cos S' d\phi' + \sin S' \sin\phi' d\theta'),$$

$$d\Sigma \qquad = \frac{\cot\Phi}{na^2\sqrt{1-e^2}}\frac{d\Omega}{d\Phi}\,dt \qquad + \operatorname{cosec}\Phi\,(\sin S' d\phi' - \cos S' \sin\phi' d\theta'),$$

$$d\Theta \qquad = \frac{\operatorname{cosec}\Phi}{na^2\sqrt{1-e^2}}\frac{d\Omega}{d\Phi}\,dt \qquad + \cot\Phi\,(\sin S' d\phi' - \cos S' \sin\phi' d\theta'),$$

where $\Omega = \Omega\ (r, \mathfrak{p}, \Sigma, \Theta, \Phi)$. It will be recollected that the value of S' is $= \Theta - \sigma'$. It may be noticed that

$$d\Sigma - \cos\Phi\, d\Theta = \sin\Phi\,(\sin S'\, d\phi' - \cos S' \sin\phi'\, d\theta').$$

The system just obtained is, except as regards the terms involving $d\phi'$ and $d\theta'$, precisely similar in its form to that in which the planet is referred to a fixed plane, or where $\Omega = \Omega\ (r, \mathfrak{p}, \sigma, \theta, \phi)$, and this is of course as it should be.

We have now

$$\mathfrak{p} = \varpi + f,$$
$$r = a \operatorname{elqr}\ (e, g),$$
$$f = \operatorname{elta}\ (e, g),$$

so that the position of the planet is determined by means of the elements $a, e, g, \varpi, \Sigma, \Theta, \Phi$. To find the variations of these elements, substituting for r its value in terms of f, the first, third, and fourth equations are

$$d\frac{a(1-e^2)}{1+e\cos f} = 0,$$

$$d\frac{nae\sin f}{\sqrt{1-e^2}} = \frac{d\Omega}{dr}\,dt,$$

$$d\,na^2\sqrt{1-e^2} = \frac{d\Omega}{d\mathfrak{p}}\,dt,$$

which give

$$e\sin f\,df - \frac{2e+(1+e^2)\cos f}{1-e^2}\,de + \frac{1+e\cos f}{a}\,da = 0,$$

$$e\cos f\,df + \sin f\,de = \frac{\sqrt{1-e^2}}{na}\frac{d\Omega}{dr}\,dt + \frac{na^2\sqrt{1-e^2}}{e\sin f}\frac{d\Omega}{d\mathfrak{p}}\,dt,$$

$$-\,de + \frac{1-e^2}{2ae}\,da = \frac{\sqrt{1-e^2}}{na^2 e}\frac{d\Omega}{d\mathfrak{p}}\,dt,$$

and we thence obtain

$$da = \frac{2e\sin f}{n\sqrt{1-e^2}}\frac{d\Omega}{dr}\,dt + \frac{2(1+e\cos f)^2}{na(1-e^2)^{\frac{3}{2}}}\frac{d\Omega}{d\mathfrak{p}}\,dt,$$

$$de = \frac{\sqrt{1-e^2}\sin f}{na}\frac{d\Omega}{dr}\,dt + \frac{e+2\cos f+e\cos^2 f}{na^2\sqrt{1-e^2}}\frac{d\Omega}{d\mathfrak{p}}\,dt,$$

$$df = \frac{n\sqrt{1-e^2}\cos f}{nae}\frac{d\Omega}{dr}\,dt - \frac{(2+e\cos f)\sin f}{na^2\sqrt{1-e^2}}\frac{d\Omega}{d\mathfrak{p}}\,dt,$$

the last of which equations, combined with

$$df = \frac{(1+e\cos f)^2}{(1-e^2)^{\frac{3}{2}}}\,dg + \frac{(2+e\cos f)\sin f}{1-e^2}\,de,$$

gives

$$dg = \frac{(1-e^2)(-2e+\cos f+e\cos^2 f)}{nae(1+e\cos f)}\frac{d\Omega}{dr}\,dt - \frac{(2+e\cos f)\sin f}{na^2 e}\frac{d\Omega}{d\mathfrak{p}}\,dt.$$

The fourth equation of the formulæ for the variations, viz., $d\mathfrak{p}=0$, gives $0 = d\varpi + df$, and therefore $d\varpi = -df$, that is,

$$d\varpi = -\frac{\sqrt{1-e^2}\cos f}{nae}\frac{d\Omega}{dr}\,dt + \frac{(2+e\cos f)\sin f}{na^2\sqrt{1-e^2}}\frac{d\Omega}{d\mathfrak{p}}\,dt,$$

and the complete system becomes therefore

$$da = \frac{2e\sin f}{n\sqrt{1-e^2}}\frac{d\Omega}{dr}\,dt + \frac{2(1+e\cos f)^2}{na(1-e^2)^{\frac{3}{2}}}\frac{d\Omega}{d\mathfrak{p}}\,dt,$$

$$de = \frac{\sqrt{1-e^2}\sin f}{na}\frac{d\Omega}{dr}\,dt + \frac{e+2\cos f+e\cos^2 f}{na^2\sqrt{1-e^2}}\frac{d\Omega}{d\mathfrak{p}}\,dt,$$

$$dg = \frac{(1-e^2)(-2e+\cos f+e\cos^2 f)}{nae(1+e\cos f)}\frac{d\Omega}{dr}\,dt - \frac{(2+e\cos f)\sin f}{na^2 e}\frac{d\Omega}{d\mathfrak{p}}\,dt,$$

$$d\varpi = -\frac{\sqrt{1-e^2}\sin f}{nae}\frac{d\Omega}{dr}\,dt + \frac{(2+e\cos f)\sin f}{na^2 e}\frac{d\Omega}{d\mathfrak{p}}\,dt,$$

$$d\Phi = \frac{-\cot\Phi}{na^2\sqrt{1-e^2}}\frac{d\Omega}{d\Sigma}dt - \frac{\operatorname{cosec}\Phi}{na^2\sqrt{1-e^2}}\frac{d\Omega}{d\Theta}dt - \qquad (\cos S'\,d\phi' + \sin S'\sin\phi'\,d\theta'),$$

$$d\Sigma = \frac{\cot\Phi}{na^2\sqrt{1-e^2}}\frac{d\Omega}{d\Phi}dt \qquad + \operatorname{cosec}\Phi\,(\sin S'\,d\phi' - \cos S'\sin\phi'\,d\theta'),$$

$$d\Theta = \frac{\operatorname{cosec}\Phi}{na^2\sqrt{1-e^2}}\frac{d\Omega}{d\Phi}dt \qquad + \cot\Phi\,(\sin S'\,d\phi' - \cos S'\sin\phi'\,d\theta'),$$

where, as before, $\Omega = \Omega(r, \mathfrak{p}, \Sigma, \Theta, \Phi)$. This is the first form of the expressions for the variations of the elements.

But we may in the disturbing function Ω replace r, $\mathfrak{p}$, by their values in terms of a, e, g, ϖ, and if on the right-hand side Ω has the last preceding value, we have

$$\frac{d\Omega}{da}da + \frac{d\Omega}{de}de + \frac{d\Omega}{dg}dg + \frac{d\Omega}{d\varpi}d\varpi + \frac{d\Omega}{d\Sigma}d\Sigma + \frac{d\Omega}{d\Theta}d\Theta + \frac{d\Omega}{d\Phi}d\Phi$$

$$= \frac{d\Omega}{dr}dr + \frac{d\Omega}{d\mathfrak{p}}d\mathfrak{p} + \frac{d\Omega}{d\Sigma}d\Sigma + \frac{d\Omega}{d\Theta}d\Theta + \frac{d\Omega}{d\Phi}d\Phi,$$

where on the left-hand side $\Omega = \Omega(a, e, g, \varpi, \Sigma, \Theta, \Phi)$; and the expressions for the differentials dr and $d\mathfrak{p}$, are

$$dr = \frac{1-e^2}{1+e\cos f}da - a\cos f\,de + \frac{ae\sin f}{\sqrt{1-e^2}}dg,$$

$$d\mathfrak{p} = \frac{(2+e\cos f)\sin f}{1-e^2}de + \frac{(1+e\cos f)^2}{(1-e)^{\frac{3}{2}}}dg + d\varpi,$$

and we have therefore

$$\frac{d\Omega}{da} = \frac{1-e^2}{1+e\cos f}\frac{d\Omega}{dr},$$

$$\frac{d\Omega}{de} = -a\cos f\frac{d\Omega}{dr} + \frac{(2+e\cos f)\sin f}{1-e^2}\frac{d\Omega}{d\mathfrak{p}},$$

$$\frac{d\Omega}{dg} = \frac{ae\sin f}{\sqrt{1-e^2}}\frac{d\Omega}{dr} + \frac{(1+e\cos f)^2}{(1-e^2)^{\frac{3}{2}}}\frac{d\Omega}{d\mathfrak{p}},$$

$$\frac{d\Omega}{d\varpi} = \frac{d\Omega}{d\mathfrak{p}},$$

$$\frac{d\Omega}{d\Phi} = \frac{d\Omega}{d\Phi},$$

$$\frac{d\Omega}{d\Sigma} = \frac{d\Omega}{d\Sigma},$$

$$\frac{d\Omega}{d\Theta} = \frac{d\Omega}{d\Theta},$$

where on the left-hand side $\Omega = \Omega(a, e, g, \varpi, \Sigma, \Theta, \Phi)$, and we thence obtain for the variations the new system of formulæ,

$$da = + \frac{2}{na}\frac{d\Omega}{dg}dt,$$

$$de = + \frac{1-e^2}{na^2e}\frac{d\Omega}{dg}dt - \frac{\sqrt{1-e^2}}{na^2e}\frac{d\Omega}{d\varpi}dt,$$

$$dg = -\frac{2}{na}\frac{d\Omega}{da}dt - \frac{1-e^2}{na^2e}\frac{d\Omega}{de}dt,$$

$$d\varpi = + \frac{\sqrt{1-e^2}}{na^2e}\frac{d\Omega}{de}dt,$$

$$d\Phi = \frac{-\cot\Phi}{na^2\sqrt{1-e^2}}\frac{d\Omega}{d\Sigma}dt - \frac{\operatorname{cosec}\Phi}{na^2\sqrt{1-e^2}}\frac{d\Omega}{d\Theta}dt - (\cos S'd\phi' + \sin S' \sin\phi' d\theta'),$$

$$d\Sigma = + \frac{\cot\Phi}{na^2\sqrt{1-e^2}}\frac{d\Omega}{d\Phi}dt + \operatorname{cosec}\Phi\,(\sin S'd\phi' - \cos S' \sin\phi' d\theta'),$$

$$d\Theta = + \frac{\operatorname{cosec}\Phi}{na^2\sqrt{1-e^2}}\frac{d\Omega}{d\Phi}dt + \cot\Phi\,(\sin S'd\phi' - \cos S' \sin\phi' d\theta'),$$

where, as before, $\Omega = \Omega\,(a, e, g, \varpi, \Sigma, \Theta, \Phi)$. This is the second form of the expressions for the variations of the elements. It is hardly necessary to remark, that if in either system of formulæ we omit the terms involving $d\theta'$ and $d\phi'$, and in the place of Σ, Θ, Φ, write σ, θ, ϕ, we have the formulæ for the variation of the elements when the orbit of the planet is referred to a fixed plane, and the disturbing function is given under the form $\Omega = \Omega\,(r, \text{þ}, \sigma, \theta, \phi)$, or $\Omega = \Omega\,(a, e, g, \varpi, \sigma, \theta, \phi)$.

The demonstration of the two preceding systems forms, as before remarked, the object of the present Memoir. But it is proper to give also the systems for the variations of the elements in the form in which they would have been obtained, if the notion of the departure had not been introduced into the investigation. To do this I revert to a preceding system of equations, which may be written

$$dr = 0,$$

$$d\theta = \frac{\operatorname{cosec}\phi}{na^2\sqrt{1-e^2}}\frac{d\Omega}{d\phi}dt,$$

$$dz = \frac{-\cot\phi}{na^2\sqrt{1-e^2}}\frac{d\Omega}{d\phi}dt,$$

$$d\frac{nae\sin f}{\sqrt{1-e^2}} = \frac{d\Omega}{dr}dt,$$

$$d\, na^2\sqrt{1-e^2} = \frac{d\Omega}{dz}dt,$$

$$d\phi = \frac{-\operatorname{cosec}\phi}{na^2\sqrt{1-e^2}}\frac{d\Omega}{d\theta}dt + \frac{\cot\phi}{na^2\sqrt{1-e^2}}\frac{d\Omega}{dz}dt\left(=\frac{\cot z}{na^2\sqrt{1-e^2}}\frac{d\Omega}{d\phi}dt\right),$$

where $\Omega = \Omega\,(r, z, \theta, \phi)$.

Substituting in these equations for r, z, the values $\dfrac{a(1-e^2)}{1+e\cos f}$ and $\varpi + f$, we obtain

$$da = \frac{2e\sin f}{n\sqrt{1-e^2}}\frac{d\Omega}{dr}dt + \frac{2(1+e\cos f)^2}{na(1-e^2)^{\frac{3}{2}}}\frac{d\Omega}{dz}dt,$$

$$de = \frac{\sqrt{1-e^2}\sin f}{na}\frac{d\Omega}{dr}dt + \frac{e+2\cos f+e\cos^2 f}{na^2\sqrt{1-e^2}}\frac{d\Omega}{dz}dt,$$

$$dg = \frac{(1-e^2)(-2e+\cos f+e\cos^2 f)}{nae(1+e\cos f)}\frac{d\Omega}{dr}dt - \frac{(2+e\cos f)\sin f}{na^2e}\frac{d\Omega}{dz}dt,$$

$$d\varpi = -\frac{\sqrt{1-e^2}\cos f}{nae}\frac{d\Omega}{dr}dt + \frac{(2+e\cos f)\sin f}{na^2\sqrt{1-e^2}}\frac{d\Omega}{dz}dt - \frac{\cot\phi}{na^2\sqrt{1-e^2}}\frac{d\Omega}{d\phi}dt,$$

$$d\phi = \frac{\cot\phi}{na^2\sqrt{1-e^2}}\frac{d\Omega}{dz}dt - \frac{\operatorname{cosec}\phi}{na^2\sqrt{1-e^2}}\frac{d\Omega}{d\theta}dt,$$

$$d\theta = \frac{\operatorname{cosec}\phi}{na^2\sqrt{1-e^2}}\frac{d\Omega}{d\phi}dt,$$

where $\Omega = \Omega\,(r, z, \theta, \phi)$, as before; and which is the first system for the variations of the six elements, a, e, g, ϖ, θ, ϕ.

But if in the disturbing function we replace r, z by their values, then if on the left-hand side $\Omega = \Omega\,(r, z, \theta, \phi)$, as before, we find

$$\frac{1-e^2}{1+e\cos f}\frac{d\Omega}{dr} = \frac{d\Omega}{dr},$$

$$-a\cos f\frac{d\Omega}{dr} + \frac{(2+e\cos f)\sin f}{1-e^2}\frac{d\Omega}{dz} = \frac{d\Omega}{de},$$

$$\frac{ae\sin f}{\sqrt{1-e^2}}\frac{d\Omega}{dr} + \frac{(1+e\cos f)^2}{(1-e^2)^{\frac{3}{2}}}\frac{d\Omega}{dz} = \frac{d\Omega}{dg},$$

$$\frac{d\Omega}{d\phi} = \frac{d\Omega}{d\phi},$$

$$\frac{d\Omega}{dz} = \frac{d\Omega}{d\varpi},$$

$$\frac{d\Omega}{d\theta} = \frac{d\Omega}{d\theta},$$

where on the right-hand side $\Omega = \Omega\,(a, e, g, \varpi, \theta, \phi)$.

The formulæ for the variations thus become

$$
\begin{aligned}
da &= \frac{2}{na}\frac{d\Omega}{dg}dt,\\
de &= \frac{1-e^2}{na^2e}\frac{d\Omega}{dg}dt - \frac{\sqrt{1-e^2}}{na^2e}\frac{d\Omega}{d\varpi}dt,\\
dg &= -\frac{2}{na}\frac{d\Omega}{da}dt - \frac{1-e^2}{na^2e}\frac{d\Omega}{de}dt,\\
d\varpi &= \frac{\sqrt{1-e^2}}{na^2e}\frac{d\Omega}{de}dt - \frac{\cot\phi}{na\sqrt{1-e^2}}\frac{d\Omega}{d\phi}dt,\\
d\phi &= \frac{\cot\phi}{na^2\sqrt{1-e^2}}\frac{d\Omega}{d\varpi}dt - \frac{\operatorname{cosec}\phi}{na^2\sqrt{1-e^2}}\frac{d\Omega}{d\theta}dt,\\
d\theta &= +\frac{\operatorname{cosec}\phi}{na^2\sqrt{1-e^2}}\frac{d\Omega}{d\phi}dt,
\end{aligned}
$$

where, as before, $\Omega = \Omega\ (a,\ e,\ g,\ \varpi,\ \theta,\ \phi)$. This is the second system of formulæ for the variations of the six elements a, e, g, ϖ, θ, ϕ.

The last-mentioned system may be easily deduced from Jacobi's canonical system of formulæ, viz. putting

$\mathfrak{A}$, the constant of vis viva,
$\mathfrak{B}$, the constant of areas,
$\mathfrak{C}$, the constant of the reduced area,
$\mathfrak{F}$, the constant attached to the time,
$\mathfrak{G}$, the angular distance of pericentre from node,
$\mathfrak{H}$, the longitude of node;

then the canonical system is

$$
\begin{aligned}
d\mathfrak{A} &= \frac{d\Omega}{d\mathfrak{F}}dt,\\
d\mathfrak{B} &= \frac{d\Omega}{d\mathfrak{G}}dt,\\
d\mathfrak{C} &= \frac{d\Omega}{d\mathfrak{H}}dt,\\
d\mathfrak{F} &= -\frac{d\Omega}{d\mathfrak{A}}dt,\\
d\mathfrak{G} &= -\frac{d\Omega}{d\mathfrak{B}}dt,\\
d\mathfrak{H} &= -\frac{d\Omega}{d\mathfrak{C}}dt,
\end{aligned}
$$

and the expressions for $\mathfrak{A}$, &c. in terms of the elements a, e, g, ϖ, θ, ϕ, are

$$\begin{aligned}
\mathfrak{A} &= -\tfrac{1}{2}n^2a^2,\\
\mathfrak{B} &= na^2\sqrt{1-e^2},\\
\mathfrak{C} &= na^2\sqrt{1-e^2}\cos\phi,\\
\mathfrak{F} &= \frac{1}{n}g - t,\\
\mathfrak{G} &= \varpi,\\
\mathfrak{H} &= \theta,
\end{aligned}$$

and the transformation can be effected without the slightest difficulty.

I shall conclude with the demonstration of a formula which occurs implicitly in Hansen's Lunar Theory, and which may probably be useful for other purposes.

Write

ρ, the radius vector, τ for t,

ψ, the true anomaly, τ for t,

that is, let ρ, ψ, be what r, f, become when the time t, in so far as it enters explicitly in g, and not through the variable elements, is replaced by an arbitrary quantity, τ.

And suppose, in like manner,

γ, the mean anomaly, τ for t,

λ, the departure, τ for t,

and let $l\rho$ denote the logarithm of ρ; then we have, attending only to the variation of t, in so far as it enters through the variable elements,

$$\begin{aligned}
df &= \frac{(2+e\cos f)\sin f}{1-e^2}\,de + \frac{(1+e\cos f)^2}{(1-e^2)^{\frac{3}{2}}}\,dg,\\
d\psi &= \frac{(2+e\cos\psi)\sin\psi}{1-e^2}\,de + \frac{(1+e\cos\psi)^2}{(1-e^2)^{\frac{3}{2}}}\,d\gamma,\\
dl\rho &= \frac{1}{a}\,da - \frac{(1+e\cos\psi)\cos\psi}{1-e^2}\,de + \frac{(1+e\cos\psi)\,e\sin\psi}{(1-e^2)^{\frac{3}{2}}}\,d\gamma.
\end{aligned}$$

We hence deduce,

$$\begin{aligned}
&dl\rho + \frac{\rho e\sin\psi}{a(1-e^2)}(df - d\psi)\\
&= \frac{1}{a}\,da + \frac{e\sin\psi\sin f(2+e\cos f) - \cos\psi - 2e - e^2\cos\psi}{(1-e^2)(1+e\cos\psi)}\,de + \frac{e\sin\psi\,(1+e\cos f)^2}{(1-e^2)^{\frac{3}{2}}(1+e\cos\psi)}\,dg,
\end{aligned}$$

the coefficient of $d\gamma$ being zero. And substituting for da, de, dg their values, viz.

$$da = \frac{2e\sin f}{n\sqrt{1-e^2}}\frac{d\Omega}{dr}dt + \frac{2(1+e\cos f)^2}{na(1-e^2)^{\frac{3}{2}}}\frac{d\Omega}{d\mathfrak{p}}dt,$$

$$de = \frac{\sqrt{1-e^2}\sin f}{na}\frac{d\Omega}{dr}dt + \frac{e+2\cos f+e\cos^2 f}{na^2\sqrt{1-e^2}}\frac{d\Omega}{d\mathfrak{p}}dt,$$

$$dg = \frac{(1-e^2)(-2e+\cos f+e\cos^2 f)}{nae(1+e\cos f)}\frac{d\Omega}{dr}dt + \frac{(2+e\cos f)\sin f}{na^2 e}\frac{d\Omega}{d\mathfrak{p}}dt,$$

the equation becomes

$$dl\rho + \frac{\rho e\sin\psi}{a(1-e^2)}(df - d\psi)$$

$$= \frac{-\sqrt{1-e^2}}{na(1+e\cos\psi)}\sin(f-\psi)\frac{d\Omega}{dr}dt$$

$$+ \frac{1}{na^2\sqrt{1-e^2}(1+e\cos\psi)}\left\{2+e\cos\psi - \cos(f-\psi)(2+e\cos f)\right\}\frac{d\Omega}{d\mathfrak{p}}dt.$$

But we have $\mathfrak{p} = \varpi + f$, $\lambda = \varpi + \psi$, and, consequently, $f - \psi = \mathfrak{p} - \lambda$, and the equation becomes

$$dl\rho + \frac{\rho e\sin\psi}{a(1-e^2)}(d\mathfrak{p} - d\lambda)$$

$$= \frac{-\sqrt{1-e^2}}{na(1+e\cos\psi)}\sin(\mathfrak{p}-\lambda)\frac{d\Omega}{dr}dt$$

$$+ \frac{1}{na^2\sqrt{1-e^2}(1+e\cos\psi)}\left\{2+e\cos\psi - \cos(\mathfrak{p}-\lambda)(2+e\cos f)\right\}\frac{d\Omega}{d\mathfrak{p}}dt,$$

which is the equation referred to; the expression on the right-hand side, omitting the factor dt, is, in fact, the portion not involving the arbitrary functions Π, Γ, of Hansen's function R (*Fund.* p. 43), viz. it is in Hansen's notation,

$$-\left\{\frac{\rho}{r}\cos(v,-\lambda) - 1 + \frac{\rho}{a(1-e^2)}[\cos(v,-\lambda)-1]\right\}\frac{an}{\sqrt{1-e^2}}\frac{d\Omega}{dv}$$

$$-\ \frac{\rho}{r}\sin(v,-\lambda)\frac{an}{\sqrt{1-e^2}}r\frac{d\Omega}{dr},$$

where a, n, e, r, ρ, v, λ, Ω (Hansen), correspond to a, n, e, r, ρ, $\mathfrak{p}$, λ, $n^{-2}a^{-3}\Omega$, of the present Memoir.

2, *Stone Buildings, W. C.*, 3 *March*, 1858.

213.

ON THE DEVELOPMENT OF THE DISTURBING FUNCTION IN THE LUNAR THEORY.

[From the *Memoirs of the Royal Astronomical Society*, vol. XXVII., (1859), pp. 69—95. Read November 12, 1858.]

THE development of the disturbing function for the lunar theory is effected in a very elegant manner in Hansen's *Fundamenta Nova*, and it requires only a single easy step to exhibit the result in a perfectly explicit form, and to compare it with those of other geometers. To do this is the immediate object of the present memoir, and the mode of development is a mere reproduction of that made use of by Hansen. But the memoir is written with a view to the development of and application to the lunar theory, of the theory contained in my "Memoir on the Problem of Disturbed Elliptic Motion," *ante* pp. 1—29, [212], and the notation adopted (differing from Hansen's very slightly) is consequently that of the memoir just referred to.

Taking, as usual, Ω to denote the disturbing function with the sign employed by Lagrange ($\Omega = -R$, if R be the disturbing function of the *Mécanique Céleste*), then

$$\Omega = m' \left\{ \frac{1}{(r^2 + r'^2 - 2rr' \cos H)^{\frac{1}{2}}} - \frac{r}{r'^2} \cos H \right\},$$

where we have

m', the mass of the sun,
r , the radius vector of the moon,
r' , the radius vector of the sun,
H, the angular distance of the sun and moon,

the earth being, of course, taken as the centre of motion; (Hansen's Ω is the above value divided by $M + m$, where M and m are the masses of the earth and moon respectively; that is, the disturbing function here represented by Ω is Hansen's Ω multiplied into $M + m$).

Write also

f , the true anomaly of the moon,

f' , the true anomaly of the sun,

$\mho$, the distance of moon's pericentre from ascending node of moon's orbit,

$\mho'$, the distance of sun's pericentre from same node,

Φ , the inclination of moon's orbit to that of the sun,

where the orbits referred to are the true or instantaneous orbits. Then the angular distance H is the third side of a spherical triangle, the other two sides whereof are $f+\mho$, $f'+\mho'$, the included angle between them being Φ, that is, we have

$$\cos H = \cos(f+\mho)\cos(f'+\mho') + \cos\Phi \sin(f+\mho)\sin(f'+\mho')$$

and in this equation, if we write

g , the mean anomaly of the moon,

a , the semi-axis major of the moon's orbit,

e , the eccentricity;

and in like manner,

g', the mean anomaly of the sun,

a', the semi-axis major of the sun's orbit,

e', the eccentricity;

then r, f, and r', f', are respectively given functions of a, e, g, and a', e', g', viz., we have

$$r = a \text{ elqr } (e, g),$$
$$f = \text{ elta } (e, g),$$
$$r' = a' \text{ elqr } (e', g'),$$
$$f' = \text{ elta } (e', g'),$$

and introducing, instead of the inclination, the quantity

$$\eta \,(= \sin \tfrac{1}{2}\Phi), \text{ the sine of the semi-inclination;}$$

the disturbing function Ω becomes a function of a, e, g, $\mho$, a', e', g', $\mho'$, η, and the required development is a development in multiple cosines of g, g', $\mho$, $\mho'$, the coefficients being of course functions of the remaining quantities a, e, a', e', η. The single symbols $\mho$, $\mho'$ (which denote the distances of the pericentres of the lunar and solar orbits from the mutual node) will be retained throughout the memoir; but if we write

ϖ, the departure of moon's pericentre,

Σ, the departure of moon's ascending node;

these departures being measured on the moon's orbit; and, in like manner, measured on the sun's orbit from a *departure point* on this orbit, but called for distinction "longitudes," instead of departures,

ϖ', the longitude of the sun's pericentre,

Θ, the longitude of the moon's ascending node;

then we have

$$\mho = \varpi - \Sigma,$$

$$\mho' = \varpi' - \Theta;$$

and the disturbing function Ω, so far as it depends on the position of the moon, is a function of the seven elements a, e, g, ϖ, η, Σ, Θ, and it contains also the quantities a', e', g', ϖ', which relate to the sun.

Proceeding now to develope Ω in ascending powers of $\frac{r}{r'}$, we have

$$\Omega = m' \left\{ \frac{1}{r'} + \frac{r^2}{r'^3} [\tfrac{3}{2} \cos^2 H - \tfrac{1}{2}] + \frac{r^3}{r'^4} [\tfrac{5}{2} \cos^3 H - \tfrac{3}{2} \cos H] + \frac{r^4}{r'^5} [\tfrac{35}{8} \cos^4 H - \tfrac{15}{4} \cos^2 H + \tfrac{3}{8}] + \&c. \right\}$$

where, however, the last term is neglected in the sequel, and since we are only concerned with the differential coefficients of Ω in regard to the lunar elements, the first term $m' \frac{1}{r'}$ which depends only on the solar elements may also be neglected; we have

$$\begin{aligned} \cos H = &\quad \cos^2 \tfrac{1}{2}\Phi \quad \cos(f - f' + \mho - \mho') \\ &+ \sin^2 \tfrac{1}{2}\Phi \quad \cos(f + f' + \mho + \mho'), \end{aligned}$$

and thence

$$\begin{aligned} \cos^2 H = &\quad \cos^4 \tfrac{1}{2}\Phi \quad \cos^2(f - f' + \mho - \mho') \\ &+ 2\cos^2 \tfrac{1}{2}\Phi \sin^2 \tfrac{1}{2}\Phi \quad \cos(f - f' + \mho - \mho') \cos(f + f' + \mho + \mho') \\ &+ \sin^4 \tfrac{1}{2}\Phi \quad \cos^2(f + f' + \mho + \mho'), \end{aligned}$$

$$\begin{aligned} \cos^3 H = &\quad \cos^6 \tfrac{1}{2}\Phi \quad \cos^3(f - f' + \mho - \mho') \\ &+ 3\cos^4 \tfrac{1}{2}\Phi \sin^2 \tfrac{1}{2}\Phi \quad \cos^2(f - f' + \mho - \mho') \cos(f + f' + \mho + \mho') \\ &+ 3\cos^2 \tfrac{1}{2}\Phi \sin^4 \tfrac{1}{2}\Phi \quad \cos(f - f' + \mho - \mho') \cos^2(f + f' + \mho + \mho') \\ &+ \sin^6 \tfrac{1}{2}\Phi \quad \cos^3(f + f' + \mho + \mho'), \end{aligned}$$

and converting the powers of the cosines of $f - f' + \mho - \mho'$, $f + f' + \mho + \mho'$ into multiple cosines, and expressing the coefficients in terms of η $(= \sin \tfrac{1}{2}\Phi)$ and neglecting η^6, we have

$$\begin{aligned}
\cos H = & \quad -\eta^2 \quad \cos\ f - f' + \varpi - \varpi' \\
& \quad +\eta^2 \quad \cos\ f + f' + \varpi + \varpi',
\end{aligned}$$

$$\begin{aligned}
\cos^2 H = & \ \tfrac{1}{2} - \eta^2 + \eta^4 \\
& + \tfrac{1}{2} - \eta^2 + \tfrac{1}{2}\eta^4 \quad \cos\ 2f - 2f' + 2\varpi - 2\varpi' \\
& + \quad \eta^2 - \eta^4 \quad \cos\ 2f \quad + 2\varpi \\
& + \quad \eta^2 - \eta^4 \quad \cos\ 2f' \quad + 2\varpi' \\
& + \quad \tfrac{1}{2}\eta^4 \quad \cos\ 2f + 2f' + 2\varpi + 2\varpi',
\end{aligned}$$

$$\begin{aligned}
\cos^3 H = & \ \tfrac{3}{4} - \tfrac{9}{4}\eta^2 + \tfrac{15}{4}\eta^4 \quad \cos\ f - f' + \varpi - \varpi' \\
& + \tfrac{1}{4} - \tfrac{3}{4}\eta^2 + \tfrac{3}{4}\eta^4 \quad \cos\ 3f - 3f' + 3\varpi - 3\varpi' \\
& + \quad \tfrac{3}{2}\eta^2 - 3\eta^4 \quad \cos\ f + f' + \varpi + \varpi' \\
& + \quad \tfrac{3}{4}\eta^2 - \tfrac{3}{2}\eta^4 \quad \cos\ 3f - f' + 3\varpi - \varpi' \\
& + \quad \tfrac{3}{4}\eta^2 - \tfrac{3}{2}\eta^4 \quad \cos\ f - 3f' + \varpi - 3\varpi' \\
& + \quad \tfrac{3}{4}\eta^4 \quad \cos\ 3f + f' + 3\varpi + \varpi' \\
& + \quad \tfrac{3}{4}\eta^4 \quad \cos\ f + 3f' + \varpi + 3\varpi'.
\end{aligned}$$

Hence

$\Omega = m'\dfrac{r^2}{r'^3}$ multiplied into

$$\left\{\begin{aligned}
& \tfrac{1}{4} - \tfrac{3}{2}\eta^2 + \tfrac{3}{2}\eta^4 \\
& + \tfrac{3}{4} - \tfrac{3}{2}\eta^2 + \tfrac{3}{4}\eta^4 \quad \cos\ 2f - 2f' + 2\varpi - 2\varpi' \\
& + \quad \tfrac{3}{2}\eta^2 - \tfrac{3}{2}\eta^4 \quad \cos\ 2f \quad + 2\varpi \\
& + \quad \tfrac{3}{2}\eta^2 - \tfrac{3}{2}\eta^4 \quad \cos\ 2f' + 2\varpi' \\
& + \quad \tfrac{3}{4}\eta^4 \quad \cos\ 2f + 2f' + 2\varpi + 2\varpi
\end{aligned}\right.$$

$+\, m'\dfrac{r^3}{r'^4}$ multiplied into

$$\left\{\begin{aligned}
& \tfrac{3}{8} - \tfrac{33}{8}\eta^2 + \tfrac{75}{8}\eta^4 \quad \cos\ f - f' + \varpi - \varpi' \\
& + \tfrac{5}{8} - \tfrac{15}{8}\eta^2 + \tfrac{15}{8}\eta^4 \quad \cos\ 3f - 3f' + 3\varpi - 3\varpi' \\
& + \quad \tfrac{9}{4}\eta^2 - \tfrac{15}{2}\eta^4 \quad \cos\ f + f' + \varpi + \varpi' \\
& + \quad \tfrac{15}{8}\eta^2 - \tfrac{15}{4}\eta^4 \quad \cos\ 3f - f' + 3\varpi - \varpi' \\
& + \quad \tfrac{15}{8}\eta^2 - \tfrac{15}{4}\eta^4 \quad \cos\ f - 3f' + \varpi - 3\varpi' \\
& + \quad \tfrac{15}{8}\eta^4 \quad \cos\ 3f + f' + 3\varpi + \varpi' \\
& + \quad \tfrac{15}{8}\eta^4 \quad \cos\ f + 3f' + \varpi + 3\varpi',
\end{aligned}\right.$$

but the last two terms of the part multiplied by $m'\frac{r^3}{r'^4}$ (which are besides of the fourth order in η) are neglected in the sequel.

Now i, i' denoting integer numbers, and writing down only the general terms (the summatory sign $\Sigma_{-\infty}^{\infty}$ being in each case understood), we may put

$$\frac{r^2}{a^2}\cos 2f = Q_c^i \cos ig, \quad \frac{r^2}{a^2}\sin 2f = Q_s^i \sin ig,$$

$$\frac{r^2}{a^2} = P^i \cos ig,$$

$$\frac{a'^3}{r'^3}\cos 2f' = G_c^{i'} \cos i'g', \quad \frac{a'^3}{r'^3}\sin 2f' = G_s^{i'} \sin i'g',$$

$$\frac{a'^3}{r'^3} = K^{i'} \cos i'g',$$

$$\frac{r^3}{a^3}\cos f = A_c^i \cos ig, \quad \frac{r^3}{a^3}\sin f = A_s^i \sin ig,$$

$$\frac{r^3}{a^3}\cos 3f = B_c^i \cos ig, \quad \frac{r^3}{a^3}\sin 3f = B_s^i \sin ig,$$

$$\frac{a'^4}{r'^4}\cos f' = C_c^{i'} \cos i'g', \quad \frac{a'^4}{r'^4}\sin f' = C_s^{i'} \sin i'g',$$

$$\frac{a'^4}{r'^4}\cos 3f' = D_c^{i'} \cos i'g', \quad \frac{a'^4}{r'^4}\sin 3f' = D_s^{i'} \sin i'g',$$

where

$$Q_c^i = Q_c^{-i}, \quad G_c^{i'} = G_c^{-i'}, \quad A_c^i = A_c^{-i}, \text{ \&c.}$$

$$Q_s^i = -Q_s^{-i}, \quad G_s^{i'} = -G_s^{-i'}, \quad A_s^i = -A_s^{-i}, \text{ \&c.}$$

$$P^i = P^{-i}, \quad K^{i'} = K^{-i'}.$$

Then by a known rule for the multiplication of doubly infinite sine or cosine series, and after an easy transformation of the original form of the coefficients, we obtain, for instance,

$$\frac{r^2}{r'^3}\cos(2f - 2f' + 2\varpi - 2\varpi') = \frac{a^2}{a'^3}(Q_c^i + Q_s^i)(G_c^{i'} - G_s^{i'})\cos(ig + i'g' + 2\varpi - 2\varpi')$$

where only the general term is written down, but the indices i, i', each of them extend from $-\infty$ to $+\infty$, zero included, or the summatory sign $\Sigma_{-\infty}^{\infty}\,\Sigma_{-\infty}^{\infty}$ is to be understood; and similarly for the other terms of Ω. And we may write

$$\Omega = \Omega_1 + \Omega_2 + \Omega_3 + \Omega_4 + \Omega_5 + \Omega_6 + \Omega_7 + \Omega_8 + \Omega_9 + \Omega_{10},$$

where

$$\left\{\begin{array}{lllllll}
\Omega_1 = m'\dfrac{a^2}{a'^3} & \tfrac{1}{4} - \tfrac{3}{2}\,\eta^2 + \tfrac{3}{2}\,\eta^4 & P^i & K^{i'} & \cos & ig + i'g' & , \\[2ex]
\Omega_2 = m'\dfrac{a^2}{a'^3} & \tfrac{3}{4} - \tfrac{3}{2}\,\eta^2 + \tfrac{3}{4}\,\eta^4 & Q_c^i + Q_s^i & G_c^{i'} - G_s^{i'} & \cos & ig + i'g' + 2\varpi - 2\varpi' & , \\[2ex]
\Omega_3 = m'\dfrac{a^2}{a'^3} & \tfrac{3}{2}\,\eta^2 - \tfrac{3}{2}\,\eta^4 & Q_c^i + Q_s^i & K^{i'} & \cos & ig + i'g' + 2\varpi & , \\[2ex]
\Omega_4 = m'\dfrac{a^2}{a'^3} & \tfrac{3}{2}\,\eta^2 - \tfrac{3}{2}\,\eta^4 & P^i & G_c^{i'} + G_s^{i'} & \cos & ig + i'g' \qquad + 2\varpi' & , \\[2ex]
\Omega_5 = m'\dfrac{a^2}{a'^3} & \tfrac{3}{4}\,\eta^4 & Q_c^i + Q_s^i & G_c^{i'} + G_s^{i'} & \cos & ig + i'g' + 2\varpi + 2\varpi' & ,
\end{array}\right.$$

$$\left\{\begin{array}{lllllll}
\Omega_6 = m'\dfrac{a^3}{a'^4} & \tfrac{3}{8} - \tfrac{33}{8}\,\eta^2 + \tfrac{75}{8}\,\eta^4 & A_c^i + A_s^i & C_c^{i'} - C_s^{i'} & \cos & ig + i'g' + \varpi - \varpi' & , \\[2ex]
\Omega_7 = m'\dfrac{a^3}{a'^4} & \tfrac{5}{8} - \tfrac{15}{8}\,\eta^2 + \tfrac{15}{8}\,\eta^4 & B_c^i + B_s^i & D_c^{i'} - D_s^{i'} & \cos & ig + i'g' + 3\varpi - 3\varpi' & , \\[2ex]
\Omega_8 = m'\dfrac{a^3}{a'^4} & \tfrac{9}{4}\,\eta^2 - \tfrac{15}{2}\,\eta^4 & A_c^i + A_s^i & C_c^{i'} + C_s^{i'} & \cos & ig + i'g' + \varpi + \varpi' & , \\[2ex]
\Omega_9 = m'\dfrac{a^3}{a'^4} & \tfrac{15}{8}\,\eta^2 - \tfrac{15}{4}\,\eta^4 & B_c^i + B_s^i & C_c^{i'} - C_s^{i'} & \cos & ig + i'g' + 3\varpi - \varpi' & , \\[2ex]
\Omega_{10} = m'\dfrac{a^3}{a'^4} & \tfrac{15}{8}\,\eta^2 - \tfrac{15}{4}\,\eta^4 & A_c^i + A_s^i & D_c^{i'} - D_s^{i'} & \cos & ig + i'g' + \varpi - 3\varpi' & .
\end{array}\right.$$

The values of P^i, Q_c^i, Q_s^i, &c. are made to depend ultimately upon the development in multiple cosines of the square of the radius vector and the development in multiple sines of the true anomaly. And the actual values of P^i, Q_c^i, Q_s^i, &c., expanded in powers of e or e', are given pp. 174—179 of the work above referred to. I have verified all these values by a different process, and have discovered only a single inaccuracy, which, however, is rather an important one, as it affects the evection, viz., the value of Q_s^1 should be $-\tfrac{3}{2}e + \tfrac{23}{24}e^3 + \text{\&c.}$ instead of (Hansen) $-\tfrac{3}{2}e + \tfrac{33}{24}e^3 + \text{\&c.}$ The formation of the sums or differences $Q_c^i + Q_s^i$ is of course perfectly easy, and we thus obtain the actual developed expression of the disturbing function, which I represent under the following form, viz.:—

$$\Omega = \Omega_1 + \Omega_2 + \Omega_3 + \Omega_4 + \Omega_5 + \Omega_6 + \Omega_7 + \Omega_8 + \Omega_9 + \Omega_{10},$$

where

$$\Omega_1 = m' \frac{a^2}{a'^3}\left(\frac{1}{4} - \frac{3}{2}\eta^2 + \frac{3}{2}\eta^4\right) \text{ multiplied into}$$

Lunar part infrà.	$\frac{1773}{256} e'^5$	cos	$-5g'$
	$\frac{77}{16} e'^4$		$-4g'$
	$\frac{53}{16} e'^3$		$-3g'$
	$\frac{9}{4} e'^2 + \frac{7}{4} e'^4$		$-2g'$
	$\frac{3}{2} e' + \frac{27}{16} e'^3$		$-\ \ g'$
	* $1 + \frac{3}{2} e'^2 + \frac{15}{8} e'^4 + \frac{35}{16} e'^6$		$0g'$
	$\frac{3}{2} e' + \frac{27}{16} e'^3$		$+\ \ g'$
	$\frac{9}{4} e'^2 + \frac{7}{4} e'^4$		$+2g'$
	$\frac{53}{16} e'^3$		$+3g'$
	$\frac{231}{48} e'^4$		$+4g'$
	$\frac{1773}{256} e'^5$		$+5g'$
	* Exact value is $(1 - e'^2)^{-\frac{3}{2}}$		
$-\frac{9}{160} e^6$	Solar part suprà.		$-6g$
$-\frac{25}{384} e^5$			$-5g$
$-\frac{1}{12} e^4 + \frac{1}{15} e^6$			$-4g$
$-\frac{1}{8} e^3 + \frac{9}{128} e^5$			$-3g$
$-\frac{1}{4} e^2 + \frac{1}{12} e^4 - \frac{1}{96} e^6$			$-2g$
$-e + \frac{1}{8} e^3 - \frac{1}{192} e^5$			$-\ \ g$
$1 + \frac{3}{2} e^2$ (exact)			$0g$
$-e + \frac{1}{8} e^3 - \frac{1}{192} e^5$			$+\ \ g$
$-\frac{1}{4} e^2 + \frac{1}{12} e^4 - \frac{1}{96} e^6$			$+2g$
$-\frac{1}{8} e^3 + \frac{9}{128} e^5$			$+3g$
$-\frac{1}{12} e^4 + \frac{1}{15} e^6$			$+4g$
$-\frac{25}{384} e^5$			$+5g$
$-\frac{9}{160} e^6$			$+6g$

$\Omega_2 = m' \frac{a^2}{a'^3}\left(\frac{3}{4} - \frac{3}{2}\eta^2 + \frac{3}{4}\eta^4\right)$ multiplied into

Lunar part infrà.	$\frac{228347}{3840} e'^5$	cos	$-7g'$	
	$\frac{533}{16} e'^4$		$-6g'$	
	$\frac{845}{48} e'^3$		$-5g'$	
	$\frac{17}{2} e'^2 - \frac{115}{6} e'^4$		$-4g'$	
	$\frac{7}{2} e' - \frac{123}{16} e'^3$		$-3g'$	$-2\varpi'$
	$1 - \frac{5}{2} e'^2 + \frac{13}{16} e'^4$		$-2g'$	
	$-\frac{1}{2} e' + \frac{1}{16} e'^3$		$-g'$	
	o (exact)		$0g'$	
	$\frac{1}{48} e'^3$		$+g'$	
	$\frac{1}{24} e'^4$		$+2g'$	
$-\frac{11}{720} e^6$	Solar part suprà.		$-4g$	
$-\frac{17}{640} e^5$			$-3g$	
$-\frac{1}{16} e^4 - \frac{11}{480} e^6$			$-2g$	
$-\frac{7}{24} e^3 - \frac{47}{384} e^5$			$-g$	
$\frac{5}{2} e^2$ (exact)			$0g$	
$-3e + \frac{13}{8} e^3 + \frac{5}{192} e^5$			$+g$	
$1 - \frac{5}{2} e^2 + \frac{23}{16} e^4 - \frac{65}{288} e^6$			$+2g$	$+2\varpi$
$e - \frac{19}{8} e^3 + \frac{107}{64} e^5$			$+3g$	
$e^2 - \frac{5}{2} e^4 + \frac{101}{48} e^6$			$+4g$	
$\frac{25}{24} e^3 - \frac{1075}{384} e^5$			$+5g$	
$\frac{9}{8} e^4 - \frac{261}{80} e^6$			$+6g$	
$\frac{2401}{1920} e^5$			$+7g$	
$\frac{64}{45} e^6$			$+8g$	

$2g - 2g' + 2\varpi - 2\varpi'$ variation.

$g - 2g' + 2\varpi - 2\varpi'$ evection.

$$\Omega_3 = m' \frac{a^2}{a'^3}\left(\frac{3}{2}\eta^2 - \frac{3}{2}\eta^4\right) \text{ multiplied into}$$

Lunar part infrà.	$\frac{1773}{256} e'^5$	cos	$-5g'$	
	$\frac{77}{16} e'^4$		$-4g'$	
	$\frac{53}{16} e'^3$		$-3g'$	
	$\frac{9}{4} e'^2 + \frac{7}{4} e'^4$		$-2g'$	
	$\frac{3}{2} e' + \frac{27}{16} e'^3$		$-g'$	
	$* 1 + \frac{3}{2} e'^2 + \frac{15}{8} e'^4 + \frac{35}{16} e'^6$		$0g'$	
	$\frac{3}{2} e' + \frac{27}{16} e'^3$		$+g'$	
	$\frac{9}{4} e'^2 + \frac{7}{4} e'^4$		$+2g'$	
	$\frac{53}{16} e'^3$		$+3g'$	
	$\frac{231}{48} e'^4$		$+4g'$	
	$\frac{1773}{256} e'^5$		$+5g'$	
	*Exact value is $(1 - e'^2)^{-\frac{3}{2}}$			
$-\frac{11}{720} e^6$	Solar part suprà.		$-4g$	
$-\frac{17}{640} e^5$			$-3g$	
$-\frac{1}{16} e^4 - \frac{11}{480} e^6$			$-2g$	
$-\frac{7}{24} e^3 - \frac{47}{384} e^5$			$-g$	
$\frac{5}{2} e^2$ (exact)			$0g$	
$-3e + \frac{13}{8} e^3 + \frac{5}{192} e^5$			$+g$	
$1 - \frac{5}{2} e^2 + \frac{23}{16} e^4 - \frac{65}{288} e^6$			$+2g$	$+2\varpi$
$e - \frac{19}{8} e^3 + \frac{107}{64} e^5$			$+3g$	
$e^2 - \frac{5}{2} e^4 + \frac{101}{48} e^6$			$+4g$	
$\frac{25}{24} e^3 - \frac{1075}{384} e^5$			$+5g$	
$\frac{9}{8} e^4 - \frac{261}{80} e^6$			$+6g$	
$\frac{2401}{1920} e^5$			$+7g$	
$\frac{64}{45} e^6$			$+8g$	

$$\Omega_4 = m'\frac{a^2}{a'^3}\left(\frac{3}{2}\eta^2 - \frac{3}{2}\eta^4\right) \text{ multiplied into}$$

Lunar part infrà.	$\frac{1}{24}e'^4$	cos	$-2g'$	
	$\frac{1}{48}e'^3$		$-\ g'$	
	o (exact)		$0\ g'$	
	$-\frac{1}{2}e' + \frac{1}{16}e'^3$		$+\ g'$	
	$1 - \frac{5}{2}e'^2 + \frac{13}{16}e'^4$		$+2g'$	
	$\frac{7}{2}e' - \frac{123}{16}e'^3$		$+3g'$	$+2\varpi'$
	$\frac{17}{2}e'^2 - \frac{115}{6}e'^4$		$+4g'$	
	$\frac{845}{48}e'^3$		$+5g'$	
	$\frac{533}{16}e'^4$		$+6g$	
	$\frac{228347}{3840}e'^5$		$+7g'$	
$-\frac{9}{160}e^6$	Solar part suprà.		$-6g$	
$-\frac{25}{384}e^5$			$-5g$	
$-\frac{1}{12}e^4 + \frac{1}{15}e^6$			$-4g$	
$-\frac{1}{8}e^3 + \frac{9}{128}e^5$			$-3g$	
$-\frac{1}{4}e^2 + \frac{1}{12}e^4 - \frac{1}{96}e^6$			$-2g$	
$-e + \frac{1}{8}e^3 - \frac{1}{192}e^5$			$-\ g$	
$1 + \frac{3}{2}e^2$ (exact)			$0\ g$	
$-e + \frac{1}{8}e^3 - \frac{1}{192}e^5$			$+\ g$	
$-\frac{1}{4}e^2 + \frac{1}{12}e^4 - \frac{1}{96}e^6$			$+2g$	
$-\frac{1}{8}e^3 + \frac{9}{128}e^5$			$+3g$	
$-\frac{1}{12}e^4 + \frac{1}{15}e^6$			$+4g$	
$-\frac{25}{384}e^5$			$+5g$	
$-\frac{9}{160}e^6$			$+6g$	

$$\Omega_5 = m' \frac{a^2}{a'^3}\left(\frac{3}{4}\eta^4\right) \text{ multiplied into}$$

Lunar part infrà.	$\frac{1}{24} e'^4$	cos	$-2g'$	$+2\varpi'$
	$\frac{1}{48} e'^3$		$-\ g'$	
	0 (exact)		$0g'$	
	$-\frac{1}{2} e' + \frac{1}{16} e'^3$		$+\ g'$	
	$1 - \frac{5}{2} e'^2 + \frac{13}{16} e'^4$		$+2g'$	
	$\frac{7}{2} e' - \frac{123}{16} e'^3$		$+3g'$	
	$\frac{17}{2} e'^2 - \frac{115}{6} e'^4$		$+4g'$	
	$\frac{845}{48} e'^3$		$+5g'$	
	$\frac{533}{16} e'^4$		$+6g'$	
	$\frac{228347}{3840} e'^5$		$+7g'$	
$-\frac{11}{720} e^6$	Solar part suprà.		$-4g$	$+2\varpi$
$-\frac{17}{640} e^5$			$-3g$	
$-\frac{1}{16} e^4 - \frac{11}{480} e^6$			$-2g$	
$-\frac{7}{24} e^3 - \frac{47}{384} e^5$			$-\ g$	
$\frac{5}{2} e^2$ (exact)			$0g$	
$-3e + \frac{13}{8} e^3 + \frac{5}{192} e^5$			$+\ g$	
$1 - \frac{5}{2} e^2 + \frac{23}{16} e^4 - \frac{65}{288} e^6$			$+2g$	
$e - \frac{19}{8} e^3 + \frac{107}{64} e^5$			$+3g$	
$e^2 - \frac{5}{2} e^4 + \frac{101}{48} e^6$			$+4g$	
$\frac{25}{24} e^3 - \frac{1075}{384} e^5$			$+5g$	
$\frac{9}{8} e^4 - \frac{261}{80} e^6$			$+6g$	
$\frac{2401}{1920} e^5$			$+7g$	
$\frac{64}{45} e^6$			$+8g$	

$$\Omega_6 = m'\frac{a^3}{a'^4}\left(\frac{3}{8} - \frac{33}{8}\eta^2 + \frac{75}{8}\eta^4\right) \text{ multiplied into}$$

Lunar part infrà.	$\frac{77}{6}e'^3$	cos	$-4g'$	
	$\frac{53}{8}e'^2$		$-3g'$	
	$3e' + \frac{11}{4}e'^3$		$-2g'$	
	$1 + 2e'^2$		$-\ \ g'$	$-\varpi'$
	* $e' + \frac{5}{2}e'^3$		$0g'$	
	$\frac{11}{8}e'^2$		$+\ \ g'$	
	$\frac{23}{12}e'^3$		$+2g'$	
	* Exact value is $e'(1-e'^2)^{-\frac{5}{2}}$			
$\frac{7}{128}e^4$	Solar part suprà.		$-3g$	
$\frac{1}{6}e^3$			$-2g$	
$\frac{11}{8}e^2 + \frac{7}{48}e^4$			$-\ \ g$	
$-\frac{5}{2}e - \frac{15}{8}e^3$ (exact)			$0g$	
$1 + 2e^2 - \frac{41}{64}e^4$			$+\ \ g$	$+\varpi$
$-\frac{1}{2}e + e^3$			$+2g$	
$-\frac{3}{8}e^2 + \frac{11}{16}e^4$			$+3g$	
$-\frac{7}{24}e^3$			$+4g$	
$-\frac{95}{384}e^4$			$+5g$	

$g - g' + \varpi - \varpi'$ parallactic equation.

$$\Omega_7 = m' \frac{a^3}{a'^4}\left(\frac{5}{8} - \frac{15}{8}\eta^2 + \frac{15}{8}\eta^4\right) \text{ multiplied into}$$

Lunar part infrà.	$\frac{163}{4}e'^3$	cos	$-6g'$	$-3\varpi'$
	$\frac{127}{8}e'^2$		$-5g'$	
	$5e' - 22e'^3$		$-4g'$	
	$1 - 6e'^2$		$-3g'$	
	$-e' + \frac{5}{4}e'^3$		$-2g'$	
	$\frac{1}{8}e'^2$		$-\ g'$	
	0 (exact)		$0g'$	
$\frac{75}{128}e^4$	Solar part suprà.		$-\ g$	$+3\varpi$
$-\frac{35}{8}e^3$ (exact)			$0g$	
$\frac{57}{8}e^2 - \frac{65}{16}e^4$			$+\ g$	
$-\frac{9}{2}e + \frac{33}{4}e^3$			$+2g$	
$1 - 6e^2 + \frac{591}{64}e^4$			$+3g$	
$\frac{3}{2}e - \frac{57}{8}e^3$			$+4g$	
$\frac{15}{8}e^2 - \frac{135}{16}e^4$			$+5g$	
$\frac{9}{4}e^3$			$+6g$	
$\frac{343}{128}e^4$			$+7g$	

$$\Omega_8 = m' \frac{a^3}{a'^4} \left(\frac{9}{4} \eta^2 - \frac{15}{2} \eta^4 \right) \text{ multiplied into}$$

Lunar part infrà.	$\frac{23}{12} e'^3$	cos	$-2g'$	$+\varpi'$
	$\frac{11}{8} e'^2$		$-\ g'$	
	$* e' + \frac{5}{2} e'^3$		$0 g'$	
	$1 + 2e'^2$		$+\ g'$	
	$3e' + \frac{11}{4} e'^3$		$+2g'$	
	$\frac{53}{8} e'^2$		$+3g'$	
	$\frac{77}{6} e'^3$		$+4g'$	
	* Exact value is $e'(1 - e'^2)^{-\frac{5}{2}}$			
$\frac{7}{128} e^4$	Solar part suprà.		$-3g$	$+\varpi$
$\frac{1}{6} e^3$			$-2g$	
$\frac{11}{8} e^2 + \frac{7}{48} e^4$			$-\ g$	
$-\frac{5}{2} e - \frac{15}{8} e^3$ (exact)			$0 g$	
$1 + 2e^2 - \frac{41}{64} e^4$			$+\ g$	
$-\frac{1}{2} e + e^3$			$+2g$	
$-\frac{3}{8} e^2 + \frac{11}{16} e^4$			$+3g$	
$-\frac{7}{24} e^3$			$+4g$	
$-\frac{95}{344} e^4$			$+5g$	

$$\Omega_9 = m' \frac{a^3}{a'^4}\left(\frac{15}{8}\eta^2 - \frac{15}{4}\eta^4\right) \text{ multiplied into}$$

Lunar part infrà.	$\frac{77}{6}e'^3$	cos	$-4g'$	$-\varpi'$
	$\frac{53}{8}e'^2$		$-3g'$	
	$3e' + \frac{11}{4}e'^3$		$-2g'$	
	$1 + 2e'^2$		$-g'$	
	$^*e' + \frac{5}{2}e'^3$		$0g'$	
	$\frac{11}{8}e'^2$		$+g'$	
	$\frac{23}{12}e'^3$		$+2g'$	
	* Exact value is $e'(1-e'^2)^{-\frac{5}{2}}$			
$\frac{75}{128}e^4$	Solar part suprà.		$-g$	$+3\varpi$
$-\frac{35}{8}e^3$ (exact)			$0g$	
$\frac{57}{8}e^2 - \frac{65}{16}e^4$			$+g$	
$-\frac{9}{2}e + \frac{33}{4}e^3$			$+2g$	
$1 - 6e^2 + \frac{591}{64}e^4$			$+3g$	
$\frac{3}{2}e - \frac{57}{8}e^3$			$+4g$	
$\frac{15}{8}e^2 - \frac{135}{16}e^4$			$+5g$	
$\frac{9}{4}e^3$			$+6g$	
$\frac{343}{128}e^4$			$+7g$	

$$\Omega_{10} = m' \frac{a^3}{a'^4}\left(\frac{15}{8}\eta^2 - \frac{15}{4}\eta^4\right) \text{ multiplied into}$$

Lunar part infrà.	$\frac{163}{4} e'^3$	cos	$-6g'$	
	$\frac{127}{8} e'^2$		$-5g'$	
	$5e' - 22e'^3$		$-4g'$	
	$1 - 6e'^2$		$-3g'$	$-3\varpi'$
	$-e' + \frac{5}{4} e'^3$		$-2g'$	
	$\frac{1}{8} e'^2$		$-\ g'$	
	0 (exact)		$0g'$	
$\frac{7}{128} e^4$	Solar part suprà.		$-3g$	
$\frac{1}{6} e^3$			$-2g$	
$\frac{11}{8} e^2 + \frac{7}{48} e^4$			$-\ g$	
$-\frac{5}{2} e - \frac{15}{8} e^3$ (exact)			$0g$	
$1 + 2e^2 - \frac{41}{64} e^4$			$+\ g$	$+\varpi$
$-\frac{1}{2} e + e^3$			$+2g$	
$-\frac{3}{8} e^2 + \frac{11}{16} e^4$			$+3g$	
$-\frac{7}{24} e^3$			$+4g$	
$-\frac{95}{384} e^4$			$+5g$	

The arrangement of the tables hardly requires any explanation; each line of the upper half of a table is to be read in combination with each line of the lower half of the same table. Thus a term of Ω_2 is

$$m'\frac{a^2}{a'^3}(\tfrac{3}{4}-\tfrac{3}{2}\eta^2+\tfrac{3}{4}\eta^4)(-\tfrac{7}{24}e^3-\tfrac{47}{384}e^5)(1-\tfrac{5}{2}e'^2+\tfrac{13}{16}e'^4)\cos(-g-2g'+2\varpi-2\varpi').$$

It is to be observed that in the table for Ω_1 (where the arguments depend only on the mean anomalies), the several terms (other than the constant term) occur in pairs of equal terms, having respectively the arguments $ig+i'g'$ and $-ig-i'g'$; such equal terms are to be united together, or, what is the same thing, the coefficients must be multiplied by 2. Thus we have the two terms

$$m'\frac{a^2}{a'^3}(\tfrac{1}{4}-\tfrac{3}{2}\eta^2+\tfrac{3}{2}\eta^4)(-\tfrac{9}{160}e^6)(\tfrac{1773}{256}e'^5)\qquad\cos(-6g-5g')$$

$$\text{Ditto}\qquad\cos(\ 6g+5g')$$

or, what is the same thing, the term is

$$m'\frac{a^2}{a'^3}(\tfrac{1}{4}-\tfrac{3}{2}\eta^2+\tfrac{3}{2}\eta^4)(-\tfrac{9}{160}e^6)(\tfrac{1773}{256}e'^5)\,2\cos(\ 6g+5g').$$

This is the case even when either i or i' vanishes; but, as already noticed, it is not the case for the constant term where i and i' both vanish. There are not any such equal pairs in the tables for Ω_2, &c., where all the arguments contain a part independent of the mean anomalies. The quantity ϖ' occurs in the upper or solar half of the table; but it is to be recollected that $\varpi'(=\varpi'-\Theta)$ involves Θ, which is an element of the moon's orbit.

The peculiar form in which the coefficients are exhibited, viz. as the product of a term depending on the inclination, a term depending on the eccentricity of the moon's orbit, and a term depending on the eccentricity of the sun's orbit is not to be considered as a want of completion of the development; it is, on the contrary, an important advantage.

[The Errata in the Tables, noticed *Mem. R. Ast. Soc.* vol. XXVIII. p. 216, have been corrected. The values of P^i, $Q_c^{\,i}$, $Q_s^{\,i}$, &c. used in the construction of the Tables *ante* p. 298 may also be obtained from the Tables of the Developments of Functions in The Theory of Elliptic Motion, 216.]

Addition.

I deduce from the preceding an expression for the disturbing function in a form similar to and easily comparable with the forms given by Sir J. W. Lubbock in his work *On the Theory of the Moon* &c. (London, 1834), pp. 30—35, and by Pontécoulant in the *Théorie Analytique du Système du Monde*, t. IV. (Paris, 1846), pp. 58—61. The several notations are as follows, viz.:

Lubbock.	Pontécoulant.	Suprà.
ξ,	ϕ,	g
ξ',	ϕ',	g'
τ,	ξ,	$g-g'+\varpi-\varpi'$
η,	η,	$g+\varpi$
γ,	γ,	$2\eta\sqrt{1-\eta^2}\div(1-2\eta^2)$

(γ, the tangent of the inclination, is employed by the above-named two authors instead of η, the sine of the semi-inclination, the relation between these two quantities gives to γ^4 or η^4,

$$\gamma^2 = 4\eta^2 + 12\eta^4,$$

$$\gamma^4 = 16\eta^4,$$

and conversely,

$$\eta^2 = \tfrac{1}{4}\gamma^2 - \tfrac{3}{16}\gamma^4,$$

$$\eta^4 = \tfrac{1}{16}\gamma^4,$$

which are useful for replacing one of these quantities by the other). The disturbing function is taken by Lubbock with the contrary sign, and he includes in the before-mentioned omitted term depending only on the sun's radius vector, that is, R (Lubbock) $= -m'\frac{1}{r'} - \Omega$; he uses also, in reference to the sun, subscript strokes instead of accents (so that, in referring to his notation, $\xi_{,}$ should properly be here used in the place of ξ', but this difference is obviously immaterial). Pontécoulant's disturbing function is taken with the same sign as in the present memoir, or we have R (Pontécoulant) $= \Omega$. Lubbock and Pontécoulant give in the principal terms the part involving $m'\frac{a^4}{a'^5}$, which depends on the square of the parallax, and which was disregarded in the preceding development. I have in the sequel inserted these terms from Lubbock's expression. We have, in fact:

$\Omega = m' \frac{a^2}{a'^3}$ multiplied into

		Lubbock. Arguments.	Lubbock. Nos.	Pontécoulant. Arguments.	Pontécoulant. Nos.	Arguments ut suprà.
$\frac{1}{4}+\frac{3}{8}e^2+\frac{3}{8}e'^2+\frac{9}{16}e^2e'^2+\frac{15}{32}e'^4$ $-\frac{3}{2}\eta^2-\frac{9}{4}\eta^2e^2-\frac{9}{4}\eta^2e'^2+\frac{3}{2}\eta^4+\frac{9}{64}\frac{a^2}{a'^2}$			0		0	
* $+\frac{3}{4}-\frac{15}{8}e^2-\frac{15}{8}e'^2+\frac{69}{64}e^4+\frac{75}{16}e^2e'^2+\frac{39}{64}e'^4$ $-\frac{3}{2}\eta^2+\frac{15}{4}\eta^2e^2+\frac{15}{4}\eta^2e'^2+\frac{3}{4}\eta^4+\frac{5}{16}\frac{a^2}{a'^2}$	cos	2τ	1	2ξ	30	$2g-2g' \;\; +2\varpi-2\varpi'$
$-\frac{1}{2}e+\frac{1}{16}e^3-\frac{3}{4}ee'^2+3\eta^2e \quad -\frac{9}{16}\frac{a^2}{a'^2}e$		ξ	2	ϕ	1	g
† $-\frac{9}{4}e+\frac{39}{32}e^3+\frac{45}{8}ee'^2+\frac{9}{2}\eta^2e \quad -\frac{5}{4}\frac{a^2}{a'^2}e$		$2\tau-\xi$	3	$2\xi-\phi$	31	$g-2g'$ } $+2\varpi-2\varpi'$
$+\frac{3}{4}e-\frac{57}{32}e^3-\frac{15}{8}ee'^2-\frac{3}{2}\eta^2e$		$2\tau+\xi$	4	$2\xi+\phi$	32	$3g-2g'$
$+\frac{3}{4}e'+\frac{9}{8}e^2e'+\frac{27}{32}e'^3-\frac{9}{2}\eta^2e' \quad +\frac{35}{64}\frac{a^2}{a'^2}e'$		ξ'	5	ϕ'	6	g'
$+\frac{21}{8}e'-\frac{105}{16}e^2e'-\frac{369}{64}e'^3-\frac{21}{4}\eta^2e'+\frac{45}{32}\frac{a^2}{a'^2}e'$		$2\tau-\xi'$	6	$2\xi-\phi'$	33	$2g-3g'$ } $+2\varpi-2\varpi'$
$-\frac{3}{8}e'+\frac{15}{16}e^2e'+\frac{3}{64}e'^3+\frac{3}{4}\eta^2e' \quad +\frac{15}{32}\frac{a^2}{a'^2}e'$		$2\tau+\xi'$	7	$2\xi+\phi'$	34	$2g-g'$
$-\frac{1}{8}e^2+\frac{1}{24}e^4-\frac{3}{16}e^2e'^2+\frac{3}{4}\eta^2e^2$		2ξ	8	2ϕ	2	$2g$
$+\frac{15}{8}e^2-\frac{75}{16}e^2e'^2 \quad -\frac{15}{4}\eta^2e^2$		$2\tau-2\xi$	9	$2\xi-2\phi$	35	$-2g'$ } $+2\varpi-2\varpi'$
$+\frac{3}{4}e^2-\frac{15}{8}e^4-\frac{15}{8}e^2e'^2-\frac{3}{2}\eta^2e^2$		$2\tau+2\xi$	10	$2\xi+2\phi$	36	$4g-2g'$
$-\frac{3}{4}ee'+\frac{3}{32}e^3e'-\frac{27}{32}ee'^3+\frac{9}{2}\eta^2ee'$		$\xi+\xi'$	11	$\phi+\phi'$	9	$g+g'$
$-\frac{63}{8}ee'+\frac{273}{64}e^3e'+\frac{1107}{64}ee'^3+\frac{63}{4}\eta^2ee'$		$2\tau-\xi-\xi'$	12	$2\xi-\phi-\phi'$	37	$g-3g'$ } $+2\varpi-2\varpi'$
$-\frac{3}{8}ee'+\frac{57}{64}e^3e'+\frac{3}{64}ee'^3+\frac{3}{4}\eta^2ee'$		$2\tau+\xi+\xi'$	13	$2\xi+\phi+\phi'$	40	$3g-g'$
$-\frac{3}{4}ee'+\frac{3}{32}e^3e'-\frac{27}{32}ee'^3+\frac{9}{2}\eta^2ee'$		$\xi-\xi'$	14	$\phi-\phi'$	8	$g-g'$
$+\frac{9}{8}ee'-\frac{39}{64}e^3e'-\frac{9}{64}ee'^3-\frac{9}{4}\eta^2ee'$		$2\tau-\xi+\xi'$	15	$2\xi-\phi+\phi'$	38	$g-g'$ } $+2\varpi-2\varpi'$
$+\frac{21}{8}ee'-\frac{399}{64}e^3e'-\frac{369}{64}ee'^3-\frac{21}{4}\eta^2ee'$		$2\tau+\xi-\xi'$	16	$2\xi+\phi-\phi'$	39	$3g-3g'$
$+\frac{9}{8}e'^2+\frac{27}{16}e^2e'^2+\frac{7}{8}e'^4-\frac{27}{4}\eta^2e'^2$		$2\xi'$	17	$2\phi'$	7	$2g'$
$+\frac{51}{8}e'^2-\frac{255}{16}e^2e'^2-\frac{115}{8}e'^4-\frac{51}{4}\eta^2e'^2$		$2\tau-2\xi'$	18	$2\xi-2\phi'$	41	$2g-4g' \;\; +2\varpi-2\varpi'$

* Variation. † Evection.

Ω (*continued*) $= m' \dfrac{a^2}{a'^3}$ multiplied into

		Lubbock.		Pontécoulant.		Arguments ut suprà.	
		Arguments.	Nos.	Arguments.	Nos.		
$-\frac{1}{16}e^3$	cos	3ξ	20	3ϕ	3	$3g$	
$-\frac{7}{32}e^3$		$2\tau-3\xi$	21	$2\xi-3\phi$	43	$-g-2g'$	$+2\varpi-2\varpi'$
$+\frac{25}{32}e^3$		$2\tau+3\xi$	22	$2\xi+3\phi$	44	$5g-2g'$	
$-\frac{3}{16}e^2e'$		$2\xi+\xi'$	23	$2\phi+\phi'$	11	$2g+g'$	
$+\frac{105}{16}e^2e'$		$2\tau-2\xi-\xi'$	24	$2\xi-2\phi-\phi'$	45	$-3g'$	$+2\varpi-2\varpi'$
$-\frac{3}{8}e^2e'$		$2\tau+2\xi+\xi'$	25	$2\xi+2\phi+\phi'$	48	$4g-g'$	
$-\frac{3}{16}e^2e'$		$2\xi-\xi'$	26	$2\phi-\phi'$	10	$2g-g'$	
$-\frac{15}{16}e^2e'$		$2\tau-2\xi+\xi'$	27	$2\xi-2\phi+\phi'$	46	$-g'$	$+2\varpi-2\varpi'$
$+\frac{21}{8}e^2e'$		$2\tau+2\xi-\xi'$	28	$2\xi+2\phi-\phi'$	47	$4g-3g'$	
$-\frac{9}{8}ee'^2$		$\xi+2\xi'$	29	$\phi+2\phi'$	13	$g+2g'$	
$-\frac{153}{8}ee'^2$		$2\tau-\xi-2\xi'$	30	$2\xi-\phi-2\phi'$	49	$g-4g'$	$+2\varpi-2\varpi'$
$-\frac{9}{8}ee'^2$		$\xi-2\xi'$	32	$\phi-2\phi'$	12	$g-2g'$	
$+\frac{51}{8}ee'^2$		$2\tau+\xi-2\xi'$	34	$2\xi+\phi-2\phi'$	51	$3g-4g'$	$+2\varpi-2\varpi'$
$+\frac{53}{32}e'^3$		$3\xi'$	35			$3g'$	
$+\frac{845}{64}e'^3$		$2\tau-3\xi'$	36			$2g-5g'$	$+2\varpi-2\varpi'$
$+\frac{1}{64}e'^3$		$2\tau+3\xi'$	37			$2g+g'$	
$-\frac{1}{24}e^4$		4ξ	38			$4g$	
$-\frac{3}{64}e^4$		$2\tau-4\xi$	39	$2\xi-4\phi$	52	$-2g-2g'$	$+2\varpi-2\varpi'$
$+\frac{27}{32}e^4$		$2\tau+4\xi$	40			$6g-2g'$	
$-\frac{3}{32}e^3e'$		$3\xi+\xi'$	41	$3\phi+\phi'$	17	$3g+g'$	
$-\frac{49}{64}e^3e'$		$2\tau-3\xi-\xi'$	42	$2\xi-3\phi-\phi'$	53	$-g-3g'$	$+2\varpi-2\varpi'$

Ω (*continued*) $= m'\dfrac{a^2}{a'^3}$ multiplied into

		Lubbock.		Pontécoulant.		Arguments ut suprà.
		Arguments.	Nos.	Arguments.	Nos.	
$-\frac{25}{64}e^3e'$	cos	$2\tau+3\xi+\xi'$	43			$5g-g'+2\varpi-2\varpi'$
$-\frac{3}{32}e^3e'$		$3\xi-\xi'$	44	$3\phi-\phi'$	16	$3g-g'$
$+\frac{7}{64}e^3e'$		$2\tau-3\xi+\xi'$	45	$2\xi-3\phi+\phi'$	54	$-g-g'+2\varpi-2\varpi'$
$+\frac{175}{64}e^3e'$		$2\tau+3\xi-\xi'$	46			$5g-3g'+2\varpi-2\varpi'$
$-\frac{9}{32}e^2e'^2$		$2\xi+2\xi'$	47	$2\phi+2\phi'$	15	$2g+2g'$
$+\frac{255}{16}e^2e'^2$		$2\tau-2\xi-2\xi'$	48	$2\xi-2\phi-2\phi'$	55	$-4g'+2\varpi-2\varpi'$
$-\frac{9}{32}e^2e'^2$		$2\xi-2\xi'$	50	$2\phi-2\phi'$	14	$2g-2g'$
$+\frac{51}{8}e^2e'^2$		$2\tau+2\xi-2\xi'$	52			$4g-4g'+2\varpi-2\varpi'$
$-\frac{53}{32}ee'^3$		$\xi+3\xi'$	53			$g+3g'$
$-\frac{2535}{64}ee'^3$		$2\tau-\xi-3\xi'$	54			$g-5g'+2\varpi-2\varpi'$
$+\frac{1}{64}ee'^3$		$2\tau+\xi+3\xi'$	55			$3g+g'+2\varpi-2\varpi'$
$-\frac{53}{32}ee'^3$		$\xi-3\xi'$	56			$g-3g'$
$-\frac{3}{64}ee'^3$		$2\tau-\xi+3\xi'$	57			$g+g'+2\varpi-2\varpi'$
$+\frac{845}{64}ee'^3$		$2\tau+\xi-3\xi'$	58			$3g-5g'+2\varpi-2\varpi'$
$+\frac{77}{32}e'^4$		$4\xi'$	59			$4g'$
$+\frac{1599}{64}e'^4$		$2\tau-4\xi'$	60			$2g-6g'+2\varpi-2\varpi'$
$+\frac{1}{32}e'^4$		$2\tau+4\xi'$	61			$2g+2g'+2\varpi-2\varpi'$
$+\frac{3}{2}\eta^2-\frac{15}{4}\eta^2e^2+\frac{9}{4}\eta^2e'^2-\frac{3}{2}\eta^4$		2η	62	2η	18	$2g+2\varpi$
$+\frac{3}{2}\eta^2+\frac{9}{4}\eta^2e^2-\frac{15}{4}\eta^2e'^2-\frac{3}{2}\eta^4$		$2\tau-2\eta$ rev.	63	$2\xi-2\eta$ rev.	57	$2g'+2\varpi'$
$-\frac{9}{2}\eta^2e$		$\xi-2\eta$ rev.	65	$\phi-2\eta$ rev.	19	$g+2\varpi$
$+\frac{3}{2}\eta^2e$		$\xi+2\eta$	66	$\phi+2\eta$	20	$3g+2\varpi$

Ω (*continued*) $= m' \dfrac{a^2}{a'^3}$ multiplied into

		Lubbock.		Pontécoulant.		Arguments ut suprà.	
		Arguments.	Nos.	Arguments.	Nos.		
$-\frac{3}{2}\eta^2 e$	cos	$2\tau - \xi - 2\eta$ rev.	67	$2\xi - \phi - 2\eta$ rev.	59	$g + 2g'$	$+2\varpi'$
$-\frac{3}{2}\eta^2 e$		$2\tau + \xi - 2\eta$ rev.	69	$2\xi + \phi - 2\eta$ rev.	61	$-g + 2g'$	
$+\frac{9}{4}\eta^2 e'$		$\xi' - 2\eta$ rev.	71	$\phi' - 2\eta$ rev.	21	$2g - g'$	$+2\varpi$
$+\frac{9}{4}\eta^2 e'$		$\xi' + 2\eta$	72	$\phi' + 2\eta$	22	$2g + g'$	
$+\frac{21}{4}\eta^2 e'$		$2\tau - \xi' - 2\eta$ rev.	73	$2\xi - \phi' - 2\eta$ rev.	63	$3g'$	$+2\varpi'$
$-\frac{3}{4}\eta^2 e'$		$2\tau + \xi' - 2\eta$ rev.	75	$2\xi + \phi' - 2\eta$ rev.	65	g'	
$+\frac{15}{4}\eta^2 e^2$		$2\xi - 2\eta$ rev.	77	$2\phi - 2\eta$ rev.	23	$0g$	$+2\varpi$
$+\frac{3}{2}\eta^2 e^2$		$2\xi + 2\eta$	78	$2\phi + 2\eta$	24	$4g$	
$-\frac{3}{8}\eta^2 e^2$		$2\tau - 2\xi - 2\eta$ rev.	79	$2\xi - 2\phi - 2\eta$ rev.	67	$2g + 2g'$	$+2\varpi'$
$-\frac{3}{8}\eta^2 e^2$		$2\tau + 2\xi - 2\eta$ rev.	81	$2\xi + 2\phi - 2\eta$ rev.	69	$-2g + 2g'$	
$-\frac{27}{4}\eta^2 e e'$		$\xi + \xi' - 2\eta$ rev.	83	$\phi + \phi' - 2\eta$ rev.	27	$g - g'$	$+2\varpi$
$+\frac{9}{4}\eta^2 e e'$		$\xi + \xi' + 2\eta$	84	$\phi + \phi' + 2\eta$	28	$3g + g'$	
$-\frac{21}{4}\eta^2 e e'$		$2\tau - \xi - \xi' - 2\eta$ rev.	85			$g + 3g'$	$+2\varpi'$
$+\frac{3}{4}\eta^2 e e'$		$2\tau + \xi + \xi' - 2\eta$ rev.	87			$-g + g'$	
$-\frac{27}{4}\eta^2 e e'$		$\xi - \xi' - 2\eta$ rev.	89	$\phi - \phi' - 2\eta$ rev.	25	$g + g'$	$+2\varpi$
$+\frac{9}{4}\eta^2 e e'$		$\xi - \xi' + 2\eta$	90	$\phi - \phi' + 2\eta$	26	$3g - g'$	
$+\frac{3}{4}\eta^2 e e'$		$2\tau - \xi + \xi' - 2\eta$ rev.	91			$g + g'$	$+2\varpi'$
$-\frac{21}{4}\eta^2 e e'$		$2\tau + \xi - \xi' - 2\eta$ rev.	93			$-g + 3g'$	
$+\frac{27}{8}\eta^2 e'^2$		$2\xi' - 2\eta$ rev.	95			$2g - 2g'$	$+2\varpi$
$+\frac{27}{8}\eta^2 e'^2$		$2\xi' + 2\eta$	96			$2g + 2g'$	
$+\frac{51}{4}\eta^2 e'^2$		$2\tau - 2\xi' - 2\eta$ rev.	97			$4g'$	$+2\varpi'$

Ω (*continued*) $= m' \dfrac{a^3}{a'^4}$ multiplied into

		Lubbock.		Pontécoulant.		Arguments ut suprà.	
		Arguments.	Nos.	Arguments.	Nos.		
* $\frac{3}{8}+\frac{3}{4}e^2+\frac{3}{4}e'^2-\frac{33}{8}\eta^2$	cos	τ	101	ξ	70	$g-g'$	
$-\frac{15}{16}e$		$\tau-\xi$	102	$\xi-\phi$	71	$-g'$	
$-\frac{3}{16}e$		$\tau+\xi$	103	$\xi+\phi$	72	$2g-g'$	
$+\frac{9}{8}e'$		$\tau-\xi'$	104	$\xi-\phi'$	73	$g-2g'$	
$+\frac{3}{8}e'$		$\tau+\xi'$	105	$\xi+\phi'$	74	g	
$+\frac{33}{64}e^2$		$\tau-2\xi$	106	$\xi-2\phi$	75	$-g-g'$	
$-\frac{9}{64}e^2$		$\tau+2\xi$	107	$\xi+2\phi$	76	$3g-g'$	$+\varpi-\varpi'$
$-\frac{45}{16}ee'$		$\tau-\xi-\xi'$	108	$\xi-\phi-\phi'$	77	$-2g'$	
$-\frac{3}{16}ee'$		$\tau+\xi+\xi'$	109	$\xi+\phi+\phi'$	79	$2g$	
$-\frac{15}{16}ee'$		$\tau-\xi+\xi'$	110	$\xi-\phi+\phi'$	78	$0g+0g'$	
$-\frac{9}{16}ee'$		$\tau+\xi-\xi'$	111			$2g-2g'$	
$+\frac{159}{64}e'^2$		$\tau-2\xi'$	112			$g-3g'$	
$+\frac{33}{64}e'^2$		$\tau+2\xi'$	113			$g+g'$	
$+\frac{9}{4}\eta^2$		$\tau-2\eta$ rev.	114			$g+g'$	$+\varpi+\varpi'$
$+\frac{15}{8}\eta^2$		$\tau+2\eta$	115			$3g-3g'$	$+3\varpi-\varpi'$
$+\frac{5}{8}-\frac{15}{4}e^2-\frac{15}{4}e'^2-\frac{15}{8}\eta^2$		3τ	116	3ξ	80	$3g-3g'$	
$-\frac{45}{16}e$		$3\tau-\xi$	117	$3\xi-\phi$	81	$2g-3g'$	
$+\frac{15}{16}e$		$3\tau+\xi$	118	$3\xi+\phi$	82	$4g-3g'$	
$+\frac{25}{8}e'$		$3\tau-\xi'$	119	$3\xi-\phi'$	83	$3g-4g'$	$+3\varpi-3\varpi'$
$-\frac{5}{8}e'$		$3\tau+\xi'$	120	$3\xi+\phi'$	84	$3g-2g'$	
$+\frac{285}{64}e^2$		$3\tau-2\xi$	121	$3\xi-2\phi$	85	$g-3g'$	

* Parallactic equation.

Ω (*continued*) $= m'\dfrac{a^3}{a'^4}$ multiplied into

		Lubbock.		Pontécoulant.		Arguments ut suprà.	
		Arguments.	Nos.	Arguments.	Nos.		
$+\frac{75}{64}e^2$	cos	$3\tau+2\xi$	122			$5g-3g'$	$+3\varpi-3\varpi'$
$-\frac{225}{16}ee'$		$3\tau-\ \xi-\xi'$	123			$2g-4g'$	
$-\frac{15}{16}ee'$		$3\tau+\ \xi+\xi'$	124			$4g-2g'$	
$+\frac{45}{16}ee'$		$3\tau-\ \xi+\xi'$	125			$2g-2g'$	
$+\frac{75}{16}ee'$		$3\tau+\ \xi-\xi'$	126			$4g-4g'$	
$+\frac{635}{64}e'^2$		$3\tau-2\xi'$	127			$3g-5g'$	
$+\frac{5}{64}e'^2$		$3\tau+2\xi'$	128			$3g-\ g'$	
$+\frac{15}{8}\eta^2$		$3\tau-2\eta$	129			$g-3g'$	$+\ \varpi-3\varpi'$

where the abbreviation *rev.* denotes that the argument to which it is attached has its sign reversed in the third column of arguments: thus Arg. 63 (Lubbock), it is $-(2\tau-2\eta)$ which is equal to $2g'+2\varpi'$. As the formula contains only cosines, there is, of course, no change in the sign of the coefficient.

On comparing with Lubbock's value, I find some differences, which are as follows:—

It will be recollected that R (Lubbock) $= -m'\dfrac{1}{r'} - \Omega$, so that the signs of the coefficients of Ω are to be reversed in order to deduce the corresponding coefficients of R. The Nos. refer to Lubbock's arguments, and the exterior factor $m'\dfrac{a^2}{a'^3}$ or $m'\dfrac{a^3}{a'^4}$ is disregarded[1].

Arg. 1. Lubbock's coefficient (viz. the coefficient in R) is

$$-\tfrac{3}{4}\left(1-\tfrac{5}{2}e^2-\tfrac{5}{2}e'^2+\tfrac{23}{16}e^4+\tfrac{25}{4}e^2e'^2+\tfrac{13}{16}e'^4+\tfrac{5}{12}\frac{a^2}{a'^2}\right)\cos^4\tfrac{1}{2}\iota$$

[1] I have been favoured with a note from Sir J. W. Lubbock, confirming my values of the coefficients for the arguments 8, 18, 58, 101 and 123.—Added 15th Feb. 1859, A. C.

which, substituting for $\cos^4 \frac{1}{2}\iota$ its value $1 - \frac{1}{2}\gamma^2 + \frac{7}{16}\gamma^4 - \&c.$, and developing to the fourth order, gives

$$-\tfrac{3}{4}\left(1 - \tfrac{5}{2}e^2 - \tfrac{5}{2}e'^2 + \tfrac{23}{16}e^4 + \tfrac{25}{4}e^2e'^2 + \tfrac{13}{16}e'^4 - \tfrac{1}{2}\gamma^2 + \tfrac{5}{4}\gamma^2 e^2 + \tfrac{5}{4}\gamma^2 e'^2 + \tfrac{7}{16}\gamma^4 + \tfrac{5}{12}\frac{a^2}{a'^2}\right)$$

which, in fact, agrees with the value given suprà. I have only referred to this term in order to make the reduction.

Arg. 8, Lubbock's coefficient is

$$-\tfrac{1}{8}(1 - \tfrac{1}{3}e^2 + \tfrac{3}{2}e'^2 - \tfrac{3}{2}\gamma^2)\, e^2$$

the exterior sign should be + instead of −. The term is given with the correct sign, Pontécoulant, Arg. 2.

Arg. 18, Lubbock's coefficient is

$$-\tfrac{51}{8}(1 - \tfrac{5}{2}e^2 - \tfrac{115}{51}e'^2 - \tfrac{1}{2}\gamma^2)\cos^4 \tfrac{1}{2}\iota$$

which, developed to γ^2, would be

$$-\tfrac{51}{8}(1 - \tfrac{5}{2}e^2 - \tfrac{115}{51}e'^2 - \gamma^2).$$

I make it

$$-\tfrac{51}{8}(1 - \tfrac{5}{2}e^2 - \tfrac{115}{51}e'^2 - \tfrac{1}{2}\gamma^2).$$

The remaining differences are

No. of Arg. Lubbock.	Lubbock's coefficient.	Coefficient from Development suprà.
58	$+\frac{45}{64}e e'^3$	$-\frac{845}{64}e e'^3$
59	$+\frac{591}{64}e'^4$	$-\frac{77}{32}e'^4$
60	$-\frac{2453}{128}e'^4$	$-\frac{1599}{64}e'^4$
61	$+\frac{741}{128}e'^4$	$-\frac{1}{32}e'^4$
*62	$-\frac{3}{8}\left(1 - \frac{5}{2}e^2 + \frac{3}{2}e'^2\right)\gamma^2$	$-\frac{3}{8}\left(1 - \frac{5}{2}e^2 + \frac{3}{2}e'^2 - \gamma^2\right)\gamma^2$

No. of Arg. Lubbock.	Lubbock's coefficient.	Coefficient from Development suprà.
*63	$-\frac{3}{8}\left(1+\frac{3}{2}e^2-\frac{5}{2}e'^2+\frac{1}{8}\gamma^2\right)\gamma^2$	$-\frac{3}{8}\left(1+\frac{3}{2}e^2-\frac{5}{2}e'^2-\gamma^2\right)\gamma^2$
95	$-\frac{27}{16}\gamma^2e'^2$	$-\frac{27}{32}\gamma^2e'^2$
96	$-\frac{27}{16}\gamma^2e'^2$	$-\frac{27}{32}\gamma^2e'^2$
*101	$-\frac{3}{8}\left(1+3e^2+3e'^2-\frac{11}{4}\gamma^2\right)$	$-\frac{3}{8}\left(1+2e^2+2e'^2-\frac{11}{4}\gamma^2\right)$
115	$-\frac{15}{128}\gamma^2$	$-\frac{15}{32}\gamma^2$
123	$-\frac{225}{16}ee'$	$+\frac{225}{16}ee'$

The greater part of the discordant terms do not occur in Pontécoulant's development, which is not carried so far, and the only differences which I find in the coefficients of Pontécoulant's R $(=\Omega)$ are, as regards the arguments 18, 57, 70, corresponding respectively to Lubbock's arguments 62, 63, 101, included in the preceding table, and for which Pontécoulant's coefficients, correcting for the change of sign, correspond with those given by Lubbock. But I see no room for a mistake in the preceding investigation as regards the coefficients of these three terms; the terms in η of the coefficients of 62 and 63 are simply the quantity $(\frac{3}{2}\eta^2-\frac{3}{2}\eta^4)$, which forms the exterior factor of Ω_3 and Ω_4 respectively, and which, putting for η^2 and η^4, their values, is equal to $\frac{3}{8}(1-\gamma^2)\gamma^2$, and as regards the coefficient of 101, the portion $1+2e^2+2e'^2$ of this coefficient is obtained by the mere multiplication of the factors $1+2e^2-$&c. and $1+2e'^2$ set opposite to $g+\varpi$ and $0g'-\varpi'$ respectively in the table for Ω_6.

2, *Stone Buildings, W.C.*, 7*th July*, 1858.

214.

THE FIRST PART OF A MEMOIR ON THE DEVELOPMENT OF THE DISTURBING FUNCTION IN THE LUNAR AND PLANETARY THEORIES.

[From the *Memoirs of the Royal Astronomical Society*, vol. XXVIII. (1860), pp. 187—215. Read November 10, 1858.]

THE development, as is well known, depends upon that of the reciprocal of the distance of the two planets: and Hansen's Memoir "Entwickelung der negativen und ungeraden Potenzen der Quadratwurzel der Function $r^2 + r'^2 - 2rr'(\cos U \cos U' + \sin U \sin U' \cos J)$," *Abh. der K. Sächs. Ges. zu Leipzig*, t. II., pp. 286—376 (1854), contains a formula which is truly fundamental, viz. the expression of the coefficient of the general term

$$\frac{r^n}{r'^{n+1}} \cos(jU + j'U')$$

of the development of the reciprocal of the distance as expressed in the above-mentioned form, where r, r' are the radius vectors of the inferior and superior planets respectively, and U, U' are the angular distances from the mutual node. In the lunar theory, where the higher powers of $\frac{r}{r'}$ are neglected, we have in this manner a small number of terms each of which is to be separately developed in multiple cosines of the mean anomalies. This can be effected as in the *Fundamenta Nova*, and my "Memoir on the Development of the Disturbing Function in the Lunar Theory," *R. Ast. Soc. Mem.*, t. XXVII., 1859, [213], which is a mere completion of Hansen's process([1]). In fact if f, f' are the true anomalies, and ϖ, ϖ' the distances

[1] I take the opportunity of mentioning the memoir of Hansen's which immediately precedes that above referred to, viz. "Entwickelung des Products einer Potenz des Radius Vectors mit dem Sinus oder Cosinus eines Vielfachen der wahren Anomalie in Reihen die nach den Sinussen oder Cosinussen der Vielfachen der wahren der excentrischen oder mittleren Anomalie fortschreiten," t. II., pp. 183—281 (1853).

of the pericentres from the mutual node, then we have $U = f + \varpi$, $U' = f' + \varpi'$, and the general term is

$$\frac{r^n}{r'^{n+1}} \cos(jf + j'f' + j\varpi + j'\varpi')$$

where r, f are given functions of the mean anomaly g, and r', f' are the like functions of the mean anomaly g'. And the development depends upon those of

$$r^n \frac{\cos}{\sin} jf, \qquad r'^{-n-1} \frac{\cos}{\sin} j'f',$$

which (if we consider as well negative as positive values of the index n) are each of the form

$$r^n \frac{\cos}{\sin} jf,$$

and when the developments of these expressions are known, we obtain at once by the mere addition and subtraction of the coefficients of the cosines and sines of the different multiples of g and g', the development of

$$\frac{r^n}{r'^{n+1}} \cos(jf + j'f' + j\varpi + j'\varpi')$$

in the tabular form employed in my memoir just referred to. In the planetary theory we must unite together the terms containing the different powers of $\frac{r}{r'}$ so as to form the entire coefficient $D(j, j')$ of $\cos(jU + j'U')$; if then we write

$$r = a(1 + x), \; r' = a'(1 + x'),$$

and develope the coefficient in powers of x, x' we have the general term

$$x^a x'^{a} \cos(jU + j'U')$$

which admits of development in multiple cosines of the mean anomalies, in the same manner precisely as the before-mentioned general term

$$\frac{r^n}{r'^{n+1}} \cos(jU + j'U')$$

in the lunar theory. It is proper to remark that this method is really identical with that commonly made use of in the planetary theory: the only difference is, that by Hansen's fundamental formula, we have the complete expression of the coefficient $D(j, j')$ developed in powers of sin or of $\tan \frac{1}{2}\phi$, instead of (as in the ordinary methods) the first two or three terms of this development.

The required development of

$$x^{\alpha} x'^{\alpha'} \cos(jU + j'U')$$

or, what is the same thing,

$$x^{\alpha} x'^{\alpha'} \cos(jf + j'f' + j\varpi + j'\varpi')$$

depends on the developments of

$$x^{\alpha} \begin{matrix}\cos\\ \sin\end{matrix} jf, \quad x'^{\alpha'} \begin{matrix}\cos\\ \sin\end{matrix} j'f'.$$

These are functions of the same form, and we may consider only

$$x^{\alpha} \begin{matrix}\cos\\ \sin\end{matrix} jf.$$

The value of x is $\left(\frac{r}{a} - 1\right)$ and we could of course calculate

$$\left(\frac{r}{a} - 1\right)^{\alpha} \begin{matrix}\cos\\ \sin\end{matrix} jf$$

by the methods of the *Fundamenta Nova* or the memoir of Hansen's referred to in the foot-note. But if we write $f = g + y$ (y is the equation of the centre), then the required expressions depend on

$$x^{\alpha} \begin{matrix}\cos\\ \sin\end{matrix} jy$$

which are actually calculated as far as e^7 for $\alpha = 0, 1, 2 \ldots 7$, and j an undetermined symbol, by Leverrier in the *Annales de l'Observatoire de Paris*, t. I. pp. 346—348 (1855). Hence, by the mere substitution, in Leverrier's formula, of the numerical values of j and j', and by the addition and subtraction of the coefficients of the cosines or sines of the different multiples of g and g', we may obtain the development of

$$x^{\alpha} x'^{\alpha'} \cos(jf + j'f' + j\varpi + j'\varpi')$$

in a tabular form similar to that employed in my memoir already referred to.

I have thought it desirable to put together the various results above referred to, and to investigate by a different process the expression for Hansen's coefficient, which in his memoir is obtained by means of a long series of transformations which it is not very easy to follow, and is not exhibited in quite the most simple form. And this is what I have done in the present first part of a memoir on the development of the disturbing function. My object has been to exhibit, in as complete a form as possible, the preliminary development in multiple cosines of the true anomalies; and to indicate the process of the ulterior development in multiple cosines of the mean anomalies.

I.

Let $\mathfrak{M}$ be the inferior, $\mathfrak{M}'$ the superior of the two planets (in the lunar theory $\mathfrak{M}$ is the moon, $\mathfrak{M}'$ the sun) and let the quantities relating to the two bodies be, for $\mathfrak{M}$,

r, the radius-vector,
U, the distance from node,
$\mho$, the distance of pericentre from node,
f, the true anomaly,
g, the mean anomaly,
a, the mean distance,
e, the eccentricity,

and for $\mathfrak{M}'$ the accented letters, r', &c., in the like significations.

The node referred to is the ascending node of the orbit of $\mathfrak{M}$ upon that of $\mathfrak{M}'$, and I write also

Φ, the inclination of the orbit of $\mathfrak{M}$ to that of $\mathfrak{M}'$,
η, $=\sin \frac{1}{2}\Phi$.

The disturbing function is in the first instance given as a function of r, r', H, where

$$\cos H = \cos U \cos U' + \sin U \sin U' \cos \Phi,$$

that is, as a function of r, r', U, U', Φ, or, what is the same thing, of r, r', U, U', η. And the preliminary development is a development in multiple cosines of U, U'. We have then

$$U = f + \mho,$$
$$U' = f' + \mho',$$

and finally

$$r = a \operatorname{elqr}(e, g),$$
$$f = \operatorname{elta}(e, g),$$
$$r' = a' \operatorname{elqr}(e', g'),$$
$$f' = \operatorname{elta}(e', g'),$$

and the ulterior development is a development in multiple cosines of g, g', $\mho$, $\mho'$, the coefficients involving, as before, η, and also a, e, a', e'. But as usual it is not attempted to carry the development further, by introducing in the coefficients in place of the relative quantities $\mho$, $\mho'$, η, the remaining elements of the two orbits, which, if it were necessary to use them, would be

for $\mathfrak{M}$,

θ, the longitude of node,
σ, the departure of node,
ϕ, the inclination,
ϖ, the departure of pericentre,

and for $\mathfrak{M}'$, the like accented quantities, the orbit of reference being any fixed or moveable orbit whatever.

II.

The expression for the disturbing function on $\mathfrak{M}$ (that is, when the superior planet disturbs the inferior) is

$$\mathfrak{M}'\left\{-\frac{r\cos H}{r'^2}+\frac{1}{\sqrt{r^2+r'^2-2rr'\cos H}}\right\}$$

and that for the disturbing function on $\mathfrak{M}'$ (that is, when the inferior planet disturbs the superior) is

$$\mathfrak{M}\left\{-\frac{r'\cos H}{r^2}+\frac{1}{\sqrt{r^2+r'^2-2rr'\cos H}}\right\}$$

where the disturbing function is taken with Lagrange's sign ($=-R$, if R be the disturbing function of the *Mécanique Céleste*).

But we may in the first instance consider the development of the reciprocal of the distance of the two planets

$$=\frac{1}{\sqrt{r^2+r'^2-2rr'\cos H}}.$$

The preceding expression for $\cos H$ may be written

$$\cos H=\cos U\cos U'+\sin U\sin U'(1-2\eta^2),$$

or in either of the two forms

$$\cos H=\cos(U-U')-2\eta^2\sin U\sin U',$$

$$\cos H=(1-\eta^2)\cos(U-U')+\eta^2\cos(U+U').$$

Now imagine the function developed in ascending powers of $\frac{r}{r'}$, the coefficient of $\frac{r^n}{r'^{n+1}}$ will contain $\cos^n H$, $\cos^{n-2} H,\ldots$ to $\cos H$ or 1 according as n is even or odd; and if we then substitute for $\cos H$ the last given expression, and express the different powers of $\cos H$ in multiple cosines of $U-U'$ and $U+U'$, and make the final expression contain the cosines of opposite arguments each with the same coefficient, it is easy to see that the form of the general term is

$$\frac{r^n}{r'^{n+1}}C_n(j,\ j')\cos(jU+j'U'),$$

where j, j', each of them extend through the values n, $n-2,\ldots -n$, and where

$$C_n(-j,\ -j')=C_n(j,\ j').$$

Thus in particular for $n=0$, 1, 2, 3, the combinations (j, j') belonging to the several arguments are,

0, 0

1, 1	−1, 1
1, −1	−1, −1

2, 2	0, 2	−2, 2
2, 0	0, 0	−2, 0
2, −2	0, −2	−2, −2

3, 3	1, 3	−1, 3	−3, 3
3, 1	1, 1	−1, 1	−3, 1
3, −1	1, −1	−1, −1	−3, −1
3, −3	1, −3	−1, −3	−3, −3

But as the coefficients C satisfy the condition $C(-j, -j') = C(j, j')$, the two terms $C(j, j')\cos(jU+j'U')$ and $C(-j, -j')\cos(-jU-j'U')$ are equal to each other, and they may be combined together into a single term. The general term may consequently be written

$$\frac{r^n}{r'^{n+1}} 2C_n(j, j')\cos(jU+j'U'),$$

where j has only the values n, $n-2$,...1 or 0, and j' has as before the values n, $n-2$,... $-n$; except (which occurs only when n is even) for $j=0$, when j' has only the values n, $n-2$,...0: and in the particular case $j=j'=0$, the last-mentioned expression for the general term must be multiplied by $\frac{1}{2}$, or, what is the same thing, the factor 2 must be omitted. In particular for $n=0$, 1, 2, 3, 4, the combinations (j, j') belonging to the several arguments are

0, 0

1, 1	
1, −1	

2, 2	0, 2	
2, 0	0, 0	
2, −2		

3, 3	1, 3		
3, 1	1, 1		
3, −1	1, −1		
3, −3	1, −3		

4, 4	2, 4	0, 4		
4, 2	2, 2	0, 2		
4, 0	2, 0	0, 0		
4, −2	2, −2			
4, −4	2, −4			

I remark that in a series $K(j, j') \genfrac{}{}{0pt}{}{\cos}{\sin} jU+j'U'$, where each argument occurs positively and negatively, and $K(-j, -j') = \pm K(j, j')$ according as the series is one of cosines or of sines, we may say that the *discrete* general term is $K(j, j') \overset{\cos}{\underset{\sin}{}} (jU + j'U')$, or that $K(j, j')$ is the discrete coefficient of the cosine or sine, but if we unite together the terms with opposite arguments so as to form the general term $2K(j, j') \overset{\cos}{\underset{\sin}{}} (jU + j'U')$, then this may be called the *concrete* general term, and $2K(j, j')$ may be said to be the concrete coefficient of the cosine or sine. In a sine series the term corresponding to the argument zero vanishes; in a cosine series this is not in general the case, and the concrete term corresponding to the argument zero must be multiplied by $\frac{1}{2}$.

Returning to the question in hand, from the symmetry of the expression to be developed, we have

$$C_n(j, j') = C_n(j', j),$$

and it follows that the only coefficients which need be calculated are those for which j is not negative, and not less in absolute magnitude than j'; the remaining coefficients are respectively equal to coefficients which satisfy these conditions. Thus for $n = 0, 1, 2, 3, 4$, the combinations (j, j') corresponding to the coefficients in question are,

0, 0

,

1, 1	
1, −1	

,

2, 2		
2, 0	0, 0	
2, −2		

,

3, 3		
3, 1	1, 1	
3, −1	1, −1	
3, −3		

,

4, 4				
4, 2	2, 2			
4, 0	2, 0	0, 0		
4, −2	2, −2			
4, −4	2, −4			

;

and we have, for instance, $C(2, 4) = C(4, 2)$, $C(-2, -4) = C(2, 4) = C(4, 2)$, &c. Under the preceding restriction, viz. j not negative, and not less in absolute magnitude than j', the expression for $C_n(j, j')$ (deduced from the formulæ of Hansen's Memoir) is as follows; viz. putting as usual $\Pi x = 1.2.3 \ldots x$, and also $\Pi_1(x - \frac{1}{2}) = \frac{1}{2} . \frac{3}{2} . \frac{5}{2} \ldots (x - \frac{1}{2})$, and representing the hypergeometric series

$$1 + \frac{\alpha . \beta}{1 . \gamma} x + \frac{\alpha . \alpha + 1 . \beta . \beta + 1}{1 . \ 2 . \ \gamma . \gamma + 1} x^2 + \&c.$$

by $F(\alpha, \beta, \gamma, x)$, we have

$$C(j, j') = \frac{2^{j+j'} \Pi_1 (\frac{1}{2}(n + j) - \frac{1}{2}) \Pi_1 (\frac{1}{2}(n + j') - \frac{1}{2})}{\Pi \frac{1}{2}(n - j) \Pi \frac{1}{2}(n - j') \Pi (j + j')}$$

$$\times \eta^{j+j'} (1 - \eta^2)^{\frac{1}{2}(j-j')} F(-n + j, n + j + 1, j + j' + 1, \eta^2)$$

which, it is to be noticed, is a rational and integral function of η, the highest power being η^{2n}, and the lowest power or order of the coefficient being $\eta^{j+j'}$. I reserve the demonstration of this formula for a separate paragraph.

Let the discrete general term involving $\cos(jU+j'U')$ be represented by

$$D(j, j')\cos(jU+j'U');$$

we have, in like manner as for the coefficients C, $D(-j, -j')=D(j, j')$, $D(j', j)=D(j, j')$, and consequently $D(j, j')$ will be known, if we know its value when the before-mentioned conditions are satisfied; viz., if j be not negative and not less in absolute magnitude than j'; and collecting together the different terms which involve the cosine in question, we find at once, the conditions being satisfied,

$$D(j, j')=\frac{r^j}{r'^{j+1}}C_j(j, j')+\frac{r^{j+2}}{r'^{j+3}}C_{j+2}(j, j')+\&c....$$

so that the value of $D(j, j')$ is known. A transformation of this expression will be given in the sequel.

III.

The *concrete* general term is

$$2D(j, j')\cos(jU+j'U');$$

in particular the concrete terms involving the arguments $U-U'$, $U+U'$, are

$$2D(1,-1)\cos(U-U'),$$
$$2D(1,\ \ 1)\cos(U+U'),$$

or, substituting for the D coefficients their values, the terms are

$$2\left\{\frac{r}{r'^2}C_1(1,-1)+\frac{r^3}{r'^4}C_3(1,-1)+\ldots\right\}\cos(U-U'),$$

$$2\left\{\frac{r}{r'^2}C_1(1,\ \ 1)+\frac{r^3}{r'^4}C_3(1,\ \ 1)+\ldots\right\}\cos(U+U').$$

So far I have considered the reciprocal of the radius vector; but if we consider, instead, the disturbing functions, the only difference will be that we must add the term

$$-\frac{r}{r'^2}\cos H,$$

or

$$-\frac{r}{r^2}\cos H,$$

according as the superior planet disturbs the inferior one, or the inferior planet the superior one. The value of $\cos H$ is

$$= \quad (1-\eta^2)\cos(U-U') + \quad \eta^2 \quad \cos(U+U'),$$

$$= 2\,C_1(1,-1)\cos(U-U') + 2\,C_1(1,\ 1)\cos(U+U'),$$

observing that

$$C_1(1,-1) = \tfrac{1}{2}(1-\eta^2), \qquad C_1(1,\ 1) = \tfrac{1}{2}\eta^2\,;$$

and the terms to be added are consequently, when the superior planet disturbs the inferior,

$$-2\frac{r}{r'^2}\,C_1(1,-1)\cos(U-U')$$

$$-2\frac{r}{r'^2}\,C_1(1,\ \ 1)\cos(U+U'),$$

the effect of which is simply to destroy the same terms contained with the opposite sign in the reciprocal of the distance;

and when the inferior planet disturbs the superior one, the terms to be added are

$$-2\frac{r'}{r^2}\,C_1(1,-1)\cos(U-U')$$

$$-2\frac{r'}{r^2}\,C_1(1,\ \ 1)\cos(U+U'),$$

which are not equal to any terms in the reciprocal of the distance. We may write as follows:

Reciprocal of Distance is,

Disturbing function ÷ Mass of Disturbing Planet is,

		U	$+U'$
$D(0, 0)$	cos	0	0
$+2D(1, -1)$		1	-1
[1] $\left[-2\dfrac{r}{r'^2}C(1, -1)\right]$		1	-1
[2] $\left[-2\dfrac{r'}{r^2}C(1, -1)\right]$		1	-1
$+2D(1, 1)$		1	1
[1] $\left[-2\dfrac{r}{r'^2}C(1, 1)\right]$		1	1
[2] $\left[-2\dfrac{r'}{r^2}C(1, 1)\right]$		1	1
$+2D(2, -2)$		2	-2
$+2D(2, 0)$		2	0
$+2D(0, 2)$		0	2
$+2D(2, 2)$		2	2
$+2D(3, -3)$		3	-3
$+2D(3, -1)$		3	-1
$+2D(1, -3)$		1	-3
$+2D(3, 1)$		3	1
$+2D(1, 3)$		1	3
$+2D(3, 3)$		3	3
$+2D(4, -4)$		4	-4
$+2D(4, -2)$		4	-2
$+2D(2, -4)$		2	-4
$+2D(4, 0)$		4	0
$+2D(0, 4)$		0	4
$+2D(4, 2)$		4	2
$+2D(2, 4)$		2	4
$+2D(4, 4)$		4	4
+ &c., &c.			

[1] Only to be inserted for the disturbing function when the superior planet disturbs the inferior, and having the effect of destroying portions of the coefficients of the next preceding terms.

[2] Only to be inserted for the disturbing function when the inferior planet disturbs the superior one.

For the lunar theory, to the extent to which it is necessary to carry the development, and to which it is carried in Hansen's *Fundamenta Nova*, and my memoir before referred to, we might simply write,

Disturbing function ÷ Sun's Mass is

			U	$+U'$
	$\frac{1}{r'} C_0(0, \quad 0)$	cos	0	0
(1)	$+ \frac{r^2}{r'^3} C_2(0, \quad 0)$		0	0
(6)	$+2\frac{r^3}{r'^4} C_3(1, -1)$		1	-1
(8)	$+2\frac{r^3}{r'^4} C_3(1, \quad 1)$		1	1
(2)	$+2\frac{r^2}{r'^3} C_2(2, -2)$		2	-2
(3)	$+2\frac{r^2}{r'^3} C_2(2, \quad 0)$		2	0
(4)	$+2\frac{r^2}{r'^3} C_2(0, \quad 2)$		0	2
(5)	$+2\frac{r^2}{r'^3} C_2(2, \quad 2)$		2	2
(7)	$+2\frac{r^3}{r'^4} C_3(3, \quad 3)$		3	-3
(9)	$+2\frac{r^3}{r'^4} C_3(3, -1)$		3	-1
(10)	$+2\frac{r^3}{r'^4} C_3(1, -3)$		1	-3
	$+2\frac{r^3}{r'^4} C_3(3, \quad 1)$		3	1
	$+2\frac{r^3}{r'^4} C_3(1, \quad 3)$		1	3

where the prefixed numbers are those of Hansen's ten parts $\Omega_1, \Omega_2 \ldots \Omega_{10}$; or omitting the first term which depends only on the sun's radius vector, and the last two terms which are ultimately neglected, and substituting for the coefficients C their values we have,

Disturbing function ÷ Sun's Mass is

				U	$+U'$
(1)	$\frac{1}{4}-\frac{3}{2}\eta^2+\frac{3}{2}\eta^4$	$\frac{r^2}{r'^3}$	cos	0	0
(2)	$\frac{3}{4}-\frac{3}{2}\eta^2+\frac{3}{4}\eta^4$	$\frac{r^2}{r'^3}$		2	− 2
(3)	$\frac{3}{2}\eta^2-\frac{3}{2}\eta^4$	$\frac{r^2}{r'^3}$		2	0
(4)	$\frac{3}{2}\eta^2-\frac{3}{2}\eta^4$	$\frac{r^2}{r'^3}$		0	2
(5)	$\frac{3}{4}\eta^4$	$\frac{r^3}{r'^4}$		2	2
(6)	$\frac{3}{8}-\frac{33}{8}\eta^2+\frac{75}{8}\eta^4$	$\frac{r^3}{r'^4}$		1	− 1
(7)	$\frac{5}{8}-\frac{15}{8}\eta^2+\frac{15}{8}\eta^4$	$\frac{r^3}{r'^4}$		3	− 3
(8)	$\frac{9}{4}\eta^2-\frac{15}{2}\eta^4$	$\frac{r^3}{r'^4}$		1	1
(9)	$\frac{15}{8}\eta^2-\frac{15}{4}\eta^4$	$\frac{r^3}{r'^4}$		3	− 1
(10)	$\frac{15}{8}\eta^2-\frac{15}{4}\eta^4$	$\frac{r^3}{r'^4}$		1	− 3

If the same form be adopted for the planetary theory, the expressions for the leading coefficients will be,

$C_0(0,\ 0)$,	$D(0,\ 0)=$	$\frac{1}{r'}$	1
2		$+\frac{r^2}{r'^3}$	$\frac{1}{4}(1-\ 6\eta^2+\ 6\eta^4)$
4		$+\frac{r^4}{r'^5}$	$\frac{9}{64}(1-20\eta^2+\ 90\eta^4-\ 140\eta^6+\ 70\eta^8)$
6		$+\frac{r^6}{r'^7}$	$\frac{25}{256}(1-42\eta^2+\ 420\eta^4-1680\eta^6+3150\eta^8\ldots)$
8		$+\frac{r^8}{r'^9}$	$\frac{1225}{16384}(1-72\eta^2+1260\eta^4-9240\eta^6-34650\eta^8\ldots)$
⋮			
$2p$		$+\frac{r^{2p}}{r'^{2p+1}}$	$\frac{\Pi_1(p-\frac{1}{2})\,\Pi_1(p-\frac{1}{2})}{\Pi p\,\Pi p}F(-2p,\ 2p+1,\ 1,\ \eta^2)$;

$C_1(1,\ -1),\qquad D(1,\ -1) = \frac{r}{r'^2}\ \tfrac{1}{2}(1-\eta^2)$

$3\qquad + \frac{r^3}{r'^4}\ \tfrac{3}{16}(1-\eta^2)(1-10\,\eta^2+\ 15\,\eta^4)$

$5\qquad + \frac{r^5}{r'^6}\ \tfrac{15}{128}(1-\eta^2)(1-28\,\eta^2+168\,\eta^4-\ 336\,\eta^6+\ 210\,\eta^8)$

$7\qquad + \frac{r^7}{r'^8}\ \tfrac{175}{2048}(1-\eta^2)(1-54\,\eta^2+675\,\eta^4-3100\,\eta^6+7425\,\eta^8)$

$\vdots$

$2p+1\qquad + \frac{r^{2p+1}}{r'^{2p+2}}\ \frac{\Pi_1(p+\frac{1}{2})\,\Pi_1(p-\frac{1}{2})}{\Pi(p+1)\,\Pi p}(1-\eta^2)\,F(-2p,\ 2p+3,\ 1,\ \eta^2);$

$C_1(1,\ 1),\qquad D(1,\ 1) = \frac{r}{r'^2}\ \tfrac{1}{2}\eta^2$

$3\qquad + \frac{r^3}{r'^4}\ \tfrac{9}{8}\eta^2(1-\tfrac{10}{3}\eta^2+\ \tfrac{5}{3}\eta^4)$

$5\qquad + \frac{r^5}{r'^6}\ \tfrac{225}{128}\eta^2(1-\tfrac{28}{3}\eta^2+\ 28\,\eta^4-\tfrac{168}{5}\eta^6)$

$7\qquad + \frac{r^7}{r'^8}\ \tfrac{1225}{512}\eta^2(1-18\,\eta^2+\tfrac{225}{2}\eta^4-330\,\eta^6)$

$\vdots$

$2p+1\qquad + \frac{r^{2p+1}}{r'^{2p+2}}\ 2\,\frac{\Pi_1(p+\frac{1}{2})\,\Pi_1(p+\frac{1}{2})}{\Pi p\,\Pi p}\eta^2\,F(-2p,\ 2p+3,\ 3,\ \eta^2);$

$C_2(2,\ -2),\qquad D(2,\ -2) = \frac{r^2}{r'^3}\ \tfrac{3}{8}(1-\eta^2)^2$

$4\qquad + \frac{r^4}{r'^5}\ \tfrac{5}{32}(1-\eta^2)^2(1-14\,\eta^2+\ 28\,\eta^4)$

$6\qquad + \frac{r^6}{r'^7}\ \tfrac{105}{1024}(1-\eta^2)^2(1-36\,\eta^2+270\,\eta^4-660\,\eta^6+495\,\eta^8)$

$\vdots$

$2p\qquad + \frac{r^{2p}}{r'^{2p+1}}\ \frac{\Pi_1(p+\frac{1}{2})\,\Pi_1(p-\frac{3}{2})}{\Pi(p+1)\,\Pi(p-1)}(1-\eta^2)^2\,F(-2p+2,\ 2p+3,\ 1,\ \eta^2);$

$C_2(2,\ 0),\qquad D(2,\ 0) = \frac{r^2}{r'^3}\ \tfrac{3}{4}\eta^2(1-\eta^2)$

$4\qquad + \frac{r^4}{r'^5}\ \tfrac{45}{32}\eta^2(1-\eta^2)(1-\tfrac{14}{3}\eta^2+\tfrac{14}{3}\eta^4)$

$6\qquad + \frac{r^6}{r'^7}\ \tfrac{525}{256}\eta^2(1-\eta^2)(1-12\,\eta^2+45\,\eta^4-66\,\eta^6\ldots)$

$\vdots$

$2p\qquad + \frac{r^{2p}}{r'^{2p+1}}\ 2\,\frac{\Pi_1(p+\frac{1}{2})\,\Pi_1(p-\frac{1}{2})}{\Pi p\,\Pi(p-1)}\eta^2(1-\eta^2)\,F(-2p+2,\ 2p+3,\ 3,\ \eta^2);$

$C_n(0,\ 2) = C_n(2,\ 0)$, $\qquad D(0,\ 2) = D(2,\ 0)$;

$C_2(2,\ 2)$, $\qquad D(2,\ 2) = \dfrac{r^2}{r'^3}\ \tfrac{3}{8}\,\eta^4$

4 $\qquad + \dfrac{r^4}{r'^5}\ \tfrac{75}{32}\,\eta^4\left(1 - \tfrac{14}{5}\,\eta^2 + \tfrac{28}{15}\,\eta^4\right)$

6 $\qquad + \dfrac{r^5}{r'^6}\ \tfrac{3675}{512}\,\eta^4\left(1 - \tfrac{36}{5}\,\eta^2 + 18\,\eta^4\right)$

⋮

$2p$ $\qquad + \dfrac{r^{2p}}{r'^{2p+1}}\ \tfrac{2}{3}\,\dfrac{\Pi_1(p+\frac{1}{2})\,\Pi_1(p+\frac{1}{2})}{\Pi(p-1)\,\Pi(p-1)}\,\eta^4\,F(-2p+2,\ 2p+3,\ 5,\ \eta^2)$;

⋮

$C_3(3,\ -3)$, $\qquad D(3,\ -3) = \dfrac{r^3}{r'^4}\ \tfrac{5}{16}(1-\eta^2)^3$

5 $\qquad + \dfrac{r^5}{r'^6}\ \tfrac{35}{256}(1-\eta^2)^3(1 - 18\eta^2 + 45\eta^4)$

⋮

$2p+1$ $\qquad + \dfrac{r^{2p+1}}{r'^{2p+2}}\ \dfrac{\Pi_1(p+\frac{3}{2})\,\Pi_1(p-\frac{3}{2})}{\Pi(p+2)\,\Pi(p-1)}\,(1-\eta^2)^3\,F(-2p+2,\ 2p+5,\ 1,\ \eta^2)$;

$C_3(3,\ -1)$, $\qquad D(3,\ -1) = \dfrac{r^3}{r'^4}\ \tfrac{15}{16}\,\eta^2(1-\eta^2)^2$

5 $\qquad + \dfrac{r^5}{r'^6}\ \tfrac{105}{64}\,\eta^2(1-\eta^2)^2\left(1 - 6\,\eta^2 + \tfrac{15}{2}\,\eta^4\right)$

⋮

$2p+1$ $\qquad + \dfrac{r^{2p+1}}{r'^{2p+2}}\ 2\,\dfrac{\Pi_1(p+\frac{1}{2})\,\Pi_1(p-\frac{1}{2})}{\Pi(p+1)\,\Pi(p-1)}\,\eta^2(1-\eta^2)^2 F(-2p+2,\ 2p+5,\ 3,\ \eta^2)$;

$C_n(1,\ -3) = C_n(3,\ -1)$, $\qquad D(1,\ -3) = D(3,\ -1)$;

$C_3(3,\ 1)$, $\qquad D(3,\ 1) = \dfrac{r^3}{r'^4}\ \tfrac{15}{16}\,\eta^4(1-\eta^2)$

5 $\qquad + \dfrac{r^5}{r'^6}\ \tfrac{525}{128}\,\eta^4(1-\eta^2)\left(1 - \tfrac{18}{5}\,\eta^2 + 3\,\eta^4\right)$

⋮

$2p+1$ $\qquad - \dfrac{r^{2p+1}}{r'^{2p+2}}\ \tfrac{2}{3}\,\dfrac{\Pi_1(p+\frac{3}{2})\,\Pi_1(p+\frac{1}{2})}{\Pi p\ \Pi(p-1)}\,\eta^4(1-\eta^2)\,F(-2p+2,\ 2p+5,\ 5,\ \eta^2)$;

$C_n(1,\ 3) = C_n(3,\ 1)$, $\qquad D(1,\ 3) = D(3,\ 1)$;

$C_3(3,\ 3)$, $\qquad D(3,\ 3) = \dfrac{r^3}{r'^4}\ \tfrac{5}{16}\,\eta^6$

5 $\qquad + \dfrac{r^5}{r'^6}\ \tfrac{245}{64}\,\eta^6\left(1 + \tfrac{18}{7}\,\eta^2 \ldots\right)$

⋮

$2p+1$ $\qquad + \dfrac{r^{2p+1}}{r'^{2p+2}}\ \tfrac{4}{45}\,\dfrac{\Pi_1(p+\frac{3}{2})\,\Pi_1(p+\frac{3}{2})}{\Pi(p-1)\,\Pi(p-1)}\,\eta^6\,F(-2p+2,\ 2p+5,\ 7,\ \eta^2)$; &c.

IV.

But for the planetary theory it is more suitable to arrange the expression of $D(j, j')$ according to the powers of η. We have, under the before-mentioned conditions, j not negative and not less in absolute magnitude than j',

$$D(j, j') = \Sigma \frac{r^{j+2\lambda}}{r'^{j+2\lambda+1}} C_{j+2\lambda}(j, j')$$

$$= \frac{r^j}{r'^{j+1}} \Sigma \left(\frac{r}{r'}\right)^{2\lambda} C_{j+2\lambda}(j, j')$$

where λ extends from 0 to ∞; and, writing in the expression of $C_n(j, j')$, $j+2\lambda$ for n, we find

$$C_{j+2\lambda}(j, j') = \frac{2^{j+j'}\,\Pi(j+\lambda-\frac{1}{2})\,\Pi_1(\frac{1}{2}(j+j')+\lambda-\frac{1}{2})}{\Pi\lambda\,\Pi(\frac{1}{2}(j-j')+\lambda)\,\Pi(j+j')} F(-2\lambda,\ 2j+2\lambda+1, j+j'+1, \eta^2),$$

or, substituting for the hypergeometric series its value,

$$C_{j+2\lambda}(j, j') = \Sigma_\theta \frac{2^{j+j'}\,\Pi_1(j+\lambda-\frac{1}{2})\,\Pi_1(\frac{1}{2}(j+j')+\lambda-\frac{1}{2})}{\Pi\lambda\,\Pi(\frac{1}{2}(j-j')+\lambda)\,\Pi(j+j')} \frac{(-)^\theta [2\lambda]^\theta [2j+2\lambda+\theta]^\theta}{[\theta]^\theta\,[j+j'+\theta]^\theta} \eta^{2\theta}$$

$$= \Sigma_\theta \frac{(-)^\theta\, 2^{j+j'}\,\Pi_1(j+\lambda-\frac{1}{2})\,\Pi_1(\frac{1}{2}(j+j')+\lambda-\frac{1}{2})\,\Pi\,2\lambda\,\Pi(2j+2\lambda+\theta)}{\Pi\,\lambda\,\Pi(\frac{1}{2}(j-j')+\lambda)\,\Pi(j+j'+\theta)\,\Pi\,\theta\,\Pi(2\lambda-\theta)\,\Pi(2j+2\lambda)} \eta^{2\theta}$$

from $\theta = 0$ to $\theta = \infty$. Substituting in the numerator for $\Pi\,2\lambda$, $2^{2\lambda}\,\Pi\,\lambda\,\Pi_1(\lambda-\frac{1}{2})$, and in the denominator for $\Pi(2j+2\lambda)$, $2^{2j+2\lambda}\,\Pi(j+\lambda)\,\Pi_1(j+\lambda-\frac{1}{2})$, and reducing, this becomes

$$C_{j+2\lambda}(j, j') = \Sigma_\theta \frac{(-)^\theta \eta^{2\theta}}{\Pi(j+j'+\theta)\,\Pi\theta} \frac{1}{2^{j-j'}} \frac{\Pi_1(\frac{1}{2}(j+j')+\lambda-\frac{1}{2})\,\Pi(2j+2\lambda+\theta)\,\Pi_1(\lambda-\frac{1}{2})}{\Pi(\frac{1}{2}(j-j')+\lambda)\,\Pi(j+\lambda)\,\Pi(2\lambda-\theta)},$$

and substituting this expression in $D(j, j')$, we find

$$D(j, j') = \frac{1}{2^{j-j'}} \eta^{j+j'} (1-\eta^2)^{\frac{1}{2}(j-j')} \frac{r^j}{r'^{j+1}} \Sigma_\theta \frac{(-)^\theta \eta^{2\theta}}{\Pi(j+j'+\theta)\,\Pi\theta}$$

$$\Sigma_\lambda \left(\frac{\Pi_1(\frac{1}{2}(j+j')+\lambda-\frac{1}{2})\,\Pi(2j+2\lambda+\theta)\,\Pi_1(\lambda-\frac{1}{2})}{\Pi(\frac{1}{2}(j-j')+\lambda)\,\Pi(j+\lambda)\,\Pi(2\lambda-\theta)} \left(\frac{r}{r'}\right)^{2\lambda} \right)$$

from $\lambda = 0$ to $\lambda = \infty$ and $\theta = 0$ to $\theta = \infty$; which is the required expression. It is to be remarked that the series in $\left(\frac{r}{r'}\right)$ which multiplies $\eta^{2\theta}$ is not in general expressible as a hypergeometric series. If, however, we attend only to the leading term in η, or write $\theta = 0$, we find for $D(j, j')$ the value

$$\frac{1}{2^{j-j'}} \eta^{j+j'} (1-\eta^2)^{\frac{1}{2}(j-j')} \frac{r^j}{r'^{j+1}} \frac{1}{\Pi(j+j')}$$

$$\Sigma_\lambda \left(\frac{\Pi_1(\frac{1}{2}(j+j')+\lambda-\frac{1}{2})\,\Pi(2j+2\lambda)\,\Pi_1(\lambda-\frac{1}{2})}{\Pi(\frac{1}{2}(j-j')+\lambda)\,\Pi(j+\lambda)\,\Pi\,2\lambda} \left(\frac{r}{r'}\right)^{2\lambda} \right)$$

which may be simplified by putting in the numerator for $\Pi(2j+2\lambda)$,

$$2^{2j+2\lambda}\,\Pi(j+\lambda)\,\Pi_1(j+\lambda-\tfrac{1}{2}),$$

and in the denominator for $\Pi(j+j')$,

$$2^{j+j'}\,\Pi\tfrac{1}{2}(j+j')\,\Pi_1(\tfrac{1}{2}(j+j')-\tfrac{1}{2}),$$

and for $\Pi 2\lambda$, $2^{2\lambda}\,\Pi\lambda\,\Pi_1(\lambda-\tfrac{1}{2})$. We thus obtain the value

$$\eta^{j+j'}(1-\eta^2)^{\frac{1}{2}(j-j')}\,\frac{r^j}{r'^{j+1}}\,\frac{1}{\Pi\frac{1}{2}(j+j')\,\Pi(\frac{1}{2}(j+j')-\frac{1}{2})}$$

$$\Sigma_\lambda\,\frac{\Pi_1(\frac{1}{2}(j+j')+\lambda-\frac{1}{2})\,\Pi_1(j+\lambda-\frac{1}{2})}{\Pi(\frac{1}{2}(j-j')+\lambda)\,\Pi\lambda}\left(\frac{r}{r'}\right)^{2\lambda},$$

which is equal to

$$\eta^{j+j'}(1-\eta^2)^{\frac{1}{2}(j-j')}\,\frac{r^j}{r'^{j+1}}\,\frac{\Pi_1(j-\frac{1}{2})}{\Pi\frac{1}{2}(j+j')\,\Pi\frac{1}{2}(j-j')}$$

$$\Sigma_\lambda\,\frac{[\frac{1}{2}(j+j')+\lambda-\frac{1}{2}]^\lambda\,[j+\lambda-\frac{1}{2}]^\lambda}{[\frac{1}{2}(j-j')+\lambda]^\lambda\,[\lambda]^\lambda}\left(\frac{r}{r'}\right)^{2\lambda},$$

or finally we have, as regards the leading term in η,

$$D(j,j')=\eta^{j+j'}(1-\eta^2)^{\frac{1}{2}(j-j')}\,\frac{r^j}{r'^{j+1}}\,\frac{\Pi_1(j-\frac{1}{2})}{\Pi\frac{1}{2}(j+j')\,\Pi\frac{1}{2}(j-j')}$$

$$F\left(\tfrac{1}{2}(j+j')+\tfrac{1}{2},\ j+\tfrac{1}{2},\ \tfrac{1}{2}(j-j')+1,\ \frac{r^2}{r'^2}\right).$$

I remark that if in general

$$r^x r'^x(r^2+r'^2-2rr'\cos\vartheta)^{-x-\frac{1}{2}}=\Sigma R_x{}^i\cos i\vartheta,\quad R_x{}^{-i}=R_x{}^i$$

then, writing $\frac{1}{2}(j-j')$ for i and $\frac{1}{2}(j+j')$ for x, we have

$$R_{\frac{1}{2}(j+j')}^{\frac{1}{2}(j-j')}=\frac{r^j}{r'^{j+1}}\,\frac{\Pi_1(j-\frac{1}{2})}{\Pi_1(\frac{1}{2}(j+j')-\frac{1}{2})\,\Pi\frac{1}{2}(j-j')}\,F\left(\tfrac{1}{2}(j+j')+\tfrac{1}{2},\ j+\tfrac{1}{2},\ \tfrac{1}{2}(j-j')+1,\ \frac{r^2}{r'^2}\right)$$

and the last-mentioned expression for $D(j,j')$, as regards the leading term in η, becomes therefore

$$D(j,j')=\eta^{j+j'}(1-\eta^2)^{\frac{1}{2}(j-j')}\,\frac{\Pi_1(\frac{1}{2}(j+j')-\frac{1}{2})}{\Pi\frac{1}{2}(j+j')}\,R_{\frac{1}{2}(j+j')}^{\frac{1}{2}(j-j')},$$

and we have in general

$$D(j,j')=\eta^{j+j'}(1-\eta^2)^{\frac{1}{2}(j-j')}\left\{\frac{\Pi_1(\frac{1}{2}(j+j')-\frac{1}{2})}{\Pi\frac{1}{2}(j+j')}\,R_{\frac{1}{2}(j+j')}^{\frac{1}{2}(j-j')}+\text{terms in }\eta\right\}.$$

As a verification, it may be noticed that for $j+j'=0$ we have $D(j,-j')=(1-\eta^2)^j(R_0{}^j+\text{terms in }\eta)$, and for $\eta=0$, $D(j,-j)=R_0{}^j$, which is right, since the two sides each denote the coefficient of $\cos j\vartheta$ in $(r^2+r'^2-2rr'\cos\vartheta)^{-\frac{1}{2}}$. There is reason to believe that the expression for $D(j,j')$ might be further reduced so as to obtain in a convenient form the coefficients of the successive powers of η; but I have not yet accomplished this.

V.

Considering now $D(j, j')$ as a given function of r and r', we must in the planetary theory write $r = a(1+x)$, $r' = a'(1+x')$, and develope in powers of x and x'. The general term is, of course,

$$\frac{a^\alpha a'^{\alpha'}}{\Pi\alpha\,\Pi\alpha}\, d_a{}^\alpha d_{a'}{}^{\alpha'}\, \overline{D(j, j')}\, x^\alpha x'^{\alpha'}$$

where $\overline{D(j, j')}$ is what $D(j, j')$ becomes when a, a' are substituted for r, r'; and, writing $f + \varpi$, $f' + \varpi'$, for U, U', we see that in the planetary theory the terms to be developed are of the form

$$x^\alpha x'^{\alpha'} \cos(jf + j'f' + j\varpi + j'\varpi'),$$

while, from what has preceded, the form for the lunar theory is

$$r^n r'^{n'} \cos(jf + j'f' + j\varpi + j'\varpi'),$$

$$\left(n' \text{ is always } = -n-1, \text{ except in the terms } \frac{r'}{r^2}\cos(U \pm U')\right)$$

the values which have to be substituted being

$$\begin{aligned} r &= a \operatorname{elqr}(e, g), & x &= \operatorname{elqr}(e, g) - 1, \\ f &= \quad \operatorname{elta}(e, g), & & \\ r' &= a' \operatorname{elqr}(e', g'), & x' &= \operatorname{elqr}(e', g') - 1, \\ f' &= \quad \operatorname{elta}(e', g'). & & \end{aligned}$$

Suppose that in the former case the developments of

$$x^\alpha \cos jf, \quad x^\alpha \sin jf, \quad x'^{\alpha'} \cos j'f', \quad x'^{\alpha'} \sin j'f'$$

and in the latter case the developments of

$$\left(\frac{r}{a}\right)^n \cos jf, \quad \left(\frac{r}{a}\right)^n \sin jf, \quad \left(\frac{r'}{a'}\right)^{n'} \cos j'f', \quad \left(\frac{r'}{a'}\right)^{n'} \sin j'f'$$

are

$$\Sigma\,[\cos]^i \cos ig, \quad \Sigma\,[\sin]^i \sin ig, \quad \Sigma\,[\cos]^{i'} \cos i'g', \quad \Sigma\,[\sin]^{i'} \sin i'g',$$

where the summations extend from i or $i' = -\infty$ to ∞, and where the coefficients $[\cos]^i$, $[\sin]^i$ satisfy

$$[\cos]^{-i} = [\cos]^i, \quad [\sin]^{-i} = -[\sin]^i,$$

and in like manner the coefficients $[\cos]^{i'}$, $[\sin]^{i'}$ satisfy

$$[\cos]^{-i'} = [\cos]^{i'}, \quad [\sin]^{-i'} = -[\sin]^{i'}.$$

It is to be observed that $[\cos]^i$, $[\sin]^i$ are functions of e, and $[\cos]^{i'}$, $[\sin]^{i'}$

functions of e'; the accents to the indices i and i' are sufficient to indicate this. Hence observing that

$$\begin{aligned}
\Sigma\,[\cos]^i \cos ig \;.\; \Sigma\,[\cos]^{i'} \cos i'g' &= \Sigma\Sigma\,[\cos]^i\,[\cos]^{i'} \cos (ig + i'g'),\\
\Sigma\,[\cos]^i \cos ig \;.\; \Sigma\,[\sin]^{i'} \sin i'g' &= \Sigma\Sigma\,[\cos]^i\,[\sin]^{i'} \sin (ig + i'g'),\\
\Sigma\,[\sin]^i \sin ig \;.\; \Sigma\,[\cos]^{i'} \cos i'g' &= \Sigma\Sigma\,[\sin]^i\,[\cos]^{i'} \sin (ig + i'g'),\\
\Sigma\,[\sin]^i \sin ig \;.\; \Sigma\,[\sin]^{i'} \sin i'g' &= -\,\Sigma\Sigma\,[\sin]^i\,[\sin]^{i'} \cos (ig + i'g'),
\end{aligned}$$

we have for the products of $x^a x'^{a'}$, or as the case may be $\left(\frac{r}{a}\right)^m \left(\frac{r'}{a'}\right)^n$ into

$$\begin{aligned}
&\cos (jf + j'f'), \text{ the values } \Sigma\Sigma\,([\cos]^i\,[\cos]^{i'} + [\sin]^i\,[\sin]^{i'}) \cos (ig + i'g'),\\
&\sin (jf + j'f'), \quad\;\; \text{"} \quad\;\; \Sigma\Sigma\,([\sin]^i\,[\cos]^{i'} + [\cos]^i\,[\sin]^{i'}) \sin (ig + i'g'),
\end{aligned}$$

and thence observing that

$$\begin{aligned}
\cos (j\varpi + j'\varpi')\,\Sigma\Sigma\, P^{i,i'} \cos (ig + i'g') &= \Sigma\Sigma\, P^{i,i'} \cos (ig + i'g' + j\varpi + j'\varpi'),\\
-\sin (j\varpi + j'\varpi')\,\Sigma\Sigma\, Q^{i,i'} \sin (ig + i'g') &= \Sigma\Sigma\, Q^{i,i'} \cos (ig + i'g' + j\varpi + j'\varpi'),
\end{aligned}$$

provided only that $P^{-i,-i'} = P^{i,-i'}$, $Q^{-i,-i'} = -\,Q^{i,i'}$, we find for the product of $x^a x'^{a'}$ or as the case may be $\left(\frac{r}{a}\right)^n \left(\frac{r'}{a'}\right)^{n'}$ into $\cos (jf + j'f' + j\varpi + j'\varpi')$ the expression

$$\Sigma\Sigma\,([\cos]^i\,[\cos]^{i'} + [\sin]^i\,[\sin]^{i'} + [\sin]^i\,[\cos]^{i'} + [\cos]^i\,[\sin]^{i'}) \cos (ig + i'g' + j\varpi + j'\varpi')$$

or, finally, the expression

$$\Sigma\Sigma\,([\cos]^i + [\sin]^i)\,([\cos]^{i'} + [\sin]^{i'}) \cos (ig + i'g' + j\varpi + j'\varpi')$$

which is the required development of the general term in multiple cosines of the mean anomalies.

VI.

Investigation of the coefficient $C_n(j, j')$.

It is possible that there might be some advantage in developing in the first instance according to the powers of $\cos H$, a process which, as it has been seen, leads very readily to the form of the general term; but the mode which I have adopted is to develope in the first instance according to the powers of η. I put therefore

$$\cos H = \cos (U - U') - 2\eta^2 \sin U \sin U',$$

and we have then for the reciprocal of the distance,

$$\{r^2 + r'^2 - 2rr' \cos (U - U') - rr' (-4 \sin U \sin U')\,\eta^2\}^{-\frac{1}{2}}$$

which is to be developed in ascending powers of $\frac{r}{r'}$, and we have for the discrete general term of the development

$$\frac{r^n}{r'^{n+1}} C_n(j, j') \cos(jU + j'U');$$

which is the definition of $C_n(j, j')$, the coefficient the value of which is sought for. It has been seen that j, j', are each of them of the same parity with n (even or odd according as n is even or odd), and it has been seen also that it is sufficient to consider the case where j is not negative and not inferior in absolute magnitude to j'.

Expanding in powers of η, and putting as before

$$\Pi x = 1.2.3 \ldots x, \quad \Pi_1(x - \tfrac{1}{2}) = \tfrac{1}{2}.\tfrac{3}{2}.\tfrac{5}{2} \ldots (x - \tfrac{1}{2})$$

the general term is

$$\frac{\Pi_1(x-\frac{1}{2})}{\Pi x} r^x r'^x \{(r^2 + r'^2 - 2rr' \cos(U - U')\}^{-x-\frac{1}{2}} (-4 \sin U \sin U')^x \eta^{2x}$$

where x extends from 0 to ∞.

The factor $(-4 \sin U \sin U')^x$ consists of a series of multiple cosines, and as usual it is assumed that the cosines to opposite arguments are made to occur with equal coefficients. The form of the general term is $\cos(\lambda U + \lambda' U')$, where λ, λ' have each of them the values x, $x - 2$, $x - 4$, ... $-x$, that is, λ, λ' are each of them of the same parity with x. Hence j, j' being as before of the same parity with each other, and ϑ being even or odd according as j, j' and x are of the same parity or of opposite parities, the development of $(-4 \sin U \sin U')^x$ will contain a series of terms $\cos[(j + \vartheta) U + (j' - \vartheta) U']$, which (since the other factor contains only multiple cosines of $U - U'$) are the only terms which give rise to a term $\cos(jU + j'U')$. I represent the discrete term of $(-4 \sin U \sin U')^x$ which contains the before-mentioned argument by

$$M_x^{\vartheta} \cos[(j + \vartheta) U + (j - \vartheta) U'].$$

On the before-mentioned assumption, j not negative and not less in absolute magnitude than j', we have $j + j'$ and $j - j'$, each of them not negative. We must have $j + \vartheta \ngtr x$ and $j' - \vartheta \ngtr x$, that is $\vartheta \ngtr (x - j)$ and $\nless -(x - j')$, consequently $x - j \nless -(x - j')$ or $2x \nless j + j'$. And this relation being assumed, ϑ extends from the inferior limit $-(x - j')$ to the superior limit $(x - j)$ by steps of two units, the extreme terms being

$$\vartheta = -(x - j'), \quad M_x^{\vartheta} \cos[(j + j' - x) U + xU'],$$

$$\vartheta = \quad (x - j), \quad M_x^{\vartheta} \cos[xU + (j + j' - x) U'],$$

the coefficient of U increasing from $j + j' - x$ to x, and that of U' diminishing from x to $j + j' - x$ by steps of two units.

Now expanding in ascending powers of $\frac{r}{r'}$, write

$$r^x r'^x \{(r^2 + r'^2 - 2rr' \cos(U - U')\}^{-x-\frac{1}{2}} = \Sigma R_x^i \cos i (U - U')$$

where i extends from $-\infty$ to $+\infty$ and $R_x^{-i} = R_x^i$; so that $R_x^i \cos i (U - U')$ is the discrete general term of the development. The term $R_x^\vartheta \cos \vartheta (U - U')$ in combination with the term $M_x^\vartheta \cos [(j + \vartheta) U + (j' - \vartheta) U']$ gives rise to $M_x^\vartheta R_x^\vartheta \cos (jU + j'U)$, and restoring the multiplier which has been omitted, and giving to ϑ and x the different admissible values, we find for the discrete general term containing $\cos (jU + j'U')$ the value

$$\Sigma \frac{\Pi_1 (x - \frac{1}{2})}{\Pi x} \eta^{2x} \Sigma (M_x^\vartheta R_x^\vartheta) \cos (jU + j'U')$$

where ϑ extends from the inferior limit $\vartheta = -(x - j')$ to the superior limit $\vartheta = x - j$, by steps of two units, and x extends from $x = \frac{1}{2}(j + j')$ to $x = \infty$. The portion of this containing $\frac{r^n}{r'^{n+1}} \cos (jU + j'U')$ is

$$\frac{r^n}{r'^{n+1}} C_n (j, j') \cos (jU + j'U')$$

and we have therefore

$$C_n (j, j') = \text{coeff. } \frac{r^n}{r'^{n+1}} \text{ in } \Sigma \frac{\Pi_1 (x - \frac{1}{2})}{\Pi x} \eta^{2x} \Sigma (M_x^\vartheta R_x^\vartheta).$$

There is some speciality in the case $j + j' = 0$, but the result just obtained subsists without variation. To find M_x^ϑ we have

$$M_x^\vartheta = \text{Discrete coeff. } \cos [(j + \vartheta) U + (j' - \vartheta) U'] \text{ in } (-4 \sin U \sin U')^x,$$

and putting $\sin U = \frac{1}{2i} \left(\upsilon - \frac{1}{\upsilon}\right)$, $\sin U' = \frac{1}{2i} \left(\upsilon' - \frac{1}{\upsilon'}\right)$ where $i = \sqrt{-1}$ we have

$$(-4 \sin U \sin U')^x = \left(\upsilon - \frac{1}{\upsilon}\right)^x \left(\upsilon' - \frac{1}{\upsilon'}\right)^x,$$

and the function on the right hand contains the term

$$(-)^{f+f'} \frac{\Pi x}{\Pi f \Pi (x - f)} \frac{\Pi x}{\Pi f' \Pi (x - f')} \upsilon^{x+2f} \upsilon'^{x-2f'},$$

or putting

$$x - 2f = j + \vartheta, \ x - 2f' = j' - \vartheta,$$

and therefore

$$f = \tfrac{1}{2}(x - j - \vartheta), \ f' = \tfrac{1}{2}(x - j' + \vartheta),$$

which give integer values for f, f', since ϑ is even or odd according as j, j', and x are of the same parity or of opposite parities, and replacing $\upsilon^{x-2f} \upsilon'^{x-2f'}$ by the *half* of its value $2 \cos [(j + \vartheta) U + (j' - \vartheta) U']$, the term is

$$(-)^{x + \frac{1}{2}(j+j')} \frac{\Pi x}{\Pi \frac{1}{2}(x - j - \vartheta) \Pi \frac{1}{2}(x + j + \vartheta)} \frac{\Pi x}{\Pi \frac{1}{2}(x - j' + \vartheta) \Pi \frac{1}{2}(x + j' - \vartheta)} \times \cos [(j + \vartheta) U + (j' - \vartheta) U'],$$

and consequently,

$$M_x^{\vartheta} = (-)^{x-\frac{1}{2}(j+j')} \frac{\Pi x}{\Pi\frac{1}{2}(x-j-\vartheta)\,\Pi\frac{1}{2}(x+j+\vartheta)} \frac{\Pi x}{\Pi\frac{1}{2}(x-j'+\vartheta)\,\Pi\frac{1}{2}(x+j'-\vartheta)},$$

which is the expression for M_x^{ϑ}.

The expression for R_x^i is found in a similar manner, viz., by substituting for $\cos(U-U')$ its exponential expression, by which means the function

$$(r^2+r'^2-2rr'\cos(U-U'))^{-x-\frac{1}{2}}$$

breaks up into a pair of factors, each of which can be separately expanded; the result, i being positive, or zero, is

$$R_x^i = \frac{r^{x+i}}{r'^{x+i+1}} \Sigma \frac{\Pi_1(x+i+m-\frac{1}{2})\,\Pi_1(x+m-\frac{1}{2})}{\Pi(i+m)\,\Pi_1(x-\frac{1}{2})\,\Pi m\,\Pi_1(x-\frac{1}{2})} \left(\frac{r}{r'}\right)^{2m}$$

from $m=0$ to $m=\infty$: this may also be written

$$R_x^i = \frac{r^{x+i}}{r'^{x+i+1}} \frac{\Pi_1(x+i-\frac{1}{2})}{\Pi i\,\Pi_1(x-\frac{1}{2})} F\left(x+i+\tfrac{1}{2},\ x+\tfrac{1}{2},\ i+1,\ \frac{r^2}{r'^2}\right);$$

writing now ϑ for i, and $x+\vartheta+2m=n$, that is $m=\frac{1}{2}(n-x-\vartheta)$ (n is of the same parity with j, j', and ϑ is even or odd according as j, j' and x are of the same parity or of opposite parities, m is therefore, as it should be, an integer), R_x^{ϑ} contains the term

$$\frac{r^n}{r'^{n+1}} \frac{\Pi_1(\frac{1}{2}(n+x+\vartheta)-\frac{1}{2})\,\Pi_1(\frac{1}{2}(n+x-\vartheta)-\frac{1}{2})}{\Pi\frac{1}{2}(n-x+\vartheta)\,\Pi_1(x-\frac{1}{2})\,\Pi\frac{1}{2}(n-x-\vartheta)\,\Pi_1(x-\frac{1}{2})},$$

$$= \frac{r^n}{r'^{n+1}} K_x^{\vartheta}$$

if for shortness

$$K_x^{\vartheta} = \frac{\Pi_1(\frac{1}{2}(n+x+\vartheta)-\frac{1}{2})\,\Pi_1(\frac{1}{2}(n+x-\vartheta)-\frac{1}{2})}{\Pi\frac{1}{2}(n-x+\vartheta)\,\Pi_1(x-\frac{1}{2})\,\Pi_1\frac{1}{2}(n-x-\vartheta)\,\Pi_1(x-\frac{1}{2})},$$

a formula which I assume to subsist as well for negative as for positive values of ϑ, so that $K_x^{-\vartheta}=K_x^{\vartheta}$. It is to be noticed that if $x>n$, then either $n-x+\vartheta$ or $n-x-\vartheta$ vanishes, and we have therefore $K_x^{\vartheta}=0$.

Substituting for R_x^{ϑ} its value we have

$$C_n(j,\ j') = \Sigma \frac{\Pi_1(x-\frac{1}{2})}{\Pi x} \eta^{2x} \Sigma (M_x^{\vartheta} K_x^{\vartheta})$$

where M_x^{ϑ}, K_x^{ϑ} have the values already obtained, and as before ϑ extends to $\vartheta=-(x-j')$ to $\vartheta=(x-j)$ by steps of two units, and x (since K_x^{ϑ} vanishes for $x<n$) extends from $x=\frac{1}{2}(j+j')$ to $x=n$.

To simplify; write $x = \frac{1}{2}(j+j') + u$; $\vartheta = -(x-j') + 2s, = (x-j') - 2t$, so that $s+t=u$, s and t being any integer values (zero not excluded) which satisfy this relation, and lastly u extends from $u=0$ to $u = n - \frac{1}{2}(j+j')$. We have

$$C_n(j,\, j') = \eta^{j+j'}\, \Sigma \eta^{2u}\, \frac{\Pi_1(\frac{1}{2}(j+j')+u-\frac{1}{2})}{\Pi(\frac{1}{2}(j+j)+u)}\, \Sigma\, \overline{M_u{}^s}\, \overline{K_u{}^s},$$

where

$$\overline{M_u{}^s} = (-)^u\, \frac{\Pi(\frac{1}{2}(j+j')+u)}{\Pi t\, \Pi(\frac{1}{2}(j+j')+s)}\, \frac{\Pi(\frac{1}{2}(j+j')+u)}{\Pi s\, \Pi(\frac{1}{2}(j+j')+t)},$$

$$\overline{K_u{}^s} = \frac{\Pi_1(\frac{1}{2}(n+j')+s-\frac{1}{2})}{\Pi(\frac{1}{2}(n-j)-t)\, \Pi_1(\frac{1}{2}(j+j')+u-\frac{1}{2})}\, \frac{\Pi_1(\frac{1}{2}(n+j)+t-\frac{1}{2})}{\Pi(\frac{1}{2}(n-j')-s)\, \Pi_1(\frac{1}{2}(j+j')+u-\frac{1}{2})},$$

and substituting these values, and observing that the result contains a factor $\dfrac{1}{\Pi(\frac{1}{2}(j+j')+u)\, \Pi_1(\frac{1}{2}(j+j')+u-\frac{1}{2})}$ which may be replaced by $\dfrac{2^{j+j'+2u}}{\Pi(j+j'+2u)}$ the result is

$$C_n(j,\, j') = 2^{j+j'}\, \eta^{j+j'}\, \Sigma_u\, \frac{(-)^u\, 2^{2u}\, \eta^{2u}}{\Pi(j+j'+2u)}$$

$$\times\, \Sigma \left\{ \frac{\Pi(\frac{1}{2}(j+j')+u)}{\Pi(\frac{1}{2}(j+j')+s)}\, \frac{\Pi(\frac{1}{2}(j+j')+u)}{\Pi(\frac{1}{2}(j+j')+t)}\, \frac{1}{\Pi s 11 t}\, \frac{\Pi_1(\frac{1}{2}(n+j')+s-\frac{1}{2})\, \Pi_1(\frac{1}{2}(n+j)+t-\frac{1}{2})}{\Pi(\frac{1}{2}(n-j')-s)\, \Pi(\frac{1}{2}(n-j)-t)} \right\}$$

which is easily transformed into

$$C_n(j,\, j') = j^{j+j'}\, \eta^{j+j'}\, \frac{\Pi_1(\frac{1}{2}(n+j)-\frac{1}{2})\, \Pi_1(\frac{1}{2}(n+j')-\frac{1}{2})}{\Pi\frac{1}{2}(n-j)\, \Pi\frac{1}{2}(n-j')\, \Pi(j+j')}\, S$$

if for shortness $S =$

$$\Sigma_u\, \frac{(-)^u\, 2^{2u}\, \eta^{2u}}{[j+j'+2u]^{2u}}\, \Sigma\, \frac{[\frac{1}{2}(j+j')+u]^s [\frac{1}{2}(n+j')+s-\frac{1}{2}]^s [\frac{1}{2}(n-j')]^s [\frac{1}{2}(j+j')+u]^t [\frac{1}{2}(n+j)+t-\frac{1}{2}]^t [\frac{1}{2}(n-j)]}{[s]^s\, [t]^t}$$

$$s+t=u,\ u=0 \text{ to } u = n - \tfrac{1}{2}(j+j');$$

the value of the sum S is

$$S = (1-\eta^2)^{\frac{1}{2}(j-j')}\, F(-n+j,\, n+j+1,\, j+j'+1,\, \eta^2),$$

and we have consequently

$$C_n(j,\, j') = \frac{2^{j+j'}\, \Pi_1(\frac{1}{2}(n+j)-\frac{1}{2})\, \Pi_1(\frac{1}{2}(n+j')-\frac{1}{2})}{\Pi\frac{1}{2}(n-j)\, \Pi\frac{1}{2}(n-j')\, \Pi(j+j')}$$

$$\times\, \eta^{j+j'}\, (1-\eta^2)^{\frac{1}{2}(j-j')}\, F(-n+j,\, n+j+1,\, j+j'+1,\, \eta^2),$$

the required expression for $C_n(j,\, j')$. It only remains to prove the formula for S.

For this purpose, observing the equation $s+t=u$, I form the equations

$$\frac{2^{2u}}{[j+j'+2u]^{2u}}=\frac{1}{[\tfrac{1}{2}(j+j')+u]^u\,[\tfrac{1}{2}(j+j')+u-\tfrac{1}{2}]^u},$$

$$[\tfrac{1}{2}(j+j')+u]^s=\frac{[\tfrac{1}{2}(j+j')+u]^u}{[\tfrac{1}{2}(j+j')+t]^t},$$

$$[\tfrac{1}{2}(j+j')+u]^t=\frac{[\tfrac{1}{2}(j+j')+u]^u}{[\tfrac{1}{2}(j+j')+s]^s},$$

and thence

$$\frac{2^{2u}}{[j+j'+2u]^{2u}}=[\tfrac{1}{2}(j+j')+u]^s\,[\tfrac{1}{2}(j+j')+u]^t$$

$$=\frac{[\tfrac{1}{2}(j+j')+u]^u}{[\tfrac{1}{2}(j+j')+u-\tfrac{1}{2}]^u}\,\frac{1}{[\tfrac{1}{2}(j+j')+s]^s\,[\tfrac{1}{2}(j+j')+t]^t},$$

and we then have (putting also $(-)^u\,\eta^{2u}=(-)^{s+t}\,\eta^{2s+2t}$),

$$S=\Sigma_u\,\frac{[\tfrac{1}{2}(j+j')+u]^u}{[\tfrac{1}{2}(j+j')+u-\tfrac{1}{2}]^u}\,\Sigma\,\frac{(-)^s\,[\tfrac{1}{2}(n-j')]^s\,[\tfrac{1}{2}(n+j')+s-\tfrac{1}{2}]^s}{[s]^s\,[\tfrac{1}{2}(j+j')+s]^s}\,\eta^{2s}$$
$$\times\,\frac{(-)^t\,[\tfrac{1}{2}(\eta-j)]^t\,[\tfrac{1}{2}(n+j)+t-\tfrac{1}{2}]^t}{[t]^t\,[\tfrac{1}{2}(j+j')+t]^t}\,\eta^{2t},$$

and it is proper to remark that the summation as regards u may be continued indefinitely; for, if $u=s+t$ be $>n-\tfrac{1}{2}(j+j')$, then one at least of the relations $s>\tfrac{1}{2}(n-j)$, $t>\tfrac{1}{2}(n-j)$, must hold good, and at least one of the factorials $[\tfrac{1}{2}(n-j)]^s$, $[\tfrac{1}{2}(n-j)]^t$, will vanish. The two factors in the second sum are the general terms of two hypergeometric series. In fact, if we put

$$\tfrac{1}{2}(n+j)+\tfrac{1}{2}=\beta,$$
$$\tfrac{1}{2}(-n+j)=\alpha,$$
$$\tfrac{1}{2}(j+j')+\tfrac{1}{2}=\epsilon,$$

and therefore

$$\tfrac{1}{2}(n+j')+\tfrac{1}{2}=\epsilon-\alpha,$$
$$\tfrac{1}{2}(-n+j')=\epsilon-\beta,$$

and if, for shortness, we use $\{\alpha\}^s$ to denote the factorial $\alpha(\alpha+1)\ldots(\alpha+s-1)$ of the increment positive unity, then we have

$$S=\Sigma_u\,\frac{\{\epsilon+\tfrac{1}{2}\}^u}{\{\epsilon\}^u}\,\Sigma\,\frac{\{\alpha\}^s\,\{\beta\}^s}{\{1\}^s\,\{\epsilon+\tfrac{1}{2}\}^s}\,\eta^{2s}\,\frac{\{\epsilon-\alpha\}^s\,\{\epsilon-\beta\}^s}{\{1\}^s\,\{\epsilon+\tfrac{1}{2}\}^s}\,\eta^{2t}$$

the two hypergeometric series being consequently

$$F(\alpha,\ \beta,\ \epsilon+\tfrac{1}{2},\ \eta^2)\qquad=1+\frac{\alpha\,.\,\beta}{1\,.\,\epsilon+\tfrac{1}{2}}\,\eta^2+\frac{\alpha\,.\,\alpha+1\,.\,\beta\,.\,\beta+1}{1\,.\,2\,.\,\epsilon+\tfrac{1}{2}\,.\,\epsilon+\tfrac{3}{2}}\,\eta^4+\text{\&c.},$$

$$F(\epsilon-\alpha,\ \epsilon-\beta,\ \epsilon+\tfrac{1}{2},\ \eta^2)=1+\frac{\epsilon-\alpha\,.\,\epsilon-\beta}{1\,.\,\epsilon+\tfrac{1}{2}}\,\eta^2+\frac{\epsilon-\alpha\,.\,\epsilon-\alpha+1\,.\,\epsilon-\beta\,.\,\epsilon-\beta+1}{1\,.\,2\,.\,\epsilon+\tfrac{1}{2}\,.\,\epsilon+\tfrac{3}{2}}\,\eta^4+\text{\&c.}$$

I assume the truth of the following remarkable theorem, [see 211] viz.:

"The series formed from the product of the two hypergeometric series,

$$F(\alpha,\ \beta,\ \epsilon+\tfrac{1}{2},\ \eta^2),\ F(\epsilon-\alpha,\ \epsilon-\beta,\ \epsilon+\tfrac{1}{2},\ \eta^2),$$

by multiplying the successive terms of the product by 1, $\frac{\epsilon+\frac{1}{2}}{\epsilon}$, $\frac{\epsilon+\frac{1}{2}\,.\,\epsilon+\frac{3}{2}}{\epsilon\ \ .\,\epsilon+1}$, &c. respectively, is

$$=(1-\eta^2)^{\alpha+\beta-\epsilon}\ F(2\alpha,\ 2\beta,\ 2\epsilon,\ \eta^2)."$$

Hence, observing that the general term of S is formed precisely in the manner in question, we have

$$S=(1-\eta^2)^{\alpha+\beta-\epsilon}\ F(2\alpha,\ 2\beta,\ 2\epsilon,\ \eta^2)$$

or, substituting for α, β, ϵ their values

$$S=(1-\eta^2)^{\frac{1}{2}(j-j')}\ F(-n+j,\ n+j+1,\ j+j'+1,\ \eta^2),$$

which is the required value.

It is clear that α, β are interchangeable with $\epsilon-\alpha$, $\epsilon-\beta$; that is, we have

$$S=(1-\eta^2)^{\alpha+\beta-\epsilon}\ F(2\alpha,\ 2\beta,\ 2\epsilon,\ \eta^2)$$
$$=(1-\eta^2)^{\epsilon-\alpha-\beta}\ F(2\epsilon-2\alpha,\ 2\epsilon-2\beta,\ 2\epsilon,\ \eta^2)$$

or

$$F(2\epsilon-2\alpha,\ 2\epsilon-2\beta,\ 2\epsilon,\ \eta^2)=(1-\eta^2)^{2\alpha+2\beta-2\epsilon}\ F(2\alpha,\ 2\beta,\ 2\epsilon,\ \eta^2),$$

which is a known property of hypergeometric series. The form

$$S=(1-\eta^2)^{-\frac{1}{2}(j-j')}F(-n+j',\ n+j'+1,\ j+j'+1,\ \eta^2)$$

is obviously less convenient than the one above mentioned, since the new expression is encumbered by a denominator $(1-\eta^2)^{\frac{1}{2}(j-j')}$, which really divides out, the finite hypergeometric series containing as a factor the square of such denominator. I have only noticed this for the verification which it affords by showing that j, j' may be interchanged.

VII.

The above expression of $C_n(j,\ j')$ is not, in its actual form, given in Hansen's memoir. The comparison with Hansen's formulæ is as follows:—The formula (36), p. 329, is

$$C\,(n-2f,\ -(n-2f-2g))$$
$$=H\cos^{2n}\tfrac{1}{2}J\tan^{2g}\tfrac{1}{2}J\ \ F(-(2n-2f-2g),\ -2f,\ 2g+1,\ -\tan^2\tfrac{1}{2}J)$$

where

$$H=\frac{\Pi\,(2n-2f)\,\Pi\,(2f+2g)}{2^{2n}\,\Pi\,(n-f)\,\Pi f\,\Pi\,(n-f-g)\,\Pi\,(f+g)\,\Pi\,2g};$$

and the formula (37), p. 330, is

$$F(-(2n-2f-2g),\ -2f,\ 2g+1,\ -\tan^2\tfrac{1}{2}J)$$
$$=\cos^{-4n+4f+4g}\tfrac{1}{2}J\ F(-(2n-2f-2g),\ 2f+2g+1,\ 2g+1,\ \sin^2\tfrac{1}{2}J).$$

Combining these, and putting $\sin\frac{1}{2}J=\eta$, we have

$$C(n-2f,\ -(n-2f-2g))$$
$$=H\eta^{2g}(1-\eta^2)^{-n+2f+g}\ F(-(2n-2f-2g),\ 2f+2g+1,\ 2g+1,\ \eta^2),$$

and then, putting

$$n-2f\qquad=j',$$
$$-n+2f+2g=j,$$

and for $C(j',j)$, $=C(j,j')$, writing $C_n(j,j')$, we have

$$C_n(j,j')=H\eta^{j+j'}(1-\eta^2)^{\frac{1}{2}(j-j')}\ F(-n+j,\ n+j+1,\ j+j'+1,\ \eta^2)$$

where

$$H=\frac{\Pi(n+j')\,\Pi(n+j)}{2^{2n}\,\Pi\frac{1}{2}(n+j')\,\Pi\frac{1}{2}(n-j')\,\Pi\frac{1}{2}(n+j)\,\Pi\frac{1}{2}(n-j)\,\Pi(j+j')},$$

or, finally, since

$$\Pi(n+j)=2^{n+j}\,\Pi\tfrac{1}{2}(n+j)\,\Pi_1(\tfrac{1}{2}(n+j)-\tfrac{1}{2}),$$
$$\Pi_1(n+j')=2^{n+j'}\,\Pi\tfrac{1}{2}(n+j')\,\Pi_1(\tfrac{1}{2}(n+j')-\tfrac{1}{2}),$$

we find

$$C_n(j,j')=\frac{2^{j+j'}\,\Pi_1(\frac{1}{2}(n+j)-\frac{1}{2})\,\Pi_1(\frac{1}{2}(n+j)-\frac{1}{2})}{\Pi\frac{1}{2}(n-j)\,\Pi\frac{1}{2}(n-j')\,\Pi(j+j')}$$
$$\times\,\eta^{j+j'}(1-\eta^2)^{\frac{1}{2}(j-j')}\ F(-n+j,\ n+j+1,\ j+j'+1,\ \eta^2),$$

which is the formula of the present memoir.

215.

A SUPPLEMENTARY MEMOIR ON THE PROBLEM OF DISTURBED ELLIPTIC MOTION.

[From the *Memoirs of the Royal Astronomical Society*, vol. XXVIII. pp. 217—234. Read January 12, 1860.]

THE present memoir contains an investigation upon fundamentally different principles of a system of formulæ given in my "Memoir on the Problem of Disturbed Elliptic Motion," *Ast. Soc. Mem.* t. XXVII. pp. 1—29 (1858), [212]; it is shown how the formulæ are deduced by means of the general equations of the *Mécanique Analytique*, from the expression for the *Vis viva* function T, in terms of the coordinates (r the radius vector, þ the departure) of the body in its instantaneous orbit, viz., the ultimate form of this function is $Tdt^2 = \frac{1}{2}(dr^2 + r^2 d\text{þ}^2)$, but T contains in the first instance terms, not identically vanishing, but which are to be equated to zero, thus furnishing equations of the problem which could not be obtained from the foregoing ultimate form of T. The investigation throws, I think, a further light upon the system of formulæ, and completes the development which I was anxious to give of the dynamical problem. I have been a great deal indebted, in the composition of the memoir, to a correspondence some time ago with Professor Donkin on the general subject.

The word orbit is used to denote a great circle of the sphere, and it is assumed that in any orbit there is an origin of longitudes; the angular position of a body in reference to the orbit is determined by the longitude and latitude. It is ultimately assumed that the longitude is measured from an origin which is what I have called a departure-point; viz., in the general case of a variable orbit the departure-point is the point of intersection of the orbit by any orthogonal trajectory of the successive positions of the orbit: this includes the case of a fixed orbit, where the departure-point is simply a fixed point. As regards points in the orbit, the word departure may be used instead of longitude. In the present memoir the origin is in the first instance taken to be, not a departure-point, but an arbitrarily varying point of the orbit.

The mutual position of any two orbits whatever, say the position of an orbit $x'y'$ and of the origin x' therein, in reference to the orbit xy and the origin x therein, is determined by

θ, the longitude of node,

σ, the departure of node,

ϕ, the inclination,

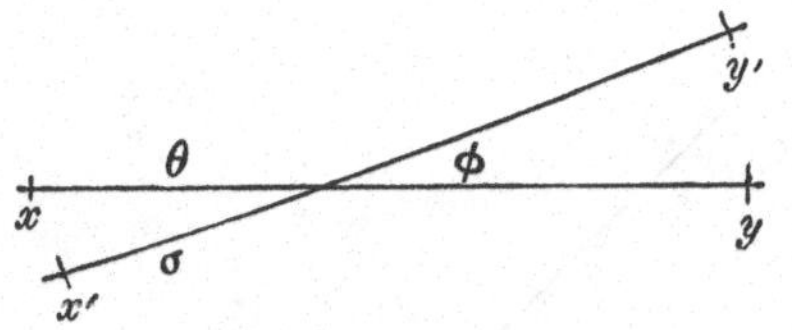

where the expression departure of node is used by way of distinction from longitude of node, the departure referring to the orbit $x'y'$ and origin x', the positions of which are to be determined, and the longitude to the orbit of reference xy and origin x. And this distinction will be preserved throughout. It should be recollected that it is not as yet assumed that the origins are departure-points.

Consider now a fixed orbit x_0y_0 and fixed origin x_0 therein, and suppose that the orbit x_1y_1 and origin x_1 therein are determined in reference to the orbit x_0y_0 and origin x_0 therein, by

θ', the longitude of node,

σ', the departure of node,

ϕ', the inclination,

the orbit x_2y_2 and origin x_2 therein, in reference to the orbit x_1y_1 and origin x_1 therein, by

Θ, the longitude of node,

Σ, the departure of node,

Φ, the inclination,

and, finally, the position of the body in reference to the orbit x_2y_2 and origin x_2 therein, by

r, the radius vector,

v, the longitude,

y, the latitude.

It is required to find the expression of T, the *Vis viva* function, or half square of the velocity. We may imagine the rectangular axes $x_0y_0z_0$, $x_1y_1z_1$, $x_2y_2z_2$, the positions of which are determined as above, and the rectangular axes $x_3y_3z_3$, that of x_3 passing through the body, that of y_3 lying in the orbit x_2y_2, 90° in advance of the body or passing through the pole of the latitude circle, and that of z_3 in the latitude circle 90° above the body. Or considering x_3, y_3, z_3, as points of the sphere, their positions in reference to the orbit x_2y_2 and origin x_2 therein are determined as follows, viz.,

for x_3, longitude is v, and latitude y,

y_3, " $v+90°$ " 0,

z_3, " v " $y+90°$.

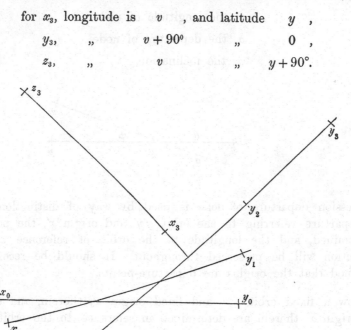

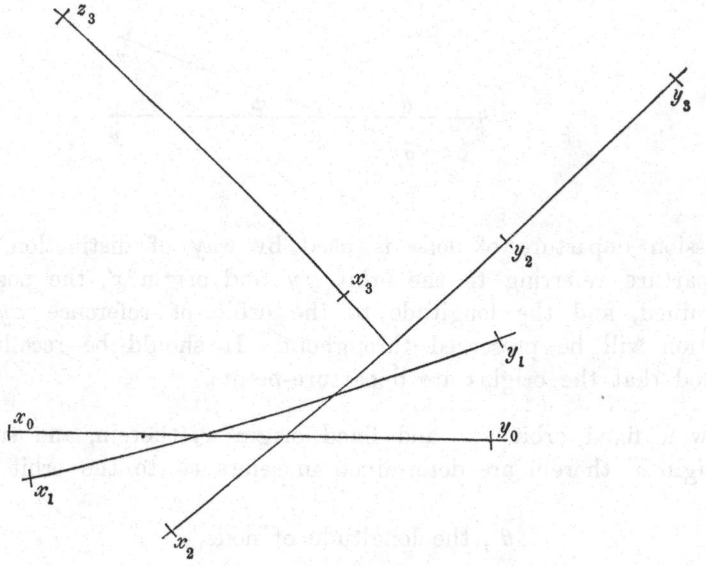

and we may suppose that (x_0, y_0, z_0), (x_1, y_1, z_1), (x_2, y_2, z_2), (x_3, y_3, z_3), are the coordinates of any point in reference to these sets of axes respectively, the coordinates of the body in reference to the axes $x_3y_3z_3$ being obviously $(r, 0, 0)$.

The displacements in the directions of the axes $x_3y_3z_3$ respectively of a point the coordinates of which are x_3, y_3, z_3, are as usual

$$dx_3 + (-Ry_3 + Qz_3)\,dt,$$
$$dy_3 + (-Pz_3 + Rx_3)\,dt,$$
$$dz_3 + (-Qx_3 + Py_3)\,dt,$$

where the expressions of the rotations P, Q, R are as follows, viz.,

$$\begin{aligned}
Pdt = {} & \sin y\, dv \\
& + \cos y\,[\sin(v-\Sigma)\sin\Phi\, d\Theta + \cos(v-\Sigma)\, d\Phi] \\
& + \sin y\,[\quad - d\Sigma + \cos\Phi\, d\Theta \quad] \\
& + \{\cos y\cos(v-\Sigma)\}\,[\sin(\Theta-\sigma')\sin\phi'\, d\theta' \cos(\Theta-\sigma')\, d\phi'] \\
& + \{\cos y\sin(v-\Sigma)\cos\Phi - \sin y\sin\Phi\}\,[\cos(\Theta-\sigma')\sin\phi'\, d\theta' \sin(\Theta-\sigma')\, d\phi'] \\
& + \{\cos y\sin(v-\Sigma)\sin\Phi + \sin y\cos\Phi\}\,[\quad -d\sigma' + \cos\phi'\, d\theta' \quad],
\end{aligned}$$

$$\begin{aligned}Qdt = & - dy \\ & + [\cos (v - \Sigma) \sin \Phi \, d\Theta - \sin (v - \Sigma) \, d\Phi] \\ & - \sin (v - \Sigma) [\sin (\Theta - \sigma') \sin \phi' \, d\theta' + \cos (\Theta - \sigma') \, d\phi'] \\ & + \cos (v - \Sigma) \cos \Phi [\cos (\Theta - \sigma') \sin \phi' \, d\theta' - \sin (\Theta - \sigma') \, d\phi'] \\ & + \cos (v - \Sigma) \sin \Phi [\quad - d\sigma' + \cos \phi' \, d\theta' \quad],\end{aligned}$$

$$\begin{aligned}Rdt = & \quad \cos y dv \\ & - \sin y [\sin (v - \Sigma) \sin \Phi \, d\Theta + \cos (v - \Sigma) \, d\Phi] \\ & + \cos y [\quad - d\Sigma + \cos \Phi \, d\Theta \quad] \\ & + \{- \sin y \cos (v - \Sigma) \} [\sin (\Theta - \sigma') \sin \phi' \, d\theta' + \cos (\Theta - \sigma') \, d\phi'] \\ & + \{- \sin y \sin (v - \Sigma) \cos \Phi - \cos y \sin \Phi\} [\cos (\Theta - \sigma') \sin \phi' \, d\theta' - \sin (\Theta - \sigma') \, d\phi'] \\ & + \{- \sin y \sin (v - \Sigma) \sin \Phi + \cos y \cos \Phi\} [\quad - d\sigma' + \cos \phi' \, d\theta' \quad].\end{aligned}$$

For the body itself we must write r, 0, 0, in the place of x_3, y_3, z_3; the displacements are therefore

$$dr, \quad rRdt, \quad -rQdt;$$

and the expression for the *Vis viva* function is

$$Tdt^2 = \tfrac{1}{2} \{dr^2 + r^2 (Q^2 + R^2) \, dt^2\}$$

where Q, R have the above-mentioned values.

Let Ω be a function of the coordinates x_0, y_0, z_0 of a point, Ω will be a function of x_3, y_3, z_3, and also of the quantities Θ, Σ, Φ, θ', σ', ϕ' which determine the positions of the axes x_3, y_3, z_3, and the complete differential of Ω will be

$$\begin{aligned}d\Omega = & \frac{d\Omega}{dx_3} \left\{dx_3 + (- Ry_3 + Qz_3) \, dt\right\} \\ & + \frac{d\Omega}{dy_3} \left\{dy_3 + (- Pz_3 + Rx_3) \, dt\right\} \\ & + \frac{d\Omega}{dz_3} \left\{dz_3 + (- Qx_3 + Py_3) \, dt\right\}.\end{aligned}$$

I assume for the present the truth of the following proposition, viz., that when Ω is a function of the coordinates of the body (in which case, as before, the values of x_3, y_3, z_3 are r, 0, 0), we have

$$\frac{d\Omega}{dx_3} = \frac{d\Omega}{dr},$$

$$\frac{d\Omega}{dy_3} = \frac{1}{r} \sec y \frac{d\Omega}{dv},$$

$$\frac{d\Omega}{dz_3} = \frac{1}{r} \frac{d\Omega}{dy}.$$

Ω is here a function of r, v, y, Θ, Σ, Φ, θ', σ', ϕ', or, as we may express it, $\Omega=\Omega(r, v, y, \Theta, \Sigma, \Phi, \theta', \sigma', \phi')$; and the expression for the complete differential, by what precedes, is

$$d\Omega=\frac{d\Omega}{dr}\,dr+Rdt\,.\sec y\,\frac{d\Omega}{dv}-Qdt\,\frac{d\Omega}{dy},$$

or, substituting for Q, R, their values, the expression is

$$\begin{aligned}
d\Omega= &\ \frac{d\Omega}{dr}\,dr\\
&+\sec y\frac{d\Omega}{dv}\left\{\begin{array}{l}
\cos y\, dv\\
-\sin y\,[\sin(v-\Sigma)\sin\Phi d\Theta+\cos(v-\Sigma)\,d\Phi]\\
+\cos y\,[\quad -d\Sigma+\cos\Phi d\Theta\qquad\qquad]\\
+\{-\sin y\cos(v-\Sigma)\qquad\qquad\quad\}\,[\sin(\Theta-\sigma')\sin\phi' d\theta'+\cos(\Theta-\sigma')\,d\phi']\\
+\{-\sin y\sin(v-\Sigma)\cos\Phi-\cos y\sin\Phi\}\,[\cos(\Theta-\sigma')\sin\phi' d\theta'-\sin(\Theta-\sigma')\,d\phi']\\
+\{-\sin y\sin(v-\Sigma)\sin\Phi+\cos y\cos\Phi\}\,[\quad -d\sigma'+\cos\phi' d\theta'\qquad\qquad]
\end{array}\right\}\\
&+\frac{d\Omega}{dy}\left\{\begin{array}{l}
-dy\\
+[\cos(v-\Sigma)\sin\Phi d\Theta-\sin(v-\Sigma)\,d\Phi]\\
\qquad\qquad -\sin(v-\Sigma)\qquad\quad [\sin(\Theta-\sigma')\sin\phi' d\theta'+\cos(\Theta-\sigma')\,d\phi']\\
\qquad\qquad +\cos(v-\Sigma)\cos\Phi\,[\cos(\Theta-\sigma')\sin\phi' d\theta'-\sin(\Theta-\sigma')\,d\phi']\\
\qquad\qquad +\cos(v-\Sigma)\sin\Phi\,[\quad -d\sigma'+\cos\phi' d\theta'\qquad\qquad]
\end{array}\right\}
\end{aligned}$$

where, on the right-hand side, the terms involving the differentials of r, v, y, are, as they should be, $\frac{d\Omega}{dr}\,dr+\frac{d\Omega}{dv}\,dv+\frac{d\Omega}{dy}\,dy$.

I have given the complete expression, as it may be useful; but for the present purpose it will be sufficient to attend to the terms involving Θ, Σ, Φ. We have

$$\begin{aligned}
&\frac{d\Omega}{d\Theta}\,d\Theta+\frac{d\Omega}{d\Sigma}\,d\Sigma+\frac{d\Omega}{d\Phi}\,d\Phi\\
&\qquad=\sec y\,\frac{d\Omega}{dv}\left\{\begin{array}{l}
-\sin y\,[\sin(v-\Sigma)\sin\Phi\, d\Theta+\cos(v-\Sigma)\,d\Phi]\\
+\cos y\,[\quad -d\Sigma+\cos\Phi d\Theta\qquad\qquad\qquad]
\end{array}\right\}\\
&\qquad\quad -\frac{d\Omega}{dy}\left\{\cos(v-\Sigma)\sin\Phi\, d\Theta-\sin(v-\Sigma)\,d\Phi\right\}.
\end{aligned}$$

and consequently

$$\frac{d\Omega}{d\Sigma} = \qquad\qquad\qquad -\frac{d\Omega}{dv},$$

$$\frac{d\Omega}{d\Theta} = \left\{\cos\Phi - \tan y \sin(v-\Sigma)\sin\Phi\right\}\frac{d\Omega}{dv} - \cos(v-\Sigma)\sin\Phi\,\frac{d\Omega}{dy},$$

$$\frac{d\Omega}{d\Phi} = \qquad -\tan y\cos(v-\Sigma)\frac{d\Omega}{dv} + \sin(v-\Sigma)\qquad\frac{d\Omega}{dy};$$

which, it is to be observed, are partial differential equations satisfied by Ω considered as a function of the form $\Omega = \Omega(r, v, y, \Theta, \Sigma, \Phi)$; Ω is, in fact, a function of three quantities only (say the coordinates x_0, y_0, z_0); and it is clear, *à priori*, that when thus expressed as a function of six quantities, it must satisfy three partial differential equations.

I write in the second equation $-\frac{d\Omega}{d\Sigma}$ for $\frac{d\Omega}{dv}$, and I then put $y=0$; we thus have

$$\cos(v-\Sigma)\frac{d\Omega}{dy} = -\cot\Phi\frac{d\Omega}{d\Sigma} - \operatorname{cosec}\Phi\frac{d\Omega}{d\Theta},$$

$$\sin(v-\Sigma)\frac{d\Omega}{dy} = \qquad\frac{d\Omega}{d\Phi}.$$

I propose, as already mentioned, to apply the formulæ to the demonstration of the results obtained in my Memoir on the Problem of Disturbed Elliptic Motion. It will be remembered that x_0y_0 and x_0 denote a fixed orbit and fixed origin therein, x_1y_1 and x_1 an arbitrarily varying orbit and origin therein. I assume, however, that x_1 is a departure-point, so that we have $-d\sigma' + \cos\phi' d\theta' = 0$. The orbit x_2y_2 will be taken to be the varying instantaneous orbit of the body itself, that is, we have constantly $y=0$; and it is assumed that x_2 is a departure-point in this orbit. To conform to my former notation, I write þ instead of v; the position of the body in the varying instantaneous orbit is consequently determined by

r, the radius vector,

þ, the departure.

The entire displacement of the body arises from the displacements dr and rdþ in the direction of the radius vector, and perpendicular to this direction, in the instantaneous orbit; that is, we have $Q=0$, $R=d$þ, and the expression for the *Vis viva* function is simply

$$Tdt^2 = \tfrac{1}{2}(dr^2 + r^2 d\text{þ}^2),$$

and, putting in the foregoing expressions of Q and R, $y=0$, and þ in the place of v, the equations $Q=0$, $Rdt=d$þ, give

$$\cos(\text{þ}-\Sigma)\sin\phi\, d\Theta - \sin(\text{þ}-\Sigma)\,d\Phi =$$

$$\sin(\text{þ}-\Sigma)\qquad[\sin(\Theta-\sigma')\sin\phi' d\theta' + \cos(\Theta-\sigma')\,d\phi']$$

$$-\cos(\text{þ}-\Sigma)\cos\Phi\,[\cos(\Theta-\sigma')\sin\phi' d\theta' - \sin(\Theta-\sigma')\,d\phi'],$$

$$d\Sigma - \cos\Phi\, d\Theta =$$

$$-\sin\Phi\,[\cos(\Theta-\sigma')\sin\phi' d\theta' - \sin(\Theta-\sigma')\,d\phi'].$$

Now considering r, þ, y as the coordinates (y being ultimately put equal to zero) the general equations of motion are

$$\frac{d}{dt}\left(\frac{dT}{d\dot{r}}\right)-\frac{dT}{dr}=\frac{dV}{dr},$$

$$\frac{d}{dt}\left(\frac{dT}{d\dot{\text{þ}}}\right)-\frac{dT}{d\text{þ}}=\frac{dV}{d\text{þ}},$$

$$\frac{d}{dt}\left(\frac{dT}{d\dot{y}}\right)-\frac{dT}{dy}=\frac{dV}{dy},$$

if for the moment $\dot{r}$, $\dot{\text{þ}}$, $\dot{y}$ denote the differential coefficients of r, þ, y: the expression of the force function V is $V=-\frac{n^2a^3}{r}+\Omega$ where n^2a^3 denotes an absolute constant, the sum of the masses, and the disturbing function Ω has a contrary sign to that of the *Mécanique Céleste.* We may in the first and second equations write at once $Tdt^2=\frac{1}{2}(dr^2+r^2d\text{þ}^2)$, and these equations thus become

$$\frac{d}{dt}\frac{dr}{dt}-r\left(\frac{d\text{þ}}{dt}\right)^2=\frac{n^2a^3}{r^2}+\frac{d\Omega}{dr},$$

$$\frac{d}{dt}\left(r^2\frac{d\text{þ}}{dt}\right)=\frac{d\Omega}{d\text{þ}}.$$

If in the third equation we take the general value of T and after the differentiation with respect to $\dot{y}$, make the substitutions $y=0$, $Q=0$, $Rdt=d\text{þ}$, we find

$$\frac{dT}{d\dot{y}}=0,$$

and the equation thus becomes

$$-\frac{dT}{dy}=\frac{d\Omega}{dy}.$$

But $Tdt^2=\frac{1}{2}\{dr^2+r^2(Q^2+R^2)\,dt^2\}$, consequently

$$-\frac{d\Omega}{dy}dt=\frac{dT}{dy}dt=r^2\left(Q\frac{dQ}{dy}+R\frac{dR}{dy}\right)dt=r^2d\text{þ}\,\frac{dR}{dy},$$

and from the value of R, putting after the differentiation $y=0$, we find

$$\begin{aligned}\frac{dR}{dy}dt=-\;&[\sin(\text{þ}-\Sigma)\sin\Phi\,d\Theta+\cos(\text{þ}-\Sigma)\,d\Phi]\\ -\cos(\text{þ}-\Sigma)\;&[\sin(\Theta-\sigma')\sin\phi'\,d\theta'+\cos(\Theta-\sigma')\,d\phi']\\ -\sin(\text{þ}-\Sigma)\cos\Phi\;&[\cos(\Theta-\sigma')\sin\phi'\,d\theta-\sin(\Theta-\sigma')\,d\phi'].\end{aligned}$$

I write $r^2 \frac{d\mathfrak{p}}{dt} = h$. We thus have

$$\frac{d}{dt}\frac{dr}{dt} - r\left(\frac{d\mathfrak{p}}{dt}\right)^2 = \frac{n^2a^3}{r^2} + \frac{d\Omega}{dr},$$

$$r^2 \frac{d\mathfrak{p}}{dt} = h,$$

and

$$dh = \frac{d\Omega}{d\mathfrak{p}}\, dt;$$

the remaining equations are

$$d\Sigma - \cos\Phi\, d\Theta = \qquad - \sin\Phi\,[\cos(\Theta - \sigma')\sin\phi'\, d\theta' - \sin(\Theta - \sigma')\, d\phi'],$$

$$\cos(\mathfrak{p} - \Sigma)\sin\Phi\, d\Theta - \sin(\mathfrak{p} - \Sigma)\, d\Phi =$$
$$\sin(\mathfrak{p} - \Sigma) \qquad [\sin(\Theta - \sigma')\sin\phi'\, d\theta' + \cos(\Theta - \sigma')\, d\phi']$$
$$- \cos(\mathfrak{p} - \Sigma)\cos\Phi\,[\cos(\Theta - \sigma')\sin\phi'\, d\theta' - \sin(\Theta - \sigma')\, d\phi'],$$

$$\sin(\mathfrak{p} - \Sigma)\sin\Phi\, d\Theta + \cos(\mathfrak{p} - \Sigma)\, d\Phi =$$
$$\frac{1}{h}\frac{d\Omega}{dy}\, dt$$
$$- \cos(\mathfrak{p} - \Sigma) \qquad [\sin(\Theta - \sigma')\sin\phi'\, d\theta' + \cos(\Theta - \sigma')\, d\phi']$$
$$- \sin(\mathfrak{p} - \Sigma)\cos\Phi\,[\cos(\Theta - \sigma')\sin\phi'\, d\theta' - \sin(\Theta - \sigma')\, d\phi'].$$

Moreover we have

$$\cos(\mathfrak{p} - \Sigma)\frac{d\Omega}{dy} = - \cot\phi\frac{d\Omega}{d\Sigma} - \operatorname{cosec}\Phi\frac{d\Omega}{d\Theta},$$

$$\sin(\mathfrak{p} - \Sigma)\frac{d\Omega}{dy} = \qquad \frac{d\Omega}{d\Phi},$$

by means of which the preceding two equations are reduced to

$$d\Theta = \frac{\operatorname{cosec}\Phi}{h}\frac{d\Omega}{d\Phi}\, dt$$
$$- \cot\Phi\,[\cos(\Theta - \sigma')\sin\phi'\, d\theta' - \sin(\Theta - \sigma')\, d\phi'],$$

$$d\Phi = - \frac{\cot\Phi}{h}\frac{d\Omega}{d\Sigma}\, dt - \frac{\operatorname{cosec}\Phi}{h}\frac{d\Omega}{d\Theta}\, dt$$
$$- \qquad [\sin(\Theta - \sigma')\sin\phi'\, d\theta' + \cos(\Theta - \sigma')\, d\phi'];$$

and we then have

$$d\Sigma = \frac{\cot\Phi}{h}\frac{d\Omega}{d\Phi}\, dt$$
$$- \operatorname{cosec}\Phi\,[\cos(\Theta - \sigma')\sin\phi'\, d\theta' - \sin(\Theta - \sigma')\, d\phi'].$$

The entire system consequently is

$$\frac{d}{dt}\left(\frac{dr}{dt}\right) - r\left(\frac{d\mathfrak{p}}{dt}\right)^2 = \frac{n^2a^3}{r^2} + \frac{d\Omega}{dr},$$

$$r^2\frac{d\mathfrak{p}}{dt} = h;$$

and

$$dh = \frac{d\Omega}{d\mathfrak{p}}dt,$$

$$d\Theta = \frac{\operatorname{cosec}\Phi}{h}\frac{d\Omega}{d\Phi}dt - \cot\Phi\,[\cos(\Theta - \sigma')\sin\phi' d\theta' - \sin(\Theta - \sigma')\,d\phi'],$$

$$d\Sigma = \frac{\cot\Phi}{h}\frac{d\Omega}{d\Phi}dt - \operatorname{cosec}\Phi\,[\cos(\Theta - \sigma')\sin\phi' d\theta' - \sin(\Theta - \sigma')\,d\phi'],$$

$$d\Phi = -\frac{\operatorname{cosec}\Phi}{h}\frac{d\Omega}{d\Theta}dt - \frac{\cot\Phi}{h}\frac{d\Omega}{d\Sigma}dt - [\sin(\Theta - \sigma')\sin\phi' d\theta' + \cos(\Theta - \sigma')\,d\phi'],$$

where $\Omega = \Omega(r, \mathfrak{p}, \Theta, \Sigma, \Phi)$. We may add the before-mentioned equation,

$$d\Sigma - \cos\Phi\, d\Theta = - \sin\Phi\,[\cos(\Theta - \sigma')\sin\phi' d\theta' - \sin(\Theta - \sigma')\,d\phi'].$$

To obtain the formulæ of my Memoir before referred to, it would only be necessary to write $h = na^2\sqrt{1 - e^2}$, and complete the solution by applying to the first equation the method of the variation of the elements; but I prefer leaving the system in its present form, as it is one which may perhaps be useful.

The formulæ have especial reference to the lunar theory, and the orbit x_0y_0 denotes the fixed ecliptic, x_1y_1 the varying ecliptic, and x_2y_2 the instantaneous orbit of the moon. But if, instead of considering with Hansen the instantaneous orbit of the moon, the position of the moon is determined (as in the theories of Clairaut, Plana, and others) by

r, the radius vector,
v, the longitude,
y, the latitude,

where the longitude is measured on the variable ecliptic, then if, in the general expressions for P, Q, R, we first neglect the terms involving $d\theta'$, $d\sigma'$, $d\phi'$, and then write θ', σ', ϕ', in the place of Θ, Σ, Φ, we find

$$\begin{aligned} Pdt = {} & \sin y\, dv \\ & + \cos y\,[\sin(v - \sigma')\sin\phi' d\theta' + \cos(v - \sigma')\,d\phi'] \\ & + \sin y\,[-d\sigma' + \cos\phi' d\theta'], \\ Qdt = {} & -dy \\ & + [\cos(v - \sigma')\sin\phi' d\theta' - \sin(v - \sigma')\,d\phi'], \end{aligned}$$

$$\begin{aligned} Rdt = \cos y\, dv & \\ - \sin y\, & [\sin (v-\sigma') \sin \phi' d\theta' + \cos (v-\sigma')\, d\phi'] \\ + \cos y\, & [-d\sigma' + \cos \phi'\, d\theta' \qquad\qquad\qquad], \end{aligned}$$

and with these values the expression for the *Vis viva* function is, as before,

$$Tdt^2 = \tfrac{1}{2} \{dr^2 + r^2 (Q^2 + R^2)\, dt^2\}.$$

The longitude may be measured from a departure-point, and then the expressions for P, Q, R, may be simplified by omitting the terms which contain $-d\sigma' + \cos \phi' d\theta'$.

Addition.

I have in the course of the memoir used some properties connected with the theory of rotation, but it is proper to give the analytical investigation.

Suppose that the rectangular coordinates (X, Y, Z) are connected with the rectangular coordinates $(X_,, Y_,, Z_,)$ by the equations

$$\begin{aligned} X &= \alpha\, X_, + \beta\, Y_, + \gamma\, Z_,, \\ Y &= \alpha' X_, + \beta' Y_, + \gamma' Z_,, \\ Z &= \alpha'' X_, + \beta'' Y_, + \gamma'' Z_,, \end{aligned}$$

then we have

$$\begin{aligned} dX &= \alpha\, dX_, + \beta\, dY_, + \gamma\, dZ_, + X_, d\alpha + Y_, d\beta + Z_, d\gamma\,, \\ dY &= \alpha' dX_, + \beta' dY_, + \gamma' dZ_, + X_, d\alpha' + Y_, d\beta' + Z_, d\gamma'\,, \\ dZ &= \alpha'' dX_, + \beta'' dY_, + \gamma'' dZ_, + X_, d\alpha'' + Y_, d\beta'' + Z_, d\gamma'', \end{aligned}$$

and then, putting

$$\begin{aligned} \mathrm{p}dt &= \gamma d\beta + \gamma' d\beta' + \gamma'' d\beta'', \\ \mathrm{q}dt &= \alpha d\gamma + \alpha' d\gamma' + \alpha'' d\gamma'', \\ \mathrm{r}dt &= \beta d\alpha + \beta' d\alpha' + \beta'' d\alpha'', \end{aligned}$$

where p, q, r are the rotations round $X_,$, $Y_,$, $Z_,$, from $Y_,$ to $Z_,$, $Z_,$ to $X_,$, and $X_,$ to $Y_,$ respectively, we have

$$\begin{aligned} \alpha dX + \alpha' dY + \alpha'' dZ &= dX_, + (-\mathrm{r} Y_, + \mathrm{q} Z_,)\, dt, \\ \beta dX + \beta' dY + \beta'' dZ &= dY_, + (-\mathrm{p} Z_, + \mathrm{r} X_,)\, dt, \\ \gamma dX + \gamma' dY + \gamma'' dZ &= dZ_, + (-\mathrm{q} X_, + \mathrm{p} Y_,)\, dt, \end{aligned}$$

where, considering XYZ as fixed axes, the left-hand sides are the displacements in the directions of the moveable axes $X_, Y_, Z_,$ arising from the variations of the coordinates $X_, Y_, Z_,$ and the variation of the position of these axes; and these displacements have therefore the values given by the right-hand sides. This is, of course, well known. It may be added that if the axes XYZ are moveable, and dX, dY, dZ denote the displacements in the directions of these axes, the formulæ are still true.

Suppose that Ω is a function of X, Y, Z; say $\Omega = \Omega(X, Y, Z)$, we have

$$d\Omega = \frac{d\Omega}{dX}dX + \frac{d\Omega}{dY}dY + \frac{d\Omega}{dZ}dZ,$$

which may be written

$$\begin{aligned} d\Omega = & \left(\alpha\frac{d\Omega}{dX} + \alpha'\frac{d\Omega}{dY} + \alpha''\frac{d\Omega}{dZ}\right)(\alpha dX + \alpha' dY + \alpha'' dZ) \\ & + \left(\beta\frac{d\Omega}{dX} + \beta'\frac{d\Omega}{dY} + \beta''\frac{d\Omega}{dZ}\right)(\beta dX + \beta' dY + \beta'' dZ) \\ & + \left(\gamma\frac{d\Omega}{dX} + \gamma'\frac{d\Omega}{dY} + \gamma''\frac{d\Omega}{dZ}\right)(\gamma dX + \gamma' dY + \gamma'' dZ), \end{aligned}$$

and the coefficients of transformation α, β, γ, &c., being independent of X, Y, Z and $X_,$, $Y_,$, $Z_,$, the first factors are $\frac{d\Omega}{dX_,}$, $\frac{d\Omega}{dY_,}$, $\frac{d\Omega}{dZ_,}$, and we have

$$\begin{aligned} d\Omega = & \frac{d\Omega}{dX_,}[dX_, + (-\mathrm{r}Y_, + \mathrm{q}Z_,)\,dt] \\ & + \frac{d\Omega}{dY_,}[dY_, + (-\mathrm{p}Z + \mathrm{r}X_,)\,dt] \\ & + \frac{d\Omega}{dZ}[dZ_, + (-\mathrm{q}X + \mathrm{p}Y_,)\,dt], \end{aligned}$$

which is a theorem assumed in the memoir.

The expressions of P, Q, R in the memoir depend upon the composition of several rotations. If, in Lamé's very convenient notation, we write

	$X_,$	$Y_,$	$Z_,$
X	α	β	γ
Y	α'	β'	γ'
Z	α''	β''	γ''

to denote the before-mentioned relation between (X, Y, Z) and $(X_{,}, Y_{,}, Z_{,})$ then the tables

	$X_{,}$	$Y_{,}$	$Z_{,}$
X	α	β	γ
Y	α'	β'	γ'
Z	α''	β''	γ''

,

	X_2	Y_2	Z_2
$X_{,}$	A	B	C
$Y_{,}$	A'	B'	C'
$Z_{,}$	A''	B''	C''

lead to the table

	X_2	Y_2	Z_2
X	$(\alpha\,,\ \beta\,,\ \gamma\,)(A,\ A',\ A'')$	$(\alpha\,,\ \beta\,,\ \gamma\,)(B,\ B',\ B'')$	$(\alpha\,,\ \beta\,,\ \gamma\,)(C,\ C',\ C'')$
Y	$(\alpha',\ \beta',\ \gamma')(A,\ A',\ A'')$	$(\alpha',\ \beta',\ \gamma')(B,\ B',\ B'')$	$(\alpha',\ \beta',\ \gamma')(C,\ C',\ C'')$
Z	$(\alpha'',\ \beta'',\ \gamma'')(A,\ A',\ A'')$	$(\alpha'',\ \beta'',\ \gamma'')(B,\ B',\ B'')$	$(\alpha'',\ \beta'',\ \gamma'')(C,\ C',\ C'')$

where for shortness $(\alpha, \beta, \gamma)(A, A', A'')$, &c., stand for $\alpha A + \beta A' + \gamma A''$, &c. these are of course the ordinary formulæ for the composition of transformations from one set of rectangular axes to another. We may say that the coefficients of transformation from (X, Y, Z) to $(X_{,}, Y_{,}, Z_{,})$ are $(\alpha, \beta, \gamma$, &c.). And in dealing with several sets of coordinates, the coefficients of transformation may be distinguished by suffixes. Thus I assume that the coefficients of transformation

from (x_0, y_0, z_0) to (x_1, y_1, z_1) are $(\alpha, \beta, \gamma, \&c.)_{01}$,
" (x_1, y_1, z_1) " (x_2, y_2, z_2) " $(\alpha, \beta, \gamma, \&c.)_{12}$,
" (x_0, y_0, z_0) " (x_2, y_2, z_2) " $(\alpha, \beta, \gamma, \&c.)_{02}$,

where $(\alpha, \beta, \gamma, \&c.)_{02}$ can be obtained as above from $(\alpha, \beta, \gamma, \&c.)_{01}$ and $(\alpha, \beta, \gamma, \&c.)_{12}$. The rotations p, q, r may be distinguished by suffixes in like manner, viz., $(\mathrm{p}, \mathrm{q}, \mathrm{r})_{01}$ will denote the rotations from (x_0, y_0, z_0) to (x_1, y_1, z_1) and so on.

There is no difficulty in obtaining first

$$\mathrm{p}_{02} = \mathrm{p}_{12} + (\alpha,\ \alpha',\ \alpha'')_{12}\ (\mathrm{p},\ \mathrm{q},\ \mathrm{r})_{01},$$
$$\mathrm{q}_{02} = \mathrm{q}_{12} + (\beta,\ \beta',\ \beta'')_{12}\ (\mathrm{p},\ \mathrm{q},\ \mathrm{r})_{01},$$
$$\mathrm{r}_{02} = \mathrm{r}_{12} + (\gamma,\ \gamma',\ \gamma'')_{12}\ (\mathrm{p},\ \mathrm{q},\ \mathrm{r})_{01},$$

and then by a repetition of the like transformation

$$p_{03} = p_{23} + (\alpha, \alpha', \alpha'')_{23} (p, q, r)_{12} + (\alpha, \alpha', \alpha'')_{13} (p, q, r)_{01},$$
$$q_{03} = q_{23} + (\beta, \beta', \beta'')_{23} (p, q, r)_{12} + (\beta, \beta', \beta'')_{13} (p, q, r)_{01},$$
$$r_{03} = r_{23} + (\gamma, \gamma', \gamma'')_{23} (p, q, r)_{12} + (\gamma, \gamma', \gamma'')_{13} (p, q, r)_{01},$$

and in like manner for any number of systems, but for the present purpose this is enough, as p_{03}, q_{03}, r_{03} are in fact the P, Q, R of the memoir. And if θ', σ', ϕ', Θ, Σ, Φ, v, y signify as before, then the coefficients of transformation from (x_0, y_0, z_0) to (x_1, y_1, z_1) are given by the table

	x_1	y_1	z_1
x_0	$\cos\sigma' \cos\theta' + \sin\sigma' \sin\theta' \cos\phi'$	$\sin\sigma' \cos\theta' - \cos\sigma' \sin\theta' \cos\phi'$	$\sin\theta' \sin\phi'$
y_0	$\cos\sigma' \sin\theta' - \sin\sigma' \cos\theta' \cos\phi'$	$\sin\sigma' \sin\theta' + \cos\sigma' \cos\theta' \cos\phi'$	$-\cos\theta' \sin\phi'$
z_0	$-\sin\sigma' \sin\phi'$	$\cos\sigma' \sin\phi'$	$\cos\phi'$

and the corresponding rotation coefficients are

$$p_{01}\, dt = -\sin\sigma' \sin\phi'\, d\theta' + \cos\sigma'\, d\phi',$$
$$q_{01}\, dt = \cos\sigma' \sin\phi'\, d\theta' + \sin\sigma'\, d\phi',$$
$$r_{01}\, dt = -d\sigma' + \cos\phi'\, d\theta'.$$

The coefficients of transformation from (x_1, y_1, z_1) to (x_2, y_2, z_2) are given by the table

	x_2	y_2	z_2
x_1	$\cos\Sigma \cos\Theta + \sin\Sigma \sin\Theta \cos\Phi$	$\sin\Sigma \cos\Theta - \cos\Sigma \sin\Theta \cos\Phi$	$\sin\Theta \sin\Phi$
y_1	$\cos\Sigma \sin\Theta - \sin\Sigma \cos\Theta \cos\Phi$	$\sin\Sigma \sin\Theta + \cos\Sigma \cos\Theta \cos\Phi$	$-\cos\Theta \sin\Phi$
z_1	$-\sin\Sigma \sin\Phi$	$\cos\Sigma \sin\Phi$	$\cos\Phi$

and the corresponding rotation coefficients are

$$p_{12}\, dt = -\sin\Sigma \sin\Phi\, d\Theta + \cos\Sigma\, d\Phi,$$
$$q_{12}\, dt = +\cos\Sigma \sin\Phi\, d\Theta + \sin\Sigma\, d\Phi,$$
$$r_{12}\, dt = -d\Sigma + \cos\Phi\, d\Theta,$$

and the coefficients of transformation from (x_2, y_2, z_2) to (x_3, y_3, z_3) are given by the table

	x_3	y_3	z_3
x_2	$\cos v \cos y$	$-\sin v$	$-\cos v \sin y$
y_2	$\sin v \cos y$	$\cos v$	$-\sin v \sin y$
z_2	$\sin y$	0	$\cos y$

and the corresponding rotation coefficients are given by

$$\mathrm{p}_{23}\, dt = -\sin y\, dv,$$
$$\mathrm{q}_{23}\, dt = -\qquad dy,$$
$$\mathrm{r}_{23}\, dt = \cos y\, dv.$$

From the last two tables it is easy to deduce the coefficients of transformation from (x_1, y_1, z_1) to (x_3, y_3, z_3), these are given by the table

	x_3	y_3
x_1	$\cos y \{\cos(v-\Sigma)\cos\Theta - \sin(v-\Sigma)\sin\Theta\cos\Phi\} + \sin y \sin\Theta\sin\Phi$	$-\sin(v-\Sigma)\cos\Theta - \cos(v-\Sigma)\sin\Theta\cos\Phi$
y_1	$\cos y \{\cos(v-\Sigma)\sin\Theta + \sin(v-\Sigma)\cos\Theta\cos\Phi\} - \sin y \cos\Theta\sin\Phi$	$-\sin(v-\Sigma)\sin\Theta + \cos(v-\Sigma)\cos\Theta\cos\Phi$
z_1	$\cos y \{\sin(v-\Sigma)\sin\Phi \qquad\qquad \} + \sin y\cos\Phi$	$\cos(v-\Sigma)\sin\Phi$

	z_3
x_1	$-\sin y \{\cos(v-\Sigma)\cos\Theta - \sin(v-\Sigma)\sin\Theta\cos\Phi\} + \cos y \sin\Theta\sin\Phi$
y_1	$-\sin y \{\cos(v-\Sigma)\sin\Theta + \sin(v-\Sigma)\cos\Theta\cos\Phi\} - \cos y\cos\Theta\sin\Phi$
z_1	$-\sin y \{\sin(v-\Sigma)\sin\Phi \qquad\qquad \} + \cos y\cos\Phi$

It is now easy to form the values of P, Q, R. We have

$$
\begin{aligned}
Pdt &= \sin y\, dv \\
&+ [\cos y \{\cos(v-\Sigma)\cos\Theta - \sin(v-\Sigma)\sin\Theta\cos\Phi\} + \sin y \sin\Theta\sin\Phi](-\sin\Sigma\sin\Phi\, d\Theta + \cos\Sigma\, d\Phi) \\
&+ [\cos y \{\cos(v-\Sigma)\sin\Theta + \sin(v-\Sigma)\cos\Theta\cos\Phi\} - \sin y\cos\Theta\sin\Sigma](\cos\Sigma\sin\Phi\, d\Theta + \sin\Sigma\, d\Phi) \\
&+ [\cos y \{\sin(v-\Sigma)\sin\Phi \qquad\qquad \} + \sin y\cos\Phi \quad](-d\Sigma + \cos\Phi\, d\Theta \qquad)
\end{aligned}
$$

$$
\begin{aligned}
&+\cos v\cos y\,(-\sin\sigma'\sin\phi'\,d\theta'+\cos\sigma'\,d\phi')\\
&+\sin v\cos y\,(\ \ \cos\sigma'\sin\phi'\,d\theta'+\sin\sigma'\,d\phi')\\
&+\qquad\ \sin y\,(-d\sigma'+\cos\phi'\,d\theta'\qquad\qquad),
\end{aligned}
$$

$$
\begin{aligned}
Qdt = &-dy\\
&+[-\sin(v-\Sigma)\cos\Theta-\cos(v-\Sigma)\sin\Theta\cos\Phi](-\sin\Sigma\sin\Phi\,d\Theta+\cos\Sigma\,d\Phi)\\
&+[-\sin(v-\Sigma)\sin\Theta+\cos(v-\Sigma)\cos\Theta\cos\Phi](\ \ \cos\Sigma\sin\Phi\,d\Theta+\sin\Sigma\,d\Phi)\\
&+[\ \ \cos(v-\Sigma)\sin\Phi\qquad\qquad\qquad\qquad\qquad](-d\Sigma+\cos\Phi\,d\Theta\qquad\qquad)\\
&\qquad\qquad-\sin v\,(-\sin\Sigma'\sin\phi'\,d\theta'+\cos\sigma'\,d\phi')\\
&\qquad\qquad+\cos v\,(\ \ \cos\Sigma'\sin\phi'\,d\theta'+\sin\sigma'\,d\phi')\\
&\qquad\qquad+\ \ 0\ \ (-d\Sigma'+\cos\phi'\,d\theta'\qquad\qquad),
\end{aligned}
$$

$$
\begin{aligned}
Rdt = &\cos y\,dv\\
&+[-\sin y\{\cos(v-\Sigma)\cos\Theta-\sin(v-\Sigma)\sin\Theta\cos\Phi\}+\cos y\sin\Theta\sin\Phi](-\sin\Sigma\sin\Phi\,d\Theta+\cos\Sigma\,d\Phi)\\
&+[-\sin y\{\cos(v-\Sigma)\sin\Theta+\sin(v-\Sigma)\cos\Theta\cos\Phi\}-\cos y\cos\Theta\sin\Phi](\ \ \cos\Sigma\sin\Phi\,d\Theta+\sin\Sigma\,d\Phi)\\
&+[-\sin y\{\sin(v-\Sigma)\sin\Phi\qquad\qquad\qquad\}+\cos y\cos\Phi\ ](-d\Sigma+\cos\Phi\,d\Theta\qquad)\\
&\qquad\qquad-\cos v\sin y\,(-\sin\sigma'\sin\phi'\,d\theta'+\cos\sigma'\,d\phi')\\
&\qquad\qquad-\sin v\sin y\,(\ \ \cos\sigma'\sin\phi'\,d\theta'+\sin\sigma'\,d\phi')\\
&\qquad\qquad+\cos y\qquad\ (-d\sigma'+\cos\phi'\,d\theta'\qquad\qquad),
\end{aligned}
$$

which are at once reducible into the before-mentioned form.

It only remains to prove the expressions for $\frac{d\Omega}{dx_3}, \frac{d\Omega}{dy_3}, \frac{d\Omega}{dz_3}$ assumed in the memoir. The relations between (x_3, y_3, z_3) and (x_2, y_2, z_2) give

$$
\begin{aligned}
\frac{d\Omega}{dx_3} &= \cos v\cos y\,\frac{d\Omega}{dx_2}+\sin v\cos y\,\frac{d\Omega}{dy_2}+\sin y\,\frac{d\Omega}{dz_2},\\
\frac{d\Omega}{dy_3} &= -\sin v\,\frac{d\Omega}{dx_2}+\cos v\,\frac{d\Omega}{dy_2},\\
\frac{d\Omega}{dz_3} &= -\cos v\sin y\,\frac{d\Omega}{dx_2}-\sin v\sin y\,\frac{d\Omega}{dy_2}+\cos y\,\frac{d\Omega}{dz_2},
\end{aligned}
$$

these formulæ apply to any point whatever the coordinates of which to the one set of axes are (x_2, y_2, z_2) and to the other set (x_3, y_3, z_3) and in obtaining them the coefficients of transformation v, y are of course considered as independent of either set of coordinates. But the coordinates (x_2, y_2, z_2) of the body are

$$
\begin{aligned}
x_2 &= r\cos y\cos v,\\
y_2 &= r\cos y\sin v,\\
z_2 &= r\sin y;
\end{aligned}
$$

and using the foregoing equations in order to introduce into the formulæ the differential coefficients $\frac{d\Omega}{dr}$, $\frac{d\Omega}{dv}$, $\frac{d\Omega}{dy}$ in the place of $\frac{d\Omega}{dx_2}$, $\frac{d\Omega}{dy_2}$, $\frac{d\Omega}{dz_2}$, we find

$$\frac{d\Omega}{dx_2} = \cos v \cos y \frac{d\Omega}{dr} - \frac{1}{r} \sin v \sec y \frac{d\Omega}{dv} - \frac{1}{r} \cos v \sin y \frac{d\Omega}{dy},$$

$$\frac{d\Omega}{dy_2} = \sin v \cos y \frac{d\Omega}{dr} + \frac{1}{r} \cos v \sec y \frac{d\Omega}{dv} - \frac{1}{r} \sin v \sin y \frac{d\Omega}{dy},$$

$$\frac{d\Omega}{dz_2} = \sin y \frac{d\Omega}{dr} + \frac{1}{r} \cos y \frac{d\Omega}{dy}.$$

Substituting these values we have the required formulæ

$$\frac{d\Omega}{dx_3} = \frac{d\Omega}{dr}, \quad \frac{d\Omega}{dy_3} = \frac{1}{r} \sec y \frac{d\Omega}{dv}, \quad \frac{d\Omega}{dz_3} = \frac{1}{r} \frac{d\Omega}{dy}$$

where on the right-hand side $\Omega = \Omega(r, v, y, \Theta, \Sigma, \Phi, \theta', \sigma', \phi')$.

2, *Stone Buildings, W.C.* 28*th Dec.*, 1858.

216.

TABLES OF THE DEVELOPMENTS OF FUNCTIONS IN THE THEORY OF ELLIPTIC MOTION.

[From the *Memoirs of the Royal Astronomical Society*, vol. XXIX. (1861), pp. 191—306. Read January 12, 1859.]

THE notation made use of is as follows; viz., r, f, a, e, g denote

r, the radius vector;
f, the true anomaly;
a, the mean distance;
e, the eccentricity;
g, the mean anomaly;

so that

$$\frac{r}{a}, = \text{elqr}\ (e, g),$$

and

$$f, = \text{elta}\ (e, g),$$

are known functions of e, g. Moreover, x denotes the periodic part of the quotient radius $\frac{r}{a}$, and y the equation of the centre, or periodic part of the true anomaly f; so that

$$\frac{r}{a} = 1 + x,$$

$$f = g + y,$$

and (x, y) are therefore also known functions of e, g.

Formulæ for the development in multiple cosines or sines, up to the terms in e^7, of the functions

$$(x^0, x^1, \ldots x^7) \frac{\cos}{\sin} jy,$$

where j is an indeterminate symbol, are given by Leverrier in the *Annales de l'Observatoire de Paris*, tom. I. (1855), pp. 346—348. It occurred to me that it would be desirable to form tables such as are contained in the present memoir, of various functions of the forms

$$x^m \frac{\cos}{\sin} jf, \quad \left(\frac{r}{a}\right)^{\pm m} \frac{\cos}{\sin} jf,$$

obtainable from Leverrier's formulæ by assigning particular integer values to the symbol j. The calculations were performed under my superintendence by Messrs Creedy and Davis; and I have to acknowledge my obligation for a grant to defray the expenses, made to me out of the Donation Fund at the disposal of the Council of the Royal Society.

A cosine series is in general represented in the form

$$\Sigma\, [\cos]^i \cos ig,$$

where i extends from $-\infty$ to $+\infty$, and the coefficients $[\cos]^i$ satisfy the condition

$$[\cos]^{-i} = [\cos]^i;$$

and in like manner a sine series is represented in the form

$$\Sigma\, [\sin]^i \sin ig,$$

where i extends from $-\infty$ to $+\infty$, and the coefficients $[\sin]^i$ satisfy the condition

$$[\sin]^{-i} = -[\sin]^i, \text{ and in particular}$$
$$[\sin]^0 = 0.$$

Thus $[\cos]^i$ is a general symbol denoting the coefficient of $\cos ig$, in a cosine series considered as involving as well negative as positive multiples of the argument g, or, what is the same thing, $[\cos]^0$ is the constant term and $[\cos]^i$ is the half coefficient of $\cos ig$, when the series is considered as containing only positive multiples of the argument. And in like manner $[\sin]^i$ is a general symbol for the coefficient of $\sin ig$, in a sine series considered as involving as well negative as positive multiples of the argument g, or, what is the same thing, $[\sin]^i$ is the half coefficient of $\sin ig$, when the series is considered as containing only positive multiples of the argument.

In the case of a pair of corresponding functions $x^m \cos jf$ and $x^m \sin jf$, or $\left(\frac{r}{a}\right)^{\pm m} \cos jf$ and $\left(\frac{r}{a}\right)^{\pm m} \sin jf$, one of them expanded in the form

$$\Sigma\, [\cos]^i \cos ig,$$

and the other in the form

$$\Sigma\, [\sin]^i \sin ig,$$

the sums and differences of the corresponding coefficients $[\cos]^i$, $[\sin]^i$, are represented by the notation

$$[\cos \pm \sin]^i,$$

and it is clear that we have

$$[\cos + \sin]^{-i} = [\cos - \sin]^i, \quad [\cos \pm \sin]^0 = [\cos]^0.$$

These coefficients $[\cos \pm \sin]^i$ are for many purposes equally useful with the coefficients $[\cos]^i$, $[\sin]^i$, and they are here tabulated accordingly.

The functions tabulated are as follows:—

$$\left(x^1,\ x^2, \ldots .\, x^7\right) \qquad , [\cos],$$

$$\left(x^0,\ x^1,\ x^2, \ldots .\, x^7\right) \begin{matrix}\cos\\ \sin\end{matrix} jf,\ j=1 \text{ to } j=7,\ [\cos], [\sin], [\cos \pm \sin],$$

$$\left(\left(\frac{r}{a}\right)^{+4}, \ldots \left(\frac{r}{a}\right)^{+1}, \log \frac{r}{a}, \left(\frac{r}{a}\right)^{-1}, \ldots \left(\frac{r}{a}\right)^{-5}\right) \qquad , [\cos],$$

$$\left(\left(\frac{r}{a}\right)^{+4}, \ldots \left(\frac{r}{a}\right)^{+1}, \qquad \left(\frac{r}{a}\right)^{-1}, \ldots \left(\frac{r}{a}\right)^{-5}\right) \begin{matrix}\cos\\ \sin\end{matrix} jf, j=1 \text{ to } j=5,\ [\cos], [\sin], [\cos \pm \sin],$$

all the developments being carried up to e^7, the limit of the formulæ from which they are deduced.

The calculations are based upon Leverrier's formulæ for

$$(x^0,\ x^1, \ldots \ldots x^7) \begin{matrix}\cos\\ \sin\end{matrix} jy,$$

which are made use of under the following forms, which I call the Datum Tables $(x^0 \ldots \ldots x^7) \begin{matrix}\cos\\ \sin\end{matrix} jy$.

Cosine Table. $\frac{1}{2}$ of	$[\ ]^0$ $\frac{e^0}{2}$	$\frac{e^2}{16}$	$\frac{e^4}{768}$	$\frac{e^6}{92160}$	$[\ ]^1$ $\frac{e}{4}$	$\frac{e^3}{96}$	$\frac{e^5}{7680}$	$\frac{e^7}{1290240}$
$x^0 \cos jy$	$+1$	$-8j^2$	$+96j^4$ $-54j^2$	$-1280j^6$ $+3920j^4$ $-3440j^2$	0	$-30j^2$	$+800j^4$ $-120j^2$	$-16800j^6$ $+30590j^4$ $-29470j^2$
$x \cos jy$		$+4$	$-48j^2$	$0j^4$ $-360j^2$	-1	$+12j^2$ $+9$	$-160j^4$ $-370j^2$ -50	$+2240j^6$ $-980j^4$ $+9695j^2$ $+245$
$x^2 \cos jy$		$+4$	$-96j^2$	$+1920j^4$ $-1320j^2$		-12	$-244j^2$ $+80$	$+16800j^4$ $-6300j^2$ -420
$x^3 \cos jy$			$+144$	$-2880j^2$		-18	$+480j^2$ $+360$	$-10080j^4$ $-22050j^2$ -3150
$x^4 \cos jy$			$+144$	$-5760j^2$			-960	$-5040j^2$ $+10080$
$x^5 \cos jy$				$+14400$			-1200	$+50400j^2$ $+37800$
$x^6 \cos jy$				$+14400$				-151200
$x^7 \cos jy$				0				-176400

Datum Table $(x_0 \ldots x^7) \cos jy \ldots$

Cosine Table. $\frac{1}{2}$ of	$[\ ]^2$			$[\ ]^2$		
	$\frac{e^2}{16}$	$\frac{e^4}{768}$	$\frac{e^6}{92160}$	$\frac{e^3}{96}$	$\frac{e^5}{7680}$	$\frac{e^7}{1290240}$
$x^0 \cos jy$	$+4j^2$	$-64j^4$ $-256j^2$	$+960j^6$ $+11820j^4$ $+4710j^2$	$+30j^2$	$-1200j^4$ $-3240j^2$	$+30240j^6$ $+243810j^4$ $+173502j^2$
$x \cos jy$	-2	$+192j^2$ $+64$	$-4800j^4$ $-8190j^2$ -1440	$-12j^2$ -9	$+240j^4$ $+2745j^2$ $+675$	$-4032j^6$ $-122220j^4$ $-225099j^2$ -35721
$x^2 \cos jy$	$+2$	$0j^2$ -96	$-480j^4$ $+1590j^2$ $+2400$	$+12$	$-840j^2$ -1080	$+10080j^4$ $+102060j^2$ $+61236$
$x^3 \cos jy$		0	$+7920j^2$ $+1440$	-6	$-240j^2$ $+540$	$+10080j^4$ $+85050j^2$ -17010
$x^4 \cos jy$		$+96$	$-1440j^2$ -8640		$+480$	$-95760j^2$ -90720
$x^5 \cos jy$			$+3600$		-600	$-10080j^2$ $+113400$
$x^6 \cos jy$			$+10800$			$+30240$
$x^7 \cos jy$			0			-105840

... Datum Table $(x^0 \dots x^7) \cos jy$ ...

Cosine Table. $\frac{1}{2}$ of	$[\ ]^4$ $\frac{e^4}{768}$	$[\ ]^4$ $\frac{e^6}{92160}$	$[\ ]^5$ $\frac{e^5}{7680}$	$[\ ]^5$ $\frac{e^7}{1290240}$	$[\ ]^6$ $\frac{e^6}{92160}$	$[\ ]^7$ $\frac{e^7}{1290240}$
$x^0 \cos jy$	$+16j^4$ $+283j^2$	$-384j^6$ $-22440j^4$ $-51528j^2$	$+400j^4$ $+3360j^2$	$-16800j^6$ $-467530j^4$ $-972790j^2$	$+64j^6$ $+8660j^4$ $+48538j^2$	$+3360j^6$ $+193130j^4$ $+828758j^2$
$x \cos jy$	$-168j^2$ -64	$+7680j^4$ $+46260j^2$ $+9216$	$-80j^4$ $-2375j^2$ -625	$+2240j^6$ $+205660j^4$ $+905135j^2$ $+153125$	$-2880j^4$ $-37890j^2$ -7776	$-448j^6$ $-82460j^4$ $-689731j^2$ -117649
$x^2 \cos jy$	$+48j^2$ $+96$	$-960j^4$ $-22380j^2$ -15360	$+1080j^2$ $+1000$	$-50400j^4$ $-540540j^2$ -262500	$+480j^4$ $+21450j^2$ $+12960$	$+23520j^4$ $+444780j^2$ $+201684$
$x^3 \cos jy$	-72	$+1440j^2$ $+11520$	$-240j^2$ -900	$+3360j^4$ $+147630j^2$ $+236250$	$-7920j^2$ -12960	$-3360j^4$ $-210630j^2$ -216090
$x^4 \cos jy$	$+24$	$+2880j^2$ $+0$	$+480$	$+35280j^2$ -84000	$+1440j^2$ $+8640$	$+65520j^2$ $+164640$
$x^5 \cos jy$		-7200	-120	$-30240j^2$ -63000	-3600	$-10080j^2$ -88200
$x^6 \cos jy$		$+4320$		$+90720$	$+720$	$+30240$
$x^7 \cos jy$		0		-35280	0	-5040

... Datum Table $(x^0 \dots x^7) \cos jy$.

Sine Table. $\frac{1}{2}$ of	$[\ ]^0$				$[\ ]^1$			
	$\frac{e^0}{2}$	$\frac{e^2}{16}$	$\frac{e^4}{768}$	$\frac{e^6}{92160}$	$\frac{e}{4}$	$\frac{e^3}{96}$	$\frac{e^5}{7680}$	$\frac{e^7}{1290240}$
$x^0 \sin jy$					$+2j$	$-24j^3$ $-6j$	$+320j^5$ $+260j^3$ $+100j$	$-4480j^7$ $-1400j^5$ $-17290j^3$ $+7490j$
$x \sin jy$						$+21j$	$-400j^3$ $-180j$	$+5040j^5$ $+35j^3$ $+6125j$
$x^2 \sin jy$						$+12j$	$-320j^3$ $+160j$	$+6720j^5$ $-420j^3$ $+3780j$
$x^3 \sin jy$							$+840j$	$-25200j^3$ $-10080j$
$x^4 \sin jy$							$+480j$	$-20160j^3$ $+25200j$
$x^5 \sin jy$								$+88200j$
$x^6 \sin jy$								$+50400j$
$x^7 \sin jy$								0

Datum Table $(x^0 \ldots x^7) \sin jy \ldots$

Sine Table. $\frac{1}{2}$ of	$[\ ]^2$			$[\ ]^3$		
	$\frac{e^2}{16}$	$\frac{e^4}{768}$	$\frac{e^6}{92160}$	$\frac{e^3}{96}$	$\frac{e^5}{7680}$	$\frac{e^7}{1290240}$
$x^0 \sin jy$	$+5j$	$-240j^3$ $-88j$	$+6000j^5$ $+8895j^3$ $+2040j$	$+8j^3$ $+26j$	$-160j^5$ $-3070j^3$ $-1290j$	$+2688j^7$ $+125160j^5$ $+257166j^3$ $+59850j$
$x \sin jy$	$-4j$	$+64j^3$ $+184j$	$-960j^5$ $-8340j^3$ $-5700j$	$-27j$	$+1400j^3$ $+2300j$	$-36624j^5$ $-207165j^3$ $-145971j$
$x^2 \sin jy$		$-72j$	$-1200j^3$ $+4260j$	$+12j$	$-160j^3$ $-1600j$	$+1344j^5$ $+39900j^3$ $+130788j$
$x^3 \sin jy$		$-96j$	$+2880j^3$ $+6120j$		$-180j$	$+55440j^3$ $+32760j$
$x^4 \sin jy$			$-8280j$		$+720j$	$-20160j^3$ $-156240j$
$x^5 \sin jy$			$-7200j$			$+37800j$
$x^6 \sin jy$			0			$+90720j$
$x^7 \sin jy$			0			0

... Datum Table $(x^0 \ldots x^7) \sin j$- ...

Sine Table. $\frac{1}{2}$ of	$[\ ]^4$ $\frac{e^4}{768}$	$\frac{e^6}{92160}$	$[\ ]^5$ $\frac{e^5}{7680}$	$\frac{e^7}{1290240}$	$[\ ]^6$ $\frac{e^6}{92160}$	$[\ ]^7$ $\frac{e^7}{1290240}$
$x^0 \sin jy$	$+120j^3$ $+206j$	$-4800j^5$ $-49200j^3$ $-21648j$	$+32j^5$ $+1790j^3$ $+2194j$	$-896j^7$ $-124600j^5$ $-939722j^3$ $-416990j$	$+1200j^5$ $+29835j^3$ $+29352j$	$+128j^7$ $+35560j^5$ $+563486j^3$ $+472730j$
$x \sin jy$	$-32j^3$ $-236j$	$+768j^5$ $+27840j^3$ $+35736j$	$-760j^3$ $-2640j$	$+34608j^5$ $+599935j^3$ $+657601j$	$-192j^5$ $-15780j^3$ $-36324j$	$-9744j^5$ $-339745j^3$ $-595231j$
$x^2 \sin jy$	$+156j$	$-7680j^3$ $-30120j$	$+160j^3$ $+2080j$	$-4032j^5$ $-237300j^3$ $-610764j$	$+5520j^3$ $+31620j$	$+1344j^5$ $+152460j^3$ $+552468j$
$x^3 \sin jy$	$-48j$	$0j^3$ $+8640j$	$-1020j$	$+35280j^3$ $+303240j$	$-960j^3$ $-19320j$	$-45360j^3$ $-384720j$
$x^4 \sin jy$		$+7200j$	$+240j$	$+6720j^3$ $+11760j$	$+7560j$	$+6720j^3$ $+193200j$
$x^5 \sin jy$		$-5760j$		$-113400j$	$-1440j$	$-63000j$
$x^6 \sin jy$		0		$+50400j$	0	$+10080j$
$x^7 \sin jy$		0		0	0	0

... Datum Table ($x^0 \ldots x^7$) $\sin jy$.

It may be remarked that with a view to the convenience of calculation, the Datum Tables give the half values of the several functions; thus, e.g., we have

$$\frac{1}{2}x^0\cos jy=\left\{+1\frac{e^0}{2}-8j^2\frac{e^2}{16}+(96j^4-54j^2)\frac{e^4}{768}+\&c.\right\}+\left\{0\frac{e}{4}-30j^2\frac{e^3}{96}+\&c.\right\}2\cos g$$

$$+\left\{4j^2\frac{e^2}{16}+\&c.\right\}2\cos 2g+\&c.$$

and so for the other functions in the two tables.

The calculation, for a given value of j, of the functions $(x^0, x^1\ldots\ldots x^7)\,{}^{\cos}_{\sin}\,jf$ is performed as follows; the numerical value of j is in the first instance substituted throughout the two Datum tables, so as to obtain the numerical values of the separate terms $\pm Aj^m$; when this is done, the terms in each compartment are added together with the proper signs, and the results written down in a single table as follows, viz. if $j=1$, and if we write down only the portion of the table which corresponds to $x^0\,{}^{\cos}_{\sin}\,y$, (the portions corresponding to $x\,{}^{\cos}_{\sin}\,y$, $x^2\,{}^{\cos}_{\sin}\,y$ &c....$x^7\,{}^{\cos}_{\sin}\,y$, being of a precisely similar form, and coming into the table in order one under the other) this is

$\frac{1}{2}x^0\,{}^{\cos}_{\sin}\,y$	$[\ ]^0$ $\frac{e^0}{2}$	$\frac{e^2}{16}$	$\frac{e^4}{768}$	$\frac{e^6}{92160}$	$[\ ]^1$ $\frac{e}{4}$	$\frac{e^3}{96}$	$\frac{e^5}{7680}$	$\frac{e^7}{1290240}$
[cos]	+ 1	− 8	+ 42	− 800	0	− 30	+ 680	− 15680
[sin]	0	0	0	0	2	− 30	+ 680	− 15680
[cos + sin]	+ 1	− 8	+ 42	− 800	+ 2	− 60	+ 1360	− 31360
[cos − sin]	Do.	”	”	”	− 2	0	0	0

$\frac{1}{2}x^0\,{}^{\cos}_{\sin}\,y$	$[\ ]^2$ $\frac{e^2}{16}$	$\frac{e^4}{768}$	$\frac{e^6}{92160}$	$[\ ]^3$ $\frac{e^3}{96}$	$\frac{e^5}{7680}$	$\frac{e^7}{1290240}$
[cos]	+ 4	− 320	+ 17490	+ 30	− 4440	+ 447552
[sin]	+ 5	− 328	+ 16935	+ 34	− 4520	+ 444864
[cos + sin]	+ 9	− 648	+ 34425	+ 64	− 8960	+ 892416
[cos − sin]	− 1	+ 8	+ 555	− 4	+ 80	+ 2688

$\frac{1}{2}x^0 \frac{\cos}{\sin} y$	[ ]^4 $\frac{e^4}{768}$	$\frac{e^6}{92160}$	[ ]^5 $\frac{e^5}{7680}$	$\frac{e^7}{1290240}$	[ ]^6 $\frac{e^6}{92160}$	[ ]^7 $\frac{e^7}{1290240}$
[cos]	+ 299	− 74352	+ 3760	− 1457120	+ 57262	+ 1025248
[sin]	+ 326	− 75648	+ 4016	− 1482208	+ 60387	+ 1071904
[cos + sin]	+ 625	− 150000	+ 7776	− 2939328	+ 117649	+ 2097152
[cos − sin]	− 27	+ 1296	− 256	+ 25088	− 3125	− 46656

where the lines [cos + sin], [cos − sin] are obtained by the addition and subtraction of the lines [cos], [sin].

The lines [cos + sin], [cos − sin], are then rearranged by reading them backwards and forwards from the terms which are in the [cos − sin] line, under [ ]^j; viz., in the present case, where $j = 1$, from the terms − 2, 0, 0, 0 under [ ]^1, as follows:—

$x^0 \frac{\cos}{\sin} f$	[ ]^0 $\frac{e}{4}$	$\frac{e^3}{96}$	$\frac{e^5}{7680}$	$\frac{e^7}{1290240}$	[ ]^1 $\frac{e^0}{2}$	$\frac{e^2}{16}$	$\frac{e^4}{768}$	$\frac{e^6}{92160}$
$\frac{1}{2}$ [cos + sin]	− 2	0	0	0	+ 1	− 8	+ 42	− 800
$\frac{1}{2}$ [cos − sin]	Do.	,,	,,	,,	0	− 1	+ 8	+ 555
[cos]	− 4	0	0	0	+ 1	− 9	+ 50	− 245
[sin]	0	0	0	0	+ 1	− 7	+ 34	− 1355

$x^0 \frac{\cos}{\sin} f$	[ ]^2 $\frac{e}{4}$	$\frac{e^3}{96}$	$\frac{e^5}{7680}$	$\frac{e^7}{1290240}$	[ ]^3 $\frac{e^2}{16}$	$\frac{e^4}{768}$	$\frac{e^6}{92160}$	[ ]^4 $\frac{e^3}{96}$	$\frac{e^5}{7680}$	$\frac{e^7}{1290240}$
$\frac{1}{2}$ [cos + sin]	+ 2	− 60	+ 1360	− 31360	+ 9	− 648	+ 34425	+ 64	− 8960	+ 892416
$\frac{1}{2}$ [cos − sin]	0	− 4	+ 80	+ 2688	0	− 27	+ 1296	0	− 256	+ 25088
[cos]	+ 2	− 64	+ 1440	− 28672	+ 9	− 675	+ 35721	+ 64	− 9216	+ 917504
[sin]	+ 2	− 56	+ 1280	− 24048	+ 9	− 621	+ 33129	+ 64	− 8704	+ 867328

$x^0 \begin{matrix}\cos\\ \sin\end{matrix} f$	$[\ ]^5$ $\frac{e^4}{768}$	$[\ ]^5$ $\frac{e^6}{92160}$	$[\ ]^6$ $\frac{e^5}{7680}$	$[\ ]^6$ $\frac{e^7}{1290240}$	$[\ ]^7$ $\frac{e^6}{92160}$	$[\ ]^8$ $\frac{e^7}{1290240}$
$\frac{1}{2}$[cos + sin]	+625	−150000	+7776	−2939328	+117649	+2097152
$\frac{1}{2}$[cos − sin]	0	−3125	0	−46656	0	0
[cos]	+625	−153125	+7776	−2985984	+117649	+2097152
[sin]	+625	−146875	+7776	−2892672	+117649	+2097152

where the lines [cos], [sin] are obtained by the addition and subtraction of the lines $\frac{1}{2}$[cos + sin], $\frac{1}{2}$[cos − sin]; and it then only remains to reduce the fractions, at the same time multiplying by 2 the lines $\frac{1}{2}$[cos + sin], $\frac{1}{2}$[cos − sin], to obtain instead the lines [cos + sin], [cos − sin]. We have thus,

$x^0 \begin{matrix}\cos\\ \sin\end{matrix} f$	$[\ ]^0$ e	e^3	e^5	e^7	$[\ ]^1$ e^0	e^2	e^4	e^6
[cos + sin]	−1	0	0	0	+1	−1	$+\frac{7}{64}$	$-\frac{5}{288}$
[cos − sin]	Do.	„	„	„	0	$-\frac{1}{8}$	$+\frac{1}{48}$	$+\frac{37}{3072}$
[cos]	−1	0	0	0	$+\frac{1}{2}$	$-\frac{9}{16}$	$+\frac{25}{384}$	$-\frac{49}{18432}$
[sin]	0	0	0	0	$+\frac{1}{2}$	$-\frac{7}{16}$	$+\frac{17}{384}$	$-\frac{271}{18432}$

$x^0 \begin{matrix}\cos\\ \sin\end{matrix} f$	$[\ ]^2$ e	e^3	e^5	e^7	$[\ ]^3$ e^2	e^4	e^6	$[\ ]^4$ e^3	e^5	e^7
[cos + sin]	+1	$-\frac{5}{4}$	$+\frac{17}{48}$	$-\frac{7}{144}$	$+\frac{9}{8}$	$-\frac{27}{16}$	$+\frac{765}{1024}$	$+\frac{4}{3}$	$-\frac{7}{3}$	$+\frac{83}{60}$
[cos − sin]	0	$-\frac{1}{12}$	$+\frac{1}{48}$	$+\frac{1}{240}$	0	$-\frac{9}{128}$	$+\frac{9}{320}$	0	$-\frac{1}{15}$	$+\frac{7}{180}$
[cos]	$+\frac{1}{2}$	$-\frac{2}{3}$	$+\frac{3}{16}$	$-\frac{1}{45}$	$+\frac{9}{16}$	$-\frac{225}{256}$	$+\frac{3969}{10240}$	$+\frac{2}{3}$	$-\frac{6}{5}$	$+\frac{32}{45}$
[sin]	$+\frac{1}{2}$	$-\frac{7}{12}$	$+\frac{1}{6}$	$-\frac{19}{720}$	$+\frac{9}{16}$	$-\frac{207}{256}$	$+\frac{3681}{10240}$	$+\frac{2}{3}$	$-\frac{17}{15}$	$+\frac{121}{180}$

$x^0 \begin{smallmatrix}\cos\\ \sin\end{smallmatrix} f$	$[\ ]^5$ e^4	$[\ ]^5$ e^6	$[\ ]^6$ e^5	$[\ ]^6$ e^7	$[\ ]^7$ e^6	$[\ ]^8$ e^7
[cos + sin]	$+\frac{625}{384}$	$-\frac{625}{192}$	$+\frac{81}{40}$	$-\frac{5103}{1120}$	$+\frac{117649}{46080}$	$+\frac{1024}{315}$
[cos − sin]	0	$-\frac{625}{9216}$	0	$-\frac{81}{1120}$	0	0
[cos]	$+\frac{625}{768}$	$-\frac{30625}{18432}$	$+\frac{81}{80}$	$-\frac{81}{35}$	$+\frac{117649}{92160}$	$+\frac{512}{315}$
[sin]	$+\frac{625}{768}$	$-\frac{29375}{18432}$	$+\frac{81}{80}$	$-\frac{2511}{1120}$	$+\frac{117649}{92160}$	$+\frac{512}{315}$

which is, in fact, part of the table $(x^0 \ldots x^7) \begin{smallmatrix}\cos\\ \sin\end{smallmatrix} f$. It is to be remarked, that when in any column the second term is zero, the third and fourth terms are each equal to one-half of the first term, and that I have in the tables represented them by the letters H, H.

The foregoing method depends, it is easy to see, on the following proposition; viz., writing

$$x^m \cos jf = \Sigma\, [\cos]_f{}^i \cos ig, \quad x^m \sin jf = \Sigma\, [\sin]_f{}^i \sin ig,$$
$$x^m \cos jy = \Sigma\, [\cos]_y{}^i \cos ig, \quad x^m \sin jy = \Sigma\, [\sin]_y{}^i \sin ig,$$

where, as before, $f = g + y$, then we have

$$[\cos + \sin]_f{}^i = [\cos + \sin]_y{}^{i-j},$$
$$[\cos - \sin]_f{}^i = [\cos - \sin]_y{}^{i+j},$$

the latter of which follows immediately from the former. To prove this, it is to be remarked that in general

$$\cos \alpha\, \Sigma\, [\cos]^i \cos ig = \quad \Sigma\, [\cos]^i \cos (ig + \alpha),$$
$$\sin \alpha\, \Sigma\, [\sin]^i \sin ig = -\Sigma\, [\sin]^i \cos (ig + \alpha),$$
$$\sin \alpha\, \Sigma\, [\cos]^i \cos ig = \quad \Sigma\, [\cos]^i \sin (ig + \alpha),$$
$$\cos \alpha\, \Sigma\, [\sin]^i \sin ig = \quad \Sigma\, [\sin]^i \sin (ig + \alpha);$$

whence, putting $\alpha = jg$,

$$\cos jf = \cos (jy + jg) = \Sigma\, [\cos + \sin]_y{}^i \cos (i + j)\, g$$
$$= \Sigma\, \tfrac{1}{2} \{[\cos + \sin]_y{}^{i-j} + [\cos + \sin]_y{}^{-i-j}\} \cos ig,$$
$$\sin jf = \sin (jy + jg) = \Sigma\, [\cos + \sin]_y{}^i \sin (i + j)\, g$$
$$= \Sigma\, \tfrac{1}{2} \{[\cos + \sin]_y{}^{i-j} - [\cos + \sin]_y{}^{-i-j}\} \sin ig;$$

and therefore

$$[\cos]_f{}^i = \tfrac{1}{2}\{[\cos+\sin]_y{}^{i-j} + [\cos+\sin]_y{}^{-i-j}\},$$

$$[\sin]_f{}^i = \tfrac{1}{2}\{[\cos+\sin]_y{}^{i-j} - [\cos+\sin]_y{}^{-i-j}\};$$

and, adding and subtracting, we have the theorem. The coefficients contained in the Datum Tables $(x^0 \ldots x^7)\,{}^{\cos}_{\sin}\, jy$ were taken to be $\tfrac{1}{2}[\cos]_y{}^i$, $\tfrac{1}{2}[\sin]_y{}^i$, in order that the calculations might in the first instance give $\tfrac{1}{2}[\cos+\sin]_f{}^i$, $\tfrac{1}{2}[\cos+\sin]_f{}^i$, and thence by an addition and subtraction, without a division by 2, give $[\cos]_f{}^i$, $[\sin]_f{}^i$.

The resulting values admit of a very simple verification; in fact, writing down the two equations

$$x^m \cos jf = \Sigma\,[\cos]^i \cos ig, \qquad x^m \sin jf = \Sigma\,[\sin]^i \sin ig,$$

or, what is the same thing, the summation now extending only from $i=1$ to $i=\infty$,

$$\tfrac{1}{2} x^m \cos jf = \tfrac{1}{2}[\cos]^0 + \Sigma'\,[\cos]^i \cos ig, \qquad \tfrac{1}{2} x^m \sin jf = \Sigma'\,[\sin]^i \sin ig\,;$$

if in the first equation, and in the second equation differentiated with respect to g, we put $g=0$, then, observing that $g=0$ gives $f=0$, $x=-e$, $\dfrac{df}{dg} = \dfrac{(1+e)^2}{(1-e^2)^{\frac{3}{2}}}$, we have

$$\tfrac{1}{2}[\cos]^0 + \Sigma'\,[\cos]^i = \tfrac{1}{2} j\,(-e)^m,$$

$$\Sigma' i\,[\sin]^i = \tfrac{1}{2} j\,(-e)^m\,(1 + 2e + \tfrac{5}{2}e^2 + 3e^3 + \tfrac{27}{8}e^4 + \tfrac{15}{4}e^5 + \tfrac{65}{16}e^6 + \tfrac{35}{8}e^7 + \ldots).$$

The values of $(x^0 \ldots x^7)\,{}^{\cos}_{\sin}\, jf$ being known, those of $\left(\dfrac{r}{a}\right)^{\pm m} {}^{\cos}_{\sin}\, jf$ can, of course, be determined by merely substituting for $\dfrac{r}{a}$ its value $(1+x)$, expanding and reducing; and a convenient mode of arranging the calculations presents itself so obviously, that no explanation is required. The verification is similar to that for $x^m\,{}^{\cos}_{\sin}\, jf$. The only difference is, that for $g=0$ we have the value of r, viz. $1-e$, in the place of the value $-e$ of x. A similar verification is applicable to the Datum Tables.

To make the arrangement of the Tables clear, I have given at the end of them some examples to show how they are to be read off.

Table (x^0, x^1, x^2, x^3, x^4, x^5, x^6, x^7).

		$[\ ]^0$				$[\ ]^1$				$[\ ]^2$		
		e^0	e^2	e^4	e^6	e	e^3	e^5	e^7	e^2	e^4	e^6
x^0		$+1$	0	0	0	0	0	0	0	0	0	0
x^1			$+\frac{1}{2}$	0	0	$-\frac{1}{2}$	$+\frac{3}{16}$	$-\frac{5}{384}$	$+\frac{7}{18432}$	$-\frac{1}{4}$	$+\frac{1}{6}$	$-\frac{1}{32}$
x^2			$+\frac{1}{2}$	0	0		$-\frac{1}{4}$	$+\frac{1}{48}$	$-\frac{1}{1536}$	$+\frac{1}{4}$	$-\frac{1}{4}$	$+\frac{5}{96}$
x^3				$+\frac{3}{8}$	0		$-\frac{3}{8}$	$+\frac{3}{32}$	$-\frac{5}{1024}$		0	$+\frac{1}{32}$
	[cos]											
x^4				$+\frac{3}{8}$	0			$-\frac{1}{4}$	$+\frac{1}{64}$		$+\frac{1}{4}$	$-\frac{3}{16}$
x^5					$+\frac{5}{16}$			$-\frac{5}{16}$	$+\frac{15}{256}$			$+\frac{5}{64}$
x^6					$+\frac{5}{16}$				$-\frac{15}{64}$			$+\frac{15}{64}$
x^7									$-\frac{35}{128}$			

Table (x^0, x^1, x^2, x^3, x^4, x^5, x^6, x^7).

		[ ]^3			[ ]^4		[ ]^5		[ ]^6	[ ]^7
		e^3	e^5	e^7	e^4	e^6	e^5	e^6	e^6	e^7
x^0	[cos]	0	0	0	0	0	0	0	0	0
x^1		$-\frac{3}{16}$	$+\frac{45}{256}$	$-\frac{567}{10240}$	$-\frac{1}{6}$	$+\frac{1}{5}$	$-\frac{125}{768}$	$+\frac{4375}{18432}$	$-\frac{27}{160}$	$-\frac{16807}{91260}$
x^2		$+\frac{1}{4}$	$-\frac{9}{32}$	$+\frac{243}{2560}$	$+\frac{1}{4}$	$-\frac{1}{3}$	$+\frac{25}{96}$	$-\frac{625}{1536}$	$+\frac{9}{32}$	$+\frac{2401}{7680}$
x^3		$-\frac{1}{8}$	$+\frac{9}{64}$	$-\frac{27}{1024}$	$-\frac{3}{16}$	$+\frac{1}{4}$	$-\frac{15}{64}$	$+\frac{375}{1024}$	$-\frac{9}{32}$	$-\frac{343}{1024}$
x^4			$+\frac{1}{8}$	$-\frac{9}{64}$	$+\frac{1}{16}$	0	$+\frac{1}{8}$	$-\frac{25}{192}$	$+\frac{3}{16}$	$+\frac{49}{192}$
x^5			$-\frac{5}{32}$	$+\frac{45}{256}$		$-\frac{5}{32}$	$-\frac{1}{32}$	$-\frac{25}{256}$	$-\frac{5}{64}$	$-\frac{35}{256}$
x^6				$+\frac{3}{64}$		$+\frac{3}{32}$		$+\frac{9}{64}$	$+\frac{1}{64}$	$+\frac{3}{64}$
x^7				$-\frac{21}{128}$				$-\frac{7}{128}$		$-\frac{1}{128}$

Table $(x^0, x^1, x^2, x^3)\ {\cos \atop \sin} f \ldots$

${\cos \atop \sin} f$		$[\ ]^0$				$[\ ]^1$			
		e	e^3	e^5	e^7	e^0	e^2	e^4	e^6
x^0	[cos + sin]	-1	0	0	0	$+1$	-1	$+\frac{7}{64}$	$-\frac{5}{288}$
	[cos − sin]	,,	,,	,,	,,	0	$-\frac{1}{8}$	$+\frac{1}{48}$	$+\frac{37}{3072}$
	[cos]	-1	0	0	0	H	$-\frac{9}{16}$	$+\frac{25}{384}$	$-\frac{49}{18432}$
	[sin]	0	0	0	0	H	$-\frac{7}{16}$	$+\frac{17}{384}$	$-\frac{271}{18432}$
x^1	[cos + sin]	$-\frac{1}{2}$	0	0	0		$+\frac{1}{2}$	$-\frac{1}{8}$	$-\frac{1}{128}$
	[cos − sin]	,,	,,	,,	,,		$+\frac{1}{4}$	$+\frac{1}{48}$	$+\frac{19}{1536}$
	[cos]	$-\frac{1}{2}$	0	0	0		$+\frac{3}{8}$	$-\frac{5}{96}$	$+\frac{7}{3072}$
	[sin]	0	0	0	0		$+\frac{1}{8}$	$-\frac{7}{96}$	$-\frac{31}{3072}$
x^2	[cos + sin]		$-\frac{1}{2}$	0	0		$+\frac{1}{2}$	$-\frac{1}{4}$	$+\frac{5}{384}$
	[cos − sin]		,,	,,	,,		$+\frac{1}{4}$	$-\frac{1}{16}$	$+\frac{5}{512}$
	[cos]		$-\frac{1}{2}$	0	0		$+\frac{3}{8}$	$-\frac{5}{32}$	$+\frac{35}{3072}$
	[sin]		0	0	0		$+\frac{1}{8}$	$-\frac{3}{32}$	$+\frac{5}{3072}$
x^3	[cos + sin]		$-\frac{3}{8}$	0	0			$+\frac{3}{8}$	$-\frac{1}{16}$
	[cos − sin]		,,	,,	,,			$+\frac{1}{4}$	$+\frac{1}{128}$
	[cos]		$-\frac{3}{8}$	0	0			$+\frac{5}{16}$	$-\frac{7}{256}$
	[sin]		0	0	0			$+\frac{1}{16}$	$-\frac{9}{256}$

N.B.—In all the Tables H, H, in the third and fourth lines of a set of four lines, denote one half of the number above them in the first line. The terms $[\cos - \sin]^0$, denoted by ,, ,, ... are the same as the terms $[\cos + \sin]^0$ immediately above them.

... *Table* $(x^0,\ x^1,\ x^2,\ x^3)\ {\cos \atop \sin} f$ ...

	$\frac{\cos}{\sin} f$	$[\ ]^2$ e	e^3	e^5	e^7	$[\ ]^3$ e^2	e^4	e^6	$[\ ]^4$ e^3	e^5	e^7
x^0	[cos + sin]	$+1$	$-\frac{5}{4}$	$+\frac{17}{48}$	$-\frac{7}{144}$	$+\frac{9}{8}$	$-\frac{27}{16}$	$+\frac{768}{1024}$	$+\frac{4}{3}$	$-\frac{7}{3}$	$+\frac{83}{60}$
	[cos − sin]	0	$-\frac{1}{12}$	$+\frac{1}{48}$	$+\frac{1}{240}$	0	$-\frac{9}{128}$	$+\frac{9}{320}$	0	$-\frac{1}{15}$	$+\frac{7}{180}$
	[cos]	H	$-\frac{2}{3}$	$+\frac{3}{16}$	$-\frac{1}{45}$	H	$-\frac{225}{256}$	$+\frac{3969}{10240}$	H	$-\frac{6}{5}$	$+\frac{32}{45}$
	[sin]	H	$-\frac{7}{12}$	$+\frac{1}{6}$	$-\frac{19}{720}$	H	$-\frac{207}{256}$	$+\frac{3681}{10240}$	H	$-\frac{17}{15}$	$+\frac{121}{180}$
x^1	[cos + sin]	$-\frac{1}{2}$	$+\frac{7}{8}$	$-\frac{29}{96}$	$+\frac{5}{144}$	$-\frac{3}{4}$	$+\frac{21}{16}$	$-\frac{327}{512}$	-1	$+\frac{23}{12}$	$-\frac{289}{240}$
	[cos − sin]	0	$+\frac{1}{8}$	$-\frac{1}{96}$	$+\frac{1}{240}$	0	$+\frac{3}{32}$	$-\frac{33}{1280}$	0	$+\frac{1}{12}$	$-\frac{29}{720}$
	[cos]	H	$+\frac{1}{2}$	$-\frac{5}{32}$	$+\frac{7}{360}$	H	$+\frac{45}{64}$	$-\frac{1701}{5120}$	H	$+1$	$-\frac{28}{45}$
	[sin]	H	$+\frac{3}{8}$	$-\frac{7}{48}$	$+\frac{11}{720}$	H	$+\frac{39}{64}$	$-\frac{1569}{5120}$	H	$+\frac{11}{12}$	$-\frac{419}{720}$
x^2	[cos + sin]		0	$-\frac{1}{12}$	$+\frac{1}{32}$	$+\frac{1}{4}$	$-\frac{7}{16}$	$+\frac{73}{512}$	$+\frac{1}{2}$	$-\frac{23}{24}$	$+\frac{257}{480}$
	[cos − sin]		0	$-\frac{1}{24}$	$+\frac{1}{480}$	0	$-\frac{1}{32}$	$-\frac{5}{256}$	0	$-\frac{1}{24}$	$-\frac{1}{480}$
	[cos]		0	$-\frac{1}{16}$	$+\frac{1}{60}$	H	$-\frac{15}{64}$	$+\frac{63}{1024}$	H	$-\frac{1}{2}$	$+\frac{4}{15}$
	[sin]		0	$-\frac{1}{48}$	$+\frac{7}{480}$	H	$-\frac{13}{64}$	$+\frac{83}{1024}$	H	$-\frac{11}{24}$	$+\frac{43}{160}$
x^3	[cos + sin]		$-\frac{3}{8}$	$+\frac{7}{16}$	$-\frac{7}{64}$		$-\frac{1}{4}$	$+\frac{51}{128}$	$-\frac{1}{8}$	$+\frac{1}{32}$	$+\frac{33}{128}$
	[cos − sin]		$-\frac{1}{8}$	$+\frac{1}{8}$	$-\frac{1}{64}$		$-\frac{1}{16}$	$+\frac{3}{32}$	0	$-\frac{1}{32}$	$+\frac{29}{384}$
	[cos]		$-\frac{1}{4}$	$+\frac{9}{32}$	$-\frac{1}{16}$		$-\frac{5}{32}$	$+\frac{63}{256}$	H	0	$+\frac{1}{6}$
	[sin]		$-\frac{1}{8}$	$+\frac{5}{32}$	$-\frac{3}{64}$		$-\frac{3}{32}$	$+\frac{39}{256}$	H	$+\frac{1}{32}$	$+\frac{35}{384}$

... *Table* $(x^0,\ x^1,\ x^2,\ x^3)\ {\cos \atop \sin} f$.

	${\cos \atop \sin} f$	$[\ ]^5$ e^4	$[\ ]^5$ e^6	$[\ ]^6$ e^5	$[\ ]^6$ e^7	$[\ ]^7$ e^7	$[\ ]^8$ e^7
x^0	[cos + sin]	$+\frac{625}{384}$	$-\frac{625}{192}$	$+\frac{81}{40}$	$-\frac{729}{160}$	$+\frac{117649}{46080}$	$+\frac{1024}{315}$
	[cos − sin]	0	$-\frac{625}{9216}$	0	$-\frac{81}{1120}$	0	0
	[cos]	H	$-\frac{30625}{18432}$	H	$-\frac{81}{35}$	H	H
	[sin]	H	$-\frac{29375}{18432}$	H	$-\frac{2511}{1120}$	H	H
x^1	[cos + sin]	$-\frac{125}{96}$	$+\frac{2125}{768}$	$-\frac{27}{16}$	$+\frac{1269}{320}$	$-\frac{16807}{7680}$	$-\frac{128}{45}$
	[cos − sin]	0	$+\frac{125}{1536}$	0	$+\frac{27}{320}$	0	0
	[cos]	H	$+\frac{4375}{3072}$	H	$+\frac{81}{40}$	H	H
	[sin]	H	$+\frac{1375}{1024}$	H	$+\frac{621}{320}$	H	H
x^2	[cos + sin]	$+\frac{25}{32}$	$-\frac{425}{256}$	$+\frac{9}{8}$	$-\frac{423}{160}$	$+\frac{2401}{1536}$	$+\frac{32}{15}$
	[cos − sin]	0	$-\frac{25}{512}$	0	$-\frac{9}{160}$	0	0
	[cos]	H	$-\frac{875}{1024}$	H	$-\frac{27}{20}$	H	H
	[sin]	H	$-\frac{825}{1024}$	H	$-\frac{207}{160}$	H	H
x^3	[cos + sin]	$-\frac{5}{16}$	$+\frac{15}{32}$	$-\frac{9}{16}$	$+\frac{9}{8}$	$-\frac{343}{384}$	$-\frac{4}{3}$
	[cos − sin]	0	$-\frac{5}{384}$	0	0	0	0
	[cos]	H	$+\frac{175}{768}$	H	H	H	H
	[sin]	H	$+\frac{185}{768}$	H	H	H	H

Table $(x^4,\ z^5,\ x^6,\ x^7) \begin{smallmatrix}\cos\\ \sin\end{smallmatrix} f \ldots$

$\begin{smallmatrix}\cos\\ \sin\end{smallmatrix} f$		$[\ ]^0$ e	$[\ ]^0$ e^3	$[\ ]^0$ e^5	$[\ ]^0$ e^7	$[\ ]^1$ e^0	$[\ ]^1$ e^2	$[\ ]^1$ e^4	$[\ ]^1$ e^6
x^4	[cos + sin]			$-\frac{3}{8}$	o			$+\frac{3}{8}$	$-\frac{1}{8}$
	[cos − sin]			"	"			$+\frac{1}{4}$	$-\frac{5}{128}$
	[cos]			$-\frac{3}{8}$	o			$+\frac{5}{16}$	$-\frac{21}{256}$
	[sin]			o	o			$+\frac{1}{16}$	$-\frac{11}{256}$
x^5	[cos + sin]			$-\frac{5}{16}$	o				$+\frac{5}{16}$
	[cos − sin]			"	"				$+\frac{15}{64}$
	[cos]			$-\frac{5}{16}$	o				$+\frac{35}{128}$
	[sin]			o	o				$+\frac{5}{128}$
x^6	[cos + sin]				$-\frac{5}{16}$				$+\frac{5}{16}$
	[cos − sin]				"				$+\frac{15}{64}$
	[cos]				$-\frac{5}{16}$				$+\frac{35}{128}$
	[sin]				o				$+\frac{5}{128}$
x^7	[cos + sin]				$-\frac{35}{128}$				
	[cos − sin]				"				
	[cos]				$-\frac{35}{128}$				
	[sin]				o				

... *Table* $(x^4, x^5, x^6, x^7) \begin{matrix}\cos\\ \sin\end{matrix} f$ ...

	$\begin{matrix}\cos\\ \sin\end{matrix} f$	$[\ ]^2$				$[\ ]^3$			$[\ ]^4$		
		e	e^3	e^5	e^7	e^2	e^4	e^6	e^3	e^5	e^7
x^4	[cos + sin]			$-\frac{1}{8}$	$+\frac{1}{64}$		$+\frac{1}{4}$	$-\frac{51}{128}$		$+\frac{5}{16}$	$-\frac{9}{16}$
	[cos − sin]			$-\frac{1}{16}$	$-\frac{1}{64}$		$+\frac{1}{16}$	$-\frac{3}{32}$		$+\frac{1}{16}$	$-\frac{5}{48}$
	[cos]			$-\frac{3}{32}$	0		$+\frac{5}{32}$	$-\frac{63}{256}$		$+\frac{3}{16}$	$-\frac{1}{3}$
	[sin]			$-\frac{1}{32}$	$+\frac{1}{64}$		$+\frac{3}{32}$	$-\frac{39}{256}$		$+\frac{1}{8}$	$-\frac{11}{48}$
x^5	[cos + sin]			$-\frac{5}{16}$	$+\frac{35}{128}$			$-\frac{5}{64}$		$-\frac{5}{32}$	$+\frac{7}{32}$
	[cos − sin]			$-\frac{5}{32}$	$+\frac{13}{128}$			$-\frac{1}{32}$		$-\frac{1}{32}$	$+\frac{1}{32}$
	[cos]			$-\frac{15}{64}$	$+\frac{3}{16}$			$-\frac{7}{128}$		$-\frac{3}{32}$	$+\frac{1}{8}$
	[sin]			$-\frac{5}{64}$	$+\frac{11}{128}$			$-\frac{3}{128}$		$-\frac{1}{16}$	$+\frac{3}{32}$
x^6	[cos + sin]				$-\frac{5}{32}$			$+\frac{15}{64}$			$+\frac{3}{16}$
	[cos − sin]				$-\frac{3}{32}$			$+\frac{3}{32}$			$+\frac{1}{16}$
	[cos]				$-\frac{1}{8}$			$+\frac{21}{128}$			$+\frac{1}{8}$
	[sin]				$-\frac{1}{32}$			$+\frac{9}{128}$			$+\frac{1}{16}$
x^7	[cos + sin]				$-\frac{35}{128}$						$-\frac{21}{128}$
	[cos − sin]				$-\frac{21}{128}$						$-\frac{7}{128}$
	[cos]				$-\frac{7}{32}$						$-\frac{7}{64}$
	[sin]				$-\frac{7}{128}$						$-\frac{7}{128}$

... *Table* $(x^4,\ x^5,\ x^6,\ x^7)\ {\cos \atop \sin} f$.

	${\cos \atop \sin} f$	$[\ ]^5$ e^4	$[\ ]^5$ e^6	$[\ ]^6$ e^5	$[\ ]^6$ e^7	$[\ ]^7$ e^6	$[\ ]^8$ e^7
x^4	[cos + sin]	$+\frac{1}{16}$	$+\frac{7}{32}$	$+\frac{3}{16}$	$-\frac{3}{64}$	$+\frac{49}{128}$	$+\frac{2}{3}$
	[cos − sin]	0	$+\frac{7}{128}$	0	$+\frac{3}{64}$	0	0
	[cos]	H	$+\frac{35}{256}$	H	0	H	H
	[sin]	H	$+\frac{21}{256}$	H	$-\frac{3}{64}$	H	H
x^5	[cos + sin]		$-\frac{9}{32}$	$-\frac{1}{32}$	$-\frac{41}{128}$	$-\frac{7}{64}$	$-\frac{1}{4}$
	[cos − sin]		$-\frac{3}{64}$	0	$-\frac{7}{128}$	0	0
	[cos]		$-\frac{21}{128}$	H	$-\frac{3}{16}$	H	H
	[sin]		$-\frac{15}{128}$	H	$-\frac{17}{128}$	H	H
x^6	[cos + sin]		$+\frac{3}{32}$		$+\frac{7}{32}$	$+\frac{1}{64}$	$+\frac{1}{16}$
	[cos − sin]		$+\frac{1}{64}$		$+\frac{1}{32}$	0	0
	[cos]		$+\frac{7}{128}$		$+\frac{1}{8}$	H	H
	[sin]		$+\frac{5}{128}$		$+\frac{3}{32}$	H	H
x^7	[cos + sin]				$-\frac{7}{128}$		$-\frac{1}{128}$
	[cos − sin]				$-\frac{1}{128}$		0
	[cos]				$-\frac{1}{32}$		H
	[sin]				$-\frac{3}{128}$		H

Table $(x^0, x^1, x^2, x^3) \begin{matrix}\cos\\ \sin\end{matrix} 2f \ldots$

	$\begin{matrix}\cos\\ \sin\end{matrix} 2f$	$[\ ]^0$			$[\ ]^1$				$[\ ]^2$			
		e^2	e^4	e^6	e	e^3	e^5	e^7	e^0	e^2	e^4	e^6
x^0	[cos + sin]	$+\frac{3}{4}$	$+\frac{1}{8}$	$+\frac{3}{64}$	-2	$+\frac{7}{4}$	$-\frac{5}{96}$	$+\frac{271}{4608}$	$+1$	-4	$+\frac{55}{16}$	$-\frac{103}{144}$
	[cos − sin]	,,	,,	,,	o	$+\frac{1}{12}$	$+\frac{5}{192}$	$+\frac{49}{7680}$	o	o	$+\frac{1}{24}$	$+\frac{1}{60}$
	[cos]	$+\frac{3}{4}$	$+\frac{1}{8}$	$+\frac{3}{64}$	H	$+\frac{11}{12}$	$-\frac{5}{384}$	$+\frac{751}{23040}$	H	H	$+\frac{167}{96}$	$-\frac{503}{1440}$
	[sin]	o	o	o	H	$+\frac{5}{6}$	$-\frac{5}{128}$	$+\frac{151}{5760}$	H	H	$+\frac{163}{96}$	$-\frac{527}{1440}$
x	[cos + sin]	$+\frac{3}{4}$	$-\frac{1}{8}$	$-\frac{3}{64}$	$-\frac{1}{2}$	$+\frac{5}{16}$	$-\frac{53}{384}$	$-\frac{203}{18432}$		$+\frac{1}{2}$	$-\frac{1}{2}$	$-\frac{1}{32}$
	[cos − sin]	,,	,,	,,	o	$-\frac{1}{16}$	$-\frac{61}{768}$	$-\frac{271}{6144}$		o	$-\frac{1}{48}$	$-\frac{17}{480}$
	[cos]	$+\frac{3}{4}$	$-\frac{1}{8}$	$-\frac{3}{64}$	H	$+\frac{1}{8}$	$-\frac{167}{1536}$	$-\frac{127}{4608}$		H	$-\frac{25}{96}$	$-\frac{1}{30}$
	[sin]	o	o	o	H	$+\frac{3}{16}$	$-\frac{15}{512}$	$+\frac{305}{18432}$		H	$-\frac{23}{96}$	$+\frac{1}{480}$
x^2	[cos + sin]	$+\frac{1}{4}$	$+\frac{1}{8}$	$+\frac{3}{64}$		$-\frac{3}{4}$	$+\frac{17}{48}$	$+\frac{19}{512}$		$+\frac{1}{2}$	-1	$+\frac{53}{96}$
	[cos − sin]	,,	,,	,,		$-\frac{1}{4}$	$+\frac{1}{96}$	$+\frac{83}{7680}$		o	$-\frac{1}{16}$	$+\frac{1}{32}$
	[cos]	$+\frac{1}{4}$	$+\frac{1}{8}$	$+\frac{3}{64}$		$-\frac{1}{2}$	$+\frac{35}{192}$	$+\frac{23}{960}$		H	$-\frac{17}{32}$	$+\frac{7}{24}$
	[sin]	o	o	o		$-\frac{1}{4}$	$+\frac{11}{64}$	$+\frac{101}{7680}$		H	$-\frac{15}{32}$	$+\frac{25}{96}$
x^3	[cos + sin]		$+\frac{1}{2}$	$-\frac{3}{64}$		$-\frac{3}{8}$	$+\frac{5}{32}$	$-\frac{49}{1024}$			$+\frac{3}{8}$	$-\frac{1}{4}$
	[cos − sin]		,,	,,		$-\frac{1}{8}$	$-\frac{1}{64}$	$-\frac{39}{1024}$			$+\frac{1}{16}$	o
	[cos]		$+\frac{1}{2}$	$-\frac{3}{64}$		$-\frac{1}{4}$	$+\frac{9}{128}$	$-\frac{11}{256}$			$+\frac{7}{32}$	$-\frac{1}{8}$
	[sin]		o	o		$-\frac{1}{8}$	$+\frac{11}{128}$	$-\frac{5}{1024}$			$+\frac{5}{32}$	$-\frac{1}{8}$

... Table $(x^0, x^1, x^2, x^3)\ {\cos \atop \sin}\ 2f$ ...

${\cos \atop \sin}\ 2f$		$[\ ]^3$				$[\ ]^4$			$[\ ]^5$		
		e	e^3	e^5	e^7	e^2	e^4	e^6	e^3	e^5	e^7
x^0	[cos + sin]	$+2$	$-\frac{27}{4}$	$+\frac{207}{32}$	$-\frac{1147}{512}$	$+\frac{13}{4}$	$-\frac{259}{24}$	$+\frac{559}{48}$	$+\frac{59}{12}$	$-\frac{3221}{192}$	$+\frac{31087}{1536}$
	[cos − sin]	0	0	$+\frac{9}{320}$	$+\frac{27}{2560}$	0	0	$+\frac{1}{45}$	0	0	$+\frac{625}{32256}$
	[cos]	H	H	$+\frac{2079}{640}$	$-\frac{1427}{1280}$	H	H	$+\frac{8401}{1440}$	H	H	$+\frac{163363}{16128}$
	[sin]	H	H	$+\frac{2061}{640}$	$-\frac{2881}{2560}$	H	H	$+\frac{8369}{1440}$	H	H	$+\frac{326101}{32256}$
x	[cos + sin]	$-\frac{1}{2}$	$+\frac{33}{16}$	$-\frac{255}{128}$	$+\frac{1081}{2048}$	$-\frac{5}{4}$	$+\frac{107}{24}$	$-\frac{229}{48}$	$-\frac{37}{16}$	$+\frac{6259}{768}$	$-\frac{59723}{6144}$
	[cos − sin]	0	0	$-\frac{3}{256}$	$-\frac{231}{10240}$	0	0	$-\frac{1}{120}$	0	0	$-\frac{125}{18432}$
	[cos]	H	H	$-\frac{513}{512}$	$+\frac{2587}{10240}$	H	H	$-\frac{1147}{480}$	H	H	$-\frac{89647}{18432}$
	[sin]	H	H	$-\frac{507}{512}$	$+\frac{1409}{5120}$	H	H	$-\frac{381}{160}$	H	H	$-\frac{44761}{9216}$
x^2	[cos + sin]		$+\frac{1}{4}$	$-\frac{13}{16}$	$+\frac{367}{512}$	$+\frac{1}{4}$	$-\frac{5}{8}$	0	$+\frac{3}{4}$	$-\frac{223}{96}$	$+\frac{2987}{1536}$
	[cos − sin]		0	$-\frac{1}{32}$	$+\frac{71}{2560}$	0	0	$-\frac{1}{48}$	0	0	$-\frac{25}{1536}$
	[cos]		H	$-\frac{27}{64}$	$+\frac{953}{2560}$	H	H	$-\frac{1}{96}$	H	H	$+\frac{1481}{1536}$
	[sin]		H	$-\frac{25}{64}$	$+\frac{441}{1280}$	H	H	$+\frac{1}{96}$	H	H	$+\frac{251}{256}$
x^3	[cos + sin]		$-\frac{3}{8}$	$+\frac{33}{32}$	$-\frac{753}{1024}$		$-\frac{1}{2}$	$+\frac{95}{64}$	$-\frac{1}{8}$	$-\frac{13}{64}$	$+\frac{1577}{1024}$
	[cos − sin]		0	$+\frac{3}{64}$	$-\frac{13}{1024}$		0	$+\frac{7}{192}$	0	0	$+\frac{95}{3072}$
	[cos]		H	$+\frac{69}{128}$	$-\frac{383}{1024}$		H	$+\frac{73}{96}$	H	H	$+\frac{2413}{3072}$
	[sin]		H	$+\frac{63}{128}$	$-\frac{185}{512}$		H	$+\frac{139}{192}$	H	H	$+\frac{1159}{1536}$

... *Table* $(x^0, x^1, x^2, x^3)\ {\cos \atop \sin}\ 2f$.

	$\cos \atop \sin$ $2f$	$[\ ]^6$ e^4	$[\ ]^6$ e^6	$[\ ]^7$ e^5	$[\ ]^7$ e^7	$[\ ]^8$ e^6	$[\ ]^9$ e^7
x^0	[cos + sin]	$+\frac{115}{16}$	$-\frac{2049}{80}$	$+\frac{9893}{960}$	$-\frac{889303}{23040}$	$+\frac{42037}{2880}$	$+\frac{367439}{17920}$
	[cos − sin]	0	0	0	0	0	0
	[cos]	H	H	H	H	H	H
	[sin]	H	H	H	H	H	H
x	[cos + sin]	$-\frac{61}{16}$	$+\frac{69}{5}$	$-\frac{4553}{768}$	$+\frac{2061389}{92160}$	$-\frac{8551}{960}$	$-\frac{26809}{2048}$
	[cos − sin]	0	0	0	0	0	0
	[cos]	H	H	H	H	H	H
	[sin]	H	H	H	H	H	H
x^2	[cos + sin]	$+\frac{25}{16}$	$-\frac{21}{4}$	$+\frac{269}{96}$	$-\frac{77143}{7680}$	$+\frac{297}{64}$	$+\frac{18749}{2560}$
	[cos − sin]	0	0	0	0	0	0
	[cos]	H	H	H	H	H	H
	[sin]	H	H	H	H	H	H
x^3	[cos + sin]	$-\frac{7}{16}$	$+\frac{3}{4}$	$-\frac{65}{64}$	$+\frac{8425}{3072}$	$-\frac{379}{192}$	$-\frac{3563}{1024}$
	[cos − sin]	0	0	0	0	0	0
	[cos]	H	H	H	H	H	H
	[sin]	H	H	H	H	H	H

Table $(x^4, x^5, x^6, x^7)\ \begin{matrix}\cos\\ \sin\end{matrix}\ 2f\ldots$

$\begin{matrix}\cos\\ \sin\end{matrix}\ 2f$		$[\ ]^0$			$[\ ]^1$				$[\ ]^2$			
		e^2	e^4	e^6	e	e^3	e^5	e^7	e^0	e^2	e^4	e^6
x^4	[cos + sin]		$+\frac{1}{4}$	$+\frac{3}{64}$			$-\frac{1}{2}$	$+\frac{5}{32}$			$+\frac{3}{8}$	$-\frac{1}{2}$
	[cos − sin]		,,	,,			$-\frac{1}{4}$	o			$+\frac{1}{16}$	$-\frac{1}{16}$
	[cos]		$+\frac{1}{4}$	$+\frac{3}{64}$			$-\frac{3}{8}$	$+\frac{5}{64}$			$+\frac{7}{32}$	$-\frac{9}{32}$
	[sin]		o	o			$-\frac{1}{8}$	$+\frac{5}{64}$			$+\frac{5}{32}$	$-\frac{7}{32}$
x^5	[cos + sin]			$+\frac{25}{64}$			$-\frac{5}{16}$	$+\frac{25}{256}$				$+\frac{5}{16}$
	[cos − sin]			,,			$-\frac{5}{32}$	$-\frac{1}{256}$				$+\frac{3}{32}$
	[cos]			$+\frac{25}{64}$			$-\frac{15}{64}$	$+\frac{3}{64}$				$+\frac{13}{64}$
	[sin]			o			$-\frac{5}{64}$	$+\frac{13}{256}$				$+\frac{7}{64}$
x^6	[cos + sin]			$+\frac{15}{64}$				$-\frac{25}{64}$				$+\frac{5}{16}$
	[cos − sin]			,,				$-\frac{15}{64}$				$+\frac{3}{32}$
	[cos]			$+\frac{15}{64}$				$-\frac{5}{16}$				$+\frac{13}{64}$
	[sin]			o				$-\frac{5}{64}$				$+\frac{7}{64}$
x^7	[cos + sin]							$-\frac{35}{128}$				
	[cos − sin]							$-\frac{21}{128}$				
	[cos]							$-\frac{7}{32}$				
	[sin]							$-\frac{7}{128}$				

... *Table* $(x^4, x^5, x^6, x^7)\begin{smallmatrix}\cos\\ \sin\end{smallmatrix} 2f$ ...

		$[\;]^3$				$[\;]^4$			$[\;]^5$		
	$\begin{smallmatrix}\cos\\ \sin\end{smallmatrix} 2f$	e	e^3	e^5	e^7	e^2	e^4	e^6	e^3	e^5	e^7
x^4	[cos + sin]			0	$-\frac{3}{16}$		$+\frac{1}{4}$	$-\frac{43}{64}$		$+\frac{1}{2}$	$-\frac{47}{32}$
	[cos − sin]			0	$-\frac{1}{32}$		0	$-\frac{1}{64}$		0	$-\frac{1}{48}$
	[cos]			0	$-\frac{7}{64}$		H	$-\frac{11}{32}$		H	$-\frac{143}{192}$
	[sin]			0	$-\frac{5}{64}$		H	$-\frac{21}{64}$		H	$-\frac{139}{192}$
x^5	[cos + sin]			$-\frac{5}{16}$	$+\frac{165}{256}$			$-\frac{15}{64}$		$-\frac{5}{32}$	$+\frac{59}{256}$
	[cos − sin]			$-\frac{1}{32}$	$+\frac{17}{256}$			$-\frac{1}{64}$		0	$-\frac{1}{256}$
	[cos]			$-\frac{11}{64}$	$+\frac{91}{256}$			$-\frac{1}{8}$		H	$+\frac{29}{256}$
	[sin]			$-\frac{9}{64}$	$+\frac{37}{128}$			$-\frac{7}{64}$		H	$+\frac{15}{128}$
x^6	[cos + sin]				$-\frac{5}{64}$			$+\frac{15}{64}$			$+\frac{21}{64}$
	[cos − sin]				$-\frac{1}{64}$			$+\frac{1}{64}$			$+\frac{1}{64}$
	[cos]				$-\frac{3}{64}$			$+\frac{1}{8}$			$+\frac{11}{64}$
	[sin]				$-\frac{1}{32}$			$+\frac{7}{64}$			$+\frac{5}{32}$
x^7	[cos + sin]				$-\frac{35}{128}$						$-\frac{21}{128}$
	[cos − sin]				$-\frac{7}{128}$						$-\frac{1}{128}$
	[cos]				$-\frac{21}{128}$						$-\frac{11}{128}$
	[sin]				$-\frac{7}{64}$						$-\frac{5}{64}$

... *Table* $(x^4,\ x^5,\ x^6,\ x^7)\ {\cos \atop \sin} 2f$.

	$\begin{matrix}\cos\\ \sin\end{matrix} 2f$	$[\ ]^6$ e^4	$[\ ]^6$ e^6	$[\ ]^7$ e^5	$[\ ]^7$ e^7	$[\ ]^8$ e^6	$[\ ]^9$ e^7
x^4	[cos + sin]	$+\frac{1}{16}$	$+\frac{9}{16}$	$+\frac{1}{4}$	$+\frac{5}{24}$	$+\frac{41}{64}$	$+\frac{43}{32}$
	[cos − sin]	0	0	0	0	0	0
	[cos]	H	H	H	H	H	H
	[sin]	H	H	H	H	H	H
x^5	[cos + sin]		$-\frac{13}{32}$	$-\frac{1}{32}$	$-\frac{163}{256}$	$-\frac{9}{64}$	$-\frac{101}{256}$
	[cos − sin]		0	0	0	0	0
	[cos]		H	H	H	H	H
	[sin]		H	H	H	H	H
x^6	[cos + sin]		$+\frac{3}{32}$		$+\frac{19}{64}$	$+\frac{1}{64}$	$+\frac{5}{64}$
	[cos − sin]		0		0	0	0
	[cos]		H		H	H	H
	[sin]		H		H	H	H
x^7	[cos + sin]				$-\frac{7}{128}$		$-\frac{1}{128}$
	[cos − sin]				0		0
	[cos]				H		H
	[sin]				H		H

Table $(x^0,\ x^1,\ x^2,\ x^3)\ {\cos \atop \sin}\ 3f \ldots$

${\cos \atop \sin}\ 3f$		$[\ ]^0$			$[\ ]^1$			$[\ ]^2$			
		e^3	e^5	e^7	e^2	e^4	e^6	e	e^3	e^5	e^7
x^0	[cos + sin]	$-\frac{1}{2}$	$-\frac{3}{16}$	$-\frac{3}{32}$	$+\frac{21}{8}$	$-\frac{31}{16}$	$-\frac{103}{1024}$	-3	$+\frac{33}{4}$	$-\frac{89}{16}$	$+\frac{163}{192}$
	[cos − sin]	,,	,,	,,	0	$-\frac{5}{128}$	$-\frac{11}{320}$	0	0	$-\frac{1}{80}$	$-\frac{17}{960}$
	[cos]	$-\frac{1}{2}$	$-\frac{3}{16}$	$-\frac{3}{32}$	H	$-\frac{253}{256}$	$-\frac{691}{10240}$	H	H	$-\frac{223}{80}$	$+\frac{133}{320}$
	[sin]	0	0	0	H	$-\frac{243}{256}$	$-\frac{339}{10240}$	H	H	$-\frac{111}{40}$	$+\frac{13}{30}$
x^1	[cos + sin]	$-\frac{3}{4}$	$+\frac{1}{32}$	$+\frac{3}{64}$	$+\frac{5}{4}$	$-\frac{61}{48}$	$+\frac{129}{512}$	$-\frac{1}{2}$	$+\frac{9}{8}$	$-\frac{125}{96}$	$+\frac{709}{1152}$
	[cos − sin]	,,	,,	,,	0	$-\frac{1}{96}$	$+\frac{59}{1280}$	0	0	$-\frac{1}{96}$	$+\frac{89}{5760}$
	[cos]	$-\frac{3}{4}$	$+\frac{1}{32}$	$+\frac{3}{64}$	H	$-\frac{41}{64}$	$+\frac{763}{5120}$	H	H	$-\frac{21}{32}$	$+\frac{1817}{5760}$
	[sin]	0	0	0	H	$-\frac{121}{192}$	$+\frac{527}{5120}$	H	H	$-\frac{31}{48}$	$+\frac{3}{10}$
x^2	[cos + sin]	$-\frac{1}{2}$	$+\frac{1}{8}$	0	$+\frac{1}{4}$	$+\frac{5}{16}$	$-\frac{85}{1536}$		-1	$+\frac{19}{12}$	$-\frac{49}{96}$
	[cos − sin]	,,	,,	,,	0	$+\frac{5}{32}$	$+\frac{53}{768}$		0	$+\frac{1}{24}$	$+\frac{7}{480}$
	[cos]	$-\frac{1}{2}$	$+\frac{1}{8}$	0	H	$+\frac{15}{64}$	$+\frac{7}{1024}$		H	$+\frac{13}{16}$	$-\frac{119}{480}$
	[sin]	0	0	0	H	$+\frac{5}{64}$	$-\frac{191}{3072}$		H	$+\frac{37}{48}$	$-\frac{21}{80}$
x^3	[cos + sin]	$-\frac{1}{8}$	$-\frac{9}{32}$	$-\frac{3}{64}$		$+\frac{3}{4}$	$-\frac{65}{128}$		$-\frac{3}{8}$	$+\frac{9}{16}$	$-\frac{61}{128}$
	[cos − sin]	,,	,,	,,		$+\frac{3}{16}$	$-\frac{1}{32}$		0	0	$-\frac{5}{128}$
	[cos]	$-\frac{1}{8}$	$-\frac{9}{32}$	$-\frac{3}{64}$		$+\frac{15}{32}$	$-\frac{69}{256}$		H	H	$-\frac{33}{128}$
	[sin]	0	0	0		$+\frac{9}{32}$	$-\frac{61}{256}$		H	H	$-\frac{7}{32}$

... *Table* $(x^0,\ x^1,\ x^2,\ x^3)\ {\cos \atop \sin} 3f$ ...

	${\cos \atop \sin} 3f$	$[\ ]^3$ e^0	e^2	e^4	e^6	$[\ ]^4$ e	e^3	e^5	e^7	$[\ ]^5$ e^2	e^4	e^6
x^0	[cos + sin]	$+1$	-9	$+\frac{1215}{64}$	$-\frac{449}{32}$	$+3$	$-\frac{39}{2}$	$+\frac{155}{4}$	$-\frac{767}{24}$	$+\frac{51}{8}$	$-\frac{593}{16}$	$+\frac{75643}{1024}$
	[cos − sin]	0	0	0	$-\frac{27}{5120}$	0	0	0	$-\frac{1}{420}$	0	0	0
	[cos]	H	H	H	$-\frac{71867}{10240}$	H	H	H	$-\frac{8949}{560}$	H	H	H
	[sin]	H	H	H	$-\frac{71813}{10240}$	H	H	H	$-\frac{26843}{1680}$	H	H	H
x^1	[cos + sin]		$+\frac{1}{2}$	$-\frac{9}{8}$	$-\frac{9}{128}$	$-\frac{1}{2}$	$+\frac{15}{4}$	$-\frac{173}{24}$	$+\frac{161}{36}$	$-\frac{7}{4}$	$+\frac{509}{48}$	$-\frac{10439}{512}$
	[cos − sin]		0	0	$-\frac{21}{2560}$	0	0	0	$-\frac{1}{144}$	0	0	0
	[cos]		H	H	$-\frac{201}{5120}$	H	H	H	$+\frac{643}{288}$	H	H	H
	[sin]		H	H	$-\frac{159}{5120}$	H	H	H	$+\frac{215}{96}$	H	H	H
x^2	[cos + sin]		$+\frac{1}{2}$	$-\frac{9}{4}$	$+\frac{399}{128}$		$+\frac{1}{2}$	$-\frac{8}{3}$	$+\frac{437}{96}$	$+\frac{1}{4}$	$-\frac{13}{16}$	$-\frac{1393}{1536}$
	[cos − sin]		0	0	$+\frac{11}{512}$		0	0	$+\frac{7}{480}$	0	0	0
	[cos]		H	H	$+\frac{1607}{1024}$		H	H	$+\frac{137}{60}$	H	H	H
	[sin]		H	H	$+\frac{1585}{1024}$		H	H	$+\frac{363}{160}$	H	H	H
x^3	[cos + sin]			$+\frac{3}{8}$	$-\frac{9}{16}$		$-\frac{3}{8}$	$+\frac{15}{8}$	$-\frac{343}{128}$		$-\frac{3}{4}$	$+\frac{469}{128}$
	[cos − sin]			0	$-\frac{1}{128}$		0	0	$-\frac{1}{128}$		0	0
	[cos]			H	$-\frac{73}{256}$		H	H	$-\frac{43}{32}$		H	H
	[sin]			H	$-\frac{71}{256}$		H	H	$-\frac{171}{128}$		H	H

... *Table* $(x^0,\ x^1,\ x^2,\ x^3)\ {\cos \atop \sin} 3f$.

	$\cos \atop \sin$ $3f$	$[\ ]^6$ e^3	$[\ ]^6$ e^5	$[\ ]^6$ e^7	$[\ ]^7$ e^4	$[\ ]^7$ e^6	$[\ ]^8$ e^5	$[\ ]^8$ e^7	$[\ ]^9$ e^6	$[\ ]^{10}$ e^7
x^0	[cos + sin]	$+\frac{47}{4}$	$-\frac{525}{8}$	$+\frac{43041}{320}$	$+\frac{2567}{128}$	$-\frac{35563}{320}$	$+\frac{2611}{80}$	$-\frac{87599}{480}$	$+\frac{263351}{5120}$	$+\frac{106469}{1344}$
	[cos − sin]	0	0	0	0	0	0	0	0	0
	[cos]	H	H	H	H	H	H	H	H	H
	[sin]	H	H	H	H	H	H	H	H	H
x^1	[cos + sin]	$-\frac{33}{8}$	$+\frac{373}{16}$	$-\frac{5919}{128}$	$-\frac{787}{96}$	$+\frac{58143}{1280}$	$-\frac{1423}{96}$	$+\frac{237373}{2880}$	$-\frac{64653}{2560}$	$-\frac{47603}{1152}$
	[cos − sin]	0	0	0	0	0	0	0	0	0
	[cos]	H	H	H	H	H	H	H	H	H
	[sin]	H	H	H	H	H	H	H	H	H
x^2	[cos + sin]	$+1$	$-\frac{37}{8}$	$+\frac{891}{160}$	$+\frac{83}{32}$	$-\frac{9871}{768}$	$+\frac{133}{24}$	$-\frac{857}{30}$	$+\frac{5431}{512}$	$+\frac{1817}{96}$
	[cos − sin]	0	0	0	0	0	0	0	0	0
	[cos]	H	H	H	H	H	H	H	H	H
	[sin]	H	H	H	H	H	H	H	H	H
x^3	[cos + sin]	$-\frac{1}{8}$	$-\frac{9}{16}$	$+\frac{627}{128}$	$-\frac{9}{16}$	$+\frac{35}{32}$	$-\frac{51}{32}$	$+\frac{367}{64}$	$-\frac{467}{128}$	$-\frac{945}{128}$
	[cos − sin]	0	0	0	0	0	0	0	0	0
	[cos]	H	H	H	H	H	H	H	H	H
	[sin]	H	H	H	H	H	H	H	H	H

Table *(x^4, x^5, x^6, x^7) $\begin{matrix}\cos\\ \sin\end{matrix}$ $3f$...

$\begin{matrix}\cos\\ \sin\end{matrix}\ 3f$		$[\ ]^0$			$[\ ]^1$			$[\ ]^2$			
		e^3	e^5	e^7	e^2	e^4	e^6	e	e^3	e^5	e^7
x^4	[cos + sin]		$-\frac{7}{16}$	$+\frac{3}{32}$		$+\frac{1}{4}$	$+\frac{9}{128}$			$-\frac{5}{8}$	$+\frac{43}{64}$
	[cos − sin]		,,	,,		$+\frac{1}{16}$	$+\frac{3}{32}$			$-\frac{1}{16}$	$+\frac{5}{192}$
	[cos]		$-\frac{7}{16}$	$+\frac{3}{32}$		$+\frac{5}{32}$	$+\frac{21}{256}$			$-\frac{11}{32}$	$+\frac{67}{192}$
	[sin]		o	o		$+\frac{3}{32}$	$-\frac{3}{256}$			$-\frac{9}{32}$	$+\frac{31}{96}$
x^5	[cos + sin]		$-\frac{5}{32}$	$-\frac{9}{64}$			$+\frac{35}{64}$			$-\frac{5}{16}$	$+\frac{45}{128}$
	[cos − sin]		,,	,,			$+\frac{7}{32}$			$-\frac{1}{32}$	$+\frac{1}{128}$
	[cos]		$-\frac{5}{32}$	$-\frac{9}{64}$			$+\frac{49}{128}$			$-\frac{11}{64}$	$+\frac{23}{128}$
	[sin]		o	o			$+\frac{21}{128}$			$-\frac{9}{64}$	$+\frac{11}{64}$
x^6	[cos + sin]			$-\frac{3}{8}$			$+\frac{15}{64}$				$-\frac{15}{32}$
	[cos − sin]			,,			$+\frac{3}{32}$				$-\frac{3}{32}$
	[cos]			$-\frac{3}{8}$			$+\frac{21}{128}$				$-\frac{9}{32}$
	[sin]			o			$+\frac{9}{128}$				$-\frac{3}{16}$
x^7	[cos + sin]			$-\frac{21}{128}$							$-\frac{35}{128}$
	[cos − sin]			,,							$-\frac{7}{128}$
	[cos]			$-\frac{21}{128}$							$-\frac{21}{128}$
	[sin]			o							$-\frac{7}{64}$

... *Table* $(x^4, x^5, x^6, x^7) \begin{matrix}\cos\\ \sin\end{matrix} 3f$ ...

	$\begin{matrix}\cos\\ \sin\end{matrix} 3f$	$[\ ]^3$ e^0	$[\ ]^3$ e^2	$[\ ]^3$ e^4	$[\ ]^3$ e^6	$[\ ]^4$ e	$[\ ]^4$ e^3	$[\ ]^4$ e^5	$[\ ]^4$ e^7	$[\ ]^5$ e^2	$[\ ]^5$ e^4	$[\ ]^5$ e^6
x^4	[cos + sin]			$+\frac{3}{8}$	$-\frac{9}{8}$			$+\frac{1}{8}$	$-\frac{25}{32}$		$+\frac{1}{4}$	$-\frac{129}{128}$
	[cos − sin]			o	$-\frac{3}{128}$			o	$-\frac{1}{96}$		o	o
	[cos]			H	$-\frac{147}{256}$			H	$-\frac{19}{48}$		H	H
	[sin]			H	$-\frac{141}{256}$			H	$-\frac{37}{96}$		H	H
x^5	[cos + sin]				$+\frac{5}{16}$			$-\frac{5}{16}$	$+\frac{75}{64}$			$-\frac{25}{64}$
	[cos − sin]				$+\frac{1}{64}$			o	$+\frac{1}{64}$			o
	[cos]				$+\frac{21}{128}$			H	$+\frac{19}{32}$			H
	[sin]				$+\frac{19}{128}$			H	$+\frac{37}{64}$			H
x^6	[cos + sin]				$+\frac{5}{16}$				o			$+\frac{15}{64}$
	[cos − sin]				$+\frac{1}{64}$				o			o
	[cos]				$+\frac{21}{128}$				o			H
	[sin]				$+\frac{19}{128}$				o			H
x^7	[cos + sin]								$-\frac{35}{128}$			
	[cos − sin]								$-\frac{1}{128}$			
	[cos]								$-\frac{9}{64}$			
	[sin]								$-\frac{17}{128}$			

... *Table* $(x^4,\ x^5,\ x^6,\ x^7)\ {\cos \atop \sin}\ 3f$.

	$\begin{matrix}\cos\\ \sin\end{matrix}\ 3f$	$[\ ]^6$ e^3	$[\ ]^6$ e^5	$[\ ]^6$ e^7	$[\ ]^7$ e^4	$[\ ]^7$ e^6	$[\ ]^8$ e^5	$[\ ]^8$ e^7	$[\ ]^9$ e^6	$[\ ]^{10}$ e^7
x^4	[cos + sin]		$+\frac{11}{16}$	$-\frac{195}{64}$	$+\frac{1}{16}$	$+\frac{33}{32}$	$+\frac{5}{16}$	$+\frac{67}{96}$	$+\frac{123}{128}$	$+\frac{451}{192}$
	[cos − sin]		0	0	0	0	0	0	0	0
	[cos]		H	H	H	H	H	H	H	H
	[sin]		H	H	H	H	H	H	H	H
x^5	[cos + sin]		$-\frac{5}{32}$	$+\frac{27}{128}$		$-\frac{17}{32}$	$-\frac{1}{32}$	$-\frac{67}{64}$	$-\frac{11}{64}$	$-\frac{73}{128}$
	[cos − sin]		0	0		0	0	0	0	0
	[cos]		H	H		H	H	H	H	H
	[sin]		H	H		H	H	H	H	H
x^6	[cos + sin]			$+\frac{15}{32}$		$+\frac{3}{32}$		$+\frac{3}{8}$	$+\frac{1}{64}$	$+\frac{3}{32}$
	[cos − sin]			0		0		0	0	0
	[cos]			H		H		H	H	H
	[sin]			H		H		H	H	H
x^7	[cos + sin]			$-\frac{21}{128}$				$-\frac{7}{128}$		$-\frac{1}{128}$
	[cos − sin]			0				0		0
	[cos]			H				H		H
	[sin]			H				H		H

Table $(x^0,\ x^1,\ x^2,\ x^3)\ {\cos \atop \sin} 4f \ldots$

	$\cos \atop \sin$ $4f$	$[\ ]^0$ e^4	$[\ ]^0$ e^6	$[\ ]^1$ e	$[\ ]^1$ e^3	$[\ ]^1$ e^5	$[\ ]^1$ e^7	$[\ ]^2$ e^2	$[\ ]^2$ e^4	$[\ ]^2$ e^6
x^0	[cos + sin]	$+\frac{5}{16}$	$+\frac{3}{16}$		$-\frac{17}{6}$	$+\frac{161}{96}$	$+\frac{907}{3840}$	$+\frac{11}{2}$	$-\frac{149}{12}$	$+\frac{325}{48}$
	[cos − sin]	,,	,,		0	$+\frac{7}{480}$	$+\frac{277}{11520}$	0	0	$+\frac{1}{720}$
	[cos]	$+\frac{5}{16}$	$+\frac{3}{16}$		H	$+\frac{203}{240}$	$+\frac{1499}{11520}$	H	H	$+\frac{1219}{360}$
	[sin]	0	0		H	$+\frac{133}{160}$	$+\frac{611}{5760}$	H	H	$+\frac{2437}{720}$
x^1	[cos + sin]	$+\frac{5}{8}$	$+\frac{3}{32}$		$-\frac{31}{16}$	$+\frac{1447}{768}$	$-\frac{4537}{30720}$	$+\frac{7}{4}$	$-\frac{53}{12}$	$+\frac{743}{192}$
	[cos − sin]	,,	,,		0	$+\frac{19}{768}$	$-\frac{793}{92160}$	0	0	$+\frac{11}{960}$
	[cos]	$+\frac{5}{8}$	$+\frac{3}{32}$		H	$+\frac{733}{768}$	$-\frac{3601}{46080}$	H	H	$+\frac{621}{320}$
	[sin]	0	0		H	$+\frac{119}{128}$	$-\frac{6409}{92160}$	H	H	$+\frac{463}{240}$
x^2	[cos + sin]	$+\frac{5}{8}$	$-\frac{5}{32}$		$-\frac{3}{4}$	$+\frac{53}{96}$	$-\frac{2123}{7680}$	$+\frac{1}{4}$	$+\frac{1}{2}$	$-\frac{49}{64}$
	[cos − sin]	,,	,,		0	$-\frac{7}{96}$	$-\frac{649}{7680}$	0	0	$-\frac{1}{64}$
	[cos]	$+\frac{5}{8}$	$-\frac{5}{32}$		H	$+\frac{23}{96}$	$-\frac{231}{1280}$	H	H	$-\frac{25}{64}$
	[sin]	0	0		H	$+\frac{5}{16}$	$-\frac{737}{7680}$	H	H	$-\frac{3}{8}$
x^3	[cos + sin]	$+\frac{5}{16}$	0		$-\frac{1}{8}$	$-\frac{43}{64}$	$+\frac{389}{1024}$		$+1$	$-\frac{7}{4}$
	[cos − sin]	,,	,,		0	$-\frac{11}{64}$	$-\frac{59}{3072}$		0	$-\frac{1}{48}$
	[cos]	$+\frac{5}{16}$	0		H	$-\frac{27}{64}$	$+\frac{277}{1536}$		H	$-\frac{85}{96}$
	[sin]	0	0		H	$-\frac{1}{4}$	$+\frac{613}{3072}$		H	$-\frac{83}{96}$

... *Table* $(x^0, x^1, x^2, x^3)\ {\cos \atop \sin}\ 4f$ ...

	$\cos \atop \sin$ $4f$	[ ]³ e	e^3	e^5	e^7	[ ]⁴ e^0	e^2	e^4	e^6
x^0	[cos + sin]	-4	$+\frac{45}{2}$	$-\frac{591}{16}$	$+\frac{5737}{256}$	$+1$	-16	$+\frac{247}{4}$	$-\frac{3355}{36}$
	[cos − sin]	0	0	0	$-\frac{9}{8960}$	0	0	0	0
	[cos]	H	H	H	$+\frac{100393}{8960}$	H	H	H	H
	[sin]	H	H	H	$+\frac{50201}{4480}$	H	H	H	H
x^1	[cos + sin]	$-\frac{1}{2}$	$+\frac{39}{16}$	$-\frac{687}{128}$	$+\frac{12355}{2048}$		$+\frac{1}{2}$	-2	$-\frac{1}{8}$
	[cos − sin]	0	0	0	$+\frac{69}{10240}$		0	0	0
	[cos]	H	H	H	$+\frac{15461}{5120}$		H	H	H
	[sin]	H	H	H	$+\frac{30853}{10240}$		H	H	H
x^2	[cos + sin]		$-\frac{5}{4}$	$+\frac{67}{16}$	$-\frac{2119}{512}$		$+\frac{1}{2}$	-4	$+\frac{245}{24}$
	[cos − sin]		0	0	$-\frac{17}{2560}$		0	0	0
	[cos]		H	H	$-\frac{2653}{1280}$		H	H	H
	[sin]		H	H	$-\frac{5289}{2560}$		H	H	H
x^3	[cos + sin]		$-\frac{3}{8}$	$+\frac{39}{32}$	$-\frac{2037}{1024}$			$+\frac{3}{8}$	-1
	[cos − sin]		0	0	$-\frac{7}{1024}$			0	0
	[cos]		H	H	$-\frac{511}{512}$			H	H
	[sin]		H	H	$-\frac{1015}{1024}$			H	H

... *Table* $(x^0, x^1, x^2, x^3)\ {\cos \atop \sin} 4f$ ...

$\cos \atop \sin$ $4f$		$[\ ]^5$ e	e^3	e^5	e^7	$[\ ]^6$ e^2	e^4	e^6	$[\ ]^7$ e^3	e^5	e^7
x^0	[cos + sin]	$+4$	$-\frac{85}{2}$	$+\frac{6845}{48}$	$-\frac{490585}{2304}$	$+\frac{21}{2}$	$-\frac{377}{4}$	$+\frac{597}{2}$	$+\frac{137}{6}$	$-\frac{18113}{96}$	$+\frac{2249741}{3840}$
	[cos − sin]	0	0	0	0	0	0	0	0	0	0
	[cos]	H	H	H	H	H	H	H	H	H	H
	[sin]	H	H	H	H	H	H	H	H	H	H
x^1	[cos + sin]	$-\frac{1}{2}$	$+\frac{95}{16}$	$-\frac{7325}{384}$	$+\frac{407635}{18432}$	$-\frac{9}{4}$	$+\frac{83}{4}$	$-\frac{4029}{64}$	$-\frac{103}{16}$	$+\frac{40967}{768}$	$-\frac{4894577}{30720}$
	[cos − sin]	0	0	0	0	0	0	0	0	0	0
	[cos]	H	H	H	H	H	H	H	H	H	H
	[sin]	H	H	H	H	H	H	H	H	H	H
x^2	[cos + sin]		$+\frac{3}{4}$	$-\frac{295}{48}$	$+\frac{8785}{512}$	$+\frac{1}{4}$	-1	$-\frac{215}{64}$	$+\frac{5}{4}$	$-\frac{779}{96}$	$+\frac{103901}{7680}$
	[cos − sin]		0	0	0	0	0	0	0	0	0
	[cos]		H	H	H	H	H	H	H	H	H
	[sin]		H	H	H	H	H	H	H	H	H
x^3	[cos + sin]		$-\frac{3}{8}$	$+\frac{95}{32}$	$-\frac{7285}{1024}$		-1	$+\frac{117}{16}$	$-\frac{1}{8}$	$-\frac{67}{64}$	$+\frac{12069}{1024}$
	[cos − sin]		0	0	0		0	0	0	0	0
	[cos]		H	H	H		H	H	H	H	H
	[sin]		H	H	H		H	H	H	H	H

... *Table* $(x^0, x^1, x^2, x^3) \begin{smallmatrix}\cos\\ \cdot\\ \sin\end{smallmatrix} 4f$.

	$\begin{smallmatrix}\cos\\ \sin\end{smallmatrix} 4f$	$[\ ]^8$ e^4	$[\ ]^8$ e^6	$[\ ]^9$ e^5	$[\ ]^9$ e^7	$[\ ]^{10}$ e^6	$[\ ]^{11}$ e^7
x^0	[cos + sin]	$+\frac{2141}{48}$	$-\frac{84857}{240}$	$+\frac{13011}{160}$	$-\frac{809813}{1280}$	$+\frac{5087}{36}$	$+\frac{19116073}{80640}$
	[cos − sin]	0	0	0	0	0	0
	[cos]	H	H	H	H	H	H
	[sin]	H	H	H	H	H	H
x^1	[cos + sin]	$-\frac{359}{24}$	$+\frac{56527}{480}$	$-\frac{7887}{256}$	$+\frac{2427367}{10240}$	$-\frac{11263}{192}$	$-\frac{9742939}{92160}$
	[cos − sin]	0	0	0	0	0	0
	[cos]	H	H	H	H	H	H
	[sin]	H	H	H	H	H	H
x^2	[cos + sin]	$+\frac{31}{8}$	$-\frac{855}{32}$	$+\frac{307}{32}$	$-\frac{172907}{2560}$	$+\frac{3995}{192}$	$+\frac{317653}{7680}$
	[cos − sin]	0	0	0	0	0	0
	[cos]	H	H	H	H	H	H
	[sin]	H	H	H	H	H	H
x^3	[cos + sin]	$-\frac{11}{16}$	$+\frac{3}{2}$	$-\frac{147}{64}$	$+\frac{10999}{1024}$	$-\frac{145}{24}$	$-\frac{42325}{3072}$
	[cos − sin]	0	0	0	0	0	0
	[cos]	H	H	H	H	H	H
	[sin]	H	H	H	H	H	H

Table $(x^4, x^5, x^6, x^7)\ {\cos \atop \sin} 4f \ldots$

	$\cos \atop \sin$ $4f$	$[\ ]^0$		$[\ ]^1$			$[\ ]^2$			$[\ ]^3$			
		e^4	e^6	e^3	e^5	e^7	e^2	e^4	e^6	e^1	e^3	e^5	e^7
x^4	[cos + sin]	$+\frac{1}{16}$	$+\frac{3}{8}$		$-\frac{5}{8}$	$+\frac{29}{64}$		$+\frac{1}{4}$	$+\frac{1}{32}$			$-\frac{3}{4}$	$+\frac{111}{64}$
	[cos − sin]	,,	,,		$-\frac{1}{8}$	$+\frac{1}{192}$		o	$+\frac{1}{32}$			o	$+\frac{1}{64}$
	[cos]	$+\frac{1}{16}$	$+\frac{3}{8}$		$-\frac{3}{8}$	$+\frac{11}{48}$		H	$+\frac{1}{32}$			H	$+\frac{7}{8}$
	[sin]	o	o		$-\frac{1}{4}$	$+\frac{43}{192}$		H	o			H	$+\frac{55}{64}$
x^5	[cos + sin]		$+\frac{11}{32}$		$-\frac{5}{32}$	$-\frac{79}{256}$			$+\frac{45}{64}$			$-\frac{5}{16}$	$+\frac{195}{256}$
	[cos − sin]		,,		$-\frac{1}{32}$	$-\frac{37}{256}$			$+\frac{3}{64}$			o	$+\frac{1}{256}$
	[cos]		$+\frac{11}{32}$		$-\frac{3}{32}$	$-\frac{29}{128}$			$+\frac{3}{8}$			H	$+\frac{49}{128}$
	[sin]		o		$-\frac{1}{16}$	$-\frac{21}{256}$			$+\frac{21}{64}$			H	$+\frac{97}{256}$
x^6	[cos + sin]		$+\frac{3}{32}$			$-\frac{33}{64}$			$+\frac{15}{64}$				$-\frac{35}{64}$
	[cos − sin]		,,			$-\frac{11}{64}$			$+\frac{1}{64}$				$-\frac{1}{64}$
	[cos]		$+\frac{3}{32}$			$-\frac{11}{32}$			$+\frac{1}{8}$				$-\frac{9}{32}$
	[sin]		o			$-\frac{11}{64}$			$+\frac{7}{64}$				$-\frac{17}{64}$
x^7	[cos + sin]					$-\frac{21}{128}$							$-\frac{35}{128}$
	[cos − sin]					$-\frac{7}{128}$							$-\frac{1}{128}$
	[cos]					$-\frac{7}{64}$							$-\frac{9}{64}$
	[sin]					$-\frac{7}{128}$							$-\frac{17}{128}$

... *Table* $(x^4,\ x^5,\ x^6,\ x^7)\ {\cos \atop \sin} 4f$ ...

	${\cos \atop \sin} 4f$	$[\ ]^4$				$[\ ]^5$				$[\ ]^6$		
		e^0	e^2	e^4	e^6	e	e^3	e^5	e^7	e^2	e^4	e^6
x^4	[cos + sin]			$+\frac{3}{8}$	-2			$+\frac{1}{4}$	$-\frac{125}{64}$		$+\frac{1}{4}$	$-\frac{45}{32}$
	[cos − sin]			0	0			0	0		0	0
	[cos]			H	H			H	H		H	H
	[sin]			H	H			H	H		H	H
x^5	[cos + sin]				$-\frac{5}{16}$			$-\frac{5}{16}$	$+\frac{475}{256}$			$-\frac{35}{64}$
	[cos − sin]				0			0	0			0
	[cos]				H			H	H			H
	[sin]				H			H	H			H
x^6	[cos + sin]				$+\frac{5}{16}$				$+\frac{5}{64}$			$+\frac{15}{64}$
	[cos − sin]				0				0			0
	[cos]				H				H			H
	[sin]				H				H			H
x^7	[cos + sin]								$-\frac{35}{128}$			
	[cos − sin]								0			
	[cos]								H			
	[sin]								H			

... *Table* $(x^4,\ x^5,\ x^6,\ x^7)\ {\cos \atop \sin} 4f$.

	${\cos \atop \sin} 4f$	$[\ ]^7$			$[\ ]^8$		$[\ ]^9$		$[\ ]^{10}$	$[\ ]^{11}$
		e^3	e^5	e^7	e^4	e^6	e^5	e^7	e^6	e^7
x^4	[cos + sin]		$+\frac{7}{8}$	$-\frac{351}{64}$	$+\frac{1}{16}$	$+\frac{13}{8}$	$+\frac{3}{8}$	$+\frac{95}{64}$	$+\frac{43}{32}$	$+\frac{719}{192}$
	[cos − sin]		o	o	o	o	o	o	o	o
	[cos]		H	H	H	H	H	H	H	H
	[sin]		H	H	H	H	H	H	H	H
x^5	[cos + sin]		$-\frac{5}{32}$	$+\frac{41}{256}$		$-\frac{21}{32}$	$-\frac{1}{32}$	$-\frac{397}{256}$	$-\frac{13}{64}$	$-\frac{199}{256}$
	[cos − sin]		o	o		o	o	o	o	o
	[cos]		H	H		H	H	H	H	H
	[sin]		H	H		H	H	H	H	H
x^6	[cos + sin]			$+\frac{39}{64}$		$+\frac{3}{32}$		$+\frac{29}{64}$	$+\frac{1}{64}$	$+\frac{7}{64}$
	[cos − sin]			o		o		o	o	o
	[cos]			H		H		H	H	H
	[sin]			H		H		H	H	H
x^7	[cos + sin]			$-\frac{21}{128}$				$-\frac{7}{128}$		$-\frac{1}{128}$
	[cos − sin]			o				o		o
	[cos]			H				H		H
	[sin]			H				H		H

Table $(x^0, x^1, x^2, x^3) \begin{matrix}\cos\\ \sin\end{matrix} 5f \ldots$

	$\begin{matrix}\cos\\ \sin\end{matrix} 5f$	[ ]^0		[ ]^1		[ ]^2			[ ]^3		
		e^5	e^7	e^4	e^6	e^3	e^5	e^7	e^2	e^4	e^6
x^0	[cos + sin]	$-\frac{3}{16}$	$-\frac{5}{32}$	$+\frac{1045}{384}$	$-\frac{229}{192}$	$-\frac{95}{12}$	$+\frac{185}{12}$	$-\frac{1295}{192}$	$+\frac{75}{8}$	$-\frac{665}{16}$	$+\frac{58515}{1024}$
	[cos − sin]	,,	,,	o	$-\frac{37}{9216}$	o	o	$+\frac{5}{4032}$	o	o	o
	[cos]	$-\frac{3}{16}$	$-\frac{5}{32}$	H	$-\frac{11029}{18432}$	H	H	$-\frac{13595}{4032}$	H	H	H
	[sin]	o	o	H	$-\frac{10955}{18432}$	H	H	$-\frac{425}{126}$	H	H	H
x^1	[cos + sin]	$-\frac{15}{32}$	$-\frac{11}{64}$	$+\frac{229}{96}$	$-\frac{7747}{3840}$	$-\frac{29}{8}$	$+\frac{205}{24}$	$-\frac{11831}{1920}$	$+\frac{9}{4}$	$-\frac{169}{16}$	$+\frac{9609}{512}$
	[cos − sin]	,,	,,	o	$-\frac{151}{7680}$	o	o	$-\frac{37}{5760}$	o	o	o
	[cos]	$-\frac{15}{32}$	$-\frac{11}{64}$	H	$-\frac{1043}{1024}$	H	H	$-\frac{3553}{1152}$	H	H	H
	[sin]	o	o	H	$-\frac{15343}{15360}$	H	H	$-\frac{277}{90}$	H	H	H
x^2	[cos + sin]	$-\frac{5}{8}$	$+\frac{1}{16}$	$+\frac{43}{32}$	$-\frac{357}{256}$	-1	$+\frac{37}{24}$	$-\frac{691}{480}$	$+\frac{1}{4}$	$+\frac{11}{16}$	$-\frac{1435}{512}$
	[cos − sin]	,,	,,	o	$+\frac{37}{1536}$	o	o	$+\frac{1}{480}$	o	o	o
	[cos]	$-\frac{5}{8}$	$+\frac{1}{16}$	H	$-\frac{2105}{3072}$	H	H	$-\frac{23}{32}$	H	H	H
	[sin]	o	o	H	$-\frac{2179}{3072}$	H	H	$-\frac{173}{240}$	H	H	H
x^3	[cos + sin]	$-\frac{15}{32}$	$+\frac{5}{32}$	$+\frac{7}{16}$	$+\frac{3}{32}$	$-\frac{1}{8}$	$-\frac{19}{16}$	$+\frac{261}{128}$		$+\frac{5}{4}$	$-\frac{531}{128}$
	[cos − sin]	,,	,,	o	$+\frac{47}{384}$	o	o	$+\frac{7}{384}$		o	o
	[cos]	$-\frac{15}{32}$	$+\frac{5}{32}$	H	$+\frac{83}{768}$	H	H	$+\frac{395}{384}$		H	H
	[sin]	o	o	H	$-\frac{11}{768}$	H	H	$+\frac{97}{96}$		H	H

... *Table* $(x^0, x^1, x^2, x^3)\ {\cos \atop \sin} 5f$ ...

${\cos \atop \sin} 5f$		$[\ ]^4$ e	e^3	e^5	e^7	$[\ ]^5$ e^0	e^2	e^4	e^6
x^0	[cos + sin]	-5	$+\frac{95}{2}$	$-\frac{1675}{12}$	$+\frac{25085}{144}$	$+1$	-25	$+\frac{9775}{64}$	$-\frac{110225}{288}$
	[cos − sin]	o	o	o	o	o	o	o	o
	[cos]	H	H	H	H	H	H	H	H
	[sin]	H	H	H	H	H	H	H	H
x^1	[cos + sin]	$-\frac{1}{2}$	$+\frac{17}{4}$	$-\frac{365}{24}$	$+\frac{8413}{288}$		$+\frac{1}{2}$	$-\frac{25}{8}$	$-\frac{25}{128}$
	[cos − sin]	o	o	o	o		o	o	o
	[cos]	H	H	H	H		H	H	H
	[sin]	H	H	H	H		H	H	H
x^2	[cos + sin]		$-\frac{3}{2}$	$+\frac{26}{3}$	$-\frac{527}{32}$		$+\frac{1}{2}$	$-\frac{25}{4}$	$+\frac{9725}{384}$
	[cos − sin]		o	o	o		o	o	o
	[cos]		H	H	H		H	H	H
	[sin]		H	H	H		H	H	H
x^3	[cos + sin]		$-\frac{3}{8}$	$+\frac{17}{8}$	$-\frac{725}{128}$			$+\frac{3}{8}$	$-\frac{25}{16}$
	[cos − sin]		o	o	o			o	o
	[cos]		H	H	H			II	H
	[sin]		H	H	H			H	H

...*Table* $(x^0,\ x^1,\ x^2,\ x^3)\ {\cos \atop \sin}\ 5f$...

${\cos \atop \sin}\ 5f$		$[\ ]^6$ e	e^3	e^5	e^7	$[\ ]^7$ e^2	e^4	e^6	$[\ ]^8$ e^3	e^5	e^7
x^0	[cos + sin]	$+5$	$-\frac{315}{4}$	$+\frac{6375}{16}$	$-\frac{59585}{64}$	$+\frac{125}{8}$	$-\frac{9605}{48}$	$+\frac{2825155}{3072}$	$+\frac{235}{6}$	$-\frac{21515}{48}$	$+\frac{187915}{96}$
	[cos − sin]	0	0	0	0	0	0	0	0	0	0
	[cos]	H	H	H	H	H	H	H	H	H	H
	[sin]	H	H	H	H	H	H	H	H	H	H
x^1	[cos + sin]	$-\frac{1}{2}$	$+\frac{69}{8}$	$-\frac{1335}{32}$	$+\frac{10003}{128}$	$-\frac{11}{4}$	$+\frac{1723}{48}$	$-\frac{242573}{1536}$	$-\frac{37}{4}$	$+\frac{10145}{96}$	$-\frac{425783}{960}$
	[cos − sin]	0	0	0	0	0	0	0	0	0	0
	[cos]	H	H	H	H	H	H	H	H	H	H
	[sin]	H	H	H	H	H	H	H	H	H	H
x^2	[cos + sin]		$+1$	$-\frac{47}{4}$	$+\frac{1553}{32}$	$+\frac{1}{4}$	$-\frac{19}{16}$	$-\frac{4295}{512}$	$+\frac{3}{2}$	$-\frac{313}{24}$	$+\frac{6977}{240}$
	[cos − sin]		0	0	0	0	0	0	0	0	0
	[cos]		H	H	H	H	H	H	H	H	H
	[sin]		H	H	H	H	H	H	H	H	H
x^3	[cos + sin]		$-\frac{3}{8}$	$+\frac{69}{16}$	$-\frac{1995}{128}$		$-\frac{5}{4}$	$+\frac{1639}{128}$	$-\frac{1}{8}$	$-\frac{53}{32}$	$+\frac{769}{32}$
	[cos − sin]		0	0	0		0	0	0	0	0
	[cos]		H	H	H		H	H	H	H	H
	[sin]		H	H	H		H	H	H	H	H

... *Table* (x^0, x^1, x^2, x^3) $\begin{matrix}\cos\\ \sin\end{matrix}\, 5f$.

	$\begin{matrix}\cos\\ \sin\end{matrix}\, 5f$	[ ]^9 e^4	[ ]^9 e^6	[ ]^{10} e^5	[ ]^{10} e^7	[ ]^{11} e^6	[ ]^{12} e^7
x^0	[cos + sin]	$+\frac{11035}{128}$	$-\frac{59127}{64}$	$+\frac{8359}{48}$	$-\frac{1033885}{576}$	$+\frac{3050417}{9216}$	$+\frac{67335}{112}$
	[cos − sin]	o	o	o	o	o	o
	[cos]	H	H	H	H	H	H
	[sin]	H	H	H	H	H	H
x^1	[cos + sin]	$-\frac{787}{32}$	$+\frac{334011}{1280}$	$-\frac{5455}{96}$	$+\frac{665623}{1152}$	$-\frac{918191}{7680}$	$-\frac{37647}{160}$
	[cos − sin]	o	o	o	o	o	o
	[cos]	H	H	H	H	H	H
	[sin]	H	H	H	H	H	H
x^2	[cos + sin]	$+\frac{173}{32}$	$-\frac{12697}{256}$	$+\frac{365}{24}$	$-\frac{13481}{96}$	$+\frac{18859}{512}$	$+\frac{12907}{160}$
	[cos − sin]	o	o	o	o	o	o
	[cos]	H	H	H	H	H	H
	[sin]	H	H	H	H	H	H
x^3	[cos + sin]	$-\frac{13}{16}$	$+\frac{63}{32}$	$-\frac{25}{8}$	$+\frac{7115}{384}$	$-\frac{3563}{384}$	$-\frac{3011}{128}$
	[cos − sin]	o	o	o	o	o	o
	[cos]	H	H	H	H	H	H
	[sin]	H	H	H	H	H	H

Table $(x^4, x^5, x^6, x^7)\ \begin{matrix}\cos\\ \sin\end{matrix}\ 5f\ldots$

$\begin{matrix}\cos\\ \sin\end{matrix}\ 5f$		$[\ ]^0$		$[\ ]^1$		$[\ ]^2$			$[\ ]^3$		
		e^5	e^7	e^4	e^6	e^3	e^5	e^7	e^2	e^4	e^6
x^4	[cos + sin]	$-\frac{3}{16}$	$-\frac{5}{32}$	$+\frac{1}{16}$	$+\frac{25}{32}$		$-\frac{13}{16}$	$+\frac{81}{64}$		$+\frac{1}{4}$	$-\frac{9}{128}$
	[cos − sin]	,,	,,	0	$+\frac{19}{128}$		0	$-\frac{1}{192}$		0	0
	[cos]	$-\frac{3}{16}$	$-\frac{5}{32}$	H	$+\frac{119}{256}$		H	$+\frac{121}{192}$		H	H
	[sin]	0	0	H	$+\frac{81}{256}$		H	$+\frac{61}{96}$		H	H
x^5	[cos + sin]	$-\frac{1}{32}$	$-\frac{25}{64}$		$+\frac{15}{32}$		$-\frac{5}{32}$	$-\frac{65}{128}$			$+\frac{55}{64}$
	[cos − sin]	,,	,,		$+\frac{5}{64}$		0	$-\frac{1}{128}$			0
	[cos]	$-\frac{1}{32}$	$-\frac{25}{64}$		$+\frac{35}{128}$		H	$-\frac{35}{128}$			H
	[sin]	0	0		$+\frac{25}{128}$		H	$-\frac{15}{64}$			H
x^6	[cos + sin]		$-\frac{1}{4}$		$+\frac{3}{32}$			$-\frac{21}{32}$			$+\frac{15}{64}$
	[cos − sin]		,,		$+\frac{1}{64}$			$-\frac{1}{32}$			0
	[cos]		$-\frac{1}{4}$		$+\frac{7}{128}$			$-\frac{11}{32}$			H
	[sin]		0		$+\frac{5}{128}$			$-\frac{5}{16}$			H
x^7	[cos + sin]		$-\frac{7}{128}$					$-\frac{21}{128}$			
	[cos − sin]		,,					$-\frac{1}{128}$			
	[cos]		$-\frac{7}{128}$					$-\frac{11}{128}$			
	[sin]		0					$-\frac{5}{64}$			

... *Table* $(x^4, x^5, x^6, x^7)\ {\cos \atop \sin} 5f$ ...

$\cos \atop \sin$ $5f$		$[\ ]^4$ e	e^3	e^5	e^7	$[\ ]^5$ e^0	e^2	e^4	e^6	$[\ ]^6$ e	e^3	e^5	e^7	$[\ ]^7$ e^2	e^4	e^6
x^4	[cos + sin]			$-\frac{7}{8}$	$+\frac{113}{32}$			$+\frac{3}{8}$	$-\frac{25}{8}$			$+\frac{3}{8}$	$-\frac{249}{64}$		$+\frac{1}{4}$	$-\frac{239}{128}$
	[cos − sin]			o	o			o	o			o	o		o	o
	[cos]			H	H			H	H			H	H		H	H
	[sin]			H	H			H	H			H	H		H	H
x^5	[cos + sin]			$-\frac{5}{16}$	$+\frac{85}{64}$				$+\frac{5}{16}$			$-\frac{5}{16}$	$+\frac{345}{128}$			$-\frac{45}{64}$
	[cos − sin]			o	o				o			o	o			o
	[cos]			H	H				H			H	H			H
	[sin]			H	H				H			H	H			H
x^6	[cos + sin]				$-\frac{5}{8}$				$+\frac{5}{16}$				$+\frac{5}{32}$			$+\frac{15}{64}$
	[cos − sin]				o				o				o			o
	[cos]				H				H				H			H
	[sin]				H				H				H			H
x^7	[cos + sin]				$-\frac{35}{128}$								$-\frac{35}{128}$			
	[cos − sin]				o								o			
	[cos]				H								H			
	[sin]				H								H			

... *Table* (x^4, x^5, x^6, x^7) $\begin{matrix}\cos\\ \sin\end{matrix}$ $5f$.

	$\begin{matrix}\cos\\ \sin\end{matrix}\ 5f$	$[\ ]^8$ e^3	$[\ ]^8$ e^5	$[\ ]^8$ e^7	$[\ ]^9$ e^4	$[\ ]^9$ e^6	$[\ ]^{10}$ e^5	$[\ ]^{10}$ e^7	$[\ ]^{11}$ e^6	$[\ ]^{12}$ e^7
x^4	[cos + sin]		$+\frac{17}{16}$	$-\frac{287}{32}$	$+\frac{1}{16}$	$+\frac{75}{32}$	$+\frac{7}{16}$	$+\frac{505}{192}$	$+\frac{229}{128}$	$+\frac{179}{32}$
	[cos − sin]		0	0	0	0	0	0	0	0
	[cos]		H	H	H	H	H	H	H	H
	[sin]		H	H	H	H	H	H	H	H
x^5	[cos + sin]		$-\frac{5}{32}$	$+\frac{5}{64}$		$-\frac{25}{32}$	$-\frac{1}{32}$	$-\frac{275}{128}$	$-\frac{15}{64}$	$-\frac{65}{64}$
	[cos − sin]		0	0		0	0	0	0	0
	[cos]		H	H		H	H	H	H	H
	[sin]		H	H		H	H	H	H	H
x^6	[cos + sin]			$+\frac{3}{4}$		$+\frac{3}{32}$		$+\frac{17}{32}$	$+\frac{1}{64}$	$+\frac{1}{8}$
	[cos − sin]			0		0		0	0	0
	[cos]			H		H		H	H	H
	[sin]			H		H		H	H	H
x^7	[cos + sin]			$-\frac{21}{128}$				$-\frac{7}{128}$		$-\frac{1}{128}$
	[cos − sin]			0				0		0
	[cos]			H				H		H
	[sin]			H				H		H

Table $(x^0, x^1, x^2, x^3) \frac{\cos}{\sin} 6f \ldots$

$\frac{\cos}{\sin} 6f$		$[\;]^0$	$[\;]^1$		$[\;]^2$		$[\;]^3$		
		e^6	e^5	e^7	e^4	e^6	e^3	e^5	e^7
x^0	[cos + sin]	$+\frac{7}{64}$	$-\frac{773}{320}$	$+\frac{5167}{7680}$	$+\frac{157}{16}$	$-\frac{1339}{80}$	$-\frac{67}{4}$	$+\frac{4053}{64}$	$-\frac{192531}{2560}$
	[cos − sin]	,,	0	$+\frac{17}{53760}$	0	0	0	0	0
	[cos]	$+\frac{7}{64}$	H	$+\frac{6031}{17920}$	H	H	H	H	H
	[sin]	0	H	$+\frac{4519}{13440}$	H	H	H	H	H
x^1	[cos + sin]	$+\frac{21}{64}$	$-\frac{1961}{768}$	$+\frac{162773}{92160}$	$+\frac{277}{48}$	$-\frac{993}{80}$	$-\frac{93}{16}$	$+\frac{6289}{256}$	$-\frac{384789}{10240}$
	[cos − sin]	,,	0	$+\frac{1091}{92160}$	0	0	0	0	0
	[cos]	$+\frac{21}{64}$	H	$+\frac{20483}{23040}$	H	H	H	H	H
	[sin]	0	H	$+\frac{26947}{30720}$	H	H	H	H	H
x^2	[cos + sin]	$+\frac{35}{64}$	$-\frac{179}{96}$	$+\frac{14689}{7680}$	$+\frac{37}{16}$	$-\frac{235}{48}$	$-\frac{5}{4}$	$+\frac{107}{32}$	$-\frac{12123}{2560}$
	[cos − sin]	,,	0	$-\frac{17}{7680}$	0	0	0	0	0
	[cos]	$+\frac{35}{64}$	H	$+\frac{917}{960}$	H	H	H	H	H
	[sin]	0	H	$+\frac{2451}{2560}$	H	H	H	H	H
x^3	[cos + sin]	$+\frac{35}{64}$	$-\frac{57}{64}$	$+\frac{739}{1024}$	$+\frac{9}{16}$	$+\frac{1}{4}$	$-\frac{1}{8}$	$-\frac{117}{64}$	$+\frac{6249}{1024}$
	[cos − sin]	,,	0	$-\frac{75}{1024}$	0	0	0	0	0
	[cos]	$+\frac{35}{64}$	H	$+\frac{83}{256}$	H	H	H	H	H
	[sin]	0	H	$+\frac{407}{1024}$	H	H	H	H	H

... *Table* $(x^0, x^1, x^2, x^3)\ {\cos \atop \sin} 6f$ ...

	$\frac{\cos}{\sin} 6f$	[ ]⁴ e^2	[ ]⁴ e^4	[ ]⁴ e^6	[ ]⁵ e	[ ]⁵ e^3	[ ]⁵ e^5	[ ]⁵ e^7
x^0	[cos + sin]	$+\frac{57}{4}$	$-\frac{829}{8}$	$+\frac{8117}{32}$	-6	$+\frac{345}{4}$	$-\frac{12605}{32}$	$+\frac{1246315}{1536}$
	[cos − sin]	o	o	o	o	o	o	o
	[cos]	H	H	H	H	H	H	H
	[sin]	H	H	H	H	H	H	H
x^1	[cos + sin]	$+\frac{11}{4}$	$-\frac{497}{24}$	$+\frac{1933}{32}$	$-\frac{1}{2}$	$+\frac{105}{16}$	$-\frac{13325}{384}$	$+\frac{1838665}{18432}$
	[cos − sin]	o	o	o	o	o	o	o
	[cos]	H	H	H	H	H	H	H
	[sin]	H	H	H	H	H	H	H
x^2	[cos + sin]	$+\frac{1}{4}$	$+\frac{7}{8}$	$-\frac{685}{96}$		$-\frac{7}{4}$	$+\frac{745}{48}$	$-\frac{72955}{1536}$
	[cos − sin]	o	o	o		o	o	o
	[cos]	H	H	H		H	H	H
	[sin]	H	H	H		H	H	H
x^3	[cos + sin]		$+\frac{3}{2}$	$-\frac{517}{64}$		$-\frac{3}{8}$	$+\frac{105}{32}$	$-\frac{13265}{1024}$
	[cos − sin]		o	o		o	o	o
	[cos]		H	H		H	H	H
	[sin]		H	H		H	H	H

... *Table* $(x^0, x^1, x^2, x^3) \begin{matrix}\cos\\ \sin\end{matrix} 6f$ ...

	$\begin{matrix}\cos\\ \sin\end{matrix} 6f$	$[\ ]^6$ e^0	e^2	e^4	e^6	$[\ ]^7$ e	e^3	e^5	e^7
x^0	[cos + sin]	$+1$	-36	$+\frac{5103}{16}$	$-\frac{19015}{16}$	$+6$	$-\frac{525}{4}$	$+\frac{29813}{32}$	$-\frac{4795063}{1536}$
	[cos − sin]	0	0	0	0	0	0	0	0
	[cos]	H	H	H	H	H	H	H	H
	[sin]	H	H	H	H	H	H	H	H
x^1	[cos + sin]		$+\frac{1}{2}$	$-\frac{9}{2}$	$-\frac{9}{32}$	$-\frac{1}{2}$	$+\frac{189}{16}$	$-\frac{30821}{384}$	$+\frac{4080685}{18432}$
	[cos − sin]		0	0	0	0	0	0	0
	[cos]		H	H	H	H	H	H	H
	[sin]		H	H	H	H	H	H	H
x^2	[cos + sin]		$+\frac{1}{2}$	-9	$+\frac{1695}{32}$		$+\frac{5}{4}$	$-\frac{959}{48}$	$+\frac{175553}{1536}$
	[cos − sin]		0	0	0		0	0	0
	[cos]		H	H	H		H	H	H
	[sin]		H	H	H		H	H	H
x^3	[cos + sin]			$+\frac{3}{8}$	$-\frac{9}{4}$		$-\frac{3}{8}$	$+\frac{189}{32}$	$-\frac{30737}{1024}$
	[cos − sin]			0	0		0	0	0
	[cos]			H	H		H	H	H
	[sin]			H	H		H	H	H

... *Table* $(x^0, x^1, x^2, x^3)\ {\cos \atop \sin}\ 6f$ ...

	$\cos \atop \sin$ $6f$	$[\ ]^8$ e^2	e^4	e^6	$[\ ]^9$ e^3	e^5	e^7
x^0	[cos + sin]	$+\frac{87}{4}$	$-\frac{3011}{8}$	$+\frac{151205}{64}$	$+\frac{247}{4}$	$-\frac{59781}{64}$	$+\frac{13947303}{2560}$
	[cos − sin]	0	0	0	0	0	0
	[cos]	H	H	H	H	H	H
	[sin]	H	H	H	H	H	H
x^1	[cos + sin]	$-\frac{13}{4}$	$+\frac{1369}{24}$	$-\frac{21969}{64}$	$-\frac{201}{16}$	$+\frac{48449}{256}$	$-\frac{10874049}{10240}$
	[cos − sin]	0	0	0	0	0	0
	[cos]	H	H	H	H	H	H
	[sin]	H	H	H	H	H	H
x^2	[cos + sin]	$+\frac{1}{4}$	$-\frac{11}{8}$	$-\frac{3317}{192}$	$+\frac{7}{4}$	$-\frac{629}{32}$	$+\frac{145449}{2560}$
	[cos − sin]	0	0	0	0	0	0
	[cos]	H	H	H	H	H	H
	[sin]	H	H	H	H	H	H
x^3	[cos + sin]		$-\frac{3}{2}$	$+\frac{1313}{64}$	$-\frac{1}{8}$	$-\frac{153}{64}$	$+\frac{44889}{1024}$
	[cos − sin]		0	0	0	0	0
	[cos]		H	H	H	H	H
	[sin]		H	H	H	H	H

... *Table* $(x^0, x^1, x^2, x^3) \begin{smallmatrix}\cos\\ \sin\end{smallmatrix} 6f$.

	$\begin{smallmatrix}\cos\\ \sin\end{smallmatrix} 6f$	$[\ ]^{10}$		$[\ ]^{11}$		$[\ ]^{12}$	$[\ ]^{13}$
		e^4	e^6	e^5	e^7	e^6	e^7
x^0	[cos + sin]	$+\frac{605}{4}$	$-\frac{16829}{8}$	$+\frac{107333}{320}$	$-\frac{33928027}{7680}$	$+\frac{110793}{160}$	$+\frac{72815971}{53760}$
	[cos − sin]	0	0	0	0	0	0
	[cos]	H	H	H	H	H	H
	[sin]	H	H	H	H	H	H
x^1	[cos + sin]	$-\frac{1805}{48}$	$+\frac{16547}{32}$	$-\frac{73961}{768}$	$+\frac{115203737}{92160}$	$-\frac{35499}{160}$	$-\frac{43634809}{92160}$
	[cos − sin]	0	0	0	0	0	0
	[cos]	H	H	H	H	H	H
	[sin]	H	H	H	H	H	H
x^2	[cos + sin]	$+\frac{115}{16}$	$-\frac{8135}{96}$	$+\frac{2173}{96}$	$-\frac{2039459}{7680}$	$+\frac{1937}{32}$	$+\frac{1111819}{7680}$
	[cos − sin]	0	0	0	0	0	0
	[cos]	H	H	H	H	H	H
	[sin]	H	H	H	H	H	H
x^3	[cos + sin]	$-\frac{15}{16}$	$+\frac{5}{2}$	$-\frac{261}{64}$	$+\frac{30707}{1024}$	$-\frac{863}{64}$	$-\frac{38507}{1024}$
	[cos − sin]	0	0	0	0	0	0
	[cos]	H	H	H	H	H	H
	[sin]	H	H	H	H	H	H

Table $(x^4, x^5, x^6, x^7) \begin{smallmatrix}\cos\\ \sin\end{smallmatrix} 6f \ldots$

$\begin{smallmatrix}\cos\\ \sin\end{smallmatrix} 6f$		$[\ ]^0$	$[\ ]^1$		$[\ ]^2$		$[\ ]^3$			$[\ ]^4$		
		e^6	e^5	e^7	e^4	e^6	e^3	e^5	e^7	e^2	e^4	e^6
x^4	[cos + sin]	$+\frac{21}{64}$	$-\frac{1}{4}$	$-\frac{25}{48}$	$+\frac{1}{16}$	$+\frac{21}{16}$		-1	$+\frac{87}{32}$		$+\frac{1}{4}$	$-\frac{15}{64}$
	[cos − sin]	,,	o	$-\frac{13}{96}$	o	o		o	o		o	o
	[cos]	$+\frac{21}{64}$	H	$-\frac{21}{64}$	H	H		H	H		H	H
	[sin]	o	H	$-\frac{37}{192}$	H	H		H	H		H	H
x^5	[cos + sin]	$+\frac{7}{64}$	$-\frac{1}{32}$	$-\frac{187}{256}$		$+\frac{19}{32}$		$-\frac{5}{32}$	$-\frac{189}{256}$			$+\frac{65}{64}$
	[cos − sin]	,,	o	$-\frac{29}{256}$		o		o	o			o
	[cos]	$+\frac{7}{64}$	H	$-\frac{27}{64}$		H		H	H			H
	[sin]	o	H	$-\frac{79}{256}$		H		H	H			H
x^6	[cos + sin]	$+\frac{1}{64}$		$-\frac{21}{64}$		$+\frac{3}{32}$			$-\frac{51}{64}$			$+\frac{15}{64}$
	[cos − sin]	,,		$-\frac{3}{64}$		o			o			o
	[cos]	$+\frac{1}{64}$		$-\frac{3}{16}$		H			H			H
	[sin]	o		$-\frac{9}{64}$		H			H			H
x^7	[cos + sin]			$-\frac{7}{128}$					$-\frac{21}{128}$			
	[cos − sin]			$-\frac{1}{128}$					o			
	[cos]			$-\frac{1}{32}$					H			
	[sin]			$-\frac{3}{128}$					H			

... *Table* (x^4, x^5, x^6, x^7) $\begin{matrix}\cos\\ \sin\end{matrix} 6f$ ...

$\begin{matrix}\cos\\ \sin\end{matrix} 6f$		$[\ ]^5$ e^3	$[\ ]^5$ e^5	$[\ ]^5$ e^7	$[\ ]^6$ e^4	$[\ ]^6$ e^6	$[\ ]^7$ e^3	$[\ ]^7$ e^5	$[\ ]^7$ e^7	$[\ ]^8$ e^4	$[\ ]^8$ e^6
x^4	[cos + sin]		-1	$+\frac{25}{4}$	$+\frac{3}{8}$	$-\frac{9}{2}$		$+\frac{1}{2}$	$-\frac{217}{32}$	$+\frac{1}{4}$	$-\frac{153}{64}$
	[cos − sin]		o	o	o	o		o	o	o	o
	[cos]		H	H	H	H		H	H	H	H
	[sin]		H	H	H	H		H	H	H	H
x^5	[cos + sin]		$-\frac{5}{16}$	$+\frac{525}{256}$		$+\frac{5}{16}$		$-\frac{5}{16}$	$+\frac{945}{256}$		$-\frac{55}{64}$
	[cos − sin]		o	o		o		o	o		o
	[cos]		H	H		H		H	H		H
	[sin]		H	H		H		H	H		H
x^6	[cos + sin]			$-\frac{45}{64}$		$+\frac{5}{16}$			$+\frac{15}{64}$		$+\frac{15}{64}$
	[cos − sin]			o		o			o		o
	[cos]			H		H			H		H
	[sin]			H		H			H		H
x^7	[cos + sin]			$-\frac{35}{128}$					$-\frac{35}{128}$		
	[cos − sin]			o					o		
	[cos]			H					H		
	[sin]			H					H		

... *Table* (x^4, x^5, x^6, x^7) $\begin{matrix}\cos\\ \sin\end{matrix}\,6f$.

	$\begin{matrix}\cos\\ \sin\end{matrix}\,6f$	$[\ ]^9$			$[\ ]^{10}$		$[\ ]^{11}$		$[\ ]^{12}$	$[\ ]^{13}$
		e^3	e^5	e^7	e^4	e^6	e^5	e^7	e^6	e^7
x^4	[cos + sin]		$+\frac{5}{4}$	$-\frac{219}{16}$	$+\frac{1}{16}$	$+\frac{51}{16}$	$+\frac{1}{2}$	$+\frac{403}{96}$	$+\frac{147}{64}$	$+\frac{191}{24}$
	[cos − sin]		o	o	o	o	o	o	o	o
	[cos]		H	H	H	H	H	H	H	H
	[sin]		H	H	H	H	H	H	H	H
x^5	[cos + sin]		$-\frac{5}{32}$	$-\frac{9}{256}$		$-\frac{29}{32}$	$-\frac{1}{32}$	$-\frac{727}{256}$	$-\frac{17}{64}$	$-\frac{329}{256}$
	[cos − sin]		o	o		o	o	o	o	o
	[cos]		H	H		H	H	H	H	H
	[sin]		H	H		H	H	H	H	H
x^6	[cos + sin]			$+\frac{57}{64}$		$+\frac{3}{32}$		$+\frac{39}{64}$	$+\frac{1}{64}$	$+\frac{9}{64}$
	[cos − sin]			o		o		o	o	o
	[cos]			H		H		H	H	H
	[sin]			H		H		H	H	H
x^7	[cos + sin]			$-\frac{21}{128}$				$-\frac{7}{128}$		$-\frac{1}{128}$
	[cos − sin]			o				o		o
	[cos]			H				H		H
	[sin]			H				H		H

Table $(x^0, x^1, x^2, x^3) \begin{smallmatrix}\cos\\ \sin\end{smallmatrix} 7f \ldots$

	$\begin{smallmatrix}\cos\\ \sin\end{smallmatrix} 7f$	$[\ ]^0$	$[\ ]^1$	$[\ ]^2$		$[\ ]^3$	
		e^7	e^6	e^5	e^7	e^4	e^6
x^0	[cos + sin]	$-\frac{1}{16}$	$+\frac{93289}{46080}$	$-\frac{329}{30}$	$+\frac{11809}{720}$	$+\frac{3227}{128}$	$-\frac{26943}{320}$
	[cos − sin]	,,	0	0	0	0	0
	[cos]	$-\frac{1}{16}$	H	H	H	H	H
	[sin]	0	H	H	H	H	H
x^1	[cos + sin]	$-\frac{7}{32}$	$+\frac{19081}{7680}$	$-\frac{187}{24}$	$+\frac{10927}{720}$	$+\frac{361}{32}$	$-\frac{55317}{1280}$
	[cos − sin]	,,	0	0	0	0	0
	[cos]	$-\frac{7}{32}$	H	H	H	H	H
	[sin]	0	H	H	H	H	H
x^2	[cos + sin]	$-\frac{7}{16}$	$+\frac{1131}{512}$	$-\frac{97}{24}$	$+\frac{4223}{480}$	$+\frac{113}{32}$	$-\frac{3177}{256}$
	[cos − sin]	,,	0	0	0	0	0
	[cos]	$-\frac{7}{16}$	H	H	H	H	H
	[sin]	0	H	H	H	H	H
x^3	[cos + sin]	$-\frac{35}{64}$	$+\frac{529}{384}$	$-\frac{23}{16}$	$+\frac{391}{192}$	$+\frac{11}{16}$	$+\frac{15}{32}$
	[cos − sin]	,,	0	0	0	0	0
	[cos]	$-\frac{35}{64}$	H	H	H	H	H
	[sin]	0	H	H	H	H	H

... *Table* $(x^0, x^1, x^2, x^3) \begin{matrix}\cos\\ \sin\end{matrix} 7f$ ...

$\begin{matrix}\cos\\ \sin\end{matrix} 7f$		$[\ ]^4$			$[\ ]^5$		
		e^3	e^5	e^7	e^2	e^4	e^6
x^0	[cos + sin]	$-\frac{91}{3}$	$+\frac{4445}{24}$	$-\frac{47299}{120}$	$+\frac{161}{8}$	$-\frac{10409}{48}$	$+\frac{2509759}{3072}$
	[cos − sin]	0	0	0	0	0	0
	[cos]	H	H	H	H	H	H
	[sin]	H	H	H	H	H	H
x^1	[cos + sin]	$-\frac{17}{2}$	$+\frac{2689}{48}$	$-\frac{3395}{24}$	$+\frac{13}{4}$	$-\frac{1721}{48}$	$+\frac{236923}{1536}$
	[cos − sin]	0	0	0	0	0	0
	[cos]	H	H	H	H	H	H
	[sin]	H	H	H	H	H	H
x^2	[cos + sin]	$-\frac{3}{2}$	$+\frac{149}{24}$	$-\frac{5897}{480}$	$+\frac{1}{4}$	$+\frac{17}{16}$	$-\frac{7671}{512}$
	[cos − sin]	0	0	0	0	0	0
	[cos]	H	H	H	H	H	H
	[sin]	H	H	H	H	H	H
x^3	[cos + sin]	$-\frac{1}{8}$	$-\frac{83}{32}$	$+\frac{1807}{128}$		$+\frac{7}{4}$	$-\frac{1781}{128}$
	[cos − sin]	0	0	0		0	0
	[cos]	H	H	H		H	H
	[sin]	H	H	H		H	H

... *Table* (x^0, x^1, x^2, x^3) $\genfrac{}{}{0pt}{}{\cos}{\sin}\, 7f$ ...

	$\genfrac{}{}{0pt}{}{\cos}{\sin}\, 7f$	[ ]^6 e^1	e^3	e^5	e^7	[ ]^7 e^0	e^2	e^4	e^6
x^0	[cos + sin]	-7	$+\frac{567}{4}$	$-\frac{14805}{16}$	$+\frac{89999}{32}$	$+1$	-49	$+\frac{37975}{64}$	$-\frac{883421}{288}$
	[cos − sin]	o	o	o	o	o	o	o	o
	[cos]	H	H	H	H	H	H	H	H
	[sin]	H	H	H	H	H	H	H	H
x^1	[cos + sin]	$-\frac{1}{2}$	$+\frac{75}{8}$	$-\frac{2199}{32}$	$+\frac{17549}{64}$		$+\frac{1}{2}$	$-\frac{49}{8}$	$-\frac{49}{128}$
	[cos − sin]	o	o	o	o		o	o	o
	[cos]	H	H	H	H		H	H	H
	[sin]	H	H	H	H		H	H	H
x^2	[cos + sin]		-2	$+\frac{101}{4}$	$-\frac{3611}{32}$		$+\frac{1}{2}$	$-\frac{49}{4}$	$+\frac{37877}{384}$
	[cos − sin]		o	o	o		o	o	o
	[cos]		H	H	H		H	H	H
	[sin]		H	H	H		H	H	H
x^3	[cos + sin]		$-\frac{3}{8}$	$+\frac{75}{16}$	$-\frac{411}{16}$			$+\frac{3}{8}$	$-\frac{49}{16}$
	[cos − sin]		o	o	o			o	o
	[cos]		H	H	H			H	H
	[sin]		H	H	H			H	H

... *Table* $(x^0, x^1, x^2, x^3) \begin{array}{c}\cos\\ \sin\end{array} 7f$ ...

$\begin{array}{c}\cos\\ \sin\end{array} 7f$		$[\ ]^8$ e	e^3	e^5	e^7	$[\ ]^9$ e^2	e^4	e^6
x^0	[cos + sin]	$+7$	-203	$+\frac{5768}{3}$	$-\frac{1255219}{144}$	$+\frac{231}{8}$	$-\frac{10381}{16}$	$+\frac{5454687}{1024}$
	[cos − sin]	o	o	o	o	o	o	o
	[cos]	H	H	H	H	H	H	H
	[sin]	H	H	H	H	H	H	H
x^1	[cos + sin]	$-\frac{1}{2}$	$+\frac{31}{2}$	$-\frac{845}{6}$	$+\frac{154651}{288}$	$-\frac{15}{4}$	$+\frac{1363}{16}$	$-\frac{344031}{512}$
	[cos − sin]	o	o	o	o	o	o	o
	[cos]	H	H	H	H	H	H	H
	[sin]	H	H	H	H	H	H	H
x^2	[cos + sin]		$+\frac{3}{2}$	$-\frac{94}{3}$	$+\frac{3791}{16}$	$+\frac{1}{4}$	$-\frac{25}{16}$	$-\frac{16155}{512}$
	[cos − sin]		o	o	o	o	o	o
	[cos]		H	H	H	H	H	H
	[sin]		H	H	H	H	H	H
x^3	[cos + sin]		$-\frac{3}{8}$	$+\frac{31}{4}$	$-\frac{3373}{64}$		$-\frac{7}{4}$	$+\frac{3945}{118}$
	[cos − sin]		o	o	o		o	o
	[cos]		H	H	H		H	H
	[sin]		H	H	H		H	H

... *Table* $(x^0,\ x^1,\ x^2,\ x^3)\ {\cos \atop \sin}\ 7f$ ...

	$\cos \atop \sin$ $7f$	$[\ ]^{10}$ e^3	$[\ ]^{10}$ e^5	$[\ ]^{10}$ e^7	$[\ ]^{11}$ e^4	$[\ ]^{11}$ e^6
x^0	[cos + sin]	$+\frac{1099}{12}$	$-\frac{84889}{48}$	$+\frac{1273433}{96}$	$+\frac{94885}{384}$	$-\frac{4151693}{960}$
	[cos − sin]	0	0	0	0	0
	[cos]	H	H	H	H	H
	[sin]	H	H	H	H	H
x^1	[cos + sin]	$-\frac{131}{8}$	$+\frac{30193}{96}$	$-\frac{436457}{192}$	$-\frac{5231}{96}$	$+\frac{3618557}{3840}$
	[cos − sin]	0	0	0	0	0
	[cos]	H	H	H	H	H
	[sin]	H	H	H	H	H
x^2	[cos + sin]	$+2$	$-\frac{677}{24}$	$+\frac{9889}{96}$	$+\frac{295}{32}$	$-\frac{34789}{256}$
	[cos − sin]	0	0	0	0	0
	[cos]	H	H	H	H	H
	[sin]	H	H	H	H	H
x^3	[cos + sin]	$-\frac{1}{8}$	$-\frac{13}{4}$	$+\frac{2361}{32}$	$-\frac{17}{16}$	$+\frac{99}{32}$
	[cos − sin]	0	0	0	0	0
	[cos]	H	H	H	H	H
	[sin]	H	H	H	H	H

Table $(x^0, x^1, x^2, x^3)\ {\cos \atop \sin} 7f\ldots$

	${\cos \atop \sin} 7f$	$[\ ]^{12}$ e^5	$[\ ]^{12}$ e^7	$[\ ]^{13}$ e^6	$[\ ]^{14}$ e^7
x^0	[cos + sin]	$+\frac{23877}{40}$	$-\frac{390873}{40}$	$+\frac{61307827}{46080}$	$+\frac{1004057}{360}$
	[cos − sin]	o	o	o	o
	[cos]	H	H	H	H
	[sin]	H	H	H	H
x^1	[cos + sin]	$-\frac{2451}{16}$	$+\frac{49413}{20}$	$-\frac{2945503}{7680}$	$-\frac{254051}{288}$
	[cos − sin]	o	o	o	o
	[cos]	H	H	H	H
	[sin]	H	H	H	H
x^2	[cos + sin]	$+\frac{257}{8}$	$-\frac{74701}{160}$	$+\frac{144373}{1536}$	$+\frac{116977}{480}$
	[cos − sin]	o	o	o	o
	[cos]	H	H	H	H
	[sin]	H	H	H	H
x^3	[cos + sin]	$-\frac{165}{32}$	$+\frac{5905}{128}$	$-\frac{7213}{384}$	$-\frac{10969}{192}$
	[cos − sin]	o	o	o	o
	[cos]	H	H	H	H
	[sin]	H	H	H	H

Table $(x^4, x^5, x^6, x^7) \begin{matrix}\cos\\ \sin\end{matrix} 7f \ldots$

	$\begin{matrix}\cos\\ \sin\end{matrix} 7f$	$[\ ]^0$	$[\ ]^1$	$[\ ]^2$		$[\ ]^3$		$[\ ]^4$			$[\ ]^5$	
		e^7	e^6	e^5	e^7	e^4	e^6	e^3	e^5	e^7	e^4	e^6
x^4	[cos + sin]	$-\frac{7}{16}$	$+\frac{73}{128}$	$-\frac{5}{16}$	$-\frac{221}{192}$	$+\frac{1}{16}$	$+\frac{63}{32}$		$-\frac{19}{16}$	$+5$	$+\frac{1}{4}$	$-\frac{59}{128}$
	[cos − sin]	,,	o	o	o	o	o		o	o	o	o
	[cos]	$-\frac{7}{16}$	H	H	H	H	H		H	H	H	H
	[sin]	o	H	H	H	H	H		H	H	H	H
x^5	[cos + sin]	$-\frac{7}{32}$	$+\frac{9}{64}$	$-\frac{1}{32}$	$-\frac{149}{128}$		$+\frac{23}{32}$		$-\frac{5}{32}$	-1		$+\frac{75}{64}$
	[cos − sin]	,,	o	o	o		o		o	o		o
	[cos]	$-\frac{7}{32}$	H	H	H		H		H	H		H
	[sin]	o	H	H	H		H		H	H		H
x^6	[cos + sin]	$-\frac{1}{16}$	$+\frac{1}{64}$		$-\frac{13}{32}$		$+\frac{3}{32}$			$-\frac{15}{16}$		$+\frac{15}{64}$
	[cos − sin]	,,	o		o		o			o		o
	[cos]	$-\frac{1}{16}$	H		H		H			H		H
	[sin]	o	H		H		H			H		H
x^7	[cos + sin]	$-\frac{1}{128}$			$-\frac{7}{128}$					$-\frac{21}{128}$		
	[cos − sin]	,,			o					o		
	[cos]	$-\frac{1}{128}$			H					H		
	[sin]	o			H					H		

... *Table* $(x^4, x^5, x^6, x^7)\ {\cos \atop \sin}\ 7f$...

	${\cos \atop \sin}\ 7f$	$[\ ]^6$		$[\ ]^7$		$[\ ]^8$				$[\ ]^9$		
		e^5	e^7	e^4	e^6	e	e^3	e^5	e^7	e^2	e^4	e^6
x^4	[cos + sin]	$-\frac{9}{8}$	$+\frac{645}{64}$	$+\frac{3}{8}$	$-\frac{49}{8}$			$+\frac{5}{8}$	$-\frac{173}{16}$		$+\frac{1}{4}$	$-\frac{381}{128}$
	[cos − sin]	o	o	o	o			o	o		o	o
	[cos]	H	H	H	H			H	H		H	H
	[sin]	H	H	H	H			H	H		H	H
x^5	[cos + sin]	$-\frac{5}{16}$	$+\frac{375}{128}$		$+\frac{5}{16}$			$-\frac{5}{16}$	$+\frac{155}{32}$			$-\frac{65}{64}$
	[cos − sin]	o	o		o			o	o			o
	[cos]	H	H		H			H	H			H
	[sin]	H	H		H			H	H			H
x^6	[cos + sin]		$-\frac{25}{32}$		$+\frac{5}{16}$				$+\frac{5}{16}$			$+\frac{15}{64}$
	[cos − sin]		o		o				o			o
	[cos]		H		H				H			H
	[sin]		H		H				H			H
x^7	[cos + sin]		$-\frac{35}{128}$						$-\frac{35}{128}$			
	[cos − sin]		o						o			
	[cos]		H						H			
	[sin]		H						H			

... *Table* $(x^4,\ x^5,\ x^6,\ x^7)\ {\cos \atop \sin}\ 7f.$

	${\cos \atop \sin}\ 7f$	$[\ ]^{10}$		$[\ ]^{11}$		$[\ ]^{12}$		$[\ ]^{13}$	$[\ ]^{14}$
		e^5	e^7	e^4	e^6	e^5	e^7	e^6	e^7
x^4	[cos + sin]	$+\frac{23}{16}$	$-\frac{3807}{192}$	$+\frac{1}{16}$	$+\frac{133}{32}$	$+\frac{9}{16}$	$+\frac{25}{4}$	$+\frac{367}{128}$	$+\frac{2093}{192}$
	[cos − sin]	0	0	0	0	0	0	0	0
	[cos]	H	H	H	H	H	H	H	H
	[sin]	H	H	H	H	H	H	H	H
x^5	[cos + sin]	$-\frac{5}{32}$	$-\frac{23}{128}$		$-\frac{33}{32}$	$-\frac{1}{32}$	$-\frac{29}{8}$	$-\frac{19}{64}$	$-\frac{203}{128}$
	[cos − sin]	0	0		0	0	0	0	0
	[cos]	H	H		H	H	H	H	H
	[sin]	H	H		H	H	H	H	H
x^6	[cos + sin]		$+\frac{33}{32}$		$+\frac{3}{32}$		$+\frac{11}{16}$	$+\frac{1}{64}$	$+\frac{5}{32}$
	[cos − sin]		0		0		0	0	0
	[cos]		H		H		H	H	H
	[sin]		H		H		H	H	H
x^7	[cos + sin]		$-\frac{21}{128}$				$-\frac{7}{128}$		$-\frac{1}{128}$
	[cos − sin]		0				0		0
	[cos]		H				H		H
	[sin]		H				H		H

Table $\left(\left(\frac{r}{a}\right)^{-5} \ldots \left(\frac{r}{a}\right)^{-1},\ \log\frac{r}{a},\ \left(\frac{r}{a}\right)^{1} \ldots \left(\frac{r}{a}\right)^{4}\right) \ldots$

		[ ]^0				[ ]^1			
		e^0	e^2	e^4	e^6	e	e^3	e^5	e^7
$\left(\frac{r}{a}\right)^{-5}$	[cos]	$+1$	$+5$	$+\frac{105}{8}$	$+\frac{105}{4}$	$+\frac{5}{2}$	$+\frac{135}{16}$	$+\frac{7285}{384}$	$+\frac{643015}{18432}$
$\left(\frac{r}{a}\right)^{-4}$		$+1$	$+3$	$+\frac{45}{8}$	$+\frac{35}{4}$	$+2$	$+\frac{17}{4}$	$+\frac{685}{96}$	$+\frac{48293}{4608}$
$\left(\frac{r}{a}\right)^{-3}$		$+1$	$+\frac{3}{2}$	$+\frac{15}{8}$	$+\frac{35}{16}$	$+\frac{3}{2}$	$+\frac{27}{16}$	$+\frac{261}{128}$	$+\frac{14309}{6144}$
$\left(\frac{r}{a}\right)^{-2}$		$+1$	$+\frac{1}{2}$	$+\frac{3}{8}$	$+\frac{5}{16}$	$+1$	$+\frac{3}{8}$	$+\frac{65}{192}$	$+\frac{2675}{9216}$
$\left(\frac{r}{a}\right)^{-1}$		$+1$	0	0	0	$+\frac{1}{2}$	$-\frac{1}{16}$	$+\frac{1}{384}$	$-\frac{1}{18432}$
$\log\frac{r}{a}$		0	$+\frac{1}{4}$	$+\frac{1}{32}$	$+\frac{1}{96}$	$-\frac{1}{2}$	$+\frac{3}{16}$	$+\frac{1}{128}$	$+\frac{127}{18432}$
$\left(\frac{r}{a}\right)^{1}$		$+1$	$+\frac{1}{2}$	0	0	$-\frac{1}{2}$	$+\frac{3}{16}$	$-\frac{5}{384}$	$+\frac{7}{18432}$
$\left(\frac{r}{a}\right)^{2}$		$+1$	$+\frac{3}{2}$	0	0	-1	$+\frac{1}{8}$	$-\frac{1}{192}$	$+\frac{1}{9216}$
$\left(\frac{r}{a}\right)^{3}$		$+1$	$+3$	$+\frac{3}{8}$	0	$-\frac{3}{2}$	$-\frac{9}{16}$	$+\frac{15}{128}$	$-\frac{35}{6144}$
$\left(\frac{r}{a}\right)^{4}$		$+1$	$+5$	$+\frac{15}{8}$	0	-2	$-\frac{9}{4}$	$+\frac{19}{96}$	$-\frac{29}{4608}$

... *Table* $\left(\left(\frac{r}{a}\right)^{-5} .. \left(\frac{r}{a}\right)^{-1},\ \log\frac{r}{a},\ \left(\frac{r}{a}\right)^{1} ... \left(\frac{r}{a}\right)^{4}\right)$ ...

		$[\ ]^2$			$[\ ]^3$			$[\ ]^4$	
		e^2	e^4	e^6	e^3	e^5	e^7	e^4	e^6
$\left(\frac{r}{a}\right)^{-5}$	[cos]	$+5$	$+\frac{155}{12}$	$+\frac{835}{32}$	$+\frac{145}{16}$	$+\frac{4715}{256}$	$+\frac{70893}{2048}$	$+\frac{745}{48}$	$+\frac{197}{8}$
$\left(\frac{r}{a}\right)^{-4}$		$+\frac{7}{2}$	$+\frac{67}{12}$	$+\frac{421}{48}$	$+\frac{23}{4}$	$+\frac{435}{64}$	$+\frac{27027}{2560}$	$+\frac{437}{48}$	$+\frac{899}{120}$
$\left(\frac{r}{a}\right)^{-3}$		$+\frac{9}{4}$	$+\frac{7}{4}$	$+\frac{141}{64}$	$+\frac{53}{16}$	$+\frac{393}{256}$	$+\frac{24753}{10240}$	$+\frac{77}{16}$	$+\frac{129}{160}$
$\left(\frac{r}{a}\right)^{-2}$		$+\frac{5}{4}$	$+\frac{1}{6}$	$+\frac{21}{64}$	$+\frac{13}{8}$	$-\frac{25}{128}$	$+\frac{393}{1024}$	$+\frac{103}{48}$	$-\frac{129}{160}$
$\left(\frac{r}{a}\right)^{-1}$		$+\frac{1}{2}$	$-\frac{1}{6}$	$+\frac{1}{48}$	$+\frac{9}{16}$	$-\frac{81}{256}$	$+\frac{729}{10240}$	$+\frac{2}{3}$	$-\frac{8}{15}$
$\log\frac{r}{a}$		$-\frac{3}{8}$	$+\frac{11}{48}$	$-\frac{3}{128}$	$-\frac{17}{48}$	$+\frac{77}{256}$	$-\frac{743}{10240}$	$-\frac{71}{192}$	$+\frac{129}{320}$
$\left(\frac{r}{a}\right)^{1}$		$-\frac{1}{4}$	$+\frac{1}{6}$	$-\frac{1}{32}$	$-\frac{3}{16}$	$+\frac{45}{256}$	$-\frac{567}{10240}$	$-\frac{1}{6}$	$+\frac{1}{5}$
$\left(\frac{r}{a}\right)^{2}$		$-\frac{1}{4}$	$+\frac{1}{12}$	$-\frac{1}{96}$	$-\frac{1}{8}$	$+\frac{9}{128}$	$-\frac{81}{5120}$	$-\frac{1}{12}$	$+\frac{1}{15}$
$\left(\frac{r}{a}\right)^{3}$		0	$-\frac{1}{4}$	$+\frac{3}{32}$	$+\frac{1}{16}$	$-\frac{45}{256}$	$+\frac{189}{2048}$	$+\frac{1}{16}$	$-\frac{3}{20}$
$\left(\frac{r}{a}\right)^{4}$		$+\frac{1}{2}$	$-\frac{7}{12}$	$+\frac{1}{8}$	$+\frac{1}{4}$	$-\frac{19}{64}$	$+\frac{261}{2560}$	$+\frac{7}{48}$	$-\frac{1}{5}$

...Table $\left(\left(\frac{r}{a}\right)^{-5} \ldots \left(\frac{r}{a}\right)^{-1}, \ \log\frac{r}{a} \ldots \left(\frac{r}{a}\right)^{1} \ldots \left(\frac{r}{a}\right)^{4}\right)$.

		$[\ ]^5$		$[\ ]^6$	$[\ ]^7$
		e^5	e^7	e^6	e^7
$\left(\frac{r}{a}\right)^{-5}$	[cos]	$+\frac{19669}{768}$	$+\frac{565135}{18432}$	$+\frac{1317}{32}$	$+\frac{1195093}{18432}$
$\left(\frac{r}{a}\right)^{-4}$		$+\frac{2701}{192}$	$+\frac{31997}{4608}$	$+\frac{1709}{80}$	$+\frac{737707}{23040}$
$\left(\frac{r}{a}\right)^{-3}$		$+\frac{1773}{256}$	$-\frac{4987}{6144}$	$+\frac{3167}{320}$	$+\frac{432091}{30720}$
$\left(\frac{r}{a}\right)^{-2}$		$+\frac{1097}{384}$	$-\frac{16621}{9216}$	$+\frac{1223}{320}$	$+\frac{47273}{9216}$
$\left(\frac{r}{a}\right)^{-1}$		$+\frac{625}{768}$	$-\frac{15625}{18432}$	$+\frac{81}{80}$	$+\frac{117649}{92160}$
$\log\frac{r}{a}$		$-\frac{523}{1280}$	$+\frac{10039}{18432}$	$-\frac{899}{1920}$	$-\frac{355081}{645120}$
$\left(\frac{r}{a}\right)^{1}$		$-\frac{125}{768}$	$+\frac{4375}{18432}$	$-\frac{27}{160}$	$-\frac{16807}{92160}$
$\left(\frac{r}{a}\right)^{2}$		$-\frac{25}{384}$	$+\frac{615}{9216}$	$-\frac{9}{160}$	$-\frac{2401}{46080}$
$\left(\frac{r}{a}\right)^{3}$		$+\frac{15}{256}$	$-\frac{875}{6144}$	$+\frac{9}{160}$	$+\frac{343}{6144}$
$\left(\frac{r}{a}\right)^{4}$		$+\frac{19}{192}$	$-\frac{725}{4608}$	$+\frac{3}{40}$	$+\frac{1421}{23040}$

Table $\left(\left(\frac{r}{a}\right)^1 \ldots \left(\frac{r}{a}\right)^4\right) \begin{matrix}\cos\\ \sin\end{matrix} f \ldots$

$\begin{matrix}\cos\\ \sin\end{matrix} f$		$[\ ]^0$ e	e^3	e^5	e^7	$[\ ]^1$ e	e^2	e^4	e^6
$\left(\frac{r}{a}\right)^1$	[cos + sin]	$-\frac{3}{2}$	0	0	0	$+1$	$-\frac{1}{2}$	$-\frac{1}{64}$	$-\frac{29}{1152}$
	[cos − sin]	,,	,,	,,	,,	0	$+\frac{1}{8}$	$+\frac{1}{24}$	$+\frac{25}{1024}$
	[cos]	$-\frac{3}{2}$	0	0	0	H	$-\frac{3}{16}$	$+\frac{5}{384}$	$-\frac{7}{18432}$
	[sin]	0	0	0	0	H	$-\frac{5}{16}$	$-\frac{11}{384}$	$-\frac{457}{18432}$
$\left(\frac{r}{a}\right)^2$	[cos + sin]	-2	$-\frac{1}{2}$	0	0	$+1$	$+\frac{1}{2}$	$-\frac{25}{64}$	$-\frac{23}{1152}$
	[cos − sin]	,,	,,	,,	,,	0	$+\frac{5}{8}$	0	$+\frac{143}{3072}$
	[cos]	-2	$-\frac{1}{2}$	0	0	H	$+\frac{9}{16}$	$-\frac{25}{128}$	$+\frac{245}{18432}$
	[sin]	0	0	0	0	H	$-\frac{1}{16}$	$-\frac{25}{128}$	$-\frac{613}{18432}$
$\left(\frac{r}{a}\right)^3$	[cos + sin]	$-\frac{5}{2}$	$-\frac{15}{8}$	0	0	$+1$	$+2$	$-\frac{41}{64}$	$-\frac{37}{576}$
	[cos − sin]	,,	,,	,,	,,	0	$+\frac{11}{8}$	$+\frac{7}{48}$	$+\frac{265}{3072}$
	[cos]	$-\frac{5}{2}$	$-\frac{15}{8}$	0	0	H	$+\frac{27}{16}$	$-\frac{95}{384}$	$+\frac{203}{18432}$
	[sin]	0	0	0	0	H	$+\frac{5}{16}$	$-\frac{151}{384}$	$-\frac{1387}{18432}$
$\left(\frac{r}{a}\right)^4$	[cos + sin]	-3	$-\frac{9}{2}$	$-\frac{3}{8}$	0	$+1$	$+4$	$-\frac{1}{64}$	$-\frac{199}{576}$
	[cos − sin]	,,	,,	,,	,,	0	$+\frac{19}{8}$	$+\frac{47}{48}$	$+\frac{115}{1024}$
	[cos]	-3	$-\frac{9}{2}$	$-\frac{3}{8}$	0	H	$+\frac{51}{16}$	$+\frac{185}{384}$	$-\frac{2149}{18432}$
	[sin]	0	0	0	0	H	$+\frac{13}{16}$	$-\frac{191}{384}$	$-\frac{4219}{18432}$

...*Table* $\left(\left(\frac{r}{a}\right)^1 \ldots \left(\frac{r}{a}\right)^4\right) \begin{matrix}\cos\\ \sin\end{matrix} f \ldots$

	$\begin{matrix}\cos\\ \sin\end{matrix} f$	$[\ ]^2$				$[\ ]^3$			$[\ ]^4$		
		e	e^3	e^5	e^7	e^2	e^4	e^6	e^3	e^5	e^7
$\left(\frac{r}{a}\right)^1$	[cos + sin]	$+\frac{1}{2}$	$-\frac{3}{8}$	$+\frac{5}{96}$	$-\frac{1}{72}$	$+\frac{3}{8}$	$-\frac{3}{8}$	$+\frac{111}{1024}$	$+\frac{1}{3}$	$-\frac{5}{12}$	$+\frac{43}{240}$
	[cos − sin]	0	$+\frac{1}{24}$	$+\frac{1}{96}$	$+\frac{1}{120}$	0	$+\frac{3}{128}$	$+\frac{3}{1280}$	0	$+\frac{1}{60}$	$-\frac{1}{720}$
	[cos]	H	$-\frac{1}{6}$	$+\frac{1}{32}$	$-\frac{1}{360}$	H	$-\frac{45}{256}$	$+\frac{567}{10240}$	H	$-\frac{1}{5}$	$+\frac{4}{45}$
	[sin]	H	$-\frac{5}{24}$	$+\frac{1}{48}$	$-\frac{1}{90}$	H	$-\frac{51}{256}$	$+\frac{543}{10240}$	H	$-\frac{13}{60}$	$+\frac{13}{144}$
$\left(\frac{r}{a}\right)^2$	[cos + sin]	0	$+\frac{1}{2}$	$-\frac{1}{3}$	$+\frac{5}{96}$	$-\frac{1}{8}$	$+\frac{1}{2}$	$-\frac{397}{1024}$	$-\frac{1}{6}$	$+\frac{13}{24}$	$-\frac{47}{96}$
	[cos − sin]	0	$+\frac{1}{6}$	$-\frac{1}{24}$	$+\frac{7}{480}$	0	$+\frac{11}{128}$	$-\frac{11}{256}$	0	$+\frac{7}{120}$	$-\frac{7}{160}$
	[cos]	0	$+\frac{1}{3}$	$-\frac{3}{16}$	$+\frac{1}{30}$	H	$+\frac{75}{256}$	$-\frac{441}{2048}$	H	$+\frac{3}{10}$	$-\frac{4}{15}$
	[sin]	0	$+\frac{1}{6}$	$-\frac{7}{48}$	$+\frac{3}{160}$	H	$+\frac{53}{256}$	$-\frac{353}{2048}$	H	$+\frac{29}{120}$	$-\frac{107}{480}$
$\left(\frac{r}{a}\right)^3$	[cos + sin]	$-\frac{1}{2}$	$+1$	$-\frac{35}{96}$	$+\frac{23}{576}$	$-\frac{3}{8}$	$+\frac{11}{16}$	$-\frac{351}{1024}$	$-\frac{7}{24}$	$+\frac{55}{96}$	$-\frac{701}{1920}$
	[cos − sin]	0	$+\frac{1}{6}$	$-\frac{1}{96}$	$+\frac{7}{960}$	0	$+\frac{7}{128}$	$-\frac{9}{640}$	0	$+\frac{13}{480}$	$-\frac{73}{5760}$
	[cos]	H	$+\frac{7}{12}$	$-\frac{3}{16}$	$+\frac{17}{720}$	H	$+\frac{95}{256}$	$-\frac{1827}{10240}$	H	$+\frac{3}{10}$	$-\frac{17}{90}$
	[sin]	H	$+\frac{5}{12}$	$-\frac{17}{96}$	$+\frac{47}{2880}$	H	$+\frac{81}{256}$	$-\frac{1683}{10240}$	H	$+\frac{131}{480}$	$-\frac{203}{1152}$
$\left(\frac{r}{a}\right)^4$	[cos + sin]	-1	$+\frac{3}{4}$	$+\frac{13}{48}$	$-\frac{83}{576}$	$-\frac{3}{8}$	$+\frac{3}{16}$	$+\frac{249}{1024}$	$-\frac{1}{6}$	$+\frac{1}{48}$	$+\frac{119}{480}$
	[cos − sin]	0	$-\frac{1}{12}$	$+\frac{1}{6}$	$-\frac{43}{960}$	0	$-\frac{9}{128}$	$+\frac{57}{640}$	0	$-\frac{11}{240}$	$+\frac{91}{1440}$
	[cos]	H	$+\frac{1}{3}$	$+\frac{7}{32}$	$-\frac{17}{180}$	H	$+\frac{15}{256}$	$+\frac{1701}{10240}$	H	$-\frac{1}{80}$	$+\frac{7}{45}$
	[sin]	H	$+\frac{5}{12}$	$+\frac{5}{96}$	$-\frac{143}{2880}$	H	$+\frac{33}{256}$	$+\frac{789}{10240}$	H	$+\frac{1}{30}$	$+\frac{133}{1440}$

... *Table* $\left(\left(\frac{r}{a}\right)^1 \ldots \left(\frac{r}{a}\right)^4\right) \begin{matrix}\cos\\ \sin\end{matrix} f.$

$\begin{matrix}\cos\\ \sin\end{matrix} f$		$[\ ]^5$ e^4	$[\ ]^5$ e^6	$[\ ]^6$ e^5	$[\ ]^6$ e^7	$[\ ]^7$ e^6	$[\ ]^8$ e^7
$\left(\frac{r}{a}\right)^1$	[cos + sin]	$+\frac{125}{384}$	$-\frac{125}{256}$	$+\frac{27}{80}$	$-\frac{189}{320}$	$+\frac{16807}{46080}$	$+\frac{128}{315}$
	[cos − sin]	0	$+\frac{125}{9216}$	0	$+\frac{27}{2240}$	0	0
	[cos]	H	$-\frac{4375}{18432}$	H	$-\frac{81}{280}$	H	H
	[sin]	H	$-\frac{4625}{18432}$	H	$-\frac{135}{448}$	H	H
$\left(\frac{r}{a}\right)^2$	[cos + sin]	$-\frac{25}{128}$	$+\frac{475}{768}$	$-\frac{9}{40}$	$+\frac{117}{160}$	$-\frac{2401}{9216}$	$-\frac{32}{105}$
	[cos − sin]	0	$+\frac{425}{9216}$	0	$+\frac{9}{224}$	0	0
	[cos]	H	$+\frac{6125}{18432}$	H	$+\frac{27}{70}$	H	H
	[sin]	H	$+\frac{5275}{18432}$	H	$+\frac{387}{1120}$	H	H
$\left(\frac{r}{a}\right)^3$	[cos + sin]	$-\frac{95}{384}$	$+\frac{205}{384}$	$-\frac{9}{40}$	$+\frac{171}{320}$	$-\frac{9947}{46080}$	$-\frac{68}{315}$
	[cos − sin]	0	$+\frac{155}{9216}$	0	$+\frac{27}{2240}$	0	0
	[cos]	H	$+\frac{5075}{18432}$	H	$+\frac{153}{560}$	H	H
	[sin]	H	$+\frac{4765}{18432}$	H	$+\frac{117}{448}$	H	H
$\left(\frac{r}{a}\right)^4$	[cos + sin]	$-\frac{31}{384}$	$-\frac{7}{128}$	$-\frac{3}{80}$	$-\frac{33}{320}$	$-\frac{539}{46080}$	$+\frac{2}{315}$
	[cos − sin]	0	$-\frac{301}{9216}$	0	$-\frac{57}{2240}$	0	0
	[cos]	H	$-\frac{805}{18432}$	H	$-\frac{9}{140}$	H	H
	[sin]	H	$-\frac{203}{18432}$	H	$-\frac{87}{2240}$	H	H

$$\ldots \textit{Table} \left(\left(\frac{r}{a}\right)^{-1} \ldots \left(\frac{r}{a}\right)^{-5}\right) \begin{matrix}\cos\\ \sin\end{matrix} f \ldots$$

	$\begin{matrix}\cos\\ \sin\end{matrix} f$	[ ]^0 e	e^3	e^5	e^7	[ ]^1 e^0	e^2	e^4	e^6
$\left(\frac{r}{a}\right)^{-1}$	[cos + sin]	$-\frac{1}{2}$	$-\frac{1}{8}$	$-\frac{1}{16}$	$-\frac{5}{128}$	$+1$	-1	$-\frac{1}{64}$	$-\frac{17}{288}$
	[cos − sin]	,,	,,	,,	,,	0	$-\frac{1}{8}$	$-\frac{1}{16}$	$-\frac{115}{3072}$
	[cos]	$-\frac{1}{2}$	$-\frac{1}{8}$	$-\frac{1}{16}$	$-\frac{5}{128}$	H	$-\frac{9}{16}$	$-\frac{5}{128}$	$-\frac{889}{18432}$
	[sin]	0	0	0	0	H	$-\frac{7}{16}$	$+\frac{3}{128}$	$-\frac{199}{18432}$
$\left(\frac{r}{a}\right)^{-2}$	[cos + sin]	0	0	0	0	$+1$	$-\frac{1}{2}$	$-\frac{1}{64}$	$-\frac{29}{1152}$
	[cos − sin]	,,	,,	,,	,,	0	$+\frac{1}{8}$	$+\frac{1}{24}$	$+\frac{25}{1024}$
	[cos]	0	0	0	0	H	$-\frac{3}{16}$	$+\frac{5}{384}$	$-\frac{7}{18432}$
	[sin]	0	0	0	0	H	$-\frac{5}{16}$	$-\frac{11}{384}$	$-\frac{457}{18432}$
$\left(\frac{r}{a}\right)^{-3}$	[cos + sin]	$+\frac{1}{2}$	$+\frac{3}{4}$	$+\frac{15}{16}$	$+\frac{35}{32}$	$+1$	$+\frac{1}{2}$	$+\frac{55}{64}$	$+\frac{1177}{1152}$
	[cos − sin]	,,	,,	,,	,,	0	$+\frac{5}{8}$	$+\frac{5}{6}$	$+\frac{3103}{3072}$
	[cos]	$+\frac{1}{2}$	$+\frac{3}{4}$	$+\frac{15}{16}$	$+\frac{35}{32}$	H	$+\frac{9}{16}$	$+\frac{325}{384}$	$+\frac{18725}{18432}$
	[sin]	0	0	0	0	H	$-\frac{1}{16}$	$+\frac{5}{384}$	$+\frac{107}{18432}$
$\left(\frac{r}{a}\right)^{-4}$	[cos + sin]	$+1$	$+\frac{5}{2}$	$+\frac{35}{8}$	$+\frac{105}{16}$	$+1$	$+2$	$+\frac{239}{64}$	$+\frac{3323}{576}$
	[cos − sin]	,,	,,	,,	,,	0	$+\frac{11}{8}$	$+\frac{49}{16}$	$+\frac{15665}{3072}$
	[cos]	$+1$	$+\frac{5}{2}$	$+\frac{35}{8}$	$+\frac{105}{16}$	H	$+\frac{27}{16}$	$+\frac{435}{128}$	$+\frac{100163}{18432}$
	[sin]	0	0	0	0	H	$+\frac{5}{16}$	$+\frac{43}{128}$	$+\frac{6173}{18432}$
$\left(\frac{r}{a}\right)^{-5}$	[cos + sin]	$+\frac{3}{2}$	$+\frac{45}{8}$	$+\frac{105}{8}$	$+\frac{1575}{64}$	$+1$	$+4$	$+\frac{647}{64}$	$+\frac{11465}{576}$
	[cos − sin]	,,	,,	,,	,,	0	$+\frac{19}{8}$	$+\frac{371}{48}$	$+\frac{17179}{1024}$
	[cos]	$+\frac{3}{2}$	$+\frac{45}{8}$	$+\frac{105}{8}$	$+\frac{1575}{64}$	H	$+\frac{51}{16}$	$+\frac{3425}{384}$	$+\frac{338051}{18432}$
	[sin]	0	0	0	0	H	$+\frac{13}{16}$	$+\frac{457}{384}$	$+\frac{28829}{18432}$

$$\ldots \textit{Table} \left(\left(\frac{r}{a}\right)^{-1} \ldots \left(\frac{r}{a}\right)^{-5}\right) \begin{matrix}\cos\\ \sin\end{matrix} f \ldots$$

$\begin{matrix}\cos\\ \sin\end{matrix} f$		$[\ ]^3$ e	e^3	e^5	e^7	$[\ ]^4$ e^2	e^4	e^6
$\left(\frac{r}{a}\right)^{-1}$	[cos + sin]	$+\frac{3}{2}$	$-\frac{7}{4}$	$+\frac{31}{96}$	$-\frac{1}{12}$	$+\frac{17}{8}$	$-\frac{47}{16}$	$+\frac{1069}{1024}$
	[cos − sin]	0	$-\frac{1}{12}$	$-\frac{1}{24}$	$-\frac{7}{240}$	0	$-\frac{9}{128}$	$-\frac{9}{320}$
	[cos]	H	$-\frac{11}{12}$	$+\frac{9}{64}$	$-\frac{9}{160}$	H	$-\frac{385}{256}$	$+\frac{5201}{10240}$
	[sin]	H	$-\frac{5}{6}$	$+\frac{35}{192}$	$-\frac{13}{480}$	H	$-\frac{367}{256}$	$+\frac{5489}{10240}$
$\left(\frac{r}{a}\right)^{-2}$	[cos + sin]	$+2$	$-\frac{3}{2}$	$+\frac{5}{24}$	$-\frac{1}{18}$	$+\frac{27}{8}$	$-\frac{27}{8}$	$+\frac{999}{1024}$
	[cos − sin]	0	$+\frac{1}{6}$	$+\frac{1}{24}$	$+\frac{1}{30}$	0	$+\frac{27}{128}$	$+\frac{27}{1280}$
	[cos]	H	$-\frac{2}{3}$	$+\frac{1}{8}$	$-\frac{1}{90}$	H	$-\frac{405}{256}$	$+\frac{5103}{10240}$
	[sin]	H	$-\frac{5}{6}$	$+\frac{1}{12}$	$-\frac{2}{45}$	H	$-\frac{459}{256}$	$+\frac{4887}{10240}$
$\left(\frac{r}{a}\right)^{-3}$	[cos + sin]	$+\frac{5}{2}$	$-\frac{1}{8}$	$+\frac{103}{96}$	$+\frac{1255}{1152}$	$+\frac{39}{8}$	-2	$+\frac{1803}{1024}$
	[cos − sin]	0	$+\frac{19}{24}$	$+\frac{43}{48}$	$+\frac{2063}{1920}$	0	$+\frac{131}{128}$	$+\frac{237}{256}$
	[cos]	H	$+\frac{1}{3}$	$+\frac{63}{64}$	$+\frac{779}{720}$	H	$-\frac{125}{256}$	$+\frac{2751}{2048}$
	[sin]	H	$-\frac{11}{24}$	$+\frac{17}{192}$	$+\frac{43}{5760}$	H	$-\frac{387}{256}$	$+\frac{855}{2048}$
$\left(\frac{r}{a}\right)^{-4}$	[cos + sin]	$+3$	$+\frac{11}{4}$	$+\frac{245}{48}$	$+\frac{463}{64}$	$+\frac{53}{8}$	$+\frac{39}{16}$	$+\frac{7041}{1024}$
	[cos − sin]	0	$+\frac{23}{12}$	$+\frac{89}{24}$	$+\frac{5663}{960}$	0	$+\frac{343}{128}$	$+\frac{2819}{640}$
	[cos]	H	$+\frac{7}{3}$	$+\frac{141}{32}$	$+\frac{197}{30}$	H	$+\frac{655}{256}$	$+\frac{57757}{10240}$
	[sin]	H	$+\frac{5}{12}$	$+\frac{67}{96}$	$+\frac{641}{960}$	H	$-\frac{31}{256}$	$+\frac{12653}{10240}$
$\left(\frac{r}{a}\right)^{-5}$	[cos + sin]	$+\frac{7}{2}$	$+\frac{15}{2}$	$+\frac{1529}{96}$	$+\frac{16207}{576}$	$+\frac{69}{8}$	$+\frac{183}{16}$	$+\frac{23865}{1024}$
	[cos − sin]	0	$+\frac{11}{3}$	$+\frac{997}{96}$	$+\frac{20279}{960}$	0	$+\frac{711}{128}$	$+\frac{8733}{640}$
	[cos]	H	$+\frac{67}{12}$	$+\frac{421}{32}$	$+\frac{8867}{360}$	H	$+\frac{2175}{256}$	$+\frac{189189}{10240}$
	[sin]	H	$+\frac{23}{12}$	$+\frac{133}{48}$	$+\frac{10099}{2880}$	H	$+\frac{753}{256}$	$+\frac{49461}{10240}$

... Table $\left(\left(\frac{r}{a}\right)^{-1} ... \left(\frac{r}{a}\right)^{-5}\right) \begin{matrix}\cos\\ \sin\end{matrix} f$ ...

	$\begin{matrix}\cos\\ \sin\end{matrix} f$	$[\ ]^4$			$[\ ]^5$	
		e^3	e^5	e^7	e^4	e^6
$\left(\frac{r}{a}\right)^{-1}$	[cos + sin]	$+\frac{71}{24}$	$-\frac{229}{48}$	$+\frac{1169}{480}$	$+\frac{523}{128}$	$-\frac{1451}{192}$
	[cos − sin]	o	$-\frac{1}{15}$	$-\frac{1}{60}$	o	$-\frac{625}{9216}$
	[cos]	H	$-\frac{387}{160}$	$+\frac{387}{320}$	H	$-\frac{70273}{18432}$
	[sin]	H	$-\frac{1129}{480}$	$+\frac{1177}{960}$	H	$-\frac{69023}{18432}$
$\left(\frac{r}{a}\right)^{-2}$	[cos + sin]	$+\frac{16}{3}$	$-\frac{20}{3}$	$+\frac{43}{15}$	$+\frac{3125}{384}$	$-\frac{3125}{256}$
	[cos − sin]	o	$+\frac{4}{15}$	$-\frac{1}{45}$	o	$+\frac{3125}{9216}$
	[cos]	H	$-\frac{16}{5}$	$+\frac{64}{45}$	H	$-\frac{109375}{18432}$
	[sin]	H	$-\frac{52}{15}$	$+\frac{13}{9}$	H	$-\frac{115625}{18432}$
$\left(\frac{r}{a}\right)^{-3}$	[cos + sin]	$+\frac{103}{12}$	$-\frac{593}{96}$	$+\frac{721}{192}$	$+\frac{5485}{384}$	$-\frac{11053}{768}$
	[cos − sin]	o	$+\frac{643}{480}$	$+\frac{2569}{2880}$	o	$+\frac{16289}{9216}$
	[cos]	H	$-\frac{387}{160}$	$+\frac{1673}{720}$	H	$-\frac{116347}{18432}$
	[sin]	H	$-\frac{451}{120}$	$+\frac{4123}{2880}$	H	$-\frac{148925}{18432}$
$\left(\frac{r}{a}\right)^{-4}$	[cos + sin]	$+\frac{77}{6}$	$-\frac{25}{48}$	$+\frac{4751}{480}$	$+\frac{2955}{128}$	$-\frac{3463}{384}$
	[cos − sin]	o	$+\frac{899}{240}$	$+\frac{2441}{480}$	o	$+\frac{48203}{9216}$
	[cos]	H	$+\frac{129}{80}$	$+\frac{899}{120}$	H	$-\frac{34909}{18432}$
	[sin]	H	$-\frac{32}{15}$	$+\frac{77}{32}$	H	$-\frac{131315}{18432}$
$\left(\frac{r}{a}\right)^{-5}$	[cos + sin]	$+\frac{437}{24}$	$+\frac{1361}{96}$	$+\frac{63341}{1920}$	$+\frac{13505}{384}$	$+\frac{1541}{128}$
	[cos − sin]	o	$+\frac{3983}{480}$	$+\frac{100849}{5760}$	o	$+\frac{113027}{9216}$
	[cos]	H	$+\frac{899}{80}$	$+\frac{36359}{1440}$	H	$+\frac{223979}{18432}$
	[sin]	H	$+\frac{1411}{480}$	$+\frac{44587}{5760}$	H	$-\frac{2075}{18432}$

$$\ldots\ Table\ \left(\left(\frac{r}{a}\right)^{-1} \ldots \left(\frac{r}{a}\right)^{-5}\right) \begin{matrix}\cos\\ \sin\end{matrix} f.$$

$\begin{matrix}\cos\\ \sin\end{matrix} f$		$[\ ]^6$ e^5	$[\ ]^6$ e^7	$[\ ]^7$ e^6	$[\ ]^8$ e^7
$\left(\frac{r}{a}\right)^{-1}$	[cos + sin]	$+\frac{899}{160}$	$-\frac{1879}{160}$	$+\frac{355081}{46080}$	$+\frac{47259}{4480}$
	[cos − sin]	0	$-\frac{81}{1120}$	0	0
	[cos]	H	$-\frac{6617}{1120}$	H	H
	[sin]	H	$-\frac{817}{140}$	H	H
$\left(\frac{r}{a}\right)^{-2}$	[cos + sin]	$+\frac{243}{20}$	$-\frac{1701}{80}$	$+\frac{823543}{46080}$	$+\frac{8192}{315}$
	[cos − sin]	0	$+\frac{243}{560}$	0	0
	[cos]	H	$-\frac{729}{70}$	H	H
	[sin]	H	$-\frac{1215}{112}$	H	H
$\left(\frac{r}{a}\right)^{-3}$	[cos + sin]	$+\frac{3669}{160}$	$-\frac{18847}{640}$	$+\frac{330911}{9216}$	$+\frac{556403}{10080}$
	[cos − sin]	0	$+\frac{2101}{896}$	0	0
	[cos]	H	$-\frac{7589}{560}$	H	H
	[sin]	H	$-\frac{71217}{4480}$	H	H
$\left(\frac{r}{a}\right)^{-4}$	[cos + sin]	$+\frac{3167}{80}$	$-\frac{8999}{320}$	$+\frac{3024637}{46080}$	$+\frac{178331}{1680}$
	[cos − sin]	0	$+\frac{16337}{2240}$	0	0
	[cos]	H	$-\frac{729}{70}$	H	H
	[sin]	H	$-\frac{7933}{448}$	H	H
$\left(\frac{r}{a}\right)^{-5}$	[cos + sin]	$+\frac{5127}{80}$	$-\frac{753}{320}$	$+\frac{5163949}{46080}$	$+\frac{3830591}{20160}$
	[cos − sin]	0	$+\frac{40263}{2240}$	0	0
	[cos]	H	$+\frac{2187}{280}$	H	H
	[sin]	H	$-\frac{22767}{2240}$	H	H

Table $\left(\left(\frac{r}{a}\right)^1 \dots \left(\frac{r}{a}\right)^4\right) \frac{\cos}{\sin} 2f \dots$

$\frac{\cos}{\sin} 2f$		$[\ ]^0$ e^2	e^4	e^6	$[\ ]^1$ e	e^3	e^5	e^7
$\left(\frac{r}{a}\right)^1$	[cos + sin]	$+\frac{3}{2}$	o	o	$-\frac{5}{2}$	$+\frac{33}{16}$	$-\frac{73}{384}$	$+\frac{881}{18432}$
	[cos − sin]	,,	,,	,,	o	$+\frac{1}{48}$	$-\frac{41}{768}$	$-\frac{1159}{30720}$
	[cos]	$+\frac{3}{2}$	o	o	H	$+\frac{25}{24}$	$-\frac{187}{1536}$	$+\frac{29}{5760}$
	[sin]	o	o	o	H	$+\frac{49}{48}$	$-\frac{35}{512}$	$+\frac{3941}{92160}$
$\left(\frac{r}{a}\right)^2$	[cos + sin]	$+\frac{5}{2}$	o	o	-3	$+\frac{13}{8}$	$+\frac{5}{192}$	$+\frac{227}{3072}$
	[cos − sin]	,,	,,	,,	o	$-\frac{7}{24}$	$-\frac{47}{384}$	$-\frac{1091}{15360}$
	[cos]	$+\frac{5}{2}$	o	o	H	$+\frac{2}{3}$	$-\frac{37}{768}$	$+\frac{11}{7680}$
	[sin]	o	o	o	H	$+\frac{23}{24}$	$+\frac{19}{256}$	$+\frac{371}{5120}$
$\left(\frac{r}{a}\right)^3$	[cos + sin]	$+\frac{15}{4}$	$+\frac{5}{8}$	o	$-\frac{7}{2}$	$+\frac{1}{16}$	$+\frac{289}{384}$	$+\frac{1645}{18432}$
	[cos − sin]	,,	,,	,,	,,	$-\frac{47}{48}$	$-\frac{151}{768}$	$-\frac{4043}{30720}$
	[cos]	$+\frac{15}{4}$	$+\frac{5}{8}$	o	H	$-\frac{11}{24}$	$+\frac{427}{1536}$	$-\frac{61}{2880}$
	[sin]	o	o	o	H	$+\frac{25}{48}$	$+\frac{243}{512}$	$+\frac{10177}{92160}$
$\left(\frac{r}{a}\right)^4$	[cos + sin]	$+\frac{21}{4}$	$+\frac{21}{8}$	o	-4	-3	$+\frac{79}{48}$	$+\frac{233}{1152}$
	[cos − sin]	,,	,,	,,	,,	$-\frac{13}{6}$	$-\frac{13}{24}$	$-\frac{989}{3840}$
	[cos]	$+\frac{21}{4}$	$+\frac{21}{8}$	o	H	$-\frac{31}{12}$	$+\frac{53}{96}$	$-\frac{637}{23040}$
	[sin]	o	o	o	H	$-\frac{5}{12}$	$+\frac{35}{32}$	$+\frac{5297}{23040}$

... *Table* $\left(\left(\frac{r}{a}\right)^1 \ldots \left(\frac{r}{a}\right)^4\right) \begin{matrix}\cos\\ \sin\end{matrix} 2f \ldots$

$\begin{matrix}\cos\\ \sin\end{matrix} 2f$		[ ]^2 e^0	[ ]^2 e^2	[ ]^2 e^4	[ ]^2 e^6	[ ]^3 e	[ ]^3 e^3	[ ]^3 e^5	[ ]^3 e^7
$\left(\frac{r}{a}\right)^1$	[cos + sin]	+ 1	$-\frac{7}{2}$	$+\frac{47}{16}$	$-\frac{215}{288}$	$+\frac{3}{2}$	$-\frac{75}{16}$	$+\frac{573}{128}$	$-\frac{3507}{2048}$
	[cos − sin]	o	o	$+\frac{1}{48}$	$-\frac{3}{160}$	o	o	$+\frac{21}{1280}$	$-\frac{123}{10240}$
	[cos]	H	H	$+\frac{71}{48}$	$-\frac{551}{1440}$	H	H	$+\frac{5751}{2560}$	$-\frac{8829}{10240}$
	[sin]	H	H	$+\frac{35}{24}$	$-\frac{131}{360}$	H	H	$+\frac{5709}{2560}$	$-\frac{4353}{5120}$
$\left(\frac{r}{a}\right)^2$	[cos + sin]	+ 1	$-\frac{5}{2}$	$+\frac{23}{16}$	$-\frac{65}{288}$	+ 1	$-\frac{19}{8}$	$+\frac{107}{64}$	$-\frac{479}{1024}$
	[cos − sin]	o	o	$-\frac{1}{16}$	$-\frac{11}{480}$	o	o	$-\frac{17}{640}$	$-\frac{7}{1024}$
	[cos]	H	H	$+\frac{11}{16}$	$-\frac{179}{1440}$	H	H	$+\frac{1053}{1280}$	$-\frac{243}{1024}$
	[sin]	H	H	$+\frac{3}{4}$	$-\frac{73}{720}$	H	H	$+\frac{1087}{1280}$	$-\frac{59}{256}$
$\left(\frac{r}{a}\right)^3$	[cos + sin]	+ 1	− 1	$-\frac{11}{16}$	$+\frac{43}{72}$	$+\frac{1}{2}$	$-\frac{3}{16}$	$-\frac{117}{128}$	$+\frac{1553}{2048}$
	[cos − sin]	o	o	$-\frac{7}{48}$	$+\frac{1}{240}$	o	o	$-\frac{69}{1280}$	$+\frac{137}{10240}$
	[cos]	H	H	$-\frac{5}{12}$	$+\frac{433}{1440}$	H	H	$-\frac{1239}{2560}$	$+\frac{3951}{10240}$
	[sin]	H	H	$-\frac{13}{48}$	$+\frac{427}{1440}$	H	H	$-\frac{1101}{2560}$	$+\frac{1907}{5120}$
$\left(\frac{r}{a}\right)^4$	[cos + sin]	+ 1	+ 1	$-\frac{43}{16}$	$+\frac{35}{36}$		$+\frac{3}{2}$	$-\frac{9}{4}$	$+\frac{267}{256}$
	[cos − sin]	o	o	$-\frac{5}{48}$	o		o	$-\frac{3}{160}$	$+\frac{3}{640}$
	[cos]	H	H	$-\frac{67}{48}$	$+\frac{35}{72}$		H	$-\frac{363}{320}$	$+\frac{1341}{2560}$
	[sin]	H	H	$-\frac{31}{24}$	$+\frac{35}{72}$		H	$-\frac{357}{320}$	$+\frac{1329}{2560}$

... *Table* $\left(\left(\frac{r}{a}\right)^1 \ldots \left(\frac{r}{a}\right)^4\right) \begin{matrix}\cos\\ \sin\end{matrix}\, 2f \ldots$

	$\begin{matrix}\cos\\ \sin\end{matrix}\, 2f$	[ ]^4			[ ]^5		
		e^2	e^4	e^6	e^3	e^5	e^7
$\left(\frac{r}{a}\right)^1$	[cos + sin]	$+2$	$-\frac{19}{3}$	$+\frac{55}{8}$	$+\frac{125}{48}$	$-\frac{6625}{768}$	$+\frac{64625}{6144}$
	[cos − sin]	0	0	$+\frac{1}{72}$	0	0	$+\frac{1625}{129024}$
	[cos]	H	H	$+\frac{31}{9}$	H	H	$+\frac{679375}{129024}$
	[sin]	H	H	$+\frac{247}{72}$	H	H	$+\frac{338875}{64512}$
$\left(\frac{r}{a}\right)^2$	[cos + sin]	$+1$	$-\frac{5}{2}$	$+\frac{101}{48}$	$+\frac{25}{24}$	$-\frac{1075}{384}$	$+\frac{8425}{3072}$
	[cos − sin]	0	0	$-\frac{11}{720}$	0	0	$-\frac{75}{7168}$
	[cos]	H	H	$+\frac{47}{45}$	H	H	$+\frac{29375}{21504}$
	[sin]	H	H	$+\frac{763}{720}$	H	H	$+\frac{925}{672}$
$\left(\frac{r}{a}\right)^3$	[cos + sin]	$+\frac{1}{4}$	$+\frac{5}{24}$	$-\frac{227}{192}$	$+\frac{5}{48}$	$+\frac{385}{768}$	$-\frac{9515}{6144}$
	[cos − sin]	0	0	$-\frac{83}{2880}$	0	0	$-\frac{2435}{129024}$
	[cos]	H	H	$-\frac{109}{180}$	H	H	$-\frac{101125}{129024}$
	[sin]	H	H	$-\frac{1661}{2880}$	H	H	$-\frac{49345}{64512}$
$\left(\frac{r}{a}\right)^4$	[cos + sin]	$-\frac{1}{4}$	$+\frac{37}{24}$	$-\frac{139}{64}$	$-\frac{1}{3}$	$+\frac{151}{96}$	$-\frac{877}{384}$
	[cos − sin]	0	0	$-\frac{17}{2880}$	0	0	$-\frac{41}{16128}$
	[cos]	H	H	$-\frac{49}{45}$	H	H	$-\frac{36875}{32256}$
	[sin]	H	H	$-\frac{3119}{2880}$	H	H	$-\frac{36793}{32256}$

... *Table* $\left(\left(\frac{r}{a}\right)^1 \ldots \left(\frac{r}{a}\right)^4\right) \begin{matrix}\cos\\ \sin\end{matrix} 2f.$

	$\begin{matrix}\cos\\ \sin\end{matrix} 2f$	[ ]^6 e^4	[ ]^6 e^6	[ ]^7 e^5	[ ]^7 e^7	[ ]^8 e^6	[ ]^9 e^7
$\left(\frac{r}{a}\right)^1$	[cos + sin]	$+\frac{27}{8}$	$-\frac{189}{16}$	$+\frac{16807}{3840}$	$-\frac{1495823}{92160}$	$+\frac{256}{45}$	$+\frac{531441}{71680}$
	[cos − sin]	o	o	o	o	o	o
	[cos]	H	H	H	H	H	H
	[sin]	H	H	H	H	H	H
$\left(\frac{r}{a}\right)^2$	[cos + sin]	$+\frac{9}{8}$	$-\frac{261}{80}$	$+\frac{2401}{1920}$	$-\frac{12005}{3072}$	$+\frac{64}{45}$	$+\frac{59049}{35840}$
	[cos − sin]	o	o	o	o	o	o
	[cos]	H	H	H	H	H	H
	[sin]	H	H	H	H	H	H
$\left(\frac{r}{a}\right)^3$	[cos + sin]	o	$+\frac{63}{80}$	$-\frac{343}{3840}$	$+\frac{102557}{92160}$	$-\frac{8}{45}$	$-\frac{19683}{71680}$
	[cos − sin]	o	o	o	o	o	o
	[cos]	o	H	H	H	H	H
	[sin]	o	H	H	H	H	H
$\left(\frac{r}{a}\right)^4$	[cos + sin]	$-\frac{3}{8}$	$+\frac{33}{20}$	$-\frac{49}{120}$	$+\frac{20531}{11520}$	$-\frac{4}{9}$	$-\frac{2187}{4480}$
	[cos − sin]	o	o	o	o	o	o
	[cos]	H	H	H	H	H	H
	[sin]	H	H	H	H	H	H

Table $\left(\left(\frac{r}{a}\right)^{-1} \ldots \left(\frac{r}{a}\right)^{-5}\right) \begin{matrix}\cos\\ \sin\end{matrix}\, 2f \ldots$

$\begin{matrix}\cos\\ \sin\end{matrix}\, 2f$		$[\ ]^0$ e^2	e^4	e^6	$[\ ]^1$ e	e^3	e^5	e^7
$\left(\frac{r}{a}\right)^{-1}$	[cos + sin]	$+\frac{1}{4}$	$+\frac{1}{8}$	$+\frac{5}{64}$	$-\frac{3}{2}$	$+\frac{17}{16}$	$+\frac{37}{384}$	$+\frac{197}{2048}$
	[cos − sin]	,,	,,	,,	0	$+\frac{1}{48}$	$+\frac{29}{768}$	$+\frac{1013}{30720}$
	[cos]	$+\frac{1}{4}$	$+\frac{1}{8}$	$+\frac{5}{64}$	H	$+\frac{13}{24}$	$+\frac{103}{1536}$	$+\frac{31}{480}$
	[sin]	0	0	0	H	$+\frac{25}{48}$	$+\frac{15}{512}$	$+\frac{971}{30720}$
$\left(\frac{r}{a}\right)^{-2}$	[cos + sin]	0	0	0	-1	$+\frac{3}{8}$	$+\frac{7}{192}$	$+\frac{295}{9216}$
	[cos − sin]	,,	,,	,,	0	$-\frac{1}{24}$	$-\frac{13}{384}$	$-\frac{389}{15360}$
	[cos]	0	0	0	H	$+\frac{1}{6}$	$+\frac{1}{768}$	$+\frac{77}{23040}$
	[sin]	0	0	0	H	$+\frac{5}{24}$	$+\frac{9}{256}$	$+\frac{1321}{46080}$
$\left(\frac{r}{a}\right)^{-3}$	[cos + sin]	0	0	0	$-\frac{1}{2}$	$+\frac{1}{16}$	$-\frac{5}{384}$	$-\frac{143}{18432}$
	[cos − sin]	,,	,,	,,	0	$+\frac{1}{48}$	$+\frac{11}{768}$	$+\frac{313}{30720}$
	[cos]	0	0	0	H	$+\frac{1}{24}$	$+\frac{1}{1536}$	$+\frac{7}{5760}$
	[sin]	0	0	0	H	$+\frac{1}{48}$	$-\frac{7}{512}$	$-\frac{827}{92160}$
$\left(\frac{r}{a}\right)^{-4}$	[cos + sin]	$+\frac{1}{4}$	$+\frac{5}{8}$	$+\frac{35}{32}$	0	$+\frac{1}{2}$	$+\frac{11}{12}$	$+\frac{1099}{768}$
	[cos − sin]	,,	,,	,,	0	$+\frac{1}{3}$	$+\frac{73}{96}$	$+\frac{2441}{1920}$
	[cos]	$+\frac{1}{4}$	$+\frac{5}{8}$	$+\frac{35}{32}$	0	$+\frac{5}{12}$	$+\frac{161}{192}$	$+\frac{3459}{2560}$
	[sin]	0	0	0	0	$+\frac{1}{12}$	$+\frac{5}{64}$	$+\frac{613}{7680}$
$\left(\frac{r}{a}\right)^{-5}$	[cos + sin]	$+\frac{3}{4}$	$+\frac{21}{8}$	$+\frac{189}{32}$	$+\frac{1}{2}$	$+\frac{33}{16}$	$+\frac{1865}{384}$	$+\frac{169229}{18432}$
	[cos − sin]	,,	,,	,,	0	$+\frac{49}{48}$	$+\frac{2545}{768}$	$+\frac{219221}{30720}$
	[cos]	$+\frac{3}{4}$	$+\frac{21}{8}$	$+\frac{189}{32}$	H	$+\frac{37}{24}$	$+\frac{6275}{1536}$	$+\frac{23497}{2880}$
	[sin]	0	0	0	H	$+\frac{25}{48}$	$+\frac{395}{512}$	$+\frac{94241}{92160}$

... *Table* $\left(\left(\frac{r}{a}\right)^{-1} \dots \left(\frac{r}{a}\right)^{-5}\right) \begin{smallmatrix}\cos\\ \sin\end{smallmatrix}\, 2f \dots$

	$\begin{smallmatrix}\cos\\ \sin\end{smallmatrix}\, 2f$	$[\ ]^2$ e^0	e^2	e^4	e^6	$[\ ]^3$ e	e^3	e^5	e^7
$\left(\frac{r}{a}\right)^{-1}$	[cos + sin]	$+1$	-4	$+\frac{47}{16}$	$-\frac{55}{144}$	$+\frac{5}{2}$	$-\frac{131}{16}$	$+\frac{887}{128}$	$-\frac{3999}{2048}$
	[cos − sin]	0	0	0	$+\frac{1}{48}$	0	0	$-\frac{9}{1280}$	$+\frac{153}{10240}$
	[cos]	H	H	H	$-\frac{13}{72}$	H	H	$+\frac{8861}{2560}$	$-\frac{9921}{10240}$
	[sin]	H	H	H	$-\frac{29}{144}$	H	H	$+\frac{8879}{2560}$	$-\frac{5037}{5120}$
$\left(\frac{r}{a}\right)^{-2}$	[cos + sin]	$+1$	$-\frac{7}{2}$	$+\frac{29}{16}$	$-\frac{53}{288}$	$+3$	$-\frac{69}{8}$	$+\frac{369}{64}$	$-\frac{1401}{1024}$
	[cos − sin]	0	0	$-\frac{1}{24}$	$-\frac{3}{80}$	0	0	$-\frac{27}{640}$	$-\frac{189}{5120}$
	[cos]	H	H	$+\frac{85}{96}$	$-\frac{319}{2880}$	H	H	$+\frac{3663}{1280}$	$-\frac{3597}{5120}$
	[sin]	H	H	$+\frac{89}{96}$	$-\frac{211}{2880}$	H	H	$+\frac{3717}{1280}$	$-\frac{213}{320}$
$\left(\frac{r}{a}\right)^{-3}$	[cos + sin]	$+1$	$-\frac{5}{2}$	$+\frac{13}{16}$	$-\frac{35}{288}$	$+\frac{7}{2}$	$-\frac{123}{16}$	$+\frac{489}{128}$	$-\frac{1763}{2048}$
	[cos − sin]	0	0	$+\frac{1}{24}$	$+\frac{7}{240}$	0	0	$+\frac{81}{1280}$	$+\frac{81}{2048}$
	[cos]	H	H	$+\frac{41}{96}$	$-\frac{133}{2880}$	H	H	$+\frac{4971}{2560}$	$-\frac{841}{2048}$
	[sin]	H	H	$+\frac{37}{96}$	$-\frac{217}{2880}$	H	H	$+\frac{4809}{2560}$	$-\frac{461}{1024}$
$\left(\frac{r}{a}\right)^{-4}$	[cos + sin]	$+1$	-1	$+\frac{17}{16}$	$+\frac{85}{72}$	$+4$	-5	$+\frac{51}{16}$	$+\frac{143}{128}$
	[cos − sin]	0	0	$+\frac{7}{16}$	$+\frac{109}{120}$	0	0	$+\frac{23}{40}$	$+\frac{1369}{1280}$
	[cos]	H	H	$+\frac{3}{4}$	$+\frac{47}{45}$	H	H	$+\frac{301}{160}$	$+\frac{2799}{2560}$
	[sin]	H	H	$+\frac{5}{16}$	$+\frac{49}{360}$	H	H	$+\frac{209}{160}$	$+\frac{61}{2560}$
$\left(\frac{r}{a}\right)^{-5}$	[cos + sin]	$+1$	$+1$	$+\frac{65}{16}$	$+\frac{139}{18}$	$+\frac{9}{2}$	$-\frac{3}{16}$	$+\frac{963}{128}$	$+\frac{22737}{2048}$
	[cos − sin]	0	0	$+\frac{67}{48}$	$+\frac{333}{80}$	0	0	$+\frac{2451}{1280}$	$+\frac{53193}{10240}$
	[cos]	H	H	$+\frac{131}{48}$	$+\frac{8557}{1440}$	H	H	$+\frac{12081}{2560}$	$+\frac{83439}{10240}$
	[sin]	H	H	$+\frac{4}{3}$	$+\frac{2563}{1440}$	H	H	$+\frac{7179}{2560}$	$+\frac{15123}{5120}$

$$\ldots \textit{Table} \left(\left(\frac{r}{a}\right)^{-1} \ldots \left(\frac{r}{a}\right)^{-5}\right) \begin{matrix}\cos\\ \sin\end{matrix}\, 2f \ldots$$

	$\begin{matrix}\cos\\ \sin\end{matrix}\, 2f$	$[\ ]^4$ e^2	$[\ ]^4$ e^4	$[\ ]^4$ e^6	$[\ ]^5$ e^3	$[\ ]^5$ e^5	$[\ ]^5$ e^7
$\left(\frac{r}{a}\right)^{-1}$	[cos + sin]	$+\frac{19}{4}$	$-\frac{121}{8}$	$+\frac{707}{48}$	$+\frac{389}{48}$	$-\frac{20267}{768}$	$+\frac{179141}{6144}$
	[cos − sin]	0	0	$-\frac{1}{90}$	0	0	$-\frac{625}{43008}$
	[cos]	H	H	$+\frac{10597}{1440}$	H	H	$+\frac{626681}{43008}$
	[sin]	H	H	$+\frac{10613}{1440}$	H	H	$+\frac{104551}{7168}$
$\left(\frac{r}{a}\right)^{-2}$	[cos + sin]	$+\frac{13}{2}$	$-\frac{55}{3}$	$+\frac{239}{16}$	$+\frac{295}{24}$	$-\frac{13745}{384}$	$+\frac{105175}{3072}$
	[cos − sin]	0	0	$-\frac{2}{45}$	0	0	$-\frac{3125}{64512}$
	[cos]	H	H	$+\frac{10723}{1440}$	H	H	$+\frac{1102775}{64512}$
	[sin]	H	H	$+\frac{10787}{1440}$	H	H	$+\frac{276475}{16128}$
$\left(\frac{r}{a}\right)^{-3}$	[cos + sin]	$+\frac{17}{2}$	$-\frac{115}{6}$	$+\frac{601}{48}$	$+\frac{845}{48}$	$-\frac{32525}{768}$	$+\frac{208225}{6144}$
	[cos − sin]	0	0	$+\frac{4}{45}$	0	0	$+\frac{15625}{129024}$
	[cos]	H	H	$+\frac{9079}{1440}$	H	H	$+\frac{2194175}{129024}$
	[sin]	H	H	$+\frac{8951}{1440}$	H	H	$+\frac{1089275}{64512}$
$\left(\frac{r}{a}\right)^{-4}$	[cos + sin]	$+\frac{43}{4}$	$-\frac{129}{8}$	$+\frac{1985}{192}$	$+\frac{145}{6}$	$-\frac{1015}{24}$	$+\frac{23581}{768}$
	[cos − sin]	0	0	$+\frac{437}{576}$	0	0	$+\frac{2701}{2688}$
	[cos]	H	H	$+\frac{799}{144}$	H	H	$+\frac{56823}{3584}$
	[sin]	H	H	$+\frac{2759}{576}$	H	H	$+\frac{159665}{10752}$
$\left(\frac{r}{a}\right)^{-5}$	[cos + sin]	$+\frac{53}{4}$	$-\frac{179}{24}$	$+\frac{977}{64}$	$+\frac{1541}{48}$	$-\frac{23479}{768}$	$+\frac{216917}{6144}$
	[cos − sin]	0	0	$+\frac{7579}{2880}$	0	0	$+\frac{467069}{129024}$
	[cos]	H	H	$+\frac{6443}{720}$	H	H	$+\frac{2511163}{129024}$
	[sin]	H	H	$+\frac{18193}{2880}$	H	H	$+\frac{1022047}{64512}$

... *Table* $\left(\left(\frac{r}{a}\right)^{-1} \dots \left(\frac{r}{a}\right)^{-5}\right) \begin{smallmatrix}\cos\\ \sin\end{smallmatrix}\, 2f$...

$\begin{smallmatrix}\cos\\ \sin\end{smallmatrix}\, 2f$		$[\ ]^6$ e^4	$[\ ]^6$ e^6	$[\ ]^7$ e^5	$[\ ]^7$ e^7	$[\ ]^8$ e^6	$[\ ]^9$ e^7
$\left(\frac{r}{a}\right)^{-1}$	[cos + sin]	$+\frac{209}{16}$	$-\frac{887}{20}$	$+\frac{78077}{3840}$	$-\frac{2228929}{30720}$	$+\frac{17807}{576}$	$+\frac{3313213}{71680}$
	[cos − sin]	0	0	0	0	0	0
	[cos]	H	H	H	H	H	H
	[sin]	H	H	H	H	H	H
$\left(\frac{r}{a}\right)^{-2}$	[cos + sin]	$+\frac{345}{16}$	$-\frac{10569}{160}$	$+\frac{69251}{1920}$	$-\frac{5394109}{46080}$	$+\frac{42037}{720}$	$+\frac{3306951}{35840}$
	[cos − sin]	0	0	0	0	0	0
	[cos]	H	H	H	H	H	H
	[sin]	H	H	H	H	H	H
$\left(\frac{r}{a}\right)^{-3}$	[cos + sin]	$+\frac{533}{16}$	$-\frac{13827}{160}$	$+\frac{228347}{3840}$	$-\frac{3071075}{18432}$	$+\frac{73369}{720}$	$+\frac{12144273}{71680}$
	[cos − sin]	0	0	0	0	0	0
	[cos]	H	H	H	H	H	H
	[sin]	H	H	H	H	H	H
$\left(\frac{r}{a}\right)^{-4}$	[cos + sin]	$+49$	-98	$+\frac{44569}{480}$	$-\frac{133707}{640}$	$+\frac{241517}{1440}$	$+\frac{2619177}{8960}$
	[cos − sin]	0	0	0	0	0	0
	[cos]	H	H	H	H	H	H
	[sin]	H	H	H	H	H	H
$\left(\frac{r}{a}\right)^{-5}$	[cos + sin]	$+\frac{555}{8}$	$-\frac{7149}{80}$	$+\frac{533617}{3840}$	$-\frac{20449667}{92160}$	$+\frac{379691}{1440}$	$+\frac{34431741}{71680}$
	[cos − sin]	0	0	0	0	0	0
	[cos]	H	H	H	H	H	H
	[sin]	H	H	H	H	H	H

Table $\left(\left(\frac{r}{a}\right)\ldots\left(\frac{r}{a}\right)^4\right)\begin{matrix}\cos\\ \sin\end{matrix}\, 3f\ldots$

$\begin{matrix}\cos\\ \sin\end{matrix} 3f$		$[\;]^0$ e^3	$[\;]^0$ e^5	$[\;]^0$ e^7	$[\;]^1$ e^2	$[\;]^1$ e^4	$[\;]^1$ e^6
$\left(\frac{r}{a}\right)^1$	[cos + sin]	$-\frac{5}{4}$	$-\frac{5}{32}$	$-\frac{3}{64}$	$+\frac{31}{8}$	$-\frac{77}{24}$	$+\frac{155}{1024}$
	[cos − sin]	„	„	„	o	$-\frac{19}{384}$	$+\frac{3}{256}$
	[cos]	$-\frac{5}{4}$	$-\frac{5}{32}$	$-\frac{3}{64}$	H	$-\frac{417}{256}$	$+\frac{167}{2048}$
	[sin]	o	o	o	H	$-\frac{1213}{768}$	$+\frac{143}{2048}$
$\left(\frac{r}{a}\right)^2$	[cos + sin]	$-\frac{5}{2}$	o	o	$+\frac{43}{8}$	$-\frac{25}{6}$	$+\frac{1069}{3072}$
	[cos − sin]	„	„	„	o	$+\frac{37}{384}$	$+\frac{487}{3840}$
	[cos]	$-\frac{5}{2}$	o	o	H	$-\frac{521}{256}$	$+\frac{2431}{10240}$
	[sin]	o	o	o	H	$-\frac{1637}{768}$	$+\frac{3397}{30720}$
$\left(\frac{r}{a}\right)^3$	[cos + sin]	$-\frac{35}{8}$	o	o	$+\frac{57}{8}$	$-\frac{65}{16}$	$-\frac{19}{1024}$
	[cos − sin]	„	„	„	o	$+\frac{75}{128}$	$+\frac{179}{640}$
	[cos]	$-\frac{35}{8}$	o	o	H	$-\frac{445}{256}$	$+\frac{1337}{10240}$
	[sin]	o	o	o	H	$-\frac{595}{256}$	$-\frac{1527}{10240}$
$\left(\frac{r}{a}\right)^4$	[cos + sin]	-7	$-\frac{7}{8}$	o	$+\frac{73}{8}$	$-\frac{91}{48}$	$-\frac{1419}{1024}$
	[cos − sin]	„	„	„	o	$+\frac{641}{384}$	$+\frac{341}{640}$
	[cos]	-7	$-\frac{7}{8}$	o	H	$-\frac{29}{256}$	$-\frac{4367}{10240}$
	[sin]	o	o	o	H	$-\frac{1369}{768}$	$-\frac{9823}{10240}$

...*Table* $\left(\left(\frac{r}{a}\right)\ldots\left(\frac{r}{a}\right)^4\right)\begin{matrix}\cos\\ \sin\end{matrix}\,3f$...

	$\begin{matrix}\cos\\ \sin\end{matrix}\,3f$	$[\ ]^2$ e	e^3	e^5	e^7	$[\ ]^3$ e^0	e^2	e^4	e^6
$\left(\frac{r}{a}\right)^1$	[cos + sin]	$-\frac{7}{2}$	$+\frac{75}{8}$	$-\frac{659}{96}$	$+\frac{1687}{1152}$	$+1$	$-\frac{17}{2}$	$+\frac{1143}{64}$	$+\frac{1805}{128}$
	[cos − sin]	0	0	$-\frac{11}{480}$	$-\frac{13}{5760}$	0	0	0	$-\frac{69}{5120}$
	[cos]	H	H	$-\frac{551}{160}$	$+\frac{4211}{5760}$	H	H	H	$-\frac{72269}{10240}$
	[sin]	H	H	$-\frac{821}{240}$	$+\frac{11}{15}$	H	H	H	$-\frac{72131}{10240}$
$\left(\frac{r}{a}\right)^2$	[cos + sin]	-4	$+\frac{19}{2}$	$-\frac{79}{12}$	$+\frac{113}{72}$	$+1$	$-\frac{15}{2}$	$+\frac{927}{64}$	$-\frac{1415}{128}$
	[cos − sin]	0	0	$+\frac{1}{120}$	$+\frac{1}{36}$	0	0	0	$-\frac{1}{5120}$
	[cos]	H	H	$-\frac{263}{80}$	$+\frac{115}{144}$	H	H	H	$-\frac{56601}{10240}$
	[sin]	H	H	$-\frac{791}{240}$	$+\frac{37}{48}$	H	H	H	$-\frac{56599}{10240}$
$\left(\frac{r}{a}\right)^3$	[cos + sin]	$-\frac{9}{2}$	$+\frac{33}{4}$	$-\frac{133}{32}$	$+\frac{11}{16}$	$+1$	-6	$+\frac{591}{64}$	$-\frac{349}{64}$
	[cos − sin]	0	0	$+\frac{13}{160}$	$+\frac{1}{30}$	0	0	0	$+\frac{137}{5120}$
	[cos]	H	H	$-\frac{163}{80}$	$+\frac{173}{480}$	H	H	H	$-\frac{27783}{10240}$
	[sin]	H	H	$-\frac{339}{160}$	$+\frac{157}{480}$	H	H	H	$-\frac{28057}{10240}$
$\left(\frac{r}{a}\right)^4$	[cos + sin]	-5	$+\frac{21}{4}$	$+\frac{17}{48}$	$-\frac{71}{72}$	$+1$	-4	$+\frac{183}{64}$	$+\frac{65}{64}$
	[cos − sin]	0	0	$+\frac{2}{15}$	$+\frac{1}{720}$	0	0	0	$+\frac{37}{1024}$
	[cos]	H	H	$+\frac{39}{160}$	$-\frac{709}{1440}$	H	H	H	$+\frac{1077}{2048}$
	[sin]	H	H	$+\frac{53}{480}$	$-\frac{79}{160}$	H	H	H	$+\frac{1003}{2048}$

$$\dots \textit{Table} \left(\left(\frac{r}{a}\right) \dots \left(\frac{r}{a}\right)^4\right) \begin{matrix}\cos\\ \sin\end{matrix} 3f \dots$$

	$\begin{matrix}\cos\\ \sin\end{matrix} 3f$	$[\ ]^4$ e	e^3	e^5	e^7	$[\ ]^5$ e^2	e^4	e^6
$\left(\frac{r}{a}\right)^1$	[cos + sin]	$+\frac{5}{2}$	$-\frac{63}{4}$	$+\frac{757}{24}$	$-\frac{1979}{72}$	$+\frac{37}{8}$	$-\frac{635}{24}$	$+\frac{54765}{1024}$
	[cos − sin]	0	0	0	$-\frac{47}{5040}$	0	0	0
	[cos]	H	H	H	$-\frac{138577}{10080}$	H	H	H
	[sin]	H	H	H	$-\frac{15387}{1120}$	H	H	H
$\left(\frac{r}{a}\right)^2$	[cos + sin]	$+2$	$-\frac{23}{2}$	$+\frac{65}{3}$	$-\frac{5317}{288}$	$+\frac{25}{8}$	$-\frac{50}{3}$	$+\frac{98875}{3072}$
	[cos − sin]	0	0	0	$-\frac{17}{10080}$	0	0	0
	[cos]	H	H	H	$-\frac{2908}{315}$	H	H	H
	[sin]	H	H	H	$-\frac{31013}{3360}$	H	H	H
$\left(\frac{r}{a}\right)^3$	[cos + sin]	$+\frac{3}{2}$	$-\frac{57}{8}$	$+11$	$-\frac{2905}{384}$	$+\frac{15}{8}$	$-\frac{135}{16}$	$+\frac{13975}{1024}$
	[cos − sin]	0	0	0	$+\frac{57}{4480}$	0	0	0
	[cos]	H	H	H	$-\frac{793}{210}$	H	H	H
	[sin]	H	H	H	$-\frac{50923}{13440}$	H	H	H
$\left(\frac{r}{a}\right)^4$	[cos + sin]	$+1$	-3	$+\frac{37}{24}$	$+\frac{251}{144}$	$+\frac{7}{8}$	$-\frac{109}{48}$	$+\frac{535}{1024}$
	[cos − sin]	0	0	0	$+\frac{79}{5040}$	0	0	0
	[cos]	H	H	H	$+\frac{277}{315}$	H	H	H
	[sin]	H	H	H	$+\frac{1451}{1680}$	H	H	H

... *Table* $\left(\left(\frac{r}{a}\right) \ldots \left(\frac{r}{a}\right)^4\right) \begin{matrix}\cos\\ \sin\end{matrix}\, 3f \ldots$

	$\begin{matrix}\cos\\ \sin\end{matrix}\, 3f$	$[\ ]^6$ e^3	$[\ ]^6$ e^5	$[\ ]^6$ e^7	$[\ ]^7$ e^4	$[\ ]^7$ e^6
$\left(\frac{r}{a}\right)^1$	[cos + sin]	$+\frac{61}{8}$	$-\frac{677}{16}$	$+\frac{56487}{640}$	$+\frac{4553}{384}$	$-\frac{84109}{1280}$
	[cos − sin]	o	o	o	o	o
	[cos]	H	H	H	H	H
	[sin]	H	H	H	H	H
$\left(\frac{r}{a}\right)^2$	[cos + sin]	$+\frac{9}{2}$	$-\frac{189}{8}$	$+\frac{3807}{80}$	$+\frac{2401}{384}$	$-\frac{127253}{3840}$
	[cos − sin]	o	o	o	o	o
	[cos]	H	H	H	H	H
	[sin]	H	H	H	H	H
$\left(\frac{r}{a}\right)^3$	[cos + sin]	$+\frac{9}{4}$	$-\frac{81}{8}$	$+\frac{2781}{160}$	$+\frac{343}{128}$	$-\frac{7889}{640}$
	[cos − sin]	o	o	o	o	o
	[cos]	H	H	H	H	H
	[sin]	H	H	H	H	H
$\left(\frac{r}{a}\right)^4$	[cos + sin]	$+\frac{3}{4}$	$-\frac{27}{16}$	$-\frac{81}{160}$	$+\frac{245}{384}$	$-\frac{147}{128}$
	[cos − sin]	o	o	o	o	o
	[cos]	H	H	H	H	H
	[sin]	H	H	H	H	H

... *Table* $\left(\left(\frac{r}{a}\right) \dots \left(\frac{r}{a}\right)^4\right) \begin{matrix}\cos\\ \sin\end{matrix} 3f.$

	$\begin{matrix}\cos\\ \sin\end{matrix} 3f$	$[\ ]^8$ e^5	$[\ ]^8$ e^7	$[\ ]^9$ e^6	$[\ ]^{10}$ e^7
$\left(\frac{r}{a}\right)^1$	[cos + sin]	$+\frac{8551}{480}$	$-\frac{288221}{2880}$	$+\frac{26809}{1024}$	$+\frac{305593}{8064}$
	[cos − sin]	0	0	0	0
	[cos]	H	H	H	H
	[sin]	H	H	H	H
$\left(\frac{r}{a}\right)^2$	[cos + sin]	$+\frac{128}{15}$	$-\frac{416}{9}$	$+\frac{59049}{5120}$	$+\frac{15625}{1008}$
	[cos − sin]	0	0	0	0
	[cos]	H	H	H	H
	[sin]	H	H	H	H
$\left(\frac{r}{a}\right)^3$	[cos + sin]	$+\frac{16}{5}$	$-\frac{76}{5}$	$+\frac{19683}{5120}$	$+\frac{3125}{672}$
	[cos − sin]	0	0	0	0
	[cos]	H	H	H	H
	[sin]	H	H	H	H
$\left(\frac{r}{a}\right)^4$	[cos + sin]	$+\frac{8}{15}$	$-\frac{26}{45}$	$+\frac{2187}{5120}$	$+\frac{625}{2016}$
	[cos − sin]	0	0	0	0
	[cos]	H	H	H	H
	[sin]	H	H	H	H

... *Table* $\left(\left(\frac{r}{a}\right)^{-1} \ldots \left(\frac{r}{a}\right)^{-5}\right) \begin{matrix}\cos\\ \sin\end{matrix} 3f$...

	$\begin{matrix}\cos\\ \sin\end{matrix} 3f$	[ ]⁰ e^3	[ ]⁰ e^5	[ ]⁰ e^7	[ ]¹ e^2	[ ]¹ e^4	[ ]¹ e^6
$\left(\frac{r}{a}\right)^{-1}$	[cos + sin]	$-\frac{1}{8}$	$-\frac{3}{32}$	$-\frac{9}{128}$	$+\frac{13}{8}$	$-\frac{41}{48}$	$-\frac{437}{3072}$
	[cos − sin]	,,	,,	,,	0	$+\frac{1}{384}$	$-\frac{11}{960}$
	[cos]	$-\frac{1}{8}$	$-\frac{3}{32}$	$-\frac{9}{128}$	H	$-\frac{109}{256}$	$-\frac{787}{10240}$
	[sin]	0	0	0	H	$-\frac{329}{768}$	$-\frac{2009}{30720}$
$\left(\frac{r}{a}\right)^{-2}$	[cos + sin]	0	0	0	$+\frac{7}{8}$	$-\frac{5}{24}$	$-\frac{29}{1024}$
	[cos − sin]	,,	,,	,,	0	$+\frac{5}{384}$	$+\frac{23}{1280}$
	[cos]	0	0	0	H	$-\frac{25}{256}$	$-\frac{53}{10240}$
	[sin]	0	0	0	H	$-\frac{85}{768}$	$-\frac{237}{10240}$
$\left(\frac{r}{a}\right)^{-3}$	[cos + sin]	0	0	0	$+\frac{3}{8}$	0	$+\frac{23}{1024}$
	[cos − sin]	,,	,,	,,	0	$-\frac{1}{128}$	$-\frac{11}{1280}$
	[cos]	0	0	0	H	$-\frac{1}{256}$	$+\frac{71}{10240}$
	[sin]	0	0	0	H	$+\frac{1}{256}$	$+\frac{159}{10240}$
$\left(\frac{r}{a}\right)^{-4}$	[cos + sin]	0	0	0	$+\frac{1}{8}$	$+\frac{1}{48}$	$+\frac{55}{3072}$
	[cos − sin]	,,	,,	,,	0	$+\frac{1}{384}$	$+\frac{1}{384}$
	[cos]	0	0	0	H	$+\frac{3}{256}$	$+\frac{21}{2048}$
	[sin]	0	0	0	H	$+\frac{7}{768}$	$+\frac{47}{6144}$
$\left(\frac{r}{a}\right)^{-5}$	[cos + sin]	$+\frac{1}{8}$	$+\frac{7}{16}$	$+\frac{63}{64}$	$+\frac{1}{8}$	$+\frac{17}{48}$	$+\frac{837}{1024}$
	[cos − sin]	,,	,,	,,	0	$+\frac{65}{384}$	$+\frac{353}{640}$
	[cos]	$+\frac{1}{8}$	$+\frac{7}{16}$	$+\frac{63}{64}$	H	$+\frac{67}{256}$	$+\frac{7009}{10240}$
	[sin]	0	0	0	H	$+\frac{71}{768}$	$+\frac{1361}{10240}$

... *Table* $\left(\left(\frac{r}{a}\right)^{-1} \ldots \left(\frac{r}{a}\right)^{-5}\right) \begin{matrix}\cos\\ \sin\end{matrix}\, 3f$ ...

$\begin{matrix}\cos\\ \sin\end{matrix}\, 3f$		[ ]^2				[ ]^3			
		e	e^3	e^5	e^7	e^0	e^2	e^4	e^6
$\left(\frac{r}{a}\right)^{-1}$	[cos + sin]	$-\frac{5}{2}$	$+\frac{13}{2}$	$-\frac{341}{96}$	$+\frac{187}{576}$	$+1$	-9	$+\frac{1143}{64}$	$-\frac{365}{32}$
	[cos − sin]	0	0	$+\frac{1}{120}$	$-\frac{1}{2880}$	0	0	0	$+\frac{9}{1024}$
	[cos]	H	H	$-\frac{567}{320}$	$+\frac{467}{2880}$	H	H	H	$-\frac{11671}{2048}$
	[sin]	H	H	$-\frac{1709}{960}$	$+\frac{13}{80}$	H	H	H	$-\frac{11689}{2048}$
$\left(\frac{r}{a}\right)^{-2}$	[cos + sin]	-2	$+\frac{9}{2}$	$-\frac{41}{24}$	$+\frac{43}{288}$	$+1$	$-\frac{17}{2}$	$+\frac{951}{64}$	$-\frac{973}{128}$
	[cos − sin]	0	0	$+\frac{1}{120}$	$+\frac{23}{1440}$	0	0	0	$+\frac{27}{5120}$
	[cos]	H	H	$-\frac{17}{20}$	$+\frac{119}{1440}$	H	H	H	$-\frac{38893}{10240}$
	[sin]	H	H	$-\frac{103}{120}$	$+\frac{1}{15}$	H	H	H	$-\frac{38947}{10240}$
$\left(\frac{r}{a}\right)^{-3}$	[cos + sin]	$-\frac{3}{2}$	$+\frac{21}{8}$	$-\frac{19}{32}$	$+\frac{23}{192}$	$+1$	$-\frac{15}{2}$	$+\frac{687}{64}$	$-\frac{535}{128}$
	[cos − sin]	0	0	$-\frac{1}{80}$	$-\frac{1}{64}$	0	0	0	$-\frac{81}{5120}$
	[cos]	H	H	$-\frac{97}{320}$	$+\frac{5}{96}$	H	H	H	$-\frac{21481}{10240}$
	[sin]	H	H	$-\frac{93}{320}$	$+\frac{13}{192}$	H	H	H	$-\frac{21319}{10240}$
$\left(\frac{r}{a}\right)^{-4}$	[cos + sin]	-1	$+\frac{5}{4}$	$-\frac{7}{48}$	$+\frac{23}{288}$	$+1$	-6	$+\frac{423}{64}$	$-\frac{125}{64}$
	[cos − sin]	0	0	$+\frac{1}{120}$	$+\frac{13}{1440}$	0	0	0	$+\frac{81}{5120}$
	[cos]	H	H	$-\frac{11}{160}$	$+\frac{2}{45}$	H	H	H	$-\frac{9919}{10240}$
	[sin]	H	H	$-\frac{37}{480}$	$+\frac{17}{480}$	H	H	H	$-\frac{10081}{10240}$
$\left(\frac{r}{a}\right)^{-5}$	[cos + sin]	$-\frac{1}{2}$	$+\frac{3}{4}$	$+\frac{61}{96}$	$+\frac{191}{144}$	$+1$	-4	$+\frac{255}{64}$	$+\frac{17}{64}$
	[cos − sin]	0	0	$+\frac{109}{480}$	$+\frac{31}{45}$	0	0	0	$+\frac{1553}{5120}$
	[cos]	H	H	$+\frac{69}{160}$	$+\frac{1451}{1440}$	H	H	H	$+\frac{2913}{10240}$
	[sin]	H	H	$+\frac{49}{240}$	$+\frac{51}{160}$	H	H	H	$-\frac{193}{10240}$

$$\ldots \textit{Table} \left(\left(\frac{r}{a}\right)^{-1} \ldots \left(\frac{r}{a}\right)^{-5}\right) \begin{matrix}\cos \\ \sin\end{matrix}\, 3f \ldots$$

	$\begin{matrix}\cos \\ \sin\end{matrix}\, 3f$	$[\ ]^4$				$[\ ]^5$		
		e	e^3	e^5	e^7	e^2	e^4	e^6
$\left(\frac{r}{a}\right)^{-1}$	[cos + sin]	$+\frac{7}{2}$	$-\frac{179}{8}$	$+\frac{2009}{48}$	$-\frac{8893}{288}$	$+\frac{67}{8}$	$-\frac{2279}{48}$	$+\frac{274345}{3072}$
	[cos − sin]	0	0	0	$+\frac{11}{1260}$	0	0	0
	[cos]	H	H	H	$-\frac{311167}{20160}$	H	H	H
	[sin]	H	H	H	$-\frac{103781}{6720}$	H	H	H
$\left(\frac{r}{a}\right)^{-2}$	[cos + sin]	$+4$	-24	$+\frac{241}{6}$	$-\frac{455}{18}$	$+\frac{85}{8}$	$-\frac{1355}{24}$	$+\frac{98525}{1024}$
	[cos − sin]	0	0	0	$+\frac{1}{315}$	0	0	0
	[cos]	H	H	H	$-\frac{15923}{1260}$	H	H	H
	[sin]	H	H	H	$-\frac{5309}{420}$	H	H	H
$\left(\frac{r}{a}\right)^{-3}$	[cos + sin]	$+\frac{9}{2}$	-24	$+\frac{545}{16}$	$-\frac{71}{4}$	$+\frac{105}{8}$	$-\frac{125}{2}$	$+\frac{94825}{1024}$
	[cos − sin]	0	0	0	$-\frac{2}{105}$	0	0	0
	[cos]	H	H	H	$-\frac{7463}{840}$	H	H	H
	[sin]	H	H	H	$-\frac{7447}{840}$	H	H	H
$\left(\frac{r}{a}\right)^{-4}$	[cos + sin]	$+5$	-22	$+\frac{607}{24}$	$-\frac{98}{9}$	$+\frac{127}{8}$	$-\frac{3065}{48}$	$+\frac{243805}{3072}$
	[cos − sin]	0	0	0	$+\frac{8}{315}$	0	0	0
	[cos]	H	H	H	$-\frac{1711}{315}$	H	H	H
	[sin]	H	H	H	$-\frac{191}{35}$	H	H	H
$\left(\frac{r}{a}\right)^{-5}$	[cos + sin]	$+\frac{11}{2}$	$-\frac{141}{8}$	$+\frac{415}{24}$	$-\frac{5021}{1152}$	$+\frac{151}{8}$	$-\frac{2809}{48}$	$+\frac{63343}{1024}$
	[cos − sin]	0	0	0	$+\frac{16319}{40320}$	0	0	0
	[cos]	H	H	H	$-\frac{19927}{10080}$	H	H	H
	[sin]	H	H	H	$-\frac{32009}{13440}$	H	H	H

... *Table* $\left(\left(\frac{r}{a}\right)^{-1} \dots \left(\frac{r}{a}\right)^{-5}\right) \begin{matrix}\cos\\ \sin\end{matrix} 3f \dots$

	$\begin{matrix}\cos\\ \sin\end{matrix} 3f$	$[\ ]^6$ e^3	$[\ ]^6$ e^5	$[\ ]^6$ e^7	$[\ ]^7$ e^4	$[\ ]^7$ e^6
$\left(\frac{r}{a}\right)^{-1}$	[cos + sin]	$+17$	$-\frac{2949}{32}$	$+\frac{57213}{320}$	$+\frac{12085}{384}$	$-\frac{32419}{192}$
	[cos − sin]	0	0	0	0	0
	[cos]	H	H	H	H	H
	[sin]	H	H	H	H	H
$\left(\frac{r}{a}\right)^{-2}$	[cos + sin]	$+\frac{47}{2}$	$-\frac{239}{2}$	$-\frac{33951}{160}$	$+\frac{17969}{384}$	$-\frac{301973}{1280}$
	[cos − sin]	0	0	0	0	0
	[cos]	H	H	H	H	H
	[sin]	H	H	H	H	H
$\left(\frac{r}{a}\right)^{-3}$	[cos + sin]	$+\frac{251}{8}$	$-\frac{4611}{32}$	$+\frac{36249}{160}$	$+\frac{8547}{128}$	$-\frac{391951}{1280}$
	[cos − sin]	0	0	0	0	0
	[cos]	H	H	H	H	H
	[sin]	H	H	H	H	H
$\left(\frac{r}{a}\right)^{-4}$	[cos + sin]	$+\frac{163}{4}$	$-\frac{2577}{16}$	$+\frac{1089}{5}$	$+\frac{35413}{384}$	$-\frac{709471}{1920}$
	[cos − sin]	0	0	0	0	0
	[cos]	H	H	H	H	H
	[sin]	H	H	H	H	H
$\left(\frac{r}{a}\right)^{-5}$	[cos + sin]	$+\frac{207}{4}$	$-\frac{2625}{16}$	$+\frac{30483}{160}$	$+\frac{47621}{384}$	$-\frac{262731}{640}$
	[cos − sin]	0	0	0	0	0
	[cos]	H	H	H	H	H
	[sin]	H	H	H	H	H

... *Table* $\left(\left(\frac{r}{a}\right)^{-1} \dots \left(\frac{r}{a}\right)^{-5}\right) \begin{smallmatrix}\cos\\ \sin\end{smallmatrix} 3f$ ...

	$\begin{smallmatrix}\cos\\ \sin\end{smallmatrix} 3f$	$[\ ]^8$		$[\ ]^9$	$[\ ]^{10}$
		e^5	e^7	e^6	e^7
$\left(\frac{r}{a}\right)^{-1}$	[cos + sin]	$+\frac{26371}{480}$	$-\frac{1710983}{5760}$	$+\frac{471527}{5120}$	$+\frac{604279}{4032}$
	[cos − sin]	0	0	0	0
	[cos]	H	H	H	H
	[sin]	H	H	H	H
$\left(\frac{r}{a}\right)^{-2}$	[cos + sin]	$+\frac{2611}{30}$	$-\frac{39893}{90}$	$+\frac{790053}{5120}$	$+\frac{532345}{2016}$
	[cos − sin]	0	0	0	0
	[cos]	H	H	H	H
	[sin]	H	H	H	H
$\left(\frac{r}{a}\right)^{-3}$	[cos + sin]	$+\frac{10531}{80}$	$-\frac{7363}{12}$	$+\frac{1258449}{5120}$	$+\frac{98715}{224}$
	[cos − sin]	0	0	0	0
	[cos]	H	H	H	H
	[sin]	H	H	H	H
$\left(\frac{r}{a}\right)^{-4}$	[cos + sin]	$+\frac{23029}{120}$	$-\frac{35614}{45}$	$+\frac{385095}{1024}$	$+\frac{44377}{63}$
	[cos − sin]	0	0	0	0
	[cos]	H	H	H	H
	[sin]	H	H	H	H
$\left(\frac{r}{a}\right)^{-5}$	[cos + sin]	$+\frac{65153}{240}$	$-\frac{2725199}{2880}$	$+\frac{2850411}{5120}$	$+\frac{2190757}{2016}$
	[cos − sin]	0	0	0	0
	[cos]	H	H	H	H
	[sin]	H	H	H	H

$$\ldots \mathit{Table}\ \left(\left(\frac{r}{a}\right)^1 \ldots \left(\frac{r}{a}\right)^4\right) \begin{matrix}\cos\\ \sin\end{matrix}\ 4f \ldots$$

$\begin{matrix}\cos\\ \sin\end{matrix}\ 4f$		$[\ ]^0$ e^4	$[\ ]^0$ e^6	$[\ ]^1$ e^3	$[\ ]^1$ e^5	$[\ ]^1$ e^7	$[\ ]^2$ e^2	$[\ ]^2$ e^4	$[\ ]^2$ e^6
$\left(\frac{r}{a}\right)^1$	[cos + sin]	$+\frac{15}{16}$	$+\frac{9}{32}$	$-\frac{229}{48}$	$+\frac{2735}{768}$	$+\frac{2719}{30720}$	$+\frac{29}{4}$	$-\frac{101}{6}$	$+\frac{681}{64}$
	[cos − sin]	,,	,,	0	$+\frac{151}{3840}$	$+\frac{1423}{92160}$	0	0	$+\frac{37}{2880}$
	[cos]	$+\frac{15}{16}$	$+\frac{9}{32}$	H	$+\frac{6913}{3840}$	$+\frac{479}{9216}$	H	H	$+\frac{15341}{2880}$
	[sin]	0	0	H	$+\frac{1127}{640}$	$+\frac{3367}{92160}$	H	H	$+\frac{1913}{360}$
$\left(\frac{r}{a}\right)^2$	[cos + sin]	$+\frac{35}{16}$	$+\frac{7}{32}$	$-\frac{179}{24}$	$+\frac{2303}{384}$	$-\frac{1031}{3072}$	$+\frac{37}{4}$	$-\frac{83}{4}$	$+\frac{2639}{192}$
	[cos − sin]	,,	,,	0	$-\frac{17}{1920}$	$-\frac{1193}{15360}$	0	0	$+\frac{5}{576}$
	[cos]	$+\frac{35}{16}$	$+\frac{7}{32}$	H	$+\frac{5749}{1920}$	$-\frac{529}{2560}$	H	H	$+\frac{3961}{576}$
	[sin]	0	0	H	$+\frac{961}{320}$	$-\frac{1981}{15360}$	H	H	$+\frac{989}{144}$
$\left(\frac{r}{a}\right)^3$	[cos + sin]	$+\frac{35}{8}$	0	$-\frac{529}{48}$	$+\frac{6385}{768}$	$-\frac{20161}{30720}$	$+\frac{23}{2}$	$-\frac{139}{6}$	$+\frac{43}{3}$
	[cos − sin]	,,	,,	0	$-\frac{1159}{3840}$	$-\frac{25297}{92160}$	0	0	$-\frac{23}{720}$
	[cos]	$+\frac{35}{8}$	0	H	$+\frac{15383}{3840}$	$-\frac{4289}{9216}$	H	H	$+\frac{10297}{1440}$
	[sin]	0	0	H	$+\frac{2757}{640}$	$-\frac{17593}{92160}$	H	H	$+\frac{10343}{1440}$
$\left(\frac{r}{a}\right)^4$	[cos + sin]	$+\frac{63}{8}$	0	$-\frac{187}{12}$	$+\frac{1769}{192}$	$-\frac{311}{7680}$	$+14$	$-\frac{137}{6}$	$+\frac{171}{16}$
	[cos − sin]	,,	,,	0	$-\frac{1091}{960}$	$-\frac{13571}{23040}$	0	0	$-\frac{71}{720}$
	[cos]	$+\frac{63}{8}$	0	H	$+\frac{3877}{960}$	$-\frac{1813}{5760}$	H	H	$+\frac{953}{180}$
	[sin]	0	0	H	$+\frac{207}{40}$	$+\frac{6319}{23040}$	H	H	$+\frac{3883}{720}$

... *Table* $\left(\left(\frac{r}{a}\right)^1 \dots \left(\frac{r}{a}\right)^4\right) \begin{matrix}\cos\\ \sin\end{matrix}\, 4f$ ...

	$\begin{matrix}\cos\\ \sin\end{matrix}\, 4f$	[ ]³				[ ]⁴			
		e	e^3	e^5	e^7	e^0	e^2	e^4	e^6
$\left(\frac{r}{a}\right)^1$	[cos + sin]	$-\frac{9}{2}$	$+\frac{399}{16}$	$-\frac{5415}{128}$	$+\frac{58251}{2048}$	$+1$	$-\frac{31}{2}$	$+\frac{239}{4}$	$-\frac{6719}{72}$
	[cos − sin]	0	0	0	$+\frac{411}{71680}$	0	0	0	0
	[cos]	H	H	H	$+\frac{509799}{35840}$	H	H	H	H
	[sin]	H	H	H	$+\frac{1019187}{71680}$	H	H	H	H
$\left(\frac{r}{a}\right)^2$	[cos + sin]	-5	$+\frac{209}{8}$	$-\frac{2783}{64}$	$+\frac{31065}{1024}$	$+1$	$-\frac{29}{2}$	$+\frac{215}{4}$	$-\frac{5993}{72}$
	[cos − sin]	0	0	0	$+\frac{209}{35840}$	0	0	0	0
	[cos]	H	H	H	$+\frac{271871}{17920}$	H	H	H	H
	[sin]	H	H	H	$+\frac{543533}{35840}$	H	H	H	H
$\left(\frac{r}{a}\right)^3$	[cos + sin]	$-\frac{11}{2}$	$+\frac{411}{16}$	$-\frac{5025}{128}$	$+\frac{53459}{2048}$	$+1$	-13	$+\frac{353}{8}$	$-\frac{1151}{18}$
	[cos − sin]	0	0	0	$-\frac{541}{71680}$	0	0	0	0
	[cos]	H	H	H	$+\frac{467631}{35840}$	H	H	H	H
	[sin]	H	H	H	$+\frac{935803}{71680}$	H	H	H	H
$\left(\frac{r}{a}\right)^4$	[cos + sin]	-6	$+\frac{93}{4}$	$-\frac{933}{32}$	$+\frac{7929}{512}$	$+1$	-11	$+\frac{253}{8}$	$-\frac{346}{9}$
	[cos − sin]	0	0	0	$-\frac{459}{17920}$	0	0	0	0
	[cos]	H	H	H	$+\frac{4329}{560}$	H	H	H	H
	[sin]	H	H	H	$+\frac{138987}{17920}$	H	H	H	H

... *Table* $\left(\left(\frac{r}{a}\right)^1 \ldots \left(\frac{r}{a}\right)^4\right) \begin{matrix}\cos\\ \sin\end{matrix} 4f$...

	$\begin{matrix}\cos\\ \sin\end{matrix} 4f$	[ ]^5 e	e^3	e^5	e^7	[ ]^6 e^2	e^4	e^6
$\left(\frac{r}{a}\right)^1$	[cos + sin]	$+\frac{7}{2}$	$-\frac{585}{16}$	$+\frac{47435}{384}$	$-\frac{3517045}{18432}$	$+\frac{33}{4}$	$-\frac{147}{2}$	$+\frac{15075}{64}$
	[cos − sin]	0	0	0	0	0	0	0
	[cos]	H	H	H	H	H	H	H
	[sin]	H	H	H	H	H	H	H
$\left(\frac{r}{a}\right)^2$	[cos + sin]	$+3$	$-\frac{239}{8}$	$+\frac{18875}{192}$	$-\frac{155175}{1024}$	$+\frac{25}{4}$	$-\frac{215}{4}$	$+\frac{10831}{64}$
	[cos − sin]	0	0	0	0	0	0	0
	[cos]	H	H	H	H	H	H	H
	[sin]	H	H	H	H	H	H	H
$\left(\frac{r}{a}\right)^3$	[cos + sin]	$+\frac{5}{2}$	$-\frac{365}{16}$	$+\frac{26845}{384}$	$-\frac{1884125}{18432}$	$+\frac{9}{2}$	-36	$+\frac{855}{8}$
	[cos − sin]	0	0	0	0	0	0	0
	[cos]	H	H	H	H	H	H	H
	[sin]	H	H	H	H	H	H	H
$\left(\frac{r}{a}\right)^4$	[cos + sin]	$+2$	$-\frac{63}{4}$	$+\frac{3989}{96}$	$-\frac{239275}{4608}$	$+3$	-21	$+\frac{435}{8}$
	[cos − sin]	0	0	0	0	0	0	0
	[cos]	H	H	H	H	H	H	H
	[sin]	H	H	H	H	H	H	H

... *Table* $\left(\left(\frac{r}{a}\right)^1 \ldots \left(\frac{r}{a}\right)^4\right) \begin{matrix}\cos\\ \sin\end{matrix}\ 4f \ldots$

	$\begin{matrix}\cos\\ \sin\end{matrix}\ 4f$	[ ]^7 e^3	[ ]^7 e^5	[ ]^7 e^7	[ ]^8 e^4	[ ]^8 e^6
$\left(\frac{r}{a}\right)^1$	[cos + sin]	$+\frac{787}{48}$	$-\frac{103937}{768}$	$+\frac{13103351}{30720}$	$+\frac{1423}{48}$	$-\frac{37724}{160}$
	[cos − sin]	0	0	0	0	0
	[cos]	H	H	H	H	H
	[sin]	H	H	H	H	H
$\left(\frac{r}{a}\right)^2$	[cos + sin]	$+\frac{269}{24}$	$-\frac{34601}{384}$	$+\frac{4312189}{15360}$	$+\frac{297}{16}$	$-\frac{13897}{96}$
	[cos − sin]	0	0	0	0	0
	[cos]	H	H	H	H	H
	[sin]	H	H	H	H	H
$\left(\frac{r}{a}\right)^3$	[cos + sin]	$+\frac{343}{48}$	$-\frac{41503}{768}$	$+\frac{4923079}{30720}$	$+\frac{32}{3}$	$-\frac{1184}{15}$
	[cos − sin]	0	0	0	0	0
	[cos]	H	H	H	H	H
	[sin]	H	H	H	H	H
$\left(\frac{r}{a}\right)^4$	[cos + sin]	$+\frac{49}{12}$	$-\frac{5243}{192}$	$+\frac{548261}{7680}$	$+\frac{16}{3}$	$-\frac{176}{5}$
	[cos − sin]	0	0	0	0	0
	[cos]	H	H	H	H	H
	[sin]	H	H	H	H	H

... *Table* $\left(\left(\frac{r}{a}\right)^1 \ldots \left(\frac{r}{a}\right)^4\right) \begin{matrix}\cos\\ \sin\end{matrix}\, 4f \ldots$

	$\begin{matrix}\cos\\ \sin\end{matrix}\, 4f$	$[\;]^9$ e^5	$[\;]^9$ e^7	$[\;]^{10}$ e^6	$[\;]^{11}$ e^7
$\left(\frac{r}{a}\right)^1$	[cos + sin]	$+\frac{64653}{1280}$	$-\frac{4051137}{10240}$	$+\frac{47603}{576}$	$+\frac{84728011}{645120}$
	[cos − sin]	o	o	o	o
	[cos]	H	H	H	H
	[sin]	H	H	H	H
$\left(\frac{r}{a}\right)^2$	[cos + sin]	$+\frac{18749}{640}$	$-\frac{1157699}{5120}$	$+\frac{25799}{576}$	$+\frac{1440343}{21504}$
	[cos − sin]	o	o	o	o
	[cos]	H	H	H	H
	[sin]	H	H	H	H
$\left(\frac{r}{a}\right)^3$	[cos + sin]	$+\frac{19683}{1280}$	$-\frac{1161297}{10240}$	$+\frac{3125}{144}$	$+\frac{19487171}{645120}$
	[cos − sin]	o	o	o	o
	[cos]	H	H	H	H
	[sin]	H	H	H	H
$\left(\frac{r}{a}\right)^4$	[cos + sin]	$+\frac{2187}{320}$	$-\frac{115911}{2560}$	$+\frac{625}{72}$	$+\frac{1771561}{161280}$
	[cos − sin]	o	o	o	o
	[cos]	H	H	H	H
	[sin]	H	H	H	H

$$\textit{Table}\ \left(\left(\frac{r}{a}\right)^{-1} \dots \left(\frac{r}{a}\right)^{-5}\right) \frac{\cos}{\sin} 4f \dots$$

$\frac{\cos}{\sin} 4f$		[ ]^0		[ ]^1			[ ]^2		
		e^4	e^6	e^3	e^5	e^7	e^2	e^4	e^6
$\left(\frac{r}{a}\right)^{-1}$	[cos + sin]	$+\frac{1}{16}$	$+\frac{1}{16}$	$-\frac{73}{48}$	$+\frac{421}{768}$	$+\frac{4231}{30720}$	$+4$	$-\frac{33}{4}$	$+\frac{331}{96}$
	[cos − sin]	,,	,,	0	$-\frac{19}{3840}$	$-\frac{1}{10240}$	0	0	$-\frac{7}{1440}$
	[cos]	$+\frac{1}{16}$	$+\frac{1}{16}$	H	$+\frac{1043}{3840}$	$+\frac{1057}{15360}$	H	H	$+\frac{2479}{1440}$
	[sin]	0	0	H	$+\frac{177}{640}$	$+\frac{2117}{30720}$	H	H	$+\frac{1243}{720}$
$\left(\frac{r}{a}\right)^{-2}$	[cos + sin]	0	0	$-\frac{17}{24}$	$+\frac{25}{384}$	$+\frac{47}{15360}$	$+\frac{11}{4}$	$-\frac{29}{6}$	$+\frac{21}{16}$
	[cos − sin]	,,	,,	0	$-\frac{7}{1920}$	$-\frac{361}{46080}$	0	0	$-\frac{1}{1440}$
	[cos]	0	0	H	$+\frac{59}{1920}$	$-\frac{11}{4608}$	H	H	$+\frac{1889}{2886}$
	[sin]	0	0	H	$+\frac{11}{320}$	$+\frac{251}{46080}$	H	H	$+\frac{1891}{2880}$
$\left(\frac{r}{a}\right)^{-3}$	[cos + sin]	0	0	$-\frac{13}{48}$	$-\frac{29}{768}$	$-\frac{197}{6144}$	$+\frac{7}{4}$	$-\frac{29}{12}$	$+\frac{1}{3}$
	[cos − sin]	,,	,,	0	$+\frac{11}{3840}$	$+\frac{407}{92160}$	0	0	$+\frac{1}{288}$
	[cos]	0	0	H	$-\frac{67}{3840}$	$-\frac{637}{46080}$	H	H	$+\frac{97}{576}$
	[sin]	0	0	H	$-\frac{13}{640}$	$-\frac{1681}{92160}$	H	H	$+\frac{95}{576}$
$\left(\frac{r}{a}\right)^{-4}$	[cos + sin]	0	0	$-\frac{1}{12}$	$-\frac{5}{192}$	$-\frac{149}{7680}$	$+1$	-1	$+\frac{1}{24}$
	[cos − sin]	,,	,,	0	$-\frac{1}{960}$	$-\frac{11}{7680}$	0	0	$-\frac{1}{360}$
	[cos]	0	0	H	$-\frac{13}{960}$	$-\frac{1}{96}$	H	H	$+\frac{7}{360}$
	[sin]	0	0	H	$-\frac{1}{80}$	$-\frac{23}{2560}$	H	H	$+\frac{1}{45}$
$\left(\frac{r}{a}\right)^{-5}$	[cos + sin]	0	0	$-\frac{1}{48}$	$-\frac{7}{768}$	$-\frac{209}{30720}$	$+\frac{1}{2}$	$-\frac{1}{3}$	0
	[cos − sin]	,,	,,	0	$+\frac{1}{3840}$	$+\frac{31}{92160}$	0	0	$+\frac{1}{720}$
	[cos]	0	0	H	$-\frac{17}{3840}$	$-\frac{149}{46080}$	H	H	$+\frac{1}{1440}$
	[sin]	0	0	H	$-\frac{3}{640}$	$-\frac{329}{92160}$	H	H	$-\frac{1}{1440}$

... *Table* $\left(\left(\frac{r}{a}\right)^{-1} \dots \left(\frac{r}{a}\right)^{-5}\right) \begin{matrix}\cos\\ \sin\end{matrix}\, 4f \dots$

	$\begin{matrix}\cos\\ \sin\end{matrix}\, 4f$	$[\ ]^3$ e	e^3	e^5	e^7	$[\ ]^4$ e^0	e^2	e^4	e^6
$\left(\frac{r}{a}\right)^{-1}$	[cos + sin]	$-\frac{7}{2}$	$+\frac{307}{16}$	$-\frac{3717}{128}$	$+\frac{30571}{2048}$	$+1$	-16	$+\frac{239}{4}$	$-\frac{3019}{36}$
	[cos − sin]	0	0	0	$-\frac{261}{71680}$	0	0	0	0
	[cos]	H	H	H	$+\frac{267431}{35840}$	H	H	H	H
	[sin]	H	H	H	$+\frac{535123}{71680}$	H	H	H	H
$\left(\frac{r}{a}\right)^{-2}$	[cos + sin]	-3	$+\frac{123}{8}$	$-\frac{1305}{64}$	$+\frac{8547}{1024}$	$+1$	$-\frac{31}{2}$	$+\frac{433}{8}$	$-\frac{9793}{144}$
	[cos − sin]	0	0	0	$+\frac{27}{35840}$	0	0	0	0
	[cos]	H	H	H	$+\frac{74793}{17920}$	H	H	H	H
	[sin]	H	H	H	$+\frac{149559}{35840}$	H	H	H	H
$\left(\frac{r}{a}\right)^{-3}$	[cos + sin]	$-\frac{5}{2}$	$+\frac{183}{16}$	$-\frac{1611}{128}$	$+\frac{8035}{2048}$	$+1$	$-\frac{29}{2}$	$+\frac{365}{8}$	$-\frac{7111}{144}$
	[cos − sin]	0	0	0	$+\frac{243}{71680}$	0	0	0	0
	[cos]	H	H	H	$+\frac{70367}{35840}$	H	H	H	H
	[sin]	H	H	H	$+\frac{140491}{71680}$	H	H	H	H
$\left(\frac{r}{a}\right)^{-4}$	[cos + sin]	-2	$+\frac{31}{4}$	$-\frac{215}{32}$	$+\frac{819}{512}$	$+1$	-13	$+\frac{283}{8}$	$-\frac{1147}{36}$
	[cos − sin]	0	0	0	$-\frac{81}{17920}$	0	0	0	0
	[cos]	H	H	H	$+\frac{7146}{8960}$	H	H	H	H
	[sin]	H	H	H	$+\frac{14373}{17920}$	H	H	H	H
$\left(\frac{r}{a}\right)^{-5}$	[cos + sin]	$-\frac{3}{2}$	$+\frac{75}{16}$	$-\frac{393}{128}$	$+\frac{1251}{2048}$	$+1$	-11	$+\frac{199}{8}$	$-\frac{655}{36}$
	[cos − sin]	0	0	0	$+\frac{243}{71680}$	0	0	0	0
	[cos]	H	H	H	$+\frac{11007}{35840}$	H	H	H	H
	[sin]	H	H	H	$+\frac{21771}{71680}$	H	H	H	H

... *Table* $\left(\left(\frac{r}{a}\right)^{-1} \dots \left(\frac{r}{a}\right)^{-5}\right) \begin{matrix}\cos\\ \sin\end{matrix}\, 4f \dots$

	$\begin{matrix}\cos\\ \sin\end{matrix}\, 4f$	$[\ ]^5$ e	e^3	e^5	e^7	$[\ ]^6$ e^2	e^4	e^6
$\left(\frac{r}{a}\right)^{-1}$	[cos + sin]	$+\frac{9}{2}$	$-\frac{757}{16}$	$+\frac{58801}{384}$	$-\frac{1316215}{6144}$	$+13$	$-\frac{459}{4}$	$+\frac{11205}{32}$
	[cos − sin]	0	0	0	0	0	0	0
	[cos]	H	H	H	H	H	H	H
	[sin]	H	H	H	H	H	H	H
$\left(\frac{r}{a}\right)^{-2}$	[cos + sin]	$+5$	$-\frac{405}{8}$	$+\frac{29485}{192}$	$-\frac{1800725}{9216}$	$+\frac{63}{4}$	$-\frac{267}{2}$	$+\frac{12255}{32}$
	[cos − sin]	0	0	0	0	0	0	0
	[cos]	H	H	H	H	H	H	H
	[sin]	H	H	H	H	H	H	H
$\left(\frac{r}{a}\right)^{-3}$	[cos + sin]	$+\frac{11}{2}$	$-\frac{833}{16}$	$+\frac{55135}{384}$	$-\frac{2975165}{18432}$	$+\frac{75}{4}$	$-\frac{595}{4}$	$+\frac{12513}{32}$
	[cos − sin]	0	0	0	0	0	0	0
	[cos]	H	H	H	H	H	H	H
	[sin]	H	H	H	H	H	H	H
$\left(\frac{r}{a}\right)^{-4}$	[cos + sin]	$+6$	$-\frac{205}{4}$	$+\frac{11935}{96}$	$-\frac{184955}{1536}$	$+22$	$-\frac{317}{2}$	$+\frac{5945}{16}$
	[cos − sin]	0	0	0	0	0	0	0
	[cos]	H	H	H	H	H	H	H
	[sin]	H	H	H	H	H	H	H
$\left(\frac{r}{a}\right)^{-5}$	[cos + sin]	$+\frac{13}{2}$	$-\frac{765}{16}$	$+\frac{37925}{384}$	$-\frac{1493005}{18432}$	$+\frac{51}{2}$	$-\frac{321}{2}$	$+\frac{2613}{8}$
	[cos − sin]	0	0	0	0	0	0	0
	[cos]	H	H	H	H	H	H	H
	[sin]	H	H	H	H	H	H	H

... *Table* $\left(\left(\frac{r}{a}\right)^{-1} \dots \left(\frac{r}{a}\right)^{-5}\right) \begin{matrix}\cos\\ \sin\end{matrix}\ 4f \dots$

	$\begin{matrix}\cos\\ \sin\end{matrix}\ 4f_{\prime}$	[ ]^7 e^3	[ ]^7 e^5	[ ]^7 e^7	[ ]^8 e^4	[ ]^8 e^6
$\left(\frac{r}{a}\right)^{-1}$	[cos + sin]	$+\frac{1471}{48}$	$-\frac{190507}{768}$	$+\frac{22796399}{30720}$	$+\frac{1027}{16}$	$-\frac{119323}{240}$
	[cos − sin]	0	0	0	0	0
	[cos]	H	H	H	H	H
	[sin]	H	H	H	H	H
$\left(\frac{r}{a}\right)^{-2}$	[cos + sin]	$+\frac{959}{24}$	$-\frac{119119}{384}$	$+\frac{13442527}{15360}$	$+\frac{2141}{24}$	$-\frac{53003}{80}$
	[cos − sin]	0	0	0	0	0
	[cos]	H	H	H	H	H
	[sin]	H	H	H	H	H
$\left(\frac{r}{a}\right)^{-3}$	[cos + sin]	$+\frac{2443}{48}$	$-\frac{284557}{768}$	$+\frac{29629663}{30720}$	$+\frac{2893}{24}$	$-\frac{40387}{48}$
	[cos − sin]	0	0	0	0	0
	[cos]	H	H	H	H	H
	[sin]	H	H	H	H	H
$\left(\frac{r}{a}\right)^{-4}$	[cos + sin]	$+\frac{763}{12}$	$-\frac{81193}{192}$	$+\frac{7623959}{7680}$	$+\frac{1273}{8}$	$-\frac{61219}{60}$
	[cos − sin]	0	0	0	0	0
	[cos]	H	H	H	H	H
	[sin]	H	H	H	H	H
$\left(\frac{r}{a}\right)^{-5}$	[cos + sin]	$+\frac{3751}{48}$	$-\frac{352919}{768}$	$+\frac{29213303}{30720}$	$+\frac{4943}{24}$	$-\frac{23591}{20}$
	[cos − sin]	0	0	0	0	0
	[cos]	H	H	H	H	H
	[sin]	H	H	H	H	H

... *Table* $\left(\left(\frac{r}{a}\right)^{-1} \dots \left(\frac{r}{a}\right)^{-5}\right) \begin{matrix}\cos\\ \sin\end{matrix}\ 4f$...

$\begin{matrix}\cos\\ \sin\end{matrix}\ 4f$		$[\ ]^9$ e^5	$[\ ]^9$ e^7	$[\ ]^{10}$ e^6	$[\ ]^{11}$ e^7
$\left(\frac{r}{a}\right)^{-1}$	[cos + sin]	$+\frac{159263}{1280}$	$-\frac{9671209}{10240}$	$+\frac{65773}{288}$	$+\frac{86564393}{215040}$
	[cos − sin]	0	0	0	0
	[cos]	H	H	H	H
	[sin]	H	H	H	H
$\left(\frac{r}{a}\right)^{-2}$	[cos + sin]	$+\frac{117099}{640}$	$-\frac{6819921}{5120}$	$+\frac{25435}{72}$	$+\frac{210276803}{322560}$
	[cos − sin]	0	0	0	0
	[cos]	H	H	H	H
	[sin]	H	H	H	H
$\left(\frac{r}{a}\right)^{-3}$	[cos + sin]	$+\frac{333513}{1280}$	$-\frac{18298713}{10240}$	$+\frac{75947}{144}$	$+\frac{131087143}{129024}$
	[cos − sin]	0	0	0	0
	[cos]	H	H	H	H
	[sin]	H	H	H	H
$\left(\frac{r}{a}\right)^{-4}$	[cos + sin]	$+\frac{115617}{320}$	$-\frac{5856453}{2560}$	$+\frac{110099}{144}$	$+\frac{27485931}{17920}$
	[cos − sin]	0	0	0	0
	[cos]	H	H	H	H
	[sin]	H	H	H	H
$\left(\frac{r}{a}\right)^{-5}$	[cos + sin]	$+\frac{627003}{1280}$	$-\frac{28615329}{10240}$	$+\frac{155681}{144}$	$+\frac{1454039059}{645120}$
	[cos − sin]	0	0	0	0
	[cos]	H	H	H	H
	[sin]	H	H	H	H

Table $\left(\left(\frac{r}{a}\right)^{-1} \ldots \left(\frac{r}{a}\right)^{4}\right) \begin{matrix}\cos\\ \sin\end{matrix} 5f \ldots$

$\begin{matrix}\cos\\ \sin\end{matrix} 5f$		$[\ ]^0$ e^6	e^7	$[\ ]^1$ e^4	e^6	$[\ ]^2$ e^3	e^5	e^7
$\left(\frac{r}{a}\right)^1$	[cos + sin]	$-\frac{21}{32}$	$-\frac{21}{64}$	$+\frac{1961}{384}$	$-\frac{4109}{1280}$	$-\frac{277}{24}$	$+\frac{575}{24}$	$-\frac{24781}{1920}$
	[cos − sin]	„	„	0	$-\frac{1091}{46080}$	0	0	$-\frac{209}{40320}$
	[cos]	$-\frac{21}{32}$	$-\frac{21}{64}$	H	$-\frac{29803}{18432}$	H	H	$-\frac{52061}{8064}$
	[sin]	0	0	H	$-\frac{146833}{92160}$	H	H	$-\frac{2032}{315}$
$\left(\frac{r}{a}\right)^2$	[cos + sin]	$-\frac{7}{4}$	$-\frac{7}{16}$	$+\frac{1131}{128}$	$-\frac{25429}{3840}$	$-\frac{97}{6}$	$+\frac{817}{24}$	$-\frac{2461}{120}$
	[cos − sin]	„	„	0	$-\frac{887}{46080}$	0	0	$-\frac{1}{105}$
	[cos]	$-\frac{7}{4}$	$-\frac{7}{16}$	H	$-\frac{61207}{18432}$	H	H	$-\frac{1149}{112}$
	[sin]	0	0	H	$-\frac{304261}{92160}$	H	H	$-\frac{17219}{1680}$
$\left(\frac{r}{a}\right)^3$	[cos + sin]	$-\frac{63}{16}$	$-\frac{21}{64}$	$+\frac{5509}{384}$	$-\frac{21763}{1920}$	$-\frac{263}{12}$	$+\frac{2135}{48}$	$-\frac{2641}{96}$
	[cos − sin]	„	„	0	$+\frac{6067}{46080}$	0	0	$+\frac{13}{2016}$
	[cos]	$-\frac{63}{16}$	$-\frac{21}{64}$	H	$-\frac{103249}{18432}$	H	H	$-\frac{6931}{504}$
	[sin]	0	0	H	$-\frac{528379}{92160}$	H	H	$-\frac{27737}{2016}$
$\left(\frac{r}{a}\right)^4$	[cos + sin]	$-\frac{63}{8}$	0	$+\frac{8501}{384}$	$-\frac{10543}{640}$	$-\frac{347}{12}$	$+\frac{2557}{48}$	$-\frac{3673}{120}$
	[cos − sin]	„	„	0	$+\frac{32251}{46080}$	0	0	$+\frac{281}{5040}$
	[cos]	$-\frac{63}{8}$	0	H	$-\frac{145369}{18432}$	H	H	$-\frac{30797}{2016}$
	[sin]	0	0	H	$-\frac{791347}{92160}$	H	H	$-\frac{154547}{10080}$

... *Table* $\left(\left(\frac{r}{a}\right)^1 \ldots \left(\frac{r}{a}\right)^4\right) \begin{matrix}\cos\\ \sin\end{matrix}\, 5f \ldots$

$\begin{matrix}\cos\\ \sin\end{matrix}\, 5f$		[ ]³			[ ]⁴			
		e^2	e^4	e^6	e	e^3	e^5	e^7
$\left(\frac{r}{a}\right)^1$	[cos + sin]	$+\frac{93}{8}$	$-\frac{417}{8}$	$+\frac{77733}{1024}$	$-\frac{11}{2}$	$+\frac{207}{4}$	$-\frac{3715}{24}$	$+\frac{58583}{288}$
	[cos − sin]	0	0	0	0	0	0	0
	[cos]	H	H	H	H	H	H	H
	[sin]	H	H	H	H	H	H	H
$\left(\frac{r}{a}\right)^2$	[cos + sin]	$+\frac{113}{8}$	-62	$+\frac{94081}{1024}$	-6	$+\frac{109}{2}$	$-\frac{484}{3}$	$+\frac{6917}{32}$
	[cos − sin]	0	0	0	0	0	0	0
	[cos]	H	H	H	H	H	H	H
	[sin]	H	H	H	H	H	H	H
$\left(\frac{r}{a}\right)^3$	[cos + sin]	$+\frac{135}{8}$	$-\frac{1119}{16}$	$+\frac{103311}{1024}$	$-\frac{13}{2}$	$+\frac{443}{8}$	$-\frac{1885}{12}$	$+\frac{238195}{1152}$
	[cos − sin]	0	0	0	0	0	0	0
	[cos]	H	H	H	H	H	H	H
	[sin]	H	H	H	H	H	H	H
$\left(\frac{r}{a}\right)^4$	[cos + sin]	$+\frac{159}{8}$	$-\frac{1191}{16}$	$+\frac{101103}{1024}$	-7	$+54$	$-\frac{3379}{24}$	$+\frac{1558}{9}$
	[cos − sin]	0	0	0	0	0	0	0
	[cos]	H	H	H	H	H	H	H
	[sin]	H	H	H	H	H	H	H

... *Table* $\left(\left(\frac{r}{a}\right)^1 \ldots \left(\frac{r}{a}\right)^4\right) \begin{matrix}\cos\\ \sin\end{matrix}\, 5f \ldots$

	$\begin{matrix}\cos\\ \sin\end{matrix}\, 5f$	[ ]⁵ e^0	[ ]⁵ e^2	[ ]⁵ e^4	[ ]⁵ e^6	[ ]⁶ e	[ ]⁶ e^3	[ ]⁶ e^5	[ ]⁶ e^7
$\left(\frac{r}{a}\right)^1$	[cos + sin]	$+1$	$-\frac{49}{2}$	$+\frac{9575}{64}$	$-\frac{441125}{1152}$	$+\frac{9}{2}$	$-\frac{561}{8}$	$+\frac{11415}{32}$	$-\frac{109167}{128}$
	[cos − sin]	0	0	0	0	0	0	0	0
	[cos]	H	H	H	H	H	H	H	H
	[sin]	H	H	H	H	H	H	H	H
$\left(\frac{r}{a}\right)^2$	[cos + sin]	$+1$	$-\frac{47}{2}$	$+\frac{8975}{64}$	$-\frac{412175}{1152}$	$+4$	$-\frac{121}{2}$	$+\frac{1213}{4}$	$-\frac{11619}{16}$
	[cos − sin]	0	0	0	0	0	0	0	0
	[cos]	H	H	H	H	H	H	H	H
	[sin]	H	H	H	H	H	H	H	H
$\left(\frac{r}{a}\right)^3$	[cos + sin]	$+1$	-22	$+\frac{7999}{64}$	$-\frac{177925}{576}$	$+\frac{7}{2}$	$-\frac{201}{4}$	$+\frac{7755}{32}$	$-\frac{9065}{16}$
	[cos − sin]	0	0	0	0	0	0	0	0
	[cos]	H	H	H	H	H	H	H	H
	[sin]	H	H	H	H	H	H	H	H
$\left(\frac{r}{a}\right)^4$	[cos + sin]	$+1$	-20	$+\frac{6695}{64}$	$-\frac{138775}{576}$	$+3$	$-\frac{159}{4}$	$+\frac{2859}{16}$	$-\frac{12591}{32}$
	[cos − sin]	0	0	0	0	0	0	0	0
	[cos]	H	H	H	H	H	H	H	H
	[sin]	H	H	H	H	H	H	H	H

$$\ldots\ Table\ \left(\left(\frac{r}{a}\right)^1 \ldots \left(\frac{r}{a}\right)^4\right) \begin{matrix}\cos\\ \sin\end{matrix}\ 5f \ldots$$

	$\begin{matrix}\cos\\ \sin\end{matrix}\ 5f$	$[\ ]^7$ e^2	e^4	e^6	$[\ ]^8$ e	e^3	e^5	e^7
$\left(\frac{r}{a}\right)^1$	[cos + sin]	$+\frac{103}{8}$	$-\frac{3941}{24}$	$+\frac{780003}{1024}$		$+\frac{359}{12}$	$-\frac{32885}{96}$	$+\frac{1453367}{960}$
	[cos − sin]	o	o	o		o	o	o
	[cos]	H	H	H		H	H	H
	[sin]	H	H	H		H	H	H
$\left(\frac{r}{a}\right)^2$	[cos + sin]	$+\frac{83}{8}$	$-\frac{259}{2}$	$+\frac{1829093}{3072}$		$+\frac{133}{6}$	$-\frac{2999}{12}$	$+\frac{263873}{240}$
	[cos − sin]	o	o	o		o	o	o
	[cos]	H	H	H		H	H	H
	[sin]	H	H	H		H	H	H
$\left(\frac{r}{a}\right)^3$	[cos + sin]	$+\frac{65}{8}$	$-\frac{4667}{48}$	$+\frac{1331743}{3072}$		$+\frac{379}{24}$	$-\frac{8255}{48}$	$+\frac{141719}{192}$
	[cos − sin]	o	o	o		o	o	o
	[cos]	H	H	H		H	H	H
	[sin]	H	H	H		H	H	H
$\left(\frac{r}{a}\right)^4$	[cos + sin]	$+\frac{49}{8}$	$-\frac{3283}{48}$	$+\frac{293853}{1024}$		$+\frac{32}{3}$	$-\frac{328}{3}$	$+\frac{6674}{15}$
	[cos − sin]	o	o	o		o	o	o
	[cos]	H	H	H		H	H	H
	[sin]	H	H	H		H	H	H

... *Table* $\left(\left(\frac{r}{a}\right)^1 \dots \left(\frac{r}{a}\right)^4\right) \begin{matrix}\cos\\ \sin\end{matrix}\, 5f.$

$\begin{matrix}\cos\\ \sin\end{matrix}\, 5f$		[ ]^9		[ ]^{10}		[ ]^{11}	[ ]^{12}
		e^4	e^6	e^5	e^7	e^6	e^7
$\left(\frac{r}{a}\right)^1$	[cos + sin]	$+\frac{7887}{128}$	$-\frac{848529}{1280}$	$+\frac{11263}{96}$	$-\frac{1402147}{1152}$	$+\frac{9742939}{46080}$	$+\frac{409821}{1120}$
	[cos − sin]	0	0	0	0	0	0
	[cos]	H	H	H	H	H	H
	[sin]	H	H	H	H	H	H
$\left(\frac{r}{a}\right)^2$	[cos + sin]	$+\frac{5431}{128}$	$-\frac{578003}{1280}$	$+\frac{1817}{24}$	$-\frac{37429}{48}$	$+\frac{5931103}{46080}$	$+\frac{236641}{1120}$
	[cos − sin]	0	0	0	0	0	0
	[cos]	H	H	H	H	H	H
	[sin]	H	H	H	H	H	H
$\left(\frac{r}{a}\right)^3$	[cos + sin]	$+\frac{3563}{128}$	$-\frac{184221}{640}$	$+\frac{4433}{96}$	$-\frac{66859}{144}$	$+\frac{3389017}{46080}$	$+\frac{101971}{896}$
	[cos − sin]	0	0	0	0	0	0
	[cos]	H	H	H	H	H	H
	[sin]	H	H	H	H	H	H
$\left(\frac{r}{a}\right)^4$	[cos + sin]	$+\frac{2187}{128}$	$-\frac{107163}{640}$	$+\frac{625}{24}$	$-\frac{71875}{288}$	$+\frac{1771561}{46080}$	$+\frac{1944}{35}$
	[cos − sin]	0	0	0	0	0	0
	[cos]	H	H	H	H	H	H
	[sin]	H	H	H	H	H	H

Table $\left(\left(\frac{r}{a}\right)^{-1} \ldots \left(\frac{r}{a}\right)^{-5}\right) \begin{matrix}\cos\\ \sin\end{matrix}\, 5f \ldots$

$\begin{matrix}\cos\\ \sin\end{matrix}\, 5f$		$[\ ]^0$ e^5	$[\ ]^0$ e^7	$[\ ]^1$ e^4	$[\ ]^1$ e^6	$[\ ]^2$ e^3	$[\ ]^2$ e^5	$[\ ]^2$ e^7
$\left(\frac{r}{a}\right)^{-1}$	[cos + sin]	$-\frac{1}{32}$	$-\frac{5}{128}$	$+\frac{167}{128}$	$-\frac{247}{960}$	$-\frac{31}{6}$	$+\frac{859}{96}$	$-\frac{18683}{6720}$
	[cos − sin]	,,	,,	0	$+\frac{151}{46080}$	0	0	$+\frac{13}{6720}$
	[cos]	$-\frac{1}{32}$	$-\frac{5}{128}$	H	$-\frac{2341}{18432}$	H	H	$-\frac{1867}{1344}$
	[sin]	0	0	H	$-\frac{12007}{92160}$	H	H	$-\frac{779}{560}$
$\left(\frac{r}{a}\right)^{-2}$	[cos + sin]	0	0	$+\frac{209}{384}$	$+\frac{43}{1280}$	$-\frac{19}{6}$	$+\frac{55}{12}$	$-\frac{77}{96}$
	[cos − sin]	,,	,,	0	$+\frac{37}{46080}$	0	0	$-\frac{1}{2016}$
	[cos]	0	0	H	$+\frac{317}{18432}$	H	H	$-\frac{809}{2016}$
	[sin]	0	0	H	$+\frac{1511}{92160}$	H	H	$-\frac{101}{252}$
$\left(\frac{r}{a}\right)^{-3}$	[cos + sin]	0	0	$+\frac{73}{384}$	$+\frac{211}{3840}$	$-\frac{43}{24}$	$+\frac{193}{96}$	$-\frac{103}{960}$
	[cos − sin]	,,	,,	0	$-\frac{47}{46080}$	0	0	$-\frac{17}{20160}$
	[cos]	0	0	H	$+\frac{497}{18432}$	H	H	$-\frac{109}{2016}$
	[sin]	0	0	H	$+\frac{2579}{92160}$	H	H	$-\frac{1073}{20160}$
$\left(\frac{r}{a}\right)^{-4}$	[cos + sin]	0	0	$+\frac{7}{128}$	$+\frac{49}{1920}$	$-\frac{11}{12}$	$+\frac{35}{48}$	$+\frac{11}{480}$
	[cos − sin]	,,	,,	0	$+\frac{19}{46080}$	0	0	$+\frac{1}{1120}$
	[cos]	0	0	H	$+\frac{239}{18432}$	H	H	$+\frac{1}{84}$
	[sin]	0	0	H	$+\frac{1157}{92160}$	H	H	$+\frac{37}{3360}$
$\left(\frac{r}{a}\right)^{-5}$	[cos + sin]	0	0	$+\frac{5}{384}$	$+\frac{1}{128}$	$-\frac{5}{12}$	$+\frac{5}{24}$	$+\frac{1}{96}$
	[cos − sin]	,,	,,	0	$-\frac{1}{9216}$	0	0	$-\frac{1}{2016}$
	[cos]	0	0	H	$+\frac{71}{18432}$	H	H	$+\frac{5}{1008}$
	[sin]	0	0	H	$+\frac{73}{18432}$	H	H	$+\frac{11}{2016}$

$$\dots Table\ \left(\left(\frac{r}{a}\right)^{-1} \dots \left(\frac{r}{a}\right)^{-5}\right) \begin{matrix}\cos\\ \sin\end{matrix}\ 5f \dots$$

	$\begin{matrix}\cos\\ \sin\end{matrix}\ 5f$	[ ]^3 e^2	e^4	e^6	[ ]^4 e	e^3	e^5	e^7
$\left(\frac{r}{a}\right)^{-1}$	[cos + sin]	$+\frac{59}{8}$	$-\frac{501}{16}$	$+\frac{39963}{1024}$	$-\frac{9}{2}$	$+\frac{337}{8}$	$-\frac{5683}{48}$	$+\frac{26119}{192}$
	[cos − sin]	o	o	o	o	o	o	o
	[cos]	H	H	H	H	H	H	H
	[sin]	H	H	H	H	H	H	H
$\left(\frac{r}{a}\right)^{-2}$	[cos + sin]	$+\frac{45}{8}$	$-\frac{177}{8}$	$+\frac{24501}{1024}$	-4	$+36$	$-\frac{565}{6}$	$+\frac{3475}{36}$
	[cos − sin]	o	o	o	o	o	o	o
	[cos]	H	H	H	H	H	H	H
	[sin]	H	H	H	H	H	H	H
$\left(\frac{r}{a}\right)^{-3}$	[cos + sin]	$+\frac{33}{8}$	$-\frac{29}{2}$	$+\frac{13281}{1024}$	$-\frac{7}{2}$	$+\frac{59}{2}$	$-\frac{3349}{48}$	$+\frac{17803}{288}$
	[cos − sin]	o	o	o	o	o	o	o
	[cos]	H	H	H	H	H	H	H
	[sin]	H	H	H	H	H	H	H
$\left(\frac{r}{a}\right)^{-4}$	[cos + sin]	$+\frac{23}{8}$	$-\frac{139}{16}$	$+\frac{6263}{1024}$	-3	$+23$	$-\frac{1145}{24}$	$+\frac{1703}{48}$
	[cos − sin]	o	o	o	o	o	o	o
	[cos]	H	H	H	H	H	H	H
	[sin]	H	H	H	H	H	H	H
$\left(\frac{r}{a}\right)^{-5}$	[cos + sin]	$+\frac{15}{8}$	$-\frac{75}{16}$	$+\frac{2535}{1024}$	$-\frac{5}{2}$	$+\frac{135}{8}$	$-\frac{715}{24}$	$+\frac{20945}{1152}$
	[cos − sin]	o	o	o	o	o	o	o
	[cos]	H	H	H	H	H	H	H
	[sin]	H	H	H	H	H	H	H

... *Table* $\left(\left(\frac{r}{a}\right)^{-1} \dots \left(\frac{r}{a}\right)^{-5}\right) \begin{matrix}\cos\\ \sin\end{matrix} 5f \dots$

$\begin{matrix}\cos\\ \sin\end{matrix} 5f$		$[\ ]^5$ e^0	e^2	e^4	e^6	$[\ ]^6$ e	e^3	e^5	e^7
$\left(\frac{r}{a}\right)^{-1}$	[cos + sin]	$+1$	-25	$+\frac{9575}{64}$	$-\frac{103325}{288}$	$+\frac{11}{2}$	-86	$+\frac{13593}{32}$	$-\frac{60877}{64}$
	[cos − sin]	0	0	0	0	0	0	0	0
	[cos]	H	H	H	H	H	H	H	H
	[sin]	H	H	H	H	H	H	H	H
$\left(\frac{r}{a}\right)^{-2}$	[cos + sin]	$+1$	$-\frac{49}{2}$	$+\frac{8999}{64}$	$-\frac{363365}{1152}$	$+6$	$-\frac{183}{2}$	$+\frac{3465}{8}$	$-\frac{29175}{32}$
	[cos − sin]	0	0	0	0	0	0	0	0
	[cos]	H	H	H	H	H	H	H	H
	[sin]	H	H	H	H	H	H	H	H
$\left(\frac{r}{a}\right)^{-3}$	[cos + sin]	$+1$	$-\frac{47}{2}$	$+\frac{8095}{64}$	$-\frac{298655}{1152}$	$+\frac{13}{2}$	$-\frac{759}{8}$	$+\frac{13509}{32}$	$-\frac{26213}{32}$
	[cos − sin]	0	0	0	0	0	0	0	0
	[cos]	H	H	H	H	H	H	H	H
	[sin]	H	H	H	H	H	H	H	H
$\left(\frac{r}{a}\right)^{-4}$	[cos + sin]	$+1$	-22	$+\frac{6935}{64}$	$-\frac{114085}{576}$	$+7$	$-\frac{383}{4}$	$+\frac{6275}{16}$	$-\frac{2751}{4}$
	[cos − sin]	0	0	0	0	0	0	0	0
	[cos]	H	H	H	H	H	H	H	H
	[sin]	H	H	H	H	H	H	H	H
$\left(\frac{r}{a}\right)^{-5}$	[cos + sin]	$+1$	-20	$+\frac{5615}{64}$	$-\frac{80455}{576}$	$+\frac{15}{2}$	$-\frac{375}{4}$	$+\frac{11055}{32}$	$-\frac{8595}{16}$
	[cos − sin]	0	0	0	0	0	0	0	0
	[cos]	H	H	H	H	H	H	H	H
	[sin]	H	H	H	H	H	H	H	H

$$\ldots \textit{Table} \left(\left(\frac{r}{a}\right)^{-1} \ldots \left(\frac{r}{a}\right)^{-5}\right) \begin{matrix}\cos\\ \sin\end{matrix}\, 5f \ldots$$

	$\begin{matrix}\cos\\ \sin\end{matrix}\, 5f$	$[\ ]^7$			$[\ ]^8$		
		e^2	e^4	e^6	e^3	e^5	e^7
$\left(\frac{r}{a}\right)^{-1}$	[cos + sin]	$+\frac{149}{8}$	$-\frac{3771}{16}$	$+\frac{3242339}{3072}$	$+\frac{1201}{24}$	$-\frac{54151}{96}$	$+\frac{4603927}{1920}$
	[cos − sin]	0	0	0	0	0	0
	[cos]	H	H	H	H	H	H
	[sin]	H	H	H	H	H	H
$\left(\frac{r}{a}\right)^{-2}$	[cos + sin]	$+\frac{175}{8}$	$-\frac{6461}{24}$	$+\frac{1183371}{1024}$	$+\frac{188}{3}$	$-\frac{4115}{6}$	$+\frac{16781}{6}$
	[cos − sin]	0	0	0	0	0	0
	[cos]	H	H	H	H	H	H
	[sin]	H	H	H	H	H	H
$\left(\frac{r}{a}\right)^{-3}$	[cos + sin]	$+\frac{203}{8}$	$-\frac{896}{3}$	$+\frac{3712093}{3072}$	$+\frac{463}{6}$	$-\frac{38771}{48}$	$+\frac{46693}{15}$
	[cos − sin]	0	0	0	0	0	0
	[cos]	H	H	H	H	H	H
	[sin]	H	H	H	H	H	H
$\left(\frac{r}{a}\right)^{-4}$	[cos + sin]	$+\frac{233}{8}$	$-\frac{5149}{16}$	$+\frac{3701999}{3072}$	$+\frac{281}{3}$	$-\frac{22135}{24}$	$+\frac{99181}{30}$
	[cos − sin]	0	0	0	0	0	0
	[cos]	H	H	H	H	H	H
	[sin]	H	H	H	H	H	H
$\left(\frac{r}{a}\right)^{-5}$	[cos + sin]	$+\frac{265}{8}$	$-\frac{16135}{48}$	$+\frac{1169805}{1024}$	$+\frac{2695}{24}$	$-\frac{24485}{24}$	$+\frac{160513}{48}$
	[cos − sin]	0	0	0	0	0	0
	[cos]	H	H	H	H	H	H
	[sin]	H	H	H	H	H	H

... *Table* $\left(\left(\frac{r}{a}\right)^{-1} \cdots \left(\frac{r}{a}\right)^{-5}\right) \begin{matrix}\cos \\ \sin\end{matrix}\ 5f.$

	$\begin{matrix}\cos \\ \sin\end{matrix}\ 5f$	$[\ ]^9$		$[\ ]^{10}$		$[\ ]^{11}$	$[\ ]^{12}$
		e^4	e^6	e^5	e^7	e^6	e^7
$\left(\frac{r}{a}\right)^{-1}$	[cos + sin]	$+\frac{14987}{128}$	$-\frac{394607}{320}$	$+\frac{11989}{48}$	$-\frac{161685}{64}$	$+\frac{22980061}{46080}$	$+\frac{2122251}{2240}$
	[cos − sin]	0	0	0	0	0	0
	[cos]	H	H	H	H	H	H
	[sin]	H	H	H	H	H	H
$\left(\frac{r}{a}\right)^{-2}$	[cos + sin]	$+\frac{19863}{128}$	$-\frac{2029257}{1280}$	$+\frac{8359}{24}$	$-\frac{983731}{288}$	$+\frac{33554587}{46080}$	$+\frac{40401}{28}$
	[cos − sin]	0	0	0	0	0	0
	[cos]	H	H	H	H	H	H
	[sin]	H	H	H	H	H	H
$\left(\frac{r}{a}\right)^{-3}$	[cos + sin]	$+\frac{25791}{128}$	$-\frac{2521323}{1280}$	$+\frac{1423}{3}$	$-\frac{1282973}{288}$	$+\frac{47722543}{46080}$	$+\frac{2391591}{1120}$
	[cos − sin]	0	0	0	0	0	0
	[cos]	H	H	H	H	H	H
	[sin]	H	H	H	H	H	H
$\left(\frac{r}{a}\right)^{-4}$	[cos + sin]	$+\frac{32907}{128}$	$-\frac{1516377}{640}$	$+\frac{7597}{12}$	$-\frac{67409}{12}$	$+\frac{66363649}{46080}$	$+\frac{1726833}{560}$
	[cos − sin]	0	0	0	0	0	0
	[cos]	H	H	H	H	H	H
	[sin]	H	H	H	H	H	H
$\left(\frac{r}{a}\right)^{-5}$	[cos + sin]	$+\frac{41355}{128}$	$-\frac{353187}{128}$	$+\frac{79711}{96}$	$-\frac{987035}{144}$	$+\frac{18100973}{9216}$	$+\frac{3906009}{896}$
	[cos − sin]	0	0	0	0	0	0
	[cos]	H	H	H	H	H	H
	[sin]	H	H	H	H	H	H

The foregoing tables are read as follows: for instance, in the Table $(x^0, x^1, x^2, x^3)\,{}^{\cos}_{\sin} f$, p. 376, *et seq.*, the third and fourth lines of the x^0-compartment show that

$$\begin{aligned}\cos f = & \quad - e \\ & + \left(\tfrac{1}{2} - \tfrac{9}{16}e^2 + \tfrac{25}{384}e^4 - \tfrac{49}{18432}e^6\right) 2\cos g \\ & + \left(\tfrac{1}{2}e - \tfrac{2}{3}e^3 + \tfrac{3}{16}e^5 - \tfrac{1}{45}e^7\right) 2\cos 2g \\ & + \quad \&c.\end{aligned}$$

$$\begin{aligned}\sin f = & \left(\tfrac{1}{2} - \tfrac{7}{16}e^2 + \tfrac{17}{384}e^4 - \tfrac{271}{18432}e^6\right) 2\sin g \\ & + \left(\tfrac{1}{2}e - \tfrac{7}{12}e^3 + \tfrac{1}{6}e^5 - \tfrac{19}{720}e^7\right) 2\sin 2g \\ & + \quad \&c.\end{aligned}$$

and the first and second lines give the sum and difference respectively of the corresponding coefficients of the cosine and sine series.

Addition, 28th Dec. 1860.—The tables have been verified by me, on the proof-sheets; the x-tables, the cosine lines of the $\left(\frac{r}{a}\right)$-tables, and a portion of the sine lines of the same tables, in the manner explained at the commencement of the memoir; but for the remainder of the sine lines it was found easier to employ the following mode of verification; viz. the equation $\frac{r}{a} = 1 + x$, gives

$$\left\{\left(\frac{r}{a}\right)^4 - 4\left(\frac{r}{a}\right)^3 + 6\left(\frac{r}{a}\right)^2 - 4\left(\frac{r}{a}\right)\right\} \sin jf = (-1 + x^4)\sin jf;$$

and as regards the terms up to e^7, the limit of the tables,

$$\left\{\left(\frac{r}{a}\right)^{-5} - 5\left(\frac{r}{a}\right)^{-4} + 10\left(\frac{r}{a}\right)^{-3} - 10\left(\frac{r}{a}\right)^{-2} + 5\left(\frac{r}{a}\right)^{-1}\right\} \sin jf = (1 - x^5 + 5x^6 - 15x^7)\sin jf,$$

which equations afford the verification referred to. In going over the earlier sheets I omitted to see that the fractions were in their least terms, and it may happen that, in some instances, they are not so. [A few reductions have been made, and I believe that the fractions are now all in their least terms.]

The expression for the true anomaly f itself has been repeatedly calculated to a much greater extent, in particular by Schubert (*Astronomie Théorique, Pet.* 1822), as far as e^{20}. The easiest way of obtaining it seems to be by means of the equation

$$\frac{df}{dg} = \left(\frac{r}{a}\right)^{-2} \sqrt{1 - e^2};$$

and, for the convenience of reference, I here give it as far as e^7, viz.,

$$f = g + \Sigma\,[\sin]^i \sin ig,$$

where, as in the other sine series, i is to be taken as well negatively as positively, and $[\sin]^{-i} = -[\sin]^i$. Or, what is the same thing,

$$f = g + \Sigma\,[\sin]^i\, 2 \sin ig,$$

where i has only positive values. And the coefficients are

	e	e^2	e^3	e^4	e^5	e^6	e^7
$[\sin]^1 =$	1		$-\frac{1}{8}$		$+\frac{5}{192}$		$+\frac{107}{9216}$
$[\sin]^2 =$		$+\frac{5}{8}$		$-\frac{11}{48}$		$+\frac{17}{384}$	
$[\sin]^3 =$			$+\frac{13}{24}$		$-\frac{43}{128}$		$+\frac{95}{1024}$
$[\sin]^4 =$				$+\frac{103}{192}$		$-\frac{451}{960}$	
$[\sin]^5 =$					$+\frac{1097}{1920}$		$-\frac{5957}{9216}$
$[\sin]^6 =$						$+\frac{1223}{1920}$	
$[\sin]^7 =$							$+\frac{47273}{64512}$

The expression for $\frac{r}{a}$, as far as e^{13}, is given in Schubert's work, above referred to; and that of $\log\left(\frac{r}{a}\right)$, as far as e^9, was calculated by Oriani, see the Introduction to Delambre's *Tables du Soleil*, Paris, 1806.

217.

A MEMOIR ON THE PROBLEM OF THE ROTATION OF A SOLID BODY.

[From the *Memoirs of the Royal Astronomical Society*, vol. XXIX. (1861), pp. 307—342. Read May 11, 1860.]

THE present memoir was written for the sake of the further elaboration of the analytical theory of the Rotation of a Solid Body, upon principles similar to those of my "Memoir on the Problem of Disturbed Elliptic Motion," *Mem. R. Ast. Soc.* vol. XXVII. pp. 1—29 (1858) [212]; the like elements are adopted, and the course of the investigation corresponds precisely to that of the memoir just referred to. The formulæ for the variations of the elements in the two problems (the motion being in each case referred to a fixed plane of reference and origin of angles therein) are found to be (as it is known they should be) identical in form; an investigation, in the present memoir, of the transformation of the system to the case of a variable plane of reference and departure-point as an origin of angles in such plane, would have been a mere repetition of that contained in the former memoir, and it was therefore unnecessary to give it. A point in the present memoir to which attention may be called is the definition of the angle g (varying uniformly with the time, but used as an element) which corresponds to the mean anomaly in elliptic motion. Besides the ultimate system of formulæ for the variations of the elements in terms of the differential coefficients of the Disturbing Function with respect to the elements, it appears to me that the intermediate formulæ for the variations in terms of the differential coefficients with respect to the coordinates (which are in the ordinary investigation altogether passed over) are not without interest, and that it is possible that they might be employed with advantage in the integration of the equations of motion for the purposes of physical astronomy.

I.

In the theory of elliptic motion, where the elements are

a, the mean distance,

e, the eccentricity,

g, the mean anomaly,

ϖ, the departure of pericentre,

θ, the longitude of node,

σ, the departure of node,

ϕ, the inclination,

and if, besides, we have

n, the mean motion ($n^2a^3 =$ sum of the masses),

and Ω denote the disturbing function taken with Lagrange's sign ($\Omega = -R$, if R be the disturbing function of the *Mécanique Céleste*), then the formulæ for the variations of the elements are

$$da = \frac{2}{na}\frac{d\Omega}{dg}dt,$$

$$de = \frac{1-e^2}{na^2e}\frac{d\Omega}{dg}dt - \frac{\sqrt{1-e^2}}{na^2e}\frac{d\Omega}{d\varpi}dt,$$

$$dg = -\frac{2}{na}\frac{d\Omega}{da}dt - \frac{1-e^2}{na^2e}\frac{d\Omega}{de}dt,$$

$$d\varpi = \frac{\sqrt{1-e^2}}{na^2e}\frac{d\Omega}{de}dt,$$

$$d\phi = -\frac{\cot\phi}{na^2\sqrt{1-e^2}}\frac{d\Omega}{d\sigma}dt - \frac{\operatorname{cosec}\phi}{na^2\sqrt{1-e^2}}\frac{d\Omega}{d\theta}dt,$$

$$d\sigma = \frac{\cot\phi}{na^2\sqrt{1-e^2}}\frac{d\Omega}{d\phi}dt,$$

$$d\theta = \frac{\operatorname{cosec}\phi}{na^2\sqrt{1-e^2}}\frac{d\Omega}{d\phi}dt,$$

where $\Omega = \Omega(a, e, g, \varpi, \phi, \sigma, \theta)$.

And if in these equations we write

$$h = -\frac{1}{a}\text{ (sum of the masses)} = -n^2a^2,$$

$$k = na^2\sqrt{1-e^2};$$

then attending to the equation $dn = -\frac{3}{2}\frac{n}{a}da$, we have

$$dh = n^2 a\, da,$$

$$dk = \tfrac{1}{2} na\sqrt{1-e^2}\, da - \frac{na^2 e}{\sqrt{1-e^2}},$$

and thence

$$\frac{d\Omega}{da} = n^2 a\frac{d\Omega}{dh} + \tfrac{1}{2}na\sqrt{1-e^2}\frac{d\Omega}{dk},$$

$$\frac{d\Omega}{de} = -\frac{n^2 a}{\sqrt{1-e^2}}\frac{d\Omega}{dk};$$

and the formulæ are very easily transformed into

$$dh = 2n\frac{d\Omega}{dg}dt,$$

$$dg = -2n\frac{d\Omega}{dh}dt,$$

$$dk = \frac{d\Omega}{d\varpi}dt,$$

$$d\varpi = \frac{d\Omega}{dk}dt,$$

$$d\phi = -\frac{\cot\phi}{k}\frac{d\Omega}{d\sigma}dt - \frac{\operatorname{cosec}\phi}{k}\frac{d\Omega}{d\theta}dt,$$

$$d\sigma = \frac{\cot\phi}{k}\frac{d\Omega}{d\phi}dt,$$

$$d\theta = \frac{\operatorname{cosec}\phi}{k}\frac{d\Omega}{d\phi}dt,$$

where $\Omega = \Omega\,(h, g, k, \varpi, \phi, \sigma, \theta)$.

This is the system of formulæ which will be obtained in the sequel for the variation of the elements in the problem of rotation, the new meanings of the symbols being explained *post*, Art. IV.

And if in either of the two problems, instead of the angles ϕ, σ, θ, which refer to a fixed plane of reference and origin of angles therein, we have the angles Φ, Σ, Θ, referring to a variable plane of reference and departure-point as an origin of angles in such plane, the position of these in respect to the fixed plane of reference and origin of angles therein being determined by

θ', the longitude of node,

σ', the departure of node,

ϕ', the inclination,

and if S' denote $\Theta - \sigma'$, then the system is

$$dh = 2n\frac{d\Omega}{dh}dt,$$

$$dg = -2n\frac{d\Omega}{dh}dt,$$

$$dk = \frac{d\Omega}{d\varpi}dt,$$

$$d\varpi = -\frac{d\Omega}{dk}dt,$$

$$d\Phi = -\frac{\cot\Phi}{k}\frac{d\Omega}{d\Sigma}dt - \frac{\operatorname{cosec}\phi}{k}\frac{d\Omega}{d\Theta}dt - (\cos S' d\phi' + \sin S' \sin\phi' d\theta'),$$

$$d\Sigma = \frac{\cot\Phi}{k}\frac{d\Omega}{d\Phi}dt + \operatorname{cosec}\Phi(\sin S' d\phi' - \cos S' \sin\phi' d\theta'),$$

$$d\Theta = \frac{\operatorname{cosec}\Phi}{k}\frac{d\Omega}{d\Phi}dt + \cot\Phi(\sin S' d\phi' - \cos S' \sin\phi' d\theta'),$$

where $\Omega = \Omega(h, g, k, \varpi, \Phi, \Sigma, \Theta, \phi', \sigma', \theta')$, or what is the same thing, $\Omega = \Omega(h, g, k, \varpi, \Phi, \Sigma, \Theta)$. As already remarked, it is not necessary to repeat in the present memoir the transformation to this set of formulæ.

II.

Considering, now, the problem of rotation, let the axes xyz denote axes fixed in space, and the axes $x_,y_,z_,$ denote the principal axes of the body; and if, to fix the ideas, xy is called the ecliptic and $x_,y_,$ the equator (the ecliptic xy being considered as a fixed circle of the sphere and the origin of longitudes x as a fixed point therein), then we may write

T, the longitude of node,
S, the departure of node,
F, the inclination.

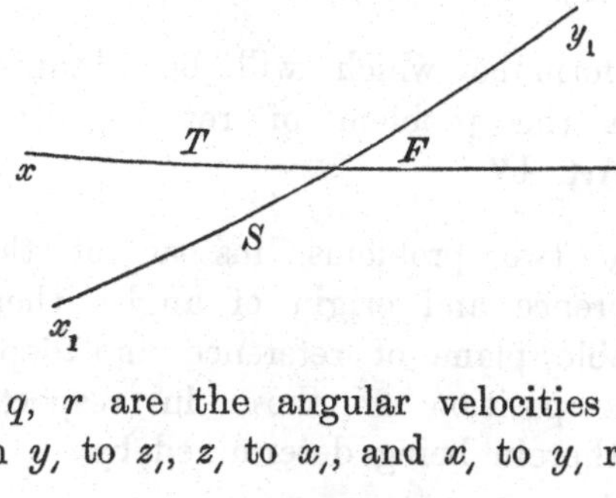

And if, as usual, p, q, r are the angular velocities round the principal axes, or axes of $x_,$, $y_,$, and $z_,$, from $y_,$ to $z_,$, $z_,$ to $x_,$, and $x_,$ to $y_,$ respectively, so that

$$pdt = -\sin S\sin F\,dT + \cos S\,dF,$$
$$qdt = \cos S\sin F\,dT + \sin S\,dF,$$
$$rdt = \cos F\,dT - dS;$$

and if A, B, C are the moments of inertia, then the *Vis Viva* function or sum of the elements of mass, each into the half square of its velocity, is

$$\mathbf{T} = \tfrac{1}{2}(Ap^2 + Bq^2 + Cr^2);$$

and if Ω be the disturbing function (taken with Lagrange's sign), then the equations of motion

$$\frac{d}{dt}\frac{d\mathbf{T}}{dT'} - \frac{d\mathbf{T}}{dT} = \frac{d\Omega}{dT}, \text{ \&c.}$$

give as usual

$$-\left[A\frac{dp}{dt}+(C-B)qr\right]\sin S\sin F+\left[B\frac{dq}{dt}+(A-C)rp\right]\cos S\sin F+\left[C\frac{dr}{dt}+(B-A)pq\right]\cos F=\frac{d\Omega}{dT},$$

$$\left[A\frac{dp}{dt}+(C-B)qr\right]\cos S \quad +\left[B\frac{dq}{dt}+(A-C)rp\right]\sin S \qquad =\frac{d\Omega}{dF},$$

$$-\left[C\frac{dr}{dt}+(B-A)pq\right] \qquad =\frac{d\Omega}{dS},$$

or, as these equations may be written,

$$A\frac{dp}{dt} + (C-B)\,qr = -\sin S\left(\operatorname{cosec} F\frac{d\Omega}{dT} + \cot F\frac{d\Omega}{dS}\right) + \cos S\frac{d\Omega}{dF},$$

$$B\frac{dq}{dt} + (A-C)\,rp = \cos S\left(\operatorname{cosec} F\frac{d\Omega}{dT} + \cot F\frac{d\Omega}{dS}\right) - \sin S\frac{d\Omega}{dF},$$

$$C\frac{dr}{dt} + (B-A)\,pq = -\frac{d\Omega}{dS};$$

which, with the equations for p, q, r, determine the motion of the body.

III.

First, to integrate the equations of motion when the disturbing forces are neglected; we have, as usual, the integral equations

$$A\,p^2 + B\,q^2 + C\,r^2 = h,$$

$$A^2p^2 + B^2q^2 + C^2r^2 = k^2;$$

where h, k are constants of integration, viz., h is the constant of *Vis Viva*, and k is the constant of the principal area.

Moreover, if the coefficients α, β, &c. are those which belong to the transformation from xyz to $x_,y_,z_,$, viz., if in the table

	$x_\prime$	$y_\prime$	$z_\prime$
x	α	β	γ
y	α'	β'	γ'
z	α''	β''	γ''

the values of the coefficients are

	$x_\prime$	$y_\prime$	$z_\prime$
x	$\cos S \cos T + \sin S \sin T \cos F$	$\sin S \cos T - \cos S \sin T \cos F$	$\sin T \sin F$
y	$\cos S \sin T - \sin S \cos T \cos F$	$\sin S \sin T + \cos S \cos T \cos F$	$-\cos T \sin F$
z	$-\sin S \sin F$	$\cos S \sin F$	$\cos F$

Then we have also, as usual, the integral equations

$$\begin{aligned} Ap\alpha + Bq\beta + Cr\gamma &= k \sin\theta \sin\phi, \\ Ap\alpha' + Bq\beta' + Cr\gamma' &= -k \cos\theta \sin\phi, \\ Ap\alpha'' + Bq\beta'' + Cr\gamma'' &= k \cos\phi\,; \end{aligned}$$

where θ, ϕ are constants of integration which determine the position of the principal plane (or invariable plane, in the undisturbed motion). Considering now a new system of axes $x_2y_2z_2$, where x_2 and z_2 are in the principal plane, and x_2 is in the first instance considered as an arbitrary fixed point therein (afterwards when the plane is treated as variable, x_2 is assumed to be a departure-point), let the position of the new set of axes in reference to the axes xyz be determined by

θ, the longitude of the node,
σ, the departure of the node,
ϕ, the inclination:

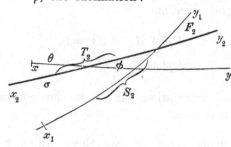

so that the relation between the coordinates xyz and $x_2y_2z_2$ is given by the table

	x_2	y_2	z_2
x	a	b	c
y	a'	b'	c'
z	a''	b''	c''

where the values of the coefficients are given by

	x_2	y_2	z_2
x	$\cos\sigma\cos\theta + \sin\sigma\sin\theta\cos\phi$	$\sin\sigma\cos\theta - \cos\sigma\sin\theta\cos\phi$	$\sin\theta\sin\phi$
y	$\cos\sigma\sin\theta - \sin\sigma\cos\theta\cos\phi$	$\sin\sigma\sin\theta + \cos\sigma\cos\theta\cos\phi$	$-\cos\theta\sin\phi$
z	$-\sin\sigma\sin\phi$	$\cos\sigma\sin\phi$	$\cos\phi$

and let the position of the axes $x_,y_,z_,$, in reference to the new axes $x_2y_2z_2$, be determined in a similar manner by

T_2, the longitude of node,

S_2, the departure of node,

F_2, the inclination;

so that we have the table

	$x_,$	$y_,$	$z_,$
x_2	α_2	β_2	γ_2
y_2	α_2'	β_2'	γ_2'
z_2	α_2''	β_2''	γ_2''

where the values of the coefficients are given by

	$x_{,}$	$y_{,}$	$z_{,}$
x_2	$\cos S_2 \cos T_2 + \sin S_2 \sin T_2 \cos F_2$	$\sin S_2 \cos T_2 - \cos S_2 \sin T_2 \cos F_2$	$\sin T_2 \sin F_2$
y_2	$\cos S_2 \sin T_2 - \sin S_2 \cos T_2 \cos F_2$	$\sin S_2 \sin T_2 + \cos S_2 \cos T_2 \cos F_2$	$-\cos T_2 \sin F_2$
z_2	$-\sin S_2 \sin F_2$	$\cos S_2 \sin F_2$	$\cos F_2$

The values of α_2, β_2, γ_2 are

$$a\alpha + a'\alpha' + a''\alpha'',$$
$$a\beta + a'\beta' + a''\beta'',$$
$$a\gamma + a'\gamma' + a''\gamma'',$$

with similar expressions for α_2', β_2', γ_2' and α_2'', β_2'', γ_2''; the last-mentioned three integrals are thus transformable into the form

$$Ap\alpha_2 + Bq\beta_2 + Cr\gamma_2 = 0,$$
$$Ap\alpha_2' + Bq\beta_2' + Cr\gamma_2' = 0,$$
$$Ap\alpha_2'' + Bq\beta_2'' + Cr\gamma_2'' = k,$$

which are, in fact, the equations which show that the plane x_2y_2 is the principal plane, or plane of maximum area.

The equations just obtained, attending to the values of α_2'', β_2'', γ_2'', give

$$Ap = -k \sin S_2 \sin F_2,$$
$$Bq = k \cos S_2 \sin F_2,$$
$$Cr = k \cos F_2;$$

and, since the expressions for p, q, r, in terms of T_2, S_2, F_2, must be similar to those in terms of T, S, F, we have

$$pdt = -\sin S_2 \sin F_2 dT_2 + \cos S_2 dF_2,$$
$$qdt = \cos S_2 \sin F_2 dT_2 + \sin S_2 dF_2,$$
$$rdt = \cos F_2 dT_2 - dS_2;$$

from which equations,

$$\sin F_2 dT_2 = (-p \sin S_2 + q \cos S_2)\, dt;$$

and substituting for $\sin S_2$, $\cos S_2$, the values $\dfrac{-Ap}{k \sin F_2}$, $\dfrac{Bq}{k \sin F_2}$, this becomes

$$k^2 \sin^2 F_2 dT_2 = k(Ap^2 + Bq^2)\, dt;$$

or, what is the same thing,

$$(k^2 - C^2r^2)\, dT_2 = k(h - Cr^2)\, dt.$$

The two equations $Ap^2+Bq^2+Cr^2=h$, $A^2p^2+B^2q^2+C^2r^2$ may be considered as determining p, q, in terms of r, and the equation $C\frac{dr}{dt}+(B-A)pq=0$, then gives

$$dt=\frac{-Cdr}{(B-A)pq},$$

whence also the equation for dT_2 becomes

$$dT_2=\frac{k(h-Cr^2)}{k^2-C^2r^2}\,\frac{-Cdr}{(B-A)pq}.$$

Instead of the time t, I consider a function $g=nt+\text{const.}$, n being for the present an arbitrary constant quantity which may be a function of the constants h and k; we have thus

$$dg=n\frac{-Cdr}{(B-A)pq},$$

and the integral equation is

$$g=n\int\frac{-Cdr}{(B-A)pq}.$$

The equation for dT_2 gives, in like manner,

$$T_2=\varpi+\int\frac{k(h-Cr^2)}{k^2-C^2r^2}\,\frac{-Cdr}{(B-A)pq};$$

where it is assumed that the integrals are each of them taken from $r=r_0$, r_0 being an arbitrary constant value, say a function of h and k. The quantity g is analogous to the mean anomaly in elliptic motion, or rather it will become so when the significations of n and r_0 are fixed; it is considered as implicitly involving a constant of integration, and, consequently, no constant of integration is added to the integral: as regards T_2, the constant of integration is ϖ, which denotes the initial value (corresponding to $r=r_0$) of the angle T_2.

IV.

Recapitulating the integral equations, we have

$$\begin{aligned}
A\,p^2+B\,q^2+C\,r^2&=h,\\
A^2p^2+B^2q^2+C^2r^2&=k^2,\\
Ap&=-k\sin S_2\sin F_2,\\
Bq&=k\cos S_2\sin F_2,\\
Cr&=k\cos F_2,\\
g&=n\int\frac{-Cdr}{(B-A)pq},
\end{aligned}$$

which equations give r, p, q, and thence S_2 and F_2 in terms of g and the constants h and k.

Moreover

$$T_2 = \varpi + \int \frac{k(h - Cr^2)}{k^2 - C^2 r^2} \frac{-Cdr}{(B-A)pq},$$

which gives T_2 in terms of g and the constants h, k, ϖ; and then T, S, F are given in terms of T_2, S_2, F_2, and the constants θ, σ, ϕ, as follows, viz., we have a spherical triangle ABC, the sides whereof a, b, c, and the opposite angles A, B, C, are respectively,

Sides	$S_2 - S$,	$T_2 - \sigma$,	$T - \theta$,
Opposite angles	ϕ ,	$180^0 - F$,	F_2 ,

as appears by the figure.

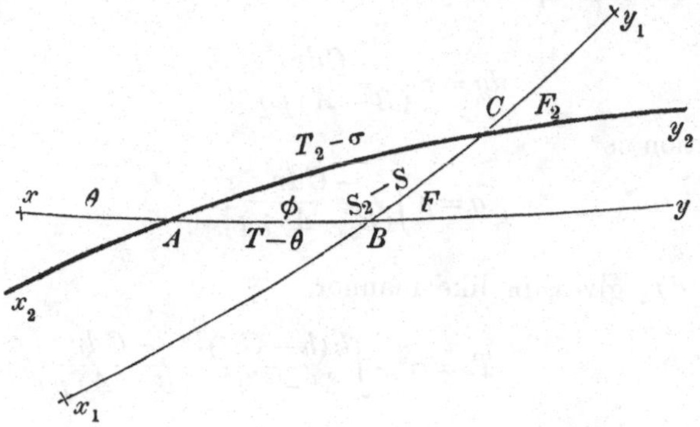

The above values of p, q, r, give

$$h - Cr^2 = \left(k^2 - C^2 r^2\right)\left(\frac{\sin^2 S_2}{A} + \frac{\cos^2 S_2}{B}\right),$$

an equation which will be useful in the sequel.

The coefficient n, and the value r_0, which is the inferior limit of the integrals, are thus far considered as arbitrary functions of h and k. As already remarked, g (which is a variable quantity, $= nt + c$) is used as an element, and the elements are h, k, g, ϖ, σ, θ, ϕ.

As to the signification of the different elements, it is proper to remark that it is not for the purposes of the present memoir necessary to completely fix the significations of the quantities n and r_0; the only conditions imposed on these quantities in the sequel are that n shall be a function of h only, and that r_0 shall be a function of h and k, satisfying an equation of the form $r_0 = f\left(\frac{h - Cr_0^2}{k^2 - C^2 r_0^2}\right)$, where f denotes an arbitrary function. The values of n and r_0 might, in accordance with these conditions, be fixed more definitively by reference to the cone rolling and sliding on the principal plane, used in the theories of Poinsot and Jacobi for the representation of the undisturbed motion, but to do this would require a further discussion of the integral equations, and it is a point which is not here entered into.

This being premised, we have (corresponding to the orbit in the theory of elliptic motion) the principal plane, with a departure point therein, the positions whereof, in respect to the fixed plane of reference, are determined by θ, the longitude of node; ϕ, the inclination; and σ, the departure of node. The precise signification of ϖ depends upon that of r_0, and the signification of r_0 being assumed to be completely determined, that of ϖ will be so likewise; ϖ is to be considered as an angle measured in the principal plane from the departure point, and determining in that plane a line (or, treating the plane as an orbit, a point) which I call the rotation pericentre, or simply the pericentre; say ϖ is the departure of the pericentre; and then g is an angle varying uniformly with the time, used for expressing in terms of the time the angle T, which determines the position, in regard to the principal plane and departure point therein, of the node of the plane of $x_{,}y_{,}$, or *equator*, upon the principal plane,—such node corresponding with the moving body in the theory of elliptic motion. We may in fact say: T, the departure of the last-mentioned node, $=\varpi$, the departure of pericentre, $+(T-\varpi)$, the rotation true anomaly of such node; $T-\varpi$ being a function of g, the rotation mean anomaly of such node. The elements are then as follows, viz.:

h, the constant of *Vis Viva*,

k, the constant of areas,

g, the rotation mean anomaly,

ϖ, the departure of rotation pericentre,

θ, the longitude of node of principal plane,

σ, the departure of ditto,

ϕ, the inclination of principal plane;

where θ, σ, ϕ determine the position of the principal plane and departure point therein, in respect to a fixed plane of reference and departure point therein; but it has been already remarked that the position of the principal plane and departure point therein might be, by the analogous quantities Θ, Σ, Φ, determined in reference to an arbitrarily varying plane of reference and departure point therein, and that the expressions for the variations of the elements would then be of the form given for this case in Art. I.

In the problem of the Rotation of the Earth, the principal plane is sensibly coincident with the plane of the equator, and on this account there is no actual physical representation of various quantities occurring in the mathematical problem. But a complete discussion of the mathematical problem does not thereby become unnecessary for the purposes of physical astronomy.

V.

If, now, the Disturbing Function is taken into account, the equations are to be integrated by the method of the variation of the elements. I use dg to denote that part of the variation of $g\,(=tdn+dc,$ if $g=nt+c)$, which depends on the variation of the constants, and in like manner for dp, dq, dr, &c. I assume, moreover, that x_2 is a departure point in the principal plane; this gives $d\sigma-\cos\phi d\theta=0$; and we see that the equations which lead to the expressions for the variations of the elements are

$$\begin{aligned}
Adp &= -\sin S\left(\operatorname{cosec} F\frac{d\Omega}{dT}+\cot F\frac{d\Omega}{dS}\right)dt+\cos S\frac{d\Omega}{dF}dt,\\
Bdq &= \cos S\left(\operatorname{cosec} F\frac{d\Omega}{dT}+\cot F\frac{d\Omega}{dS}\right)dt+\sin S\frac{d\Omega}{dF}dt,\\
Cdr &= -\frac{d\Omega}{dS}dt,\\
dT &= 0,\\
dS &= 0,\\
dF &= 0,\\
d\sigma-\cos\phi d\theta &= 0,
\end{aligned}$$

where, of course, $\Omega=\Omega\,(T,\ S,\ F)$.

The variations dh and dk are obtained from the equations

$$\begin{aligned}
\tfrac{1}{2}dh &= A\,pdp+B\,qdq+C\,rdr,\\
kdk &= A^2pdp+B^2qdq+C^2rdr,
\end{aligned}$$

expressions which will presently be resumed and reduced; S_2 and F_2 may be considered as given functions of h, k, r, and the variations dS_2 and dF_2 can be thus obtained. The expressions for T, S, F, in terms of T_2, S_2, F_2, σ, θ, ϕ, and the equations $dT=0$, $dS=0$, $dF=0$, $d\sigma-\cos\phi d\theta=0$, then lead to the expressions for $d\theta$, $d\sigma$, $d\phi$, and to an equation $-dS_2+\cos F_2dT_2=0$; dT_2 is thus also given in terms of dh, dk, dr. And this being so, the two equations

$$g=\int\frac{-Cdr}{(B-A)\,pq},$$

$$T_2=\varpi+\int\frac{k\,(h-Cr^2)}{k^2-C^2r^2}\,\frac{-Cdr}{(B-A)\,pq},$$

lead to the expressions of dg, $d\varpi$, in terms of dh, dk, dr; the form of these expressions is simplified by partially fixing, in the manner already referred to, the significations of the quantities n and r_0, considered as functions of h and k. In this manner (dh, dk, dr, being given linear functions of $\frac{d\Omega}{dT}dt$, $\frac{d\Omega}{dS}dt$, $\frac{d\Omega}{dF}dt$), we obtain

dh, dk, dg, $d\varpi$, $d\theta$, $d\sigma$, $d\phi$, all of them expressed in the same form; that is, in terms of the differential coefficients of the disturbing function Ω with respect to the coordinates T, S, F. We may then express the disturbing function Ω in terms of the elements, and transform the equations so as to obtain expressions for the variations of the elements in terms of the differential coefficients of Ω with respect to the elements.

VI.

Proceeding to carry out the foregoing plan, the equation

$$\tfrac{1}{2}dh = Apdp + Bqdq + Crdr,$$

putting for Adp, Bdq, Cdr, their values, gives

$$\tfrac{1}{2}dh = (-p\sin S + q\cos S)\left(\operatorname{cosec} F\frac{d\Omega}{dT} + \cot F\frac{d\Omega}{dS}\,dt\right)$$
$$+ (p\cos S + q\sin S)\frac{d\Omega}{dF}\,dt - r\frac{d\Omega}{dS}\,dt,$$

which may also be written

$$\tfrac{1}{2}dh = \frac{h(h - Cr^2)}{k^2 - C^2r^2}\left\{-\sin(S_2 - S)\sin F_2\frac{d\Omega}{dF}\,dt\right.$$
$$\left. + \cos(S_2 - S)\sin F_2\left(\operatorname{cosec} F\frac{d\Omega}{dT} + \cot F\frac{d\Omega}{dS}\right)dt - \cos F_2\frac{d\Omega}{dS}\,dt\right\}$$
$$+ \frac{k(B - A)pq}{k^2 - C^2r^2}\left\{\cos(S_2 - S)\sin F_2\frac{d\Omega}{dF}\,dt\right.$$
$$\left. + \sin(S_2 - S)\sin F_2\left(\operatorname{cosec} F\frac{d\Omega}{dT} + \cot F\frac{d\Omega}{dS}\right)dt\right\}$$
$$+ \frac{r(Ch - k^2)}{k^2 - C^2r^2}\frac{d\Omega}{dS}\,dt.$$

In fact, if, in this expression, we consider first the terms involving $\frac{d\Omega}{dS}dt$, the coefficient is

$$\frac{-Cr(h - Cr^2) + r(Ch - k^2)}{k^2 - C^2r^2} = \frac{-r(k^2 - C^2r^2)}{k^2 - C^2r^2} \text{ which is } = -r.$$

Next, the terms involving $\frac{d\Omega}{dF}dt$, the coefficient is

$$\left\{-(h - Cr^2)\sin(S_2 - S) + (B - A)pq\cos(S_2 - S)\right\}\frac{1}{k\sin F_2}$$
$$= \left[\left\{-(h - Cr^2)\sin S_2 + (B - A)pq\cos S_2\right\}\cos S\right.$$
$$\left. + \left\{(h - Cr^2)\cos S_2 + (B - A)pq\sin S_2\right\}\sin S\right]\frac{1}{k\sin F_2};$$

or, putting for $\sin S_2$, $\cos S_2$, their values $\dfrac{-Ap}{k\sin F_2}$, $\dfrac{Bq}{k\cos F_2}$, the coefficient is

$$\left[\left\{-(h-Cr^2)(-Ap)+(B-A)\,pq\,.\,Bq\right\}\cos S \right.$$
$$\left. +\left\{\ (h-Cr^2)\,Bq+(B-A)\,pq\,(-Ap)\right\}\sin S\right]\frac{1}{k^2\sin^2 F_2},$$

and the quantities in { } being respectively $p\,(k^2-C^2r^2)$ and $q\,(k^2-C^2r^2)$, and since $k^2-C^2r^2=k^2\sin^2 F_2$, the coefficient is $=p\cos S+q\sin S$.

In like manner, for the terms involving $\left(\operatorname{cosec} F\dfrac{d\Omega}{dT}+\cot F\dfrac{d\Omega}{dS}\right)dt$, the coefficient is

$$\left\{(h-Cr^2)\cos(S_2-S)+(B-A)\,pq\sin(S_2-S)\right\}\frac{1}{k\sin F_2}$$
$$=\left[\left\{(h-Cr^2)\cos S_2+(B-A)\,pq\sin S_2\right\}\cos S\right.$$
$$\left.+\left\{(h-Cr^2)\sin S_2-(B-A)\,pq\cos S_2\right\}\sin S\right]\frac{1}{k\sin F_2},$$

which is $=-p\sin S+q\cos S$; and the foregoing transformed expression for $\tfrac{1}{2}dh$ is thus seen to be correct.

The equation

$$kdk=A^2pdp+B^2qdq+C^2rdr,$$

substituting for Adp, Bdq, Cdr, their values, becomes

$$\begin{aligned} kdk = \ & Ap\left\{-\sin S\left(\operatorname{cosec} F\frac{d\Omega}{dT}+\cot F\frac{d\Omega}{dS}\right)dt+\cos S\frac{d\Omega}{dF}\,dt\right\} \\ &+Bq\left\{\ \cos S\left(\operatorname{cosec} F\frac{d\Omega}{dT}+\cot F\frac{d\Omega}{dS}\right)dt+\sin S\frac{d\Omega}{dF}\,dt\right\} \\ &+Cr\left\{-\frac{d\Omega}{ds}\,dt\right\}. \end{aligned}$$

But, from the equations $Ap=-k\sin S_2\sin F_2$, $Bq=k\cos S_2\sin F_2$, $Cr=k\cos F_2$, we deduce

$$\begin{aligned} -Ap\sin S+Bq\cos S &= \ \ k\cos(S_2-S)\sin F_2, \\ Ap\cos S+Bq\sin S &= -k\sin(S_2-S)\sin F_2, \\ Cr &= \ \ k\cos F_2, \end{aligned}$$

and we thence find

$$dk=-\sin(S_2-S)\sin F_2\frac{d\Omega}{dF}dt+\cos(S_2-S)\sin F_2\left(\operatorname{cosec} F\frac{d\Omega}{dT}+\cot F\frac{d\Omega}{dS}\right)dt-\cos F_2\frac{d\Omega}{dS}\,dt;$$

so that dh and dk are now determined.

VII.

The expressions for dh and dk being thus obtained, and dr being also given by the equation $Cdr = -\frac{d\Omega}{dS}dt$, we may now express dF_2 and dS_2 in terms of dh, dk, dr. In fact, the equations

$$Cr = k\cos F_2,$$

$$h - Cr^2 = (k^2 - C^2r^2)\left(\frac{\sin^2 S_2}{A} + \frac{\cos^2 S_2}{B}\right),$$

give immediately

$$k\sin F_2 dF_2 = -Cdr + \cos F_2 dk,$$

$$\sin S_2 \cos S_2 \left(\frac{1}{A} - \frac{1}{B}\right) dS_2 = \frac{C(Ch - k^2)r}{(k^2 - C^2r^2)^2}dr + \frac{\frac{1}{2}dh}{k^2 - C^2r^2} - \frac{(h - Cr^2)kdk}{(k^2 - C^2r^2)^2};$$

or, multiplying by $k^2 - C^2r^2 (= k^2\sin^2 F_2)$, and replacing $\cos S_2 \sin F_2 \sin^2 F_2$ by $-ABpq$, and dividing, the last equation becomes

$$dS_2 = \frac{-C(Ch - k^2)rdr}{(k^2 - C^2r^2)(B - A)pq} - \frac{\frac{1}{2}dh}{(B - A)pq} + \frac{(h - Cr^2)kdk}{(k^2 - C^2r^2)(B - A)pq};$$

and thence also

$$\cos F_2 dS_2 = \frac{-C^2(Ch - k^2)r^2dr}{k(k^2 - C^2r^2)(B - A)pq} - \frac{\frac{1}{2}Crdh}{k(B - A)pq} + \frac{C(h - Cr)rdk}{(k^2 - C^2r^2)(B - A)pq},$$

which is the value of dT_2, as given by the equation (not yet demonstrated)

$$dT_2 = \cos S_2 dF_2.$$

The expression for dF_2, substituting for dr and dk, their values, gives

$$\begin{aligned} k\sin F_2 dF_2 = \; & \cos F_2 \sin F_2 \cos(S_2 - S)\left(\operatorname{cosec} F\frac{d\Omega}{dT} + \cot F\frac{d\Omega}{dS}\right)dt \\ & - \cos F_2 \sin F_2 \sin(S_2 - S)\frac{d\Omega}{dF}dt \\ & - \cos^2 F_2 \frac{d\Omega}{dS}dt + \frac{d\Omega}{dS}dt; \end{aligned}$$

or, combining the last two terms, and reducing,

$$dF_2 = \frac{1}{k}\left\{\cos F_2 \cos(S_2 - S)\left(\operatorname{cosec} F\frac{d\Omega}{dT} + \cot F\frac{d\Omega}{dS}\right)dt - \cos F_2 \sin(S_2 - S)\frac{d\Omega}{dF}dt + \sin F_2 \frac{d\Omega}{dS}dt\right\}.$$

We might, in like manner, transform the expression for dS_2, but it is somewhat more simple to obtain the new form by differentiation of the equation $\tan S_2 = -\frac{Ap}{Bq}$; this, in fact, gives

$$dS_2 = \frac{AB(pdq - qdp)}{A^2p^2 + B^2q^2} = \frac{1}{k^2\sin^2 F_2}(Ap \,.\, Bdq - Bq \,.\, Adp),$$

and the value of the numerator being

$$-k\sin S_2\sin F_2\left\{\ \ \cos S\left(\operatorname{cosec} F\frac{d\Omega}{dT}+\cot F\frac{d\Omega}{dS}\right)dt+\sin S\frac{d\Omega}{dF}\,dt\right\}$$

$$-k\cos S_2\sin F_2\left\{-\sin S\left(\operatorname{cosec} F\frac{d\Omega}{dT}+\cot F\frac{d\Omega}{dS}\right)dt+\cos S\frac{d\Omega}{dF}\,dt\right\},$$

we find

$$dS_2=-\frac{1}{k\sin F_2}\left\{\sin(S_2-S)\left(\operatorname{cosec} F\frac{d\Omega}{dT}+\cot F\frac{d\Omega}{dS}\right)dt+\cos(S_2-S)\frac{d\Omega}{dF}\,dt\right\};$$

the last-mentioned expressions for dF_2 and dS_2 will be used for obtaining $d\theta$, $d\sigma$, $d\phi$.

VIII.

I form now the equations

$$\begin{aligned}-\sin S\sin F\,dT+\cos S\,dF &= pdt-\sin S_2\sin F_2\,dT_2+\cos S_2\,dF_2\\ &\quad+\alpha_2\ \ (-\sin\sigma\sin\phi\,d\theta+\cos\sigma\,d\phi)\\ &\quad+\alpha_2'\ (\ \ \cos\sigma\sin\phi\,d\theta+\sin\sigma\,d\phi)\\ &\quad+\alpha_2''(-d\sigma+\cos\phi\,d\theta\qquad\qquad),\\ \cos S\sin F\,dT+\sin S\,dF &= qdt+\cos S_2\sin F_2\,dT_2+\sin S_2\,dF_2\\ &\quad+\beta_2\ \ (-\sin\sigma\sin\phi\,d\theta+\cos\sigma\,d\phi)\\ &\quad+\beta_2'\ (\ \ \cos\sigma\sin\phi\,d\theta+\sin\sigma\,d\phi)\\ &\quad+\beta_2''(-d\sigma+\cos\phi\,d\theta\qquad\qquad),\\ -dS+\cos F\,dT &= rdt-dS_2+\cos F_2\,dT_2\\ &\quad+\gamma_2\ \ (-\sin\sigma\sin\phi\,d\theta+\cos\sigma\,d\phi)\\ &\quad+\gamma_2'\ (\ \ \cos\sigma\sin\phi\,d\theta+\sin\sigma\,d\phi)\\ &\quad+\gamma_2''(-d\sigma+\cos\phi\,d\theta\qquad\qquad);\end{aligned}$$

which will presently be useful, but which require some explanation. As the equations stand, on the left hand, dT, dS, dF, denote the entire variations of T, S, F, treating not only the constants, but also the time, as variable; but on the right hand, dT_2, dS_2, dF_2, denote the variations of T_2, S_2, F_2, depending only on the variation of the constants; if, on the left hand and the right hand respectively, dT, dS, dF, and dT_2, dS_2, dF_2, were used to denote either the entire variations of T, S, F, and T_2, S_2, F_2, or else the variations of these quantities depending only on the variations of the constants, then the terms pdt, qdt, rdt, would have to be omitted, and the equations would still be true. And it is to be noticed that, omitting the terms pdt, qdt, rdt, the equations express the relations existing between dT, dS, dF, dT_2, dS_2, dF_2, $d\theta$, $d\sigma$, $d\phi$, in virtue of the integral relations implied by the existence of the spherical triangle, the sides and angles whereof are S_2-S, $T_2-\sigma$, $T-\theta$, and ϕ, $180^\circ-F$, F_2.

For the present purpose we are to omit the terms pdt, qdt, rdt, and to write $dT = 0$, $dS = 0$, $dF = 0$. We have thus

$$\begin{aligned}
-(-\sin S_2 \sin F_2\, dT_2 + \cos S_2\, dF_2) = {} & \alpha_2\ (-\sin\sigma \sin\phi\, d\theta + \cos\sigma\, d\phi) \\
& + \alpha_2'\ (\ \ \cos\sigma \sin\phi\, d\theta + \sin\sigma\, d\phi) \\
& + \alpha_2''\,(-d\sigma + \cos\phi\, d\theta\qquad\qquad),
\end{aligned}$$

$$\begin{aligned}
-(\ \ \cos S_2 \sin F_2\, dT_2 + \sin S_2\, dF_2) = {} & \beta_2\ (-\sin\sigma \sin\phi\, d\theta + \cos\sigma\, d\phi) \\
& + \beta_2'\ (\ \ \cos\sigma \sin\phi\, d\theta + \sin\sigma\, d\phi) \\
& + \beta_2''\,(-d\sigma + \cos\phi\, d\theta\qquad\qquad),
\end{aligned}$$

$$\begin{aligned}
-(-dS_2 + \cos F_2\, dT_2) = {} & \gamma_2\ (-\sin\sigma \sin\phi\, d\theta + \cos\sigma\, d\phi) \\
& + \gamma_2'\ (\ \ \cos\sigma \sin\phi\, d\theta + \cos\sigma\, d\phi) \\
& + \gamma_2''\,(-d\sigma + \cos\phi\, d\theta\qquad\qquad).
\end{aligned}$$

Hence we have

$$\begin{aligned}
-\sin\sigma \sin\phi\, d\theta + \cos\sigma\, d\phi = {} & -\alpha_2\,(-\sin S_2 \sin F_2\, dT_2 + \cos F_2\, dS_2) \\
& -\beta_2\,(\ \ \cos S_2 \sin F_2\, dT_2 + \sin F_2\, dS_2) \\
& -\gamma_2\,(\qquad\quad \cos F_2\, dT_2 - dS_2\qquad),
\end{aligned}$$

which, attending to the values of α_2'', β_2'', γ_2'', may be written,

$$\begin{aligned}
-\sin\sigma \sin\phi\, d\theta + \cos\sigma\, d\phi = {} & -\alpha_2\,(\alpha_2''\, dT_2 + \cos S_2\, dF_2) \\
& -\beta_2\,(\beta_2''\, dT_2 + \sin S_2\, dF_2) \\
& -\gamma_2\,(\gamma_2''\, dT_2 - dS_2\qquad) \\
= {} & -(\alpha_2 \cos S_2 + \beta_2 \sin S_2)\, dF_2 + \gamma_2\, dS_2,
\end{aligned}$$

since the coefficient of dT_2 vanishes; and, in like manner,

$$\begin{aligned}
\cos\sigma \sin\phi\, d\theta + \sin\sigma\, d\phi = {} & -\alpha_2'\,(\alpha_2''\, dT_2 + \cos S_2\, dF_2) \\
& -\beta_2'\,(\beta_2''\, dT_2 + \sin S_2\, dF_2) \\
& -\gamma_2'\,(\gamma_2''\, dT_2 - dS_2\qquad) \\
= {} & -(\alpha_2' \cos S_2 + \beta_2' \sin S_2)\, dF_2 + \gamma_2'\, dS_2;
\end{aligned}$$

and so also

$$\begin{aligned}
-d\sigma + \cos\phi\, d\theta = {} & -\alpha_2''\,(\alpha_2''\, dT_2 + \cos S_2\, dF_2) \\
& -\beta_2''\,(\beta_2''\, dT_2 + \sin S_2\, dF_2) \\
& -\gamma_2''\,(\gamma_2''\, dT_2 - dS_2\qquad) \\
= {} & -dT_2 - (\alpha_2'' \cos S_2 + \beta_2'' \sin S_2)\, dF_2 + \gamma_2''\, dS_2.
\end{aligned}$$

But we have

$$\begin{aligned}
\alpha_2\ \cos S_2 + \beta_2\ \sin S_2 &= \cos T_2, & \gamma_2 &= \sin T_2 \sin F_2, \\
\alpha_2'\ \cos S_2 + \beta_2'\ \sin S_2 &= \sin T_2, & \gamma_2' &= -\cos T_2 \sin F_2, \\
\alpha_2'' \cos S_2 + \beta_2'' \sin S_2 &= 0\ , & \gamma_2'' &= \cos F_2,
\end{aligned}$$

and the foregoing equations thus become

$$-\sin\sigma\sin\phi\, d\theta+\cos\sigma\, d\phi=-\cos T_2\, dF_2+\sin T_2\sin F_2\, dS_2,$$
$$\cos\sigma\sin\phi\, d\theta+\sin\sigma\, d\phi=-\sin T_2\, dF_2-\cos T_2\sin F_2\, dS_2,$$
$$-d\sigma+\cos\phi\, d\theta\qquad=-dT_2\qquad+\cos F_2 dS_2.$$

The last equation, making x_2 a departure point, or putting $-d\sigma+\cos\phi\, d\theta=0$, gives

$$dT_2=\cos F_2 dS_2,$$

an equation above referred to. The other two equations give

$$d\phi=-\cos(T_2-\sigma)\, dF_2+\sin(T_2-\sigma)\sin F_2\, dS_2,$$
$$\sin\phi\, d\theta=-\sin(T_2-\sigma)\, dF_2-\cos(T_2-\sigma)\sin F_2\, dS_2,$$

which, with the equation,

$$d\sigma=\cos\phi\, d\theta,$$

give $d\theta$, $d\sigma$, $d\phi$, in terms of dS_2 and dF_2.

IX.

Proceeding to substitute for dS_2, dF_2, their values, we find

$$\begin{aligned}kd\phi= & -\cos(T_2-\sigma)\left\{\cos F_2\cos(S_2-S)\left(\operatorname{cosec} F\frac{d\Omega}{dT}+\cot F\frac{d\Omega}{dS}\right)dt-\cos F_2\cos(S_2-S)\frac{d\Omega}{dF}dt+\sin F_2\frac{d\Omega}{dS}dt\right\}\\ & -\sin(T_2-\sigma)\left\{\sin(S_2-S)\left(\operatorname{cosec} F\frac{d\Omega}{dT}+\cot F\frac{d\Omega}{dS}\right)dt+\cos(S_2-S)\frac{d\Omega}{dF}dt\right\},\end{aligned}$$

or reducing

$$\begin{aligned}kd\phi= & -\Big[\sin(T_2-\sigma)\sin(S_2-S)+\cos(T_2-\sigma)\cos(S_2-S)\cos F_2\Big]\left(\operatorname{cosec} F\frac{d\Omega}{dT}+\cot F\frac{d\Omega}{dS}\right)dt\\ & -\Big[\sin(T_2-\sigma)\cos(S_2-S)-\cos(T_2-\sigma)\sin(S_2-S)\cos F_2\Big]\frac{d\Omega}{dF}dt\\ & -\Big[\cos(T_2-\sigma)\sin F_2\Big]\frac{d\Omega}{dF}dt;\end{aligned}$$

and, in like manner,

$$\begin{aligned}k\sin\phi\, d\theta= & -\sin(T_2-\sigma)\left\{\cos F_2\cos(S_2-S)\left(\operatorname{cosec} F\frac{d\Omega}{dT}+\cot F\frac{d\Omega}{dS}\right)dt-\cot F_2\cos(S_2-S)\frac{d\Omega}{dF}dt+\sin F_2\frac{d\Omega}{dS}dt\right\}\\ & +\cos(T_2-\sigma)\left\{\sin(S_2-S)\left(\operatorname{cosec} F\frac{d\Omega}{dT}+\cot F\frac{d\Omega}{dS}\right)dt+\cos(S_2-S)\frac{d\Omega}{dF}dt\right\}\end{aligned}$$

or reducing

$$k \sin\phi\, d\theta = \left[\cos(T_2-\sigma)\sin(S_2-S) - \sin(T_2-\sigma)\cos(S_2-S)\cos F_2\right]\left(\operatorname{cosec} F\frac{d\Omega}{dT} + \cot F\frac{d\Omega}{dS}\right) dt$$

$$+\left[\cos(T_2-\sigma)\cos(S_2-S) + \sin(T_2-\sigma)\sin(S_2-S)\cos F_2\right]\frac{d\Omega}{dF}\, dt$$

$$+\left[\qquad -\sin(T_2-\sigma)\sin F_2 \qquad\right]\frac{d\Omega}{dS}\, dt.$$

These expressions for $d\phi$ and $d\theta$ are in a form which is convenient for some purposes, but they may be further reduced by means of the spherical triangle. In fact, in the expression for $d\phi$, the three coefficients in [ ] are respectively,

$$\begin{aligned}
\sin b \sin a + \cos b \cos a \cos C &= \sin B \sin A - \cos B \cos A \cos c \\
&= \sin F \sin\phi + \cos F \cos\phi \cos(T-\theta), \\
\sin b \cos a - \cos b \sin a \cos C &= \cos A \sin c \\
&= \cos\phi \sin(T-\theta), \\
\cos b \sin C &= \cos B \sin A + \sin B \cos A \cos c \\
&= -\cos F \sin\phi + \sin F \cos\phi \cos(T-\theta);
\end{aligned}$$

and we have thus

$$k d\phi = -\left(\sin F \sin\phi + \cos F \cos\phi \cos(T-\theta)\right)\left(\operatorname{cosec} F\frac{d\Omega}{dT} + \cot F\frac{d\Omega}{dS}\right) dt$$

$$-\cos\phi\sin(T-\theta)\,\frac{d\Omega}{dF}\, dt$$

$$-\left(-\cos F\sin\phi + \sin F\cos\phi\cos(T-\theta)\right)\frac{d\Omega}{dS}\, dt;$$

and the right-hand side is

$$-\left(\sin F \sin\phi + \cos F \cos\phi \cos(T-\theta)\right)\operatorname{cosec} F\frac{d\Omega}{dT}\, dt$$

$$-\cos\phi\sin(T-\theta)\,\frac{d\Omega}{dF}\, dt$$

$$+\operatorname{cosec} F\cos\phi\cos(T-\theta)\,\frac{d\Omega}{dS}\, dt,$$

or we have finally

$$k d\phi = -\cos\phi\cos(T-\theta)\left(\cot F\frac{d\Omega}{dT} + \operatorname{cosec} F\frac{d\Omega}{dS}\right) dt$$

$$-\sin\phi\,\frac{d\Omega}{dT}\, dt$$

$$-\cos\phi\sin(T-\theta)\,\frac{d\Omega}{dF}\, dt.$$

In like manner, in the original expression for $k \sin \phi \, d\theta$, the three coefficients in [ ] are

$$\begin{aligned}
\cos b \sin a - \sin b \cos a \cos C &= \cos B \sin c \\
&= -\cos F \sin (T-\theta), \\
\cos b \cos a + \sin b \sin a \cos C &= \cos c \\
&= \cos (T-\theta), \\
-\sin b \sin C \qquad &= -\sin B \sin c \\
&= -\sin T \sin (T-\theta),
\end{aligned}$$

and thence

$$\begin{aligned}
k \sin \phi \, d\theta = &- \cos F \sin (T-\theta) \left(\operatorname{cosec} F \frac{d\Omega}{dT} + \cot F \frac{d\Omega}{dS} \right) dt \\
&+ \qquad \cos (T-\theta) \frac{d\Omega}{dF} \, dt \\
&- \sin F \sin (T-\theta) \frac{d\Omega}{dS} \, dt,
\end{aligned}$$

and the right-hand side is

$$\begin{aligned}
&- \cot F \sin (T-\theta) \frac{d\Omega}{dT} \, dt \\
&+ \qquad \cos (T-\theta) \frac{d\Omega}{dF} \, dt \\
&- \operatorname{cosec} F \sin (T-\theta) \frac{d\Omega}{dS} \, dt,
\end{aligned}$$

or we have finally

$$\begin{aligned}
k \sin \phi \, d\theta = &- \sin (T-\theta) \left(\cot F \frac{d\Omega}{dT} + \operatorname{cosec} F \frac{d\Omega}{dS} \right) dt \\
&+ \cos (T-\theta) \frac{d\Omega}{dF} \, dt.
\end{aligned}$$

The expression for $d\sigma$ can be obtained from either of those for $d\theta$, by the equation $d\sigma = \cos \phi \, d\theta$, and we have thus the values of $d\sigma$, $d\theta$, $d\phi$.

X.

It remains to find the expressions for dg and $d\varpi$. We have

$$g = \quad n \int \frac{-C dr}{(B-A) pq},$$

$$T_2 = \varpi + \int \frac{k(h - Cr^2)}{k^2 - C^2 r^2} \frac{-C dr}{(B-A) pq},$$

where p, q, are considered as functions of r, h, k, and where, besides, n and the inferior limit r_0 of the integrals are functions of h and k. We may write

$$dg = \frac{-nC}{(B-A)pq}dr + \mathfrak{A}dh + \mathfrak{B}dk,$$

$$dT_2 = d\varpi + \frac{k(h-Cr^2)}{k^2-C^2r^2}\frac{-Cdr}{(B-A)pq}dr + \mathfrak{C}dh + \mathfrak{D}dk,$$

where $\mathfrak{A}$, $\mathfrak{B}$, $\mathfrak{C}$, $\mathfrak{D}$, are functions which contain integrals with respect to r, and there is not any algebraical relation between them, except the equation $\mathfrak{B} = -2n\mathfrak{C}$, which will be obtained presently. I retain, therefore, $\mathfrak{A}$, $\mathfrak{B}$, $\mathfrak{C}$, $\mathfrak{D}$, in the formulæ. The first equation gives the value of dg: from the second equation we have

$$d\varpi = dT_2 + \frac{Ck(h-Cr^2)}{(k^2-C^2r^2)(B-A)pq}dr - \mathfrak{C}dh - \mathfrak{D}dk,$$

or substituting for dT_2 its value $\cos F_2 dS_2$, the expression for which has been obtained above [p. 489], we have

$$d\varpi = \frac{-C^2(Ch-k^2)r^2dr}{k(k^2-C^2r^2)(B-A)pq} - \frac{\tfrac{1}{2}Crdh}{k(B-A)pq} + \frac{C(h-Cr^2)rdk}{(k^2-C^2r^2)(B-A)pq}$$
$$+\frac{Ck^2(h-Cr^2)dr}{k(k^2-C^2r^2)(B-A)pq} - \mathfrak{C}dh - \mathfrak{D}dk,$$

or reducing, this is

$$d\varpi = \frac{Chdr}{k(B-A)pq} + \left[\frac{-\tfrac{1}{2}Cr}{k(B-A)pq} - \mathfrak{C}\right]dh + \left[\frac{C(h-Cr^2)r}{(k^2-C^2r^2)(B-A)pq} - \mathfrak{D}\right]dk,$$

which might be retained in this form. I obtain, however, a different form as follows, viz., we have

$$d\varpi = \cos F_2\, dS_2 + \frac{Ck(h-Cr^2)dr}{(k^2-C^2r^2)(B-A)pq} - \mathfrak{C}dh - \mathfrak{D}dk;$$

or using the other form of dS_2 [p. 490], and substituting also for dr its value from the equation $Cdr = -\frac{d\Omega}{dS}dt$, we have

$$d\varpi = -\frac{1}{k}\cot F_2\left\{\sin(S_2-S)\left(\operatorname{cosec} F\frac{d\Omega}{dT} + \cot F\frac{d\Omega}{dS}\right)dt + \cos(S_2-S)\frac{d\Omega}{dF}dt\right\}$$
$$-\frac{k(h-Cr^2)}{(k^2-C^2r^2)(B-A)pq}\frac{d\Omega}{dS}dt - \mathfrak{C}dh - \mathfrak{D}dk.$$

We have now to prove the before mentioned equation $BC = -2n\mathfrak{C}$. We have

$$\mathfrak{B} = \frac{d}{dk}n\int\frac{-Cdr}{(B-A)pq} = n\int dr\frac{d}{dk}\frac{-C}{(B-A)pq} + \frac{dn}{dk}\int\frac{-Cdr}{(B-A)pq} + \frac{nC}{(B-A)p_0q_0}\frac{dr_0}{dk},$$

$$\mathfrak{C} = \frac{d}{dh}\int\frac{k(h-Cr^2)}{k^2-C^2r^2}\frac{-Cdr}{(B-A)pq} = \int dr\frac{d}{dh}\frac{-Ck(h-Cr^2)}{(k^2-C^2r^2)(B-A)pq} + \frac{Ck(h-Cr_0^2)}{(k^2-C^2r_0^2)(B-A)p_0q_0}\frac{dr_0}{dh},$$

where p_0, q_0, are the values of p and q corresponding to $r=r_0$. If we assume that n is a function of h only, the term multiplied by $\frac{dn}{dk}$ will disappear, and by properly determining r_0 as a function of h and k, we can, as regards the terms which contain r_0, satisfy the equation $\mathfrak{B}=-2n\mathfrak{C}$; the condition for this is

$$\frac{dr_0}{dk}=\frac{-2k(h-Cr_0^2)}{k^2-C^2r_0^2}\frac{dr_0}{dh},$$

which will be satisfied if r_0 is determined as a function of h and k by an equation of the form

$$r_0=f\left(\frac{h-Cr_0^2}{k^2-C^2r_0^2}\right),$$

where f denotes an arbitrary function.

It remains to show that, as regards the terms involving integrals, we have the same relation $\mathfrak{B}=-2n\mathfrak{C}$, and this will be the case if

$$\frac{d}{dk}\frac{-C}{(B-A)pq}=-2\frac{d}{dh}\frac{-Ck(h-Cr^2)}{(k^2-C^2r^2)(B-A)pq},$$

or, what is the same thing, if

$$\frac{d}{dk}\frac{1}{pq}=-2\frac{d}{dh}\frac{k(h-Cr^2)}{(k^2-C^2r^2)pq}=\frac{-2k}{(k^2-C^2r^2)}\frac{d}{dh}\frac{h-Cr^2}{pq},$$

in which equation p, q, are considered as functions of h, k, given by the equations

$$Ap^2+Bq^2+Cr^2=h,$$
$$A^2p^2+B^2q^2+C^2r^2=k^2.$$

We find without difficulty

$$\frac{d.pq}{dh}=\frac{-\frac{1}{2}(A^2p^2-B^2q^2)}{AB(B-A)pq},\qquad \frac{d.pq}{dk}=\frac{k(Ap^2-Bq^2)}{AB(B-A)pq},$$

and thence

$$\frac{d}{dh}\frac{h-Cr^2}{pq}=\frac{1}{pq}+\frac{\frac{1}{2}(h-Cr^2)(A^2p^2-B^2q^2)}{AB(B-A)p^3q^3}$$

$$=\frac{1}{AB(B-A)p^3q^3}[AB(B-A)p^2q^2+\tfrac{1}{2}(Ap^2+Bq^2)(A^2p^2-B^2q^2)]$$

$$=\frac{1}{AB(B-A)p^3q^3}\tfrac{1}{2}(Ap^2-Bq^2)(A^2p^2)+(B^2q^2)$$

$$=\frac{\frac{1}{2}(k^2-C^2r^2)(Ap^2-Bq^2)}{AB(B-A)p^3q^3},$$

and the right-hand side of the equation in question is therefore

$$= \frac{-k\,(Ap^2 - Bq^2)}{AB\,(B-A)\,p^3q^3},$$

which is obviously also the value of the left-hand side. Hence, under the assumed relations $\left(n \text{ a function of } h \text{ only, and } r_0 = f\left(\frac{h - Cr_0^2}{k^2 - C^2r_0^2}\right)\right)$, we have the above-mentioned equation $\mathfrak{B} = -2n\mathfrak{C}$.

XI.

It will be convenient to recapitulate the various formulæ for the variations, as follows; we have

$$\tfrac{1}{2}dh = (-p\sin S + q\cos S)\left(\operatorname{cosec} F\frac{d\Omega}{dT} + \cot F\frac{d\Omega}{dS}\right)dt$$

$$+ (\ \ p\cos S + q\sin S)\frac{d\Omega}{dF}dt - r\frac{d\Omega}{dS}dt,$$

and

$$\tfrac{1}{2}dh = \frac{k\,(h - Cr^2)}{k^2 - C^2r^2}\left\{\cos(S_2 - S)\sin F_2\left(\operatorname{cosec} F\frac{d\Omega}{dT} + \cot F\frac{d\Omega}{dS}\right)dt - \sin(S_2 - S)\sin F_2\frac{d\Omega}{dF}dt - \cos F_2\frac{d\Omega}{dS}dt\right\}$$

$$+ \frac{k\,(B-A)\,pq}{k^2 - C^2r^2}\left\{\sin(S_2 - S)\sin F_2\left(\operatorname{cosec} F\frac{d\Omega}{dT} + \cot F\frac{d\Omega}{dS}\right)dt + \cos(S_2 - S)\sin F_2\frac{d\Omega}{dF}dt\right\}$$

$$+ \frac{r\,(Ch - k^2)}{k^2 - C^2r^2}\frac{d\Omega}{dS}dt;$$

$$kdk = (-Ap\sin S + Bq\cos S)\left(\operatorname{cosec} F\frac{d\Omega}{dT} + \cot F\frac{d\Omega}{dS}\right)dt$$

$$+ (\ \ Ap\cos S + Bq\sin S)\frac{d\Omega}{dF}dt - Cr\frac{d\Omega}{dS}dt,$$

and

$$dk = \cos(S_2 - S)\sin F_2\left(\operatorname{cosec} F\frac{d\Omega}{dT} + \cot F\frac{d\Omega}{dS}\right)dt - \sin(S_2 - S)\sin F_2\frac{d\Omega}{dF}dt - \cos S_2\frac{d\Omega}{dS}dt;$$

$$kd\phi = -\cos\phi\cos(T - \theta)\left(\cot F\frac{d\Omega}{dT} + \operatorname{cosec} F\frac{d\Omega}{dS}\right)dt$$

$$-\sin\phi\frac{d\Omega}{dT}dt - \cos\phi\sin(T-\theta)\frac{d\Omega}{dF}dt,$$

and

$$kd\phi = -\left\{\sin(T_2-\sigma)\sin(S_2-S)+\cos(T_2-\sigma)\cos(S_2-S)\cos F_2\right\}\left(\operatorname{cosec} F\frac{d\Omega}{dT}+\cot F\frac{d\Omega}{dS}\right)dt$$

$$-\left\{\sin(T_2-\sigma)\cos(S_2-S)-\cos(T_2-\sigma)\sin(S_2-S)\cos F_2\right\}\frac{d\Omega}{dF}dt$$

$$-\left\{\qquad -\cos(T_2-\sigma)\sin F_2 \qquad\right\}\frac{d\Omega}{dS}dt;$$

$$k\sin\phi d\theta = -\sin(T-\theta)\left(\cot F\frac{d\Omega}{dT}+\operatorname{cosec} F\frac{d\Omega}{dS}\right)dt$$

$$+\cos(T-\theta)\frac{d\Omega}{dF}dt,$$

and

$$k\sin\phi d\theta = \left\{\cos(T_2-\sigma)\sin(S_2-S)-\sin(T_2-\sigma)\cos(S_2-S)\cos F_2\right\}\left(\operatorname{cosec} F\frac{d\Omega}{dT}+\cot F\frac{d\Omega}{dS}\right)dt$$

$$+\left\{\cos(T_2-\sigma)\cos(S_2-S)+\sin(T_2-\sigma)\sin(S_2-S)\cos F_2\right\}\frac{d\Omega}{dF}dt$$

$$+\left\{\qquad -\sin(T_2-\sigma)\sin F_2 \qquad\right\}\frac{d\Omega}{dS}dt;$$

$$d\sigma = \cos\phi d\theta;$$

$$dg = \frac{-nC}{(B-A)pq}dr \quad + \mathfrak{A}dh + \mathfrak{B}dk,$$

and

$$dg = \frac{n}{(B-A)pq}\frac{d\Omega}{dS}dt + \mathfrak{A}dh + \mathfrak{B}dk;$$

$$d\varpi = \frac{Chdr}{k(B-A)pq} + \left\{\frac{-\tfrac{1}{2}Cr}{k(B-A)pq} - \mathfrak{C}\right\}dh + \left\{\frac{C(h-Cr^2)r}{(k^2-C^2r^2)(B-A)pq} - \mathfrak{D}\right\}dk,$$

and

$$d\varpi = -\frac{1}{k}\cot F_2\left\{\sin(S_2-S)\left(\operatorname{cosec} F\frac{d\Omega}{dT}+\cot F\frac{d\Omega}{dS}\right)dt + \cos(S_2-S)\frac{d\Omega}{dF}dt\right\}$$

$$+\frac{B\sin^2 S_2 + A\cos^2 S_2}{k(B-A)\sin^2 F_2\sin S_2\cos S_2}\frac{d\Omega}{dS}dt - \mathfrak{C}dh - \mathfrak{D}dk;$$

where it will be remembered that $\mathfrak{B} = -2n\mathfrak{C}$. The different forms for the variations of the same element are, or may be, each of them useful. The first expressions for dg and $d\varpi$ respectively are to be considered as giving these variations in terms of dr, dh, dk; and the second expressions are those obtained by substituting for these quantities their values, but in the terms multiplied by the integral expressions $\mathfrak{A}$, $\mathfrak{B}$, $\mathfrak{C}$, $\mathfrak{D}$, which, on account of these multipliers, do not unite with the other terms, dh, dk are retained as standing for their values. The following equations may be added,

$$dT_2 = d\varpi - \frac{Ck(h - Cr^2)}{(k^2 - C^2r^2)(B-A)pq}\, dr + \mathfrak{C}dh + \mathfrak{D}dk,$$

$$dF_2 = \frac{-Ck\sin F_2}{k^2 - C^2r^2}\, dr + \frac{k\sin F_2 \cos F_2}{k^2 - C^2r^2}\, dk,$$

$$dS_2 = \frac{-C(Ch - k^2)\,r}{(k^2 - C^2r^2)(B-A)pq}\, dr - \frac{\frac{1}{2}dh}{(B-A)pq} + \frac{(h - Cr^2)\,kdk}{(k^2 - C^2r^2)(B-A)pq},$$

which give dF_2, dS_2, in terms of dr, dh, dk; and I call to mind, also, the equations

$$Ap = -k\sin S_2 \sin F_2, \qquad Bq = k\cos S_2 \sin F_2, \qquad Cr = k\cos F_2.$$

XII.

To find the differential coefficients of Ω with respect to the elements, I proceed as follows; considering the function first under the form $\Omega = \Omega(T, S, F)$, the total differential is

$$\frac{d\Omega}{dT}dT + \frac{d\Omega}{dS}dS + \frac{d\Omega}{dF}dF,$$

which must be equal to the total differential of Ω considered under the form $\Omega = \Omega\ (h, k, g, \varpi, \sigma, \theta, \phi)$; that is, it must be

$$= \frac{d\Omega}{dh}dh + \frac{d\Omega}{dk}dk + \frac{d\Omega}{dg}(ndt + dg) + \frac{d\Omega}{d\varpi}d\varpi + \frac{d\Omega}{d\theta}d\theta + \frac{d\Omega}{d\sigma}d\sigma + \frac{d\Omega}{d\phi}d\phi\,;$$

where, as elsewhere dg denotes only the part of the variation of g, which depends on the variation of the constants; so that the total variation of g is represented by $ndt + dg$. The value of $d\Omega$,

$$= \frac{d\Omega}{dT}dT + \frac{d\Omega}{dS}dS + \frac{d\Omega}{dF}dF,$$

is to be obtained in a form comparable with the last-mentioned expression, by means of the formulæ *suprà*, Art. VIII., which, for shortness, I represent as follows:

$$\begin{aligned} -\sin S\sin F\,dT + \cos S\,dF &= dP,\\ \cos S\sin F\,dT + \sin S\,dF &= dQ,\\ -dS + \cos F\,dT \qquad &= dR\,; \end{aligned}$$

these equations give

$$\begin{aligned} dF &= \cos S\,dP + \sin S\,dQ\,,\\ dT &= \operatorname{cosec} F\,(-\sin S\,dP + \cos S\,dQ),\\ dS &= \cot F\,(-\sin S\,dP + \cos S\,dQ) - dR\,; \end{aligned}$$

and we then have

$$d\Omega = (\cos S\, dP + \sin S\, dQ)\frac{d\Omega}{dS}$$

$$+(-\sin S\, dP + \cos S\, dQ)\left(\operatorname{cosec} F\frac{d\Omega}{dT} + \cot F\frac{d\Omega}{dS}\right)$$

$$-dR\frac{d\Omega}{dS};$$

and substituting for dP, dQ, dR, their values, the resulting expression may, for shortness, be represented by $d\Omega = d_1\Omega + d_2\Omega + d_3\Omega$, where

$$d_1\Omega = (\ p\cos S + q\sin S)\frac{d\Omega}{dF}dt$$

$$+(-p\sin S + q\cos S)\left(\operatorname{cosec} F\frac{d\Omega}{dT} + \cot F\frac{d\Omega}{dS}\right)dt$$

$$-r\frac{d\Omega}{dS}dt.$$

$$d_2\Omega = \left\{\cos(S_2 - S)\sin F_2\left(\operatorname{cosec} F\frac{d\Omega}{dT} + \cot F\frac{d\Omega}{dS}\right) - \sin(S_2 - S)\sin F_2\frac{d\Omega}{dF} - \cos F_2\frac{d\Omega}{dS}\right\}dT_2$$

$$+\left\{\sin(S_2 - S)\left(\operatorname{cosec} F\frac{d\Omega}{dT} + \cot F\frac{d\Omega}{dS}\right) + \cos(S_2 - S)\frac{d\Omega}{dF}\right\}dF_2$$

$$+\ \frac{d\Omega}{dS}dS_2,$$

which, for shortness, I represent by $d_2\Omega = XdT_2 + YdF_2 + ZdS_2$; and

$$d_3\Omega = \left\{\begin{array}{l}(\ \alpha_2\cos S + \beta_2\sin S)(-\sin\sigma\sin\phi\, d\theta + \cos\sigma\, d\phi)\\ +(\ \alpha_2'\cos S + \beta_2'\sin S)(\ \cos\sigma\sin\phi\, d\theta + \sin\sigma\, d\phi)\\ +(\ \alpha_2''\cos S + \beta_2''\sin S)(\quad -d\sigma + \cos\phi\, d\theta\quad)\end{array}\right\}\frac{d\Omega}{dF}$$

$$+\left\{\begin{array}{l}(-\alpha_2\sin S + \beta_2\cos S)(-\sin\sigma\sin\phi\, d\theta + \cos\sigma\, d\phi)\\ +(-\alpha_2'\sin S + \beta_2'\cos S)(\ \cos\sigma\sin\phi\, d\theta + \sin\sigma\, d\phi)\\ (-\alpha_2''\sin S + \beta_2''\cos S)(\quad -d\sigma + \cos\phi\, d\theta\quad)\end{array}\right\}\left(\operatorname{cosec} F\frac{d\Omega}{dT} + \cot F\frac{d\Omega}{dS}\right)$$

$$-\left\{\begin{array}{l}\gamma_2(-\sin\sigma\sin\phi\, d\theta + \cos\sigma\, d\phi)\\ +\gamma_2'(\ \cos\sigma\sin\phi\, d\theta + \sin\sigma\, d\phi)\\ +\gamma_2''(\quad -d\sigma + \cos\phi\, d\theta\quad)\end{array}\right\}\frac{d\Omega}{dS},$$

or substituting for α_2, &c., their values, and reducing, $d_3\Omega =$

$$\left\{\begin{matrix}([\cos(S_2-S)\sin(T_2-\sigma)-\sin(S_2-S)\cos(T_2-\sigma)\cos F_2]\sin\phi-\sin(S_2-S)\sin F_2\cos\phi)\,d\theta \\ +(\cos(S_2-S)\cos(T_2-\sigma)+\sin(S_2-S)\sin(T_2-\sigma)\cos F_2)\,d\phi+\sin(S_2-S)\sin F_2 d\sigma\end{matrix}\right\}\frac{d\Omega}{dF}$$

$$+\left\{\begin{matrix}([\sin(S_2-S)\sin(T_2-\sigma)+\cos(S_2-S)\cos(T_2-\sigma)\cos F_2]\sin\phi+\cos(S_2-S)\sin F_2\cos\phi)\,d\theta \\ +(\sin(S_2-S)\cos(T_2-\sigma)-\cos(S_2-S)\sin(T_2-\sigma)\cos F_2)\,d\phi-\cos(S_2-S)\cos F_2 d\sigma\end{matrix}\right\}\left(\operatorname{cosec} F\frac{d\Omega}{dT}+\cot F\frac{d\Omega}{dS}\right)$$

$$-\left\{(-\cos(T_2-\sigma)\sin F_2\sin\phi+\cos F_2\cos\phi)\,d\theta+\sin(T_2-\sigma)\sin F_2 d\phi-\cos F_2 d\sigma\right\}\frac{d\Omega}{dS}.$$

Hence, comparing $d_1\Omega+d_2\Omega+d_3\Omega$ with the other expression for $d\Omega$, and observing that dT_2, dS_2, dF_2, do not involve $d\theta$, $d\sigma$, $d\phi$, we have

$$d_1\Omega=\frac{d\Omega}{dg}\,ndt,$$

$$d_2\Omega=\frac{d\Omega}{dh}\,dh+\frac{d\Omega}{dk}\,dk+\frac{d\Omega}{dg}\,dg+\frac{d\Omega}{d\varpi}\,d\varpi,$$

$$d_3\Omega=\frac{d\Omega}{d\theta}\,d\theta+\frac{d\Omega}{d\sigma}\,d\sigma+\frac{d\Omega}{d\phi}\,d\phi.$$

XIII.

The first equation gives

$$n\frac{d\Omega}{dg}=(-p\sin S+q\cos S)\left(\operatorname{cosec} F\frac{d\Omega}{dT}+\cot F\frac{d\Omega}{dS}\right)$$
$$+(p\cos S+q\sin S)\frac{d\Omega}{dF}-r\frac{d\Omega}{dS},$$

and, comparing this with the first form of $\tfrac{1}{2}dh$, we have

$$dh=2n\frac{d\Omega}{dg}\,dt.$$

The third equation gives

$$\frac{d\Omega}{d\theta}=\left\{[\sin(S_2-S)\sin(T_2-\sigma)+\cos(S_2-S)\cos(T_2-S)\cos F_2]\sin\phi+\cos(S_2-S)\sin F_2\cos\phi\right\}\left(\operatorname{cosec} F\frac{d\Omega}{dT}+\cot F\frac{d\Omega}{dS}\right)$$
$$+\left\{[\cos(S_2-S)\sin(T_2-\sigma)-\sin(S_2-S)\cos(T_2-\sigma)\cos F_2]\sin\phi-\sin(S_2-S)\sin F_2\cos\phi\right\}\frac{d\Omega}{dF}$$
$$+\left\{\cos(T_2-\sigma)\sin F_2\sin\phi-\cos F_2\cos\phi\right\}\frac{d\Omega}{dS},$$

$$\frac{d\Omega}{d\sigma} = -\cos(S_2 - S)\cos F_2\left(\operatorname{cosec} F\frac{d\Omega}{dT} - \cot F\frac{d\Omega}{dS}\right)$$
$$+\sin(S_2 - S)\sin F_2\frac{d\Omega}{dF} + \cos F_2\frac{d\Omega}{dS},$$

$$\frac{d\Omega}{d\phi} = \left\{\sin(S_2 - S)\cos(T_2 - \sigma) - \cos(S_2 - S)\sin(T_2 - \sigma)\cos F_2\right\}\left(\operatorname{cosec} F\frac{d\Omega}{dT} + \cot F\frac{d\Omega}{dS}\right)$$
$$+\left\{\cos(S_2 - S)\cos(T_2 - \sigma) + \sin(S_2 - S)\sin(T_2 - \sigma)\cos F_2\right\}\frac{d\Omega}{dF}$$
$$-\sin(T_2 - \sigma)\sin F_2\frac{d\Omega}{dS},$$

and thence

$$\cot\phi\frac{d\Omega}{d\sigma} + \operatorname{cosec}\phi\frac{d\Omega}{d\theta} =$$
$$\left\{\sin(S_2 - S)\sin(T_2 - \sigma) + \cos(S_2 - S)\cos(T_2 - \sigma)\cos F_2\right\}\left(\operatorname{cosec} F\frac{d\Omega}{dT} + \cot F\frac{d\Omega}{dS}\right)$$
$$+\left\{\cos(S_2 - S)\sin(T_2 - \sigma) - \sin(S_2 - S)\cos(T_2 - \sigma)\cos F_2\right\}\frac{d\Omega}{dF}$$
$$+\cos(T_2 - \sigma)\sin F_2\frac{d\Omega}{dS};$$

and we have therefore

$$d\phi = -\frac{\cot\phi}{k}\frac{d\Omega}{d\sigma}dt - \frac{\operatorname{cosec}\phi}{k}\frac{d\Omega}{d\theta}dt,$$
$$d\sigma = \frac{\cot\phi}{k}\frac{d\Omega}{d\phi}dt,$$
$$d\theta = \frac{\operatorname{cosec}\phi}{k}\frac{d\Omega}{d\phi}dt.$$

The second equation, viz.

$$\frac{d\Omega}{dh}dh + \frac{d\Omega}{dk}dk + \frac{d\Omega}{dg}dg + \frac{d\Omega}{d\varpi}d\varpi = d_2\Omega = XdT_2 + YdF_2 + ZdS_2,$$

if we substitute for dg, dT_2, dF_2, dS_2 their values, becomes

$$\frac{d\Omega}{dh}dh + \frac{d\Omega}{dk}dk + \frac{d\Omega}{dg}\left(-\frac{nC}{(B-A)pq}dr + \mathfrak{A}dh + \mathfrak{B}dk\right) + \frac{d\Omega}{d\varpi}d\varpi$$
$$= X\left\{d\varpi - \frac{Ck(h - Cr^2)}{(k^2 - C^2r^2)(B-A)pq}dr + \mathfrak{C}dh + \mathfrak{D}dk\right\}$$
$$+Y\left\{-\frac{Ck\sin F_2}{k^2 - C^2r^2}dr + \frac{k\sin F_2\cos F_2}{k^2 - C^2r^2}dk\right\}$$
$$+Z\left\{-\frac{C(Ch - k^2)r}{(k^2 - C^2r^2)(B-A)pq}dr - \frac{\tfrac{1}{2}dh}{(B-A)pq} + \frac{(n - Cr^2)kdk}{(k^2 - C^2r^2)(B-A)pq}\right\}.$$

The comparison of the terms involving $d\varpi$ gives at once $\dfrac{d\Omega}{d\varpi} = X$, or, substituting for X its value,

$$\frac{d\Omega}{d\varpi} = \cos(S_2 - S)\sin F_2\left(\operatorname{cosec} F\frac{d\Omega}{dT} + \cot F\frac{d\Omega}{dS}\right)$$

$$-\sin(S_2 - S)\sin F_2\frac{d\Omega}{dF} - \cos F_2\frac{d\Omega}{dS};$$

and comparing with the expression for dk, we find

$$dk = \frac{d\Omega}{d\varpi}dt.$$

The terms involving dr give

$$n\frac{d\Omega}{dg} = \frac{k(h - Cr^2)}{k^2 - C^2r^2}X + \frac{k\sin F_2(B - A)pq}{k^2 - C^2r^2}Y + \frac{(Ch - k^2)r}{k^2 - C^2r^2}Z,$$

or, substituting for X, Y, Z, their values,

$$\begin{aligned} n\frac{d\Omega}{dg} = {} & \frac{k(h - Cr^2)}{k^2 - C^2r^2}\left\{\cos(S_2 - S)\sin F_2\left(\operatorname{cosec} F\frac{d\Omega}{dT} + \cot F\frac{d\Omega}{dS}\right) - \sin(S_2 - S)\sin F_2\frac{d\Omega}{dF} - \cos F_2\frac{d\Omega}{dS}\right\} \\ & + \frac{k(B - A)pq}{k^2 - C^2r^2}\left\{\sin(S_2 - S)\sin F_2\left(\operatorname{cosec} F\frac{d\Omega}{dT} + \cot F\frac{d\Omega}{dS}\right) + \cos(S_2 - S)\sin F_2\frac{d\Omega}{dF}\right\} \\ & + \frac{r(Ch - k^2)}{k^2 - C^2r^2}\frac{d\Omega}{dS}, \end{aligned}$$

which agrees with the second form for $\tfrac{1}{2}dh$, and gives as before

$$dh = 2n\frac{d\Omega}{dg}dt;$$

the terms involving dh give

$$\frac{d\Omega}{dh} + \mathfrak{A}\frac{d\Omega}{dg} = \mathfrak{C}X - \frac{\tfrac{1}{2}}{(B - A)pq}Z,$$

and thence

$$-2n\frac{d\Omega}{dh} = 2n\mathfrak{A}\frac{d\Omega}{dg} - 2n\mathfrak{C}X + \frac{n}{(B - A)pq}Z;$$

or, substituting for $-2n\mathfrak{C}$ its value $\mathfrak{B}$, for X the value $\dfrac{d\Omega}{d\varpi}$, and for Z its value $\dfrac{d\Omega}{dS}$, we have

$$-2n\frac{d\Omega}{dh} = 2n\mathfrak{A}\frac{d\Omega}{dg} + \mathfrak{B}\frac{d\Omega}{d\varpi} + \frac{n}{(B - A)pq}\frac{d\Omega}{dS},$$

where, on the right hand, $\frac{d\Omega}{dg}$ and $\frac{d\Omega}{d\varpi}$ may be considered as standing for given functions of $\frac{d\Omega}{dT}, \frac{d\Omega}{dS}, \frac{d\Omega}{dF}$; for the present purpose, however, multiplying by dt, and substituting for $2n\frac{d\Omega}{dg}$ and $\frac{d\Omega}{d\varpi}$ the values dh and dk, we have

$$-2n\frac{d\Omega}{dh}dt = \mathfrak{A}dh + \mathfrak{B}dk + \frac{n}{(B-A)pq}\frac{d\Omega}{dS}dt,$$

which agrees with the foregoing value of dg, or we have

$$dg = -2n\frac{d\Omega}{dh}dt.$$

The comparison of the terms involving dk gives

$$\frac{d\Omega}{dk} + \mathfrak{B}\frac{d\Omega}{dg} = \mathfrak{D}X + \frac{k\sin F_2\cos F_2}{k^2 - C^2r^2}Y + \frac{(h-Cr^2)k}{(k^2-C^2r^2)(B-A)pq}Z;$$

or, substituting and reducing,

$$-\frac{d\Omega}{dk} = -2n\mathfrak{C}\frac{d\Omega}{dg} - \mathfrak{D}\frac{d\Omega}{d\varpi} - \frac{1}{k}\cot F_2\left\{\cos(S_2-S)\frac{d\Omega}{dF} + \sin(S_2-S)\left(\operatorname{cosec} F\frac{d\Omega}{dT} + \cot F\frac{d\Omega}{dS}\right)\right\}$$
$$-\frac{(h-Cr^2)k}{(k^2-C^2r^2)(B-A)pq}\frac{d\Omega}{dS},$$

where, on the right-hand side, $\frac{d\Omega}{dg}$ and $\frac{d\Omega}{d\varpi}$ may be considered as standing for given functions of $\frac{d\Omega}{dT}, \frac{d\Omega}{dS}, \frac{d\Omega}{dF}$; for the present purpose, however, multiplying by dt and putting for $2n\frac{d\Omega}{dg}dt$ and $\frac{d\Omega}{d\varpi}dt$ the values dh and dk, we have

$$-\frac{d\Omega}{dk}dt = -\mathfrak{C}dh - \mathfrak{D}dk - \frac{1}{k}\cot F_2\left\{\cos(S_2-S)\frac{d\Omega}{dF}dt + \sin(S_2-S)\left(\operatorname{cosec} F\frac{d\Omega}{dT} + \cot F\frac{d\Omega}{dS}\right)dt\right\}$$
$$-\frac{(h-Cr^2)k}{(k^2-C^2r^2)(B-A)pq}\frac{d\Omega}{dS}dt,$$

which agrees with the foregoing value of $d\varpi$, or we have

$$d\varpi = -\frac{d\Omega}{dk}dt,$$

and we have thus the system of formulæ for the variation of the elements in the problem of rotation, given in Art. I.

218.

A THIRD MEMOIR ON THE PROBLEM OF DISTURBED ELLIPTIC MOTION.

[From the *Memoirs of the Royal Astronomical Society*, vol. XXXI, pp. 43—56. Read January 10, 1863.]

THE object of the present Memoir is to obtain the differential equations for determining

r, the radius vector,

v, the longitude,

y, the latitude,

of the disturbed body, when the last two coordinates are measured in respect to an arbitrarily varying plane (which however, to fix the ideas, is called the variable ecliptic) and the departure point or origin of longitudes therein. This is very readily effected by means of an expression for the *Vis Viva* function given in my "Supplementary Memoir on the Problem of Disturbed Elliptic Motion," *Mem. Roy. Ast. Soc.*, t. XXVIII. pp. 217—234 (1859), [215]. Neglecting the squares of the variations of the variable ecliptic, and also the products of the variations by sin y, or $\frac{dy}{dt}$, then (as might be expected) it is found that the equations for r and v are the same as for a fixed ecliptic, and the equation for y is found in a simple form, which is ultimately reduced to coincide with that obtained for the lunar theory by Laplace in the seventh book of the *Mécanique Céleste*, and which is used by him to show that the effect of the variation of the ecliptic on the latitude of the Moon (as measured from the variable ecliptic) is insensible. And it is shown conversely how the approximate formula of the Memoir may be obtained by a process similar to that made use of in the *Mécanique Céleste*.

I.

The position in respect to a fixed plane of reference and origin of longitudes therein, of the variable ecliptic and of the departure point or origin of longitudes therein, are determined by

θ', the longitude of node,

σ', the departure of node,

ϕ', the inclination,

where, by the definition of a departure point,

$$d\sigma' - \cos\phi'\, d\theta' = 0,$$

and then, in respect to the variable ecliptic and departure point or origin of longitudes therein, the position of the disturbed body is determined by

r, the radius vector,

v, the longitude,

y, the latitude;

and this being so, then (Supplementary Memoir, pp. 220, 227) the expression for the *Vis Viva* function is,

$$T = \tfrac{1}{2}\{\dot{r}^2 + r^2(Q^2 + R^2)\},$$

where

$$Q = \quad -\dot{y} \quad + [\cos(v-\sigma')\sin\phi' \,.\, \dot{\theta}' - \sin(v-\sigma')\,\dot{\phi}'],$$

$$R = \cos y \,.\, \dot{v} - \sin y\,[\sin(v-\sigma')\sin\phi' \,.\, \dot{\theta}' + \cos(v-\sigma')\,\dot{\phi}'],$$

the superscript dots being used to denote differentiation with respect to the time. The last-mentioned expressions may for shortness be denoted by

$$Q = \quad -\dot{y} + A,$$

$$R = \cos y \,.\, \dot{v} - B\sin y.$$

The equations of motion are of course,

$$\frac{d}{dt}\frac{dT}{d\dot{r}} - \frac{dT}{dr} = \frac{dV}{dr},$$

$$\frac{d}{dt}\frac{dT}{d\dot{v}} - \frac{dT}{dv} = \frac{dV}{dv},$$

$$\frac{d}{dt}\frac{dT}{d\dot{y}} - \frac{dT}{dy} = \frac{dV}{dy},$$

where $V = \dfrac{n^2a^3}{r} + \Omega$, if Ω is the disturbing function, taken with Lagrange's sign ($\Omega = -R$, if R is the disturbing function of the *Mécanique Céleste*).

To reduce these, we have in the first place

$$\frac{dT}{d\dot{r}} = \dot{r} \quad , \quad \frac{dT}{dr} = r(Q^2 + R^2),$$

$$\frac{dT}{d\dot{v}} = r^2 R \cos y, \quad \frac{dT}{dv} = r^2 \left(Q \frac{dQ}{dv} + R \frac{dR}{dv}\right) = r^2 \quad (-QB - RA \sin y),$$

$$\frac{dT}{d\dot{y}} = -r^2 Q \quad , \quad \frac{dT}{dy} = r^2 R \frac{dR}{dy} \qquad = r^2 R(-\sin y \,.\, \dot{v} - B \cos y).$$

The equations are thus reduced to

$$\frac{d^2 r}{dt^2} - r(Q^2 + R^2) + \frac{n^2 a^3}{r^2} = \frac{d\Omega}{dr},$$

$$\frac{d}{dt}(r^2 \cos y \,.\, R) + r^2 \quad (QB + RA \sin y) \quad = \frac{d\Omega}{dv},$$

$$\frac{d}{dt}(-r^2 Q \qquad) + r^2 R\left(\sin y \frac{dv}{dt} + B \cos y\right) = \frac{d\Omega}{dy},$$

and then substituting for Q, R their values, viz.

$$Q = \quad -\frac{dy}{dt} + A,$$

$$R = \cos y \frac{dv}{dt} - B \sin y,$$

we find

$$\frac{d^2 r}{dt^2} - r\left\{\cos^2 y \left(\frac{dv}{dt}\right)^2 + \left(\frac{dy}{dt}\right)^2\right\} + \frac{n^2 a^3}{r^2} = \frac{d\Omega}{dr} + \mathfrak{A},$$

$$\frac{d}{dt}\left(r^2 \cos^2 y \frac{dv}{dt}\right) \qquad = \frac{d\Omega}{dv} + \mathfrak{B},$$

$$\frac{d}{dt}\left(r^2 \frac{dy}{dt}\right) + r^2 \cos y \sin y \left(\frac{dv}{dt}\right)^2 \quad = \frac{d\Omega}{dy} + \mathfrak{C},$$

where

$$\mathfrak{A} = r\left(-2A \frac{dy}{dt} - 2B \sin y \cos y \frac{dv}{dt} + A^2 + B^2 \sin^2 y\right),$$

$$\mathfrak{B} = \frac{d}{dt}(r^2 B \sin y \cos y) + r^2 \left(B \frac{dy}{dt} - A \sin y \cos y \frac{dv}{dt} - AB \cos^2 y\right),$$

$$\mathfrak{C} = \frac{d}{dt}(r^2 A) \qquad + r^2 \left(-(\cos^2 y - \sin^2 y) B \frac{dv}{dt} + B^2 \sin y \cos y\right),$$

in which

$$A = \cos(v - \sigma') \sin \phi' \frac{d\theta'}{dt} + \sin(v - \sigma') \frac{d\phi'}{dt},$$

$$B = \sin(v - \sigma') \sin \phi' \frac{d\theta'}{dt} + \cos(v - \sigma') \frac{d\phi'}{dt};$$

θ', σ', ϕ' being given functions of t such that $d\sigma' - \cos\phi' \, d\theta' = 0$.

The foregoing equations of motion are rigorously accurate.

II.

Neglecting the terms which involve A^2, AB, B^2, we have

$$\mathfrak{A} = r\left(-2A\frac{dy}{dt} - 2B\sin y\cos y\frac{dv}{dt}\right),$$

$$\mathfrak{B} = \frac{d}{dt}(r^2 B\sin y\cos y) + r^2\left(B\frac{dy}{dt} - A\sin y\cos y\frac{dv}{dt}\right),$$

$$\mathfrak{C} = \frac{d}{dt}(r^2 A) \qquad\qquad - r^2(\cos^2 y - \sin^2 y)B\frac{dv}{dt};$$

and if we then neglect also the products of A and B by y and $\frac{dy}{dt}$, we have

$$\mathfrak{A} = 0,$$

$$\mathfrak{B} = 0,$$

$$\mathfrak{C} = \frac{d}{dt}(r^2 A) - r^2 B\frac{dv}{dt},$$

where it may be noticed that, in order to obtain this last value of $\mathfrak{C}$, the only neglected term is a term containing $B\sin^2 y$.

Now, attending to the values of A and B, we have

$$\frac{dA}{dt} = \frac{dA}{dv}\frac{dv}{dt} + \frac{d'A}{dt}$$

$$= -B\frac{dv}{dt} + \frac{d'A}{dt},$$

where here and in the sequel $\frac{d'}{dt}$ denotes differentiation in regard to t in so far only as it enters through the quantities σ', θ', ϕ', which determine the position of the variable ecliptic.

Hence

$$\mathfrak{C} = 2Ar\frac{dr}{dt} + r^2\left(-B\frac{dv}{dt} + \frac{d'A}{dt}\right) - r^2 B\frac{dv}{dt}$$

$$= 2Ar\frac{dr}{dt} + r^2\frac{d'A}{dt} - 2r^2 B\frac{dv}{dt};$$

and, as above, $\mathfrak{A} = 0$, $\mathfrak{B} = 0$.

Let r, v, y be the values obtained on the supposition that $\mathfrak{C} = 0$, and

$$r + \delta r,\ v + \delta v,\ y + \delta y,$$

the accurate values; the first and second equations show that, neglecting the products of y and $\frac{dy}{dt}$ into δy and $\frac{d\delta y}{dt}$, we have $\delta r = 0$, $\delta v = 0$; so that the values of r and v are not affected by the variation of the ecliptic. And then, substituting in the third equation $y + \delta y$ in the place of y, and for

$$\cos(y+\delta y)\sin(y+\delta y),\ = \cos y\sin y + (\cos^2 y - \sin^2 y)\,\delta y,$$

writing $\cos y \sin y + \delta y$, we have

$$\frac{d}{dt} \cdot r^2 \left(\frac{dy}{dt} + \frac{d\delta y}{dt}\right) + r^2 \left(\frac{dv}{dt}\right)^2 (\cos y \sin y + \delta y) = \frac{d\Omega}{dy} + \delta \frac{d\Omega}{dy} + \mathfrak{C};$$

or, since the terms independent of δy and $\mathfrak{C}$ must destroy each other, this is

$$\frac{d}{dt} \cdot r^2 \frac{d\delta y}{dt} + r^2 \left(\frac{dv}{dt}\right)^2 \delta y = \delta \frac{d\Omega}{dy} + \mathfrak{C};$$

or, as this may be written,

$$r^2 \frac{d^2\delta y}{dt^2} + 2r \frac{dr}{dt} \frac{d\delta y}{dt} + r^2 \left(\frac{dv}{dt}\right)^2 \delta y = \delta \frac{d\Omega}{dy} + 2Ar \frac{dr}{dt} + r^2 \frac{d'A}{dt} - 2r^2 B \frac{dv}{dt},$$

that is,

$$\frac{d^2\delta y}{dt^2} + \frac{2}{r} \frac{dr}{dt} \frac{d\delta y}{dt} + \left(\frac{dv}{dt}\right)^2 \delta y = \frac{1}{r^2} \delta \frac{d\Omega}{dy} + 2A \frac{1}{r} \frac{dr}{dt} + \frac{d'A}{dt} - 2B \frac{dv}{dt},$$

or, what is the same thing,

$$\begin{aligned}\frac{d^2\delta y}{dt^2} + n^2 (1 + \tfrac{3}{2} m^2)\, \delta y = \frac{1}{r^2} \delta \frac{d\Omega}{dy} &+ \left\{n^2 + \tfrac{3}{2} m^2 n^2 - \left(\frac{dv}{dt}\right)^2\right\} \delta y \\ &- \frac{2}{r} \frac{dr}{dt} \frac{d\delta y}{dt} + 2A \frac{1}{r} \frac{dr}{dt} + 2B \left(n - \frac{dv}{dt}\right) \\ &- 2Bn + \frac{d'A}{dt},\end{aligned}$$

where as usual $m, = \frac{n'}{n}$, is the ratio of the mean motion of the Sun to that of the Moon; the term $\frac{3}{2} m^2 n^2 \delta y$ having been added on each side of the equation in order to destroy on the right-hand side the corresponding term arising from $\frac{1}{r^2} \delta \frac{d\Omega}{dy}$. In fact, to find the approximate expression for $\frac{d\Omega}{dy}$, we have

$$\Omega = \frac{m' r^2}{r'^3} (\tfrac{3}{2} \cos^2 H - \tfrac{1}{2}),$$

where H is the angular distance of the Sun and Moon; that is, $\cos H = \cos y \cos (v - v')$; here

$$\begin{aligned}\Omega &= \frac{m' r^2}{r'^3} \{\tfrac{3}{2} \cos^2 y \cos^2 (v - v') - \tfrac{1}{2}\}, \\ \frac{d\Omega}{dy} &= -\frac{m' r^2}{r'^3} \,.\, 3 \sin y \cos y \cos^2 (v - v') \\ &= -\frac{m' r^2}{r'^3} (\tfrac{3}{2} + \tfrac{3}{2} \cos (2v - 2v')) \sin y \cos y;\end{aligned}$$

or, neglecting the periodic quantity $\cos(2v-2v')$, and writing y for $\sin y \cos y$; also putting as usual $m'=n'^2a'^3=m^2n^2a'^3$, and $r'=a'$, we have

$$\frac{d\Omega}{dy}=-m^2n^2r^2\,.\,\tfrac{3}{2}y,$$

and thence

$$\delta\frac{d\Omega}{dy}=-m^2n^2r^2\,.\,\tfrac{3}{2}\delta y$$

or

$$\frac{1}{r^2}\,\delta\,\frac{d\Omega}{dy}=-\tfrac{3}{2}m^2n^2\,\delta y.$$

Substituting this value of $\frac{1}{r^2}\,\delta\,\frac{d\Omega}{dy}$, and putting also $r=a$, $\frac{dr}{dt}=0$, $\frac{dv}{dt}=0$, the equation for δy becomes

$$\frac{d^2\delta y}{dt^2}+n^2(1+\tfrac{3}{2}m^2)\,\delta y=-2Bn+\frac{d'A}{dt}.$$

III.

To deduce the formula, seventh book of the *Mécanique Céleste,* I proceed as follows:

Putting

$$\frac{d\theta'}{dt}=\frac{1}{\cos\phi'}\frac{d\sigma'}{dt},$$

we have

$$A\cos\phi'=\frac{d\sigma'}{dt}\sin\phi'\cos(v-\sigma')-\frac{d\phi'}{dt}\cos\phi'\sin(v-\sigma'),$$

$$B\cos\phi'=\frac{d\sigma'}{dt}\sin\phi'\sin(v-\sigma')+\frac{d\phi'}{dt}\cos\phi'\cos(v-\sigma'),$$

which may be written

$$A\cos\phi'=-\sin v\frac{d}{dt}(\sin\phi'\cos\sigma')+\cos v\frac{d}{dt}(\sin\phi'\sin\sigma'),$$

$$B\cos\phi'=\cos v\frac{d}{dt}(\sin\phi'\cos\sigma')+\sin v\frac{d}{dt}(\sin\phi'\sin\sigma').$$

Laplace in effect assumes that the variations of the ecliptic are given in the form

$$\sin\phi'\sin\sigma'=-\Sigma k\sin(int+\epsilon),$$
$$\sin\phi'\cos\sigma'=\Sigma k\cos(int+\epsilon),$$

($it+\epsilon$ is there written for the argument, n being assumed $=1$) where i, k, ϵ are absolute constants, the quantities i being all very small in comparison with m^2. Substituting these values, and putting $\cos\phi'$ equal to unity, we have

$$A=-\Sigma ik\cos(v+int+\epsilon),$$
$$B=-\Sigma ik\sin(v+int+\epsilon);$$

and thence also

$$\frac{d'A}{dt} = n\Sigma\, i^2 k \sin(v + int + \epsilon),$$

and

$$2nB - \frac{d'A}{dt} = -n\Sigma\,(2i + i^2)\, k \sin(v + int + \epsilon),$$

so that the equation for δy becomes

$$\frac{d^2\delta y}{dt^2} + n^2(1 + \tfrac{3}{2}m^2)\,\delta y + n\Sigma\,(2i + i^2)\, k\sin(v + int + \epsilon) = 0;$$

or, taking as the independent variable $v\,(= nt)$ in the place of t, this is

$$\frac{d^2\delta y}{dv^2} + (1 + \tfrac{3}{2}m^2)\,\delta y + \frac{1}{n}\Sigma\,(2i + i^2)\, k\sin(v + iv + \epsilon) = 0;$$

which is, in fact, Laplace's equation, n being retained instead of being put equal to unity, and δy being the part which depends on the variation of the ecliptic, of his $s_,$.

IV.

Conversely the equation

$$\frac{d^2\delta y}{dt^2} + n^2(1 + \tfrac{3}{2}m^2)\,\delta y = -2Bn + \frac{d'A}{dt},$$

may be obtained by a process similar to Laplace's. Assuming that the Moon and Sun are each of them referred to a fixed plane of reference and origin of longitudes therein, by the coordinates

u, the reciprocal of the reduced radius vector,

v, the longitude,

s, the tangent of the latitude,

for the Moon, and by the corresponding coordinates u', v', s' for the Sun, then we have

$$\frac{d^2s}{dv^2} + s + \frac{\dfrac{ds}{dv}\dfrac{d\Omega}{dv} - su\dfrac{d\Omega}{du} - (1 + s^2)\dfrac{d\Omega}{ds}}{n^2u^2\left(1 + \dfrac{2}{n^2}\displaystyle\int\frac{d\Omega}{u^2dv}\,dv\right)} = 0.$$

Here, as before,

$$\Omega = \frac{m'r^2}{r'^3}\,(\tfrac{3}{2}\cos^2 H - \tfrac{1}{2}),$$

or, as it is now to be written,

$$\Omega = \frac{m'(1 + s^2)\,u'^3}{(1 + s'^2)^{\frac{3}{2}}\,u^2}\left\{\tfrac{3}{2}\left(\frac{\cos(v - v') + ss'}{\sqrt{1 + s^2}\,\sqrt{1 + s'^2}}\right)^2 - \tfrac{1}{2}\right\};$$

or, since the second term

$$\frac{m'u'^3}{(1+s'^2)^{\frac{3}{2}}} - \tfrac{1}{2}\frac{1+s^2}{u^2}$$

gives, as is immediately seen, no term in

$$\frac{ds}{dv}\frac{d\Omega}{dv} - su\frac{d\Omega}{du} - (1+s^2)\frac{d\Omega}{ds},$$

we may, in calculating this quantity, write

$$\Omega = \frac{m'(1+s^2)\,u'^3}{(1+s'^2)^{\frac{3}{2}}\,u^2}\cdot\tfrac{3}{2}\left(\frac{\cos(v-v')+ss'}{\sqrt{1+s^2}\sqrt{1+s'^2}}\right)^2$$

$$= \tfrac{3}{2}\frac{m'u'^3}{(1+s'^2)^{\frac{5}{2}}\,u^2}\left(\cos(v-v')+ss'\right)^2,$$

or, neglecting s'^2,

$$\Omega = \tfrac{3}{2}\,m'u'^3\left(\frac{\cos(v-v')+ss'}{u}\right)^2.$$

Hence, putting for a moment

$$\Theta = \frac{\cos(v-v')+ss'}{u},$$

we have

$$\frac{ds}{dv}\frac{d\Omega}{dv} - su\frac{d\Omega}{du} - (1+s^2)\frac{d\Omega}{ds} = 3m'u'^3\,.\,\Theta\left\{\frac{ds}{dv}\frac{d\Theta}{dv} - su\frac{d\Theta}{du} - (1+s^2)\frac{d\Theta}{ds}\right\},$$

where the factor in { } is

$$= \frac{1}{u}\left\{-\frac{ds}{dv}\sin(v-v') + s\left(\cos(v-v')+ss'\right) - (1+s^2)\,s'\right\}$$

$$= \frac{1}{u}\left\{s\cos(v-v') - \frac{ds}{dv}\sin(v-v') - s'\right\},$$

and the above-mentioned quantity is

$$= \frac{3m'u'^3}{u^2}\left\{\cos(v-v')+ss'\right\}\left\{s\cos(v-v') - \frac{ds}{dv}\sin(v-v') - s'\right\};$$

or, what is the same thing,

$$= \frac{3m'u'^3}{u^2}\cos(v-v')\left\{s\cos(v-v') - \frac{ds}{dv}\sin(v-v') - s'\right\}.$$

Considering now the Sun as moving in the variable ecliptic, its latitude is $= \sin\phi'\sin(v'-\sigma')$; that is, we have

$$s' = \sin\phi'\sin(v'-\sigma');$$

and if the Moon were in the variable ecliptic, its latitude would be $\sin\phi'\sin(v-\sigma')$; that is, the latitude, measured from the variable ecliptic, is $= s-\sin\phi'\sin(v-\sigma')$; or, putting

$s_{,}$, the Moon's latitude, measured from the variable ecliptic,

we have

$$s=\sin\phi'\sin(v-\sigma')+s_{,}.$$

Hence, disregarding the variations of ϕ', σ', we have

$$\frac{ds}{dv}=\sin\phi'\cos(v-\sigma')+\frac{ds_{,}}{dv};$$

and substituting these values of s, $\frac{ds}{dv}$, and s', we find

$$\begin{aligned}
&s\cos(v-v')-\frac{ds}{dv}\sin(v-v')-s'\\
&=\ s_{,}\cos(v-v')-\frac{ds_{,}}{dv}\sin(v-v')\\
&\quad+\sin\phi'\left\{\sin(v-\sigma')\cos(v-v')-\cos(v-\sigma')\sin(v-v')-\sin(v'-\sigma')\right\}\\
&=\ s_{,}\cos(v-v')-\frac{ds_{,}}{dv}\sin(v-v');
\end{aligned}$$

or

$$\begin{aligned}
&\frac{ds}{dv}\frac{d\Omega}{dv}-su\frac{d\Omega}{du}(1+s^2)\frac{d\Omega}{ds}\\
&\qquad=\frac{3m'u'^3}{u^2}\cos(v-v')\left\{s_{,}\cos(v-v')-\frac{ds_{,}}{dv}\sin(v-v')\right\};
\end{aligned}$$

which, neglecting the periodic terms, is

$$=\tfrac{3}{2}\frac{m'u'^3}{u^2}s_{,};$$

and then

$$\frac{\dfrac{ds}{dv}\dfrac{d\Omega}{dv}-su\dfrac{d\Omega}{du}-(1+s^2)\dfrac{d\Omega}{ds}}{h^2u^2\left(1+\dfrac{2}{h^2}\displaystyle\int\frac{d\Omega}{u^2dv}dv\right)}=\frac{1}{h^2u^2}\cdot\tfrac{3}{2}\frac{m'u'^3}{u^2}s_{,}=\tfrac{3}{2}\frac{m'u'^3}{h^2u^4}s_{,};$$

which, putting $m'=n'^2a'^3=m^2n^2a'^3$, $u'=\frac{1}{a'}$, $h^2=n^2a^4$, $u=\frac{1}{a}$, becomes

$$=\tfrac{3}{2}m^2s_{,};$$

so that the differential equation is reduced to

$$\frac{d^2s}{dv^2}+s+\tfrac{3}{2}m^2s_{,}=0.$$

But

$$s = \sin\phi' \sin(v-\sigma') + s_,,$$

$$\frac{ds}{dv} = \sin\phi' \cos(v-\sigma') + \frac{d'}{dv}\sin\phi' \sin(v-\sigma') + \frac{ds_,}{dv},$$

$$\frac{d^2s}{dv^2} = -\sin\phi' \sin(v-\sigma') + \frac{d'}{dv}\sin\phi' \cos(v-\sigma') + \frac{d}{dv}\frac{d'}{dv}\sin\phi' \sin(v-\sigma') + \frac{d^2s_,}{dv^2},$$

where $\frac{d'}{dv}$ denotes differentiation in respect to v, in so far only as it enters through ϕ' and σ' $\left(\text{these are functions of } t, \text{ which is } = \frac{1}{n}v\right)$. Hence

$$\frac{d^2s}{dv^2} + s = \frac{d^2s_,}{dv^2} + s_, + \frac{d'}{dv}\sin\phi' \cos(v-\sigma') + \frac{d}{dv}\frac{d'}{dv}\sin\phi' \sin(v-\sigma');$$

but

$$\frac{d'}{dv}\sin\phi' \sin(v-\sigma') = \frac{1}{n}\left(\sin v \frac{d}{dt}\sin\phi' \cos\sigma' - \cos v \frac{d}{dt}\sin\phi' \sin\sigma'\right)$$

$$= -\frac{1}{n}A\cos\phi'$$

$$= -\frac{1}{n}A;$$

and thence

$$\frac{d}{dv}\frac{d'}{dv}\sin\phi' \sin(v-\sigma') = -\frac{1}{n}\frac{dA}{dv} - \frac{1}{n}\frac{d'A}{dv} = \frac{1}{n}B - \frac{1}{n^2}\frac{d'A}{dt};$$

and similarly

$$\frac{d'}{dv}\sin\phi' \cos(v-\sigma') = \frac{1}{n}\left\{\cos v \frac{d}{dt}(\sin\phi' \cos\sigma') + \sin v \frac{d}{dt}\sin\phi' \sin\sigma'\right\}$$

$$= \frac{1}{n}B\cos\phi'$$

$$= \frac{1}{n}B.$$

Hence

$$\frac{d^2s}{dv^2} + s = \frac{d^2s_,}{dv^2} + s_, + \frac{2}{n}B - \frac{1}{n^2}\frac{d'A}{dt};$$

or, putting $v = nt$,

$$\frac{d^2s}{dv^2} + s = \frac{1}{n^2}\frac{d^2s_,}{dt^2} + s_, + \frac{2}{n}B - \frac{1}{n^2}\frac{d'A}{dt};$$

whence, substituting in the equation

$$\frac{d^2 s}{dv^2} + s + \tfrac{3}{2} m^2 s, = 0,$$

we have

$$\frac{d^2 s_,}{dt^2} + n^2 (1 + \tfrac{3}{2} m^2)\, s_, + 2nB - \frac{d'A}{dt} = 0\,;$$

or putting $s_, = y + \delta y$, then

$$\frac{d^2 y}{dt^2} + n^2 (1 + \tfrac{3}{2} m^2)\, y = 0\,;$$

which gives, in the approximation which is being considered, the principal term of the latitude, and then

$$\frac{d^2 \delta y}{dt^2} + n^2 (1 + \tfrac{3}{2} m^2)\, \delta y + 2nB - \frac{d'A}{dt} = 0,$$

which is the approximate equation previously obtained by the method of the present memoir.

219.

ON SOME FORMULÆ RELATING TO THE VARIATION OF THE PLANE OF A PLANET'S ORBIT.

[From the *Monthly Notices of the Royal Astronomical Society*, vol. XXI. (1861), pp. 43—47.]

IN Hansen's Memoir, "Auseinandersetzung einer zweckmässigen Methode zur Berechnung der absoluten Störungen der kleinen Planeten," *Abhand. der K. Sächs. Gesell.* t. v. (1856), are contained, § 8, some very elegant formulæ for taking account of the variation of the plane of the orbit. These, in fact, depend upon the following geometrical theorem, viz., if (in the figure) ABC is a spherical triangle; P, a point on the side AB; and PM, PN, the perpendiculars let fall from P on the other two sides AC, CB; then we have

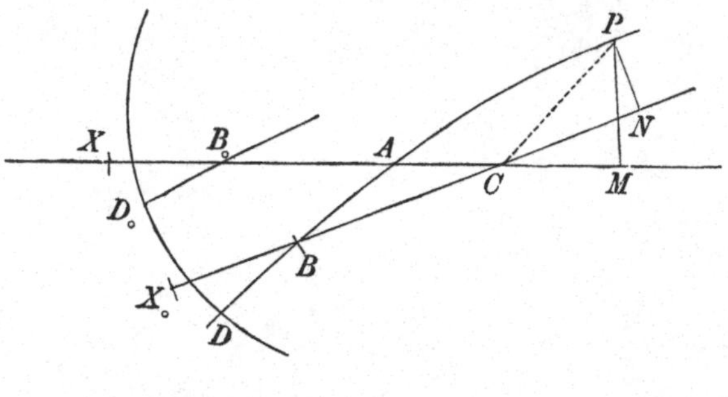

$$\cos PM \sin (BC + CM) = \cos PN \sin BN - \tan \tfrac{1}{2} C \cos BC (\sin PM + \sin PN),$$
$$\cos PM \cos (BC + CM) = \cos PN \cos BN + \tan \tfrac{1}{2} C \sin BC (\sin PM + \sin PN).$$

These equations, in fact, give

$$\cos PM \sin CM = \cos PN \sin CN - \tan \tfrac{1}{2} C (\sin PM + \sin PN),$$
$$\cos PM \cos CM = \cos PN \cos CN;$$

the latter of which is at once seen to be true, since joining the points C and P, the two sides are respectively equal to $\cos CP$. To verify the former one, write $\angle PCM = C_1$, $\angle PCN = C_2$, so that $C = C_1 - C_2$. Then, since $\cos CP = \cos PM \cos CM = \cos PN \cos CN$, $\sin PM = \sin CP \sin C_1$, $\sin PN = \sin CP \sin C_2$, the equation becomes $\cos CP (\tan CM - \tan CN) = -\tan \frac{1}{2} C \sin CP (\sin C_1 + \sin C_2)$, or since $\tan CM = \tan CP \cos C_1$, $\tan CN = \tan CP \cos C_2$, this is

$$\cos C_1 - \cos C_2 = -\tan \tfrac{1}{2} C (\sin C_1 + \sin C_2),$$

which is identically true, in virtue of the equation $C = C_1 - C_2$; and, conversely, we have the original two equations.

Suppose that XM is the ecliptic, X being the origin of longitudes, DP the instantaneous orbit, D the departure-point therein, and P the planet, DD_0 the orthogonal trajectory of the successive positions of the orbit; and writing

$\mathfrak{p}$, the departure of the planet,

v, the longitude of ditto,

y, the latitude of ditto,

θ, the longitude of node,

σ, the departure of ditto,

ϕ, the inclination;

then, in the figure, $DP = \mathfrak{p}$, $XM = v$, $PM = y$, $XA = \theta$, $DA = \sigma$, $\angle A = \phi$.

The quantities θ_0, σ_0, ϕ_0, might be considered as altogether arbitrary; but to fix the ideas it is better to assume at once that they denote

θ_0, the longitude of node,

σ_0, the departure,

ϕ_0, the inclination,

for the initial position of the orbit, viz., in the figure $XB_0 = \theta_0$, $D_0B_0 = \sigma_0$, $\angle B_0 = \phi_0$.

Take $DB = \sigma_0$, $\angle B = \phi_0$, $BX_0 = \theta_0$, this determines a travelling orbit of reference X_0N, and origin of longitudes X_0 therein; such that, with respect to this travelling orbit, the position of the planet's orbit is determined by

θ_0, the longitude of node,

σ_0, the departure of node,

ϕ_0, the inclination.

We have in the triangle ABC, $AB = \sigma - \sigma_0$, $\angle B = \phi_0$, $\angle A = 180^\circ - \phi$; and if the other parts of the triangle are represented by

$BC = \omega$,

$AC = \theta_0 - \theta + \omega + \Gamma$,

$\angle C = \Phi$;

then ω, Γ, Φ, are given in terms of $\sigma-\sigma_0$, ϕ_0, ϕ; and we have, moreover, $XC=\theta+AC$ $=\theta_0+\omega+\Gamma$, $X_0C=\sigma_0+\omega$; that is, the position of the travelling orbit X_0N, and origin of longitudes X_0 therein, are determined by

$$\begin{aligned}&\theta_0+\omega+\Gamma, \text{ the longitude of node,}\\&\sigma_0+\omega \quad\;\;\, , \text{ the departure of node,}\\&\Phi \qquad\quad\;\;\, , \text{ the inclination.}\end{aligned}$$

Suppose that in reference to this travelling orbit and origin of longitudes therein, we have

$$\begin{aligned}&v', \text{ the longitude of planet,}\\&y', \text{ the latitude of ditto,}\end{aligned}$$

viz., in the figure $X_0N=v'$ (and therefore $BN=v'-\theta_0$), $PN=y'$.

Moreover, $BC+CM=BC+AM-AC=\omega+(v-\theta)-(\theta_0-\theta+\omega+\Gamma)=v-\theta_0-\Gamma$, hence the two equations are

$$\begin{aligned}\cos y\sin(v-\theta_0-\Gamma)&=\cos y'\sin(v'-\theta_0)-\tan\tfrac{1}{2}\Phi\cos\omega\,(\sin y+\sin y'),\\\cos y\cos(v-\theta_0-\Gamma)&=\cos y'\cos(v'-\theta_0)+\tan\tfrac{1}{2}\Phi\sin\omega\,(\sin y+\sin y'),\end{aligned}$$

or, as they may also be written,

$$\begin{aligned}\cos y\sin(v-\theta_0-\Gamma)&=\cos\phi_0\sin(\text{þ}-\sigma_0)-\tan\tfrac{1}{2}\Phi\cos\omega\,(\sin y+\sin y'),\\\cos y\cos(v-\theta_0-\Gamma)&=\qquad\;\;\cos(\text{þ}-\sigma_0)+\tan\tfrac{1}{2}\Phi\sin\omega\,(\sin y+\sin y'),\end{aligned}$$

or, if we put $s=\sin y+\sin y'$, then observing that $\sin y=\sin\phi\sin(\text{þ}-\sigma)$, $\sin y'=$ $\sin\phi_0\sin(\text{þ}-\sigma_0)$, these become

$$\begin{aligned}\cos y\sin(v-\theta_0-\Gamma)&=\cos\phi_0\sin(\text{þ}-\sigma_0)-\tan\tfrac{1}{2}\Phi\,.\,s\cos\omega,\\\cos y\cos(v-\theta_0-\Gamma)&=\qquad\;\;\cos(\text{þ}-\sigma_0)+\tan\tfrac{1}{2}\Phi\,.\,s\sin\omega,\\\sin y\,\{=\sin\phi\sin(\text{þ}-\sigma)\}&=-\sin\phi_0\sin(\text{þ}-\sigma_0)+s,\end{aligned}$$

which are, in fact, Hansen's formulæ (16), p. 75, the letters corresponding as follows, viz.,

$$\begin{aligned}&v,\ \text{þ},\ y,\ \sigma,\ \sigma_0,\ \theta,\ \theta_0,\ \phi,\ \ \phi_0\ ,\ \Phi,\ \Gamma,\ \omega \text{ (suprà) to}\\&l,\ v,\ b,\ \sigma,\ h\ ,\ \theta,\ h\ ,\ i\ ,\ -k\ ,\ 2\eta,\ \Gamma,\ \omega \text{ (Hansen).}\end{aligned}$$

where, of course, the correspondence ϕ_0 to $-k$, shows that these angles are measured in a contrary direction. I had from Hansen's equations expected that the above formulæ would have contained $\sin y-\sin y'$ in place of $\sin y+\sin y'$.

2, *Stone Buildings, W. C.*, 4*th Dec*, 1860.

220.

NOTE ON A THEOREM OF JACOBI'S, IN RELATION TO THE PROBLEM OF THREE BODIES.

[From the *Monthly Notices of the Royal Astronomical Society*, vol. XXII. (1861), pp. 76—78.]

THE following theorem of Jacobi's (*Comptes Rendus*, t. III., p. 61 (1836)) has not, I think, found its way in an explicit form into any treatise of physical astronomy. The theorem is as follows, viz. "Consider the movement of a point without mass round the Sun, disturbed by a planet the orbit of which is circular. Let xyz be the rectangular coordinates of the disturbed body, the orbit of the disturbing planet being taken as the plane of xy, and the Sun as the centre of coordinates; let a' be the distance of the disturbing planet, $n't$ its longitude, m' its mass, M the mass of the Sun: then we have, *rigorously*,

$$\tfrac{1}{2}\left[\left(\frac{dx}{dt}\right)^2+\left(\frac{dy}{dt}\right)^2+\left(\frac{dz}{dt}\right)^2\right]-n'\left(x\frac{dy}{dt}-y\frac{dx}{dt}\right)$$

$$=\frac{M}{(x^2+y^2+z^2)^{\frac{1}{2}}}+m'\left\{\frac{1}{(x^2+y^2+z^2-2a'(x\cos n't+y\sin n't)+a'^2)^{\frac{1}{2}}}-\frac{x\cos n't+y\sin n't}{a'^2}\right\}+C.$$

This is therefore a new integral equation, which, in the problem of three bodies, subsists, as regards the terms independent of the eccentricity of the disturbing planet, and which is rigorous as regards all the powers of the mass of such planet. In the *Lunar Theory* the Earth must be substituted in the place of the Sun, and the Sun taken as the disturbing planet."

To prove the theorem, as expressed in polar coordinates, I take the equations of motion in the form in which I have employed them in my "Memoir on the Theory of Disturbed Elliptic Motion" (*Memoirs*, vol. XXVII. p. 1 (1859)), [212], viz.

$$\frac{d}{dt}\frac{dr}{dt} - r\cos^2 y\left(\frac{dv}{dt}\right)^2 - r\left(\frac{dy}{dt}\right)^2 + \frac{n^2a^3}{r^2} = \frac{d\Omega}{dr},$$

$$\frac{d}{dt}\left(r^2\cos^2 y\frac{dv}{dt}\right) = \frac{d\Omega}{dv},$$

$$\frac{d}{dt}\left(r^2\frac{dy}{dt}\right) + r^2\cos y\sin y = \frac{d\Omega}{dy},$$

where

$$\Omega = m'\left\{\frac{1}{\sqrt{r^2+r'^2-2rr'\cos H}} - \frac{r\cos H}{r'^2}\right\};$$

or, since $\cos H = \cos y\cos(v-v')$, and the Sun is considered as moving in a circular orbit (i.e. $r'=a'$, $v'=n't$), we have

$$\Omega = m'\left\{\frac{1}{\sqrt{r^2+a'^2-2ra'\cos y\cos(v-n't)}} - \frac{r\cos y\cos(v-n't)}{a'^2}\right\};$$

so that Ω is a function of r, v, y and of t, which last quantity enters only in the combination $v-n't$. Hence the complete differential coefficient of Ω is

$$\frac{d(\Omega)}{dt} = \frac{d'\Omega}{dt} - n'\frac{d\Omega}{dv},$$

where $\frac{d'\Omega}{dt}$ denotes, as usual, the differential coefficient in regard to the time, in so far as it enters through the coordinates r, v, y of the disturbed body.

We have, as usual,

$$\frac{d}{dt}\frac{dr^2 + r^2(\cos^2 y\, dv^2 + dy^2)}{dt^2} = \frac{d'\Omega}{dt};$$

and, from the foregoing equation,

$$\frac{d'\Omega}{dt} = \frac{d(\Omega)}{dt} + n'\frac{d\Omega}{dv}$$

$$= \frac{d(\Omega)}{dt} + n'\frac{d}{dt}\left(r^2\cos^2 y\frac{dv}{dt}\right);$$

hence, substituting this value, transposing, and integrating, we have

$$\left(\frac{dr}{dt}\right)^2 + r^2\left\{\cos^2 y\left(\frac{dv}{dt}\right)^2 + \left(\frac{dy}{dt}\right)^2\right\} - n'r^2\cos^2 y\frac{dv}{dt} = \Omega + C,$$

which is Jacobi's equation expressed in terms of the coordinates r, v, y.

M. de Pontécoulant, in his *Lunar Theory* (1846), where the solar eccentricity is neglected, writes (p. 91),

$$\int d'R = R + m\int \frac{dR}{dv}\,dt,$$

($n' = mn = m$, since n is there put equal to unity); and combining this with the equation (p. 43),

$$\frac{r^2 dv}{(1+s^2)\,dt} = h + \int \frac{dR}{dv}\,dt,$$

we have

$$\int d'R = R - mh + \frac{mr^2\,dv}{(1+s^2)\,dt};$$

and substituting this value of $\int d'R$ in the integral of *Vis Viva* (p. 41),

$$\left(\frac{dr}{dt}\right)^2 + \frac{r^2 dv^2}{(1+s^2)\,dt^2} + \frac{r^2 ds^2}{(1+s^2)^2\,dt^2} - \frac{2}{r} + \frac{1}{a} = 2\int d'R,$$

we have what is, in fact, Jacobi's equation.

221.

ON THE SECULAR ACCELERATION OF THE MOON'S MEAN MOTION.

[From the *Monthly Notices of the Royal Astronomical Society*, vol. XXII. (1862), pp. 171—231.]

THE present Memoir exhibits a new method of taking account, in the Lunar Theory, of the variation of the eccentricity of the Sun's orbit. The approximation is carried to the same extent as in Prof. Adams'. Memoir "On the Secular Variation of the Moon's Mean Motion" (*Phil. Trans.*, vol. CXLIII. (1853), pp. 397—406); and I obtain results agreeing precisely with his, viz., besides his new periodic terms in the longitude and radius vector, I obtain in the longitude the secular term

$$(-\tfrac{3}{2}m^2 + \tfrac{3771}{64}m^4)\int (e'^2 - E'^2)\, ndt,$$

and in the quotient radius, or radius vector divided by the mean distance, the secular term

$$(\tfrac{3}{4}m^2 - \tfrac{1973}{64}m^4)(e'^2 - E'^2),$$

which is, in fact, as will be shown, included implicitly in the results given in Professor Adams' Memoir. In quoting the foregoing results, I have written $e'^2 - E'^2$ in the place of $(e' + f't)^2 - e'^2 = 2e'f't$, which in the notation of the present Memoir it should have been; and I purposely refrain from here explaining the precise signification of the symbols: this is carefully done in the sequel. The method appears to me a very simple one in principle; and it possesses the advantage that it is not incorporated step by step with a lunar theory in which the eccentricity of the Sun's orbit is treated as constant; but it is added on to such a lunar theory, giving in the Moon's coordinates the supplementary terms which arise from the variation of the solar eccentricity, and thus serving as a verification of any process employed for taking account of such variation.

I have given the details of the work in a series of Annexes, 1 to 23: this appears to me the best course for presenting the investigation in a readable form.

I.

The inclination and eccentricity of the Moon's orbit, and, *à fortiori*, the variation of the position of the Ecliptic, and the Sun's latitude, are neglected; and the longitudes are measured from a fixed point in the Ecliptic. I write

n, the actual mean motion of the Moon at a given epoch;

viz., it is assumed that the mean longitude at the time t is $\epsilon + nt + n_2 t^2 +$ &c. where ϵ, n, n_2, &c. are absolute constants; and, moreover,

a, the calculated mean distance of the Moon;

that is, n^2a^3 is the sum of the masses of the Earth and Moon; a is therefore an absolute constant; and, in like manner,

n', the actual mean motion of the Sun at the same epoch,

a', the calculated mean distance of the Sun;

that is, if it were necessary to pay attention to the secular variation of the mean motion of the Sun, the assumption would be that the mean longitude was $\epsilon' + n't + n_2' t^2 +$ &c., ϵ', n', n_2', &c. being absolute constants, and $n'^2a'^3$ the sum of the masses of the Sun and Earth; a' would thus also be an absolute constant. But for the purpose of the present investigation the secular variation of the mean longitude of the Sun is neglected, or it is assumed that the mean longitude of the Sun is $\epsilon' + n't$, ϵ', n' being absolute constants; and that $n'^2a'^3$ is the sum of the masses of the Sun and Earth, a' being thus also an absolute constant.

I put also

m, the ratio of the mean motions of the Sun and Moon;

that is,

$$m = \frac{n'}{n}, \text{ or } n' = mn;$$

m is also an absolute constant.

The Sun is considered as moving in an elliptic orbit, the eccentricity whereof is $e' + \delta e'$ or $e' + f't$, e' and f' being absolute constants; the longitude of the Sun's perigee may be taken to be $\varpi' + (1 - c')\,n't$; so that the mean anomaly g' is $= n't + \epsilon - [\varpi' - (1 - c')\,n't] = c'n't + \epsilon' - \varpi'$; c', ϖ', being absolute constants; but c' is in fact treated as being $= 1$. Hence, if r', v' are the radius vector and longitude of the Sun, we have

$$r' = a' \operatorname{elqr}(e' + \delta e',\ g'),$$

$$v' = \varpi' + (1 - c')\,n't + \operatorname{elta}(e' + \delta e',\ g')$$

$$= n't + \epsilon + [\operatorname{elta}(e' + \delta e',\ g') - g'],$$

where

$$g' = c'n't + \epsilon' - \varpi'.$$

In the expression for the disturbing function the Sun's mass is taken to be $= n'^2a'^3$, or, what is the same thing, $= m^2n^2a'^3$.

Let r, v be the radius vector and longitude of the Moon; then, taking the usual approximate expression of the Disturbing Function, the equations of motion are

$$\frac{d}{dt}\frac{dr}{dt} - r\left(\frac{dv}{dt}\right)^2 + \frac{n^2a^3}{r^2} = m^2n^2a'^3\frac{r}{r'^3}\left(\tfrac{1}{2} + \tfrac{3}{2}\cos 2v - 2v'\right),$$

$$\frac{d}{dt}\frac{r^2dv}{dt} = m^2n^2a'^3\frac{r^2}{r'^3}\left(\quad - \tfrac{3}{2}\sin 2v - 2v'\right).$$

It will be convenient to assume

ρ, the quotient radius of the Moon's orbit,

ρ', the quotient radius of the Sun's orbit;

that is

$$r = \rho a, \quad r' = \rho' a'.$$

The equations of motion thus become

$$\frac{d}{dt}\frac{\delta\rho}{dt} - \rho\left(\frac{dv}{dt}\right)^2 + \frac{n^2}{\rho^3} = m^2n^2P,$$

$$\frac{d}{dt}\left(\rho^2\frac{dv}{dt}\right) = m^2n^2Q,$$

where for shortness

$$P = \frac{\rho}{\rho'^3}\left(\tfrac{1}{2} + \tfrac{3}{2}\cos 2v - 2v'\right),$$

$$Q = \frac{\rho^2}{\rho'^3}\left(\quad - \tfrac{3}{2}\sin 2v - 2v'\right),$$

in which

$$\rho' = \text{elqr}\,(e' + \delta e',\ g'),$$

$$v' = n't + \epsilon' + [\text{elta}\,(e' + \delta e',\ g') - g'].$$

I now change the notation by writing $\rho' + \delta\rho'$, $v' + \delta v'$, in the place of ρ', v', respectively, using henceforward ρ', v' to denote

$$\rho' = \qquad \text{elqr}\,(e',\ g'),$$

$$v' = n't + \epsilon' + [\text{elta}\,(e',\ g') - g'];$$

and I write also $\rho + \delta\rho$, $v + \delta v$, in the place of ρ, v, using henceforward ρ, v to denote the solutions of the equations obtained from the equations of motion by writing therein ρ', v' instead of the complete values $\rho' + \delta\rho'$, $v' + \delta v'$.

Suppose, in like manner, that the complete values of P, Q are denoted by $P + \delta P$, $Q + \delta Q$, where

$$\delta P = \frac{dP}{d\rho}\delta\rho + \frac{dP}{dv}\delta v + \frac{dP}{d\rho'}\delta\rho' + \frac{dP}{dv'}\delta v',$$

with a like value for δQ, the first powers of $\delta\rho$, δv, $\delta\rho'$, $\delta v'$ being alone attended to. Then, observing that the equations of motion are satisfied when $\delta\rho$, δv, $\delta\rho'$, $\delta v'$ are neglected, we have, it is clear,

$$\frac{d}{dt}\frac{d\delta\rho}{dt} - \delta\rho\left(\frac{dv}{dt}\right)^2 - 2\rho\frac{dv}{dt}\frac{d\delta v}{dt} - \frac{2n^2}{\rho^3}\delta\rho = m^2n^2\delta P,$$

$$\frac{d}{dt}\left(\rho^2\frac{d\delta v}{dt} + 2\rho\delta\rho\frac{dv}{dt}\right) = m^2n^2\delta Q.$$

The second of these equations gives

$$\rho^2\frac{d\delta v}{dt} + 2\rho\delta\rho\frac{dv}{dt} = m^2n^2(C + \int dQdt),$$

where the constant of integration, C, is to be so determined that δv may not contain any term of the form kt (for any such term is taken to be included in the term nt of $v+\delta v$). Multiplying the equation just obtained by $\frac{2}{\rho}\frac{dv}{dt}$, and adding it to the first equation, we have

$$\frac{d^2\delta\rho}{dt^2} + \left\{3\left(\frac{dv}{dt}\right)^2 - \frac{2n^2}{\rho^3}\right\}\delta\rho = m^2n^2\left(\delta P + \frac{2}{\rho}\frac{dv}{dt}(C + \int dQdt)\right),$$

which, with the above-mentioned integral equation, are the equations for the solution of the problem; but it will be convenient to write them under the slightly different form

$$\frac{d^2\delta\rho}{dt^2} + n^2\delta\rho = \left\{n^2 + \frac{2n^2}{\rho^3} - 3\left(\frac{dv}{dt}\right)^2\right\}\delta\rho + m^2n^2\left\{\delta P + \frac{2}{\rho}\frac{dv}{dt}\left(C + \int dQdt\right)\right\},$$

$$\frac{d\delta v}{dt} = -\frac{2}{\rho}\frac{dv}{dt}\delta\rho + \frac{m^2n^2}{\rho^2}(C + \int \delta Qdt).$$

In these equations C is determined, as above, by the condition that $\frac{d\delta v}{dt}$ may contain no constant term; the values of ρ', v', $\delta\rho'$, $\delta v'$ are of course given by the theory of elliptic motion, and those of ρ, v are given by the ordinary lunar theory, in which the eccentricity of the solar orbit is treated as a constant; and then, $\delta\rho$, δv being obtained by integrating the equations, the radius vector and longitude of the Moon are $a(\rho+\delta\rho)$ and $v+\delta v$ respectively.

We have

$$P = \frac{\rho}{\rho'^3}\left(\tfrac{1}{2} + \tfrac{3}{2}\cos 2v - 2v'\right),$$

$$Q = \frac{\rho}{\rho'^3}\left(-\tfrac{3}{2}\sin 2v - 2v'\right).$$

Moreover, by the lunar theory, observing that Plana's a is, or may be considered, identical with the a of the present Memoir, and putting also

$$\tau = nt + \epsilon - (n't + \epsilon'),$$

we have

$$
\begin{aligned}
\frac{1}{\rho} = 1 + \tfrac{1}{6} m^2 - \tfrac{3}{4}\ m^2 e'^2 & \\
- \tfrac{3}{2}\ m^2 e' & \quad \cos g' \\
+ \quad m^2 - \tfrac{5}{2} m^2 e'^2 & \quad ,, \ 2\tau \\
+ \tfrac{7}{2}\ m^2 e' & \quad ,, \ 2\tau - g' \\
- \tfrac{1}{2}\ m^2 e' & \quad ,, \ 2\tau + g' \\
- \tfrac{9}{4}\ m^2 e'^2 & \quad ,, \ 2g' \\
+ \tfrac{17}{2}\ m^2 e'^2 & \quad ,, \ 2\tau - 2g' \\
0\ m^2 e'^2 & \quad ,, \ 2\tau + 2g',
\end{aligned}
$$

$$
\begin{aligned}
v = \quad nt + \epsilon & \\
- 3\ \ m e' + 0\ m^2 e' & \quad \sin g' \\
+ \tfrac{11}{8}\ m^2 - \tfrac{55}{16} m^2 e'^2 & \quad ,, \ 2\tau \\
+ \tfrac{77}{16}\ m^2 e' & \quad ,, \ 2\tau - g' \\
- \tfrac{11}{16}\ m^2 e' & \quad ,, \ 2\tau + g' \\
- \tfrac{9}{4}\ m e'^2 + 0\ m^2 e'^2 & \quad ,, \ 2g' \\
+ \tfrac{187}{16}\ m^2 e'^2 & \quad ,, \ 2\tau - 2g' \\
0\ m^2 e'^2 & \quad ,, \ 2\tau + 2g',
\end{aligned}
$$

where the series are carried as far as m^2 and e'^2; the terms in e'^2 are given, as I shall have occasion to refer to them, but they are not used in the investigation, and, omitting them, the values are

$$
\begin{aligned}
\frac{1}{\rho} = \quad 1 + \tfrac{1}{6} m^2 & \\
- \tfrac{3}{2} m^2 e' & \quad \cos g' \\
+ \quad m^2 & \quad ,, \ 2\tau \\
+ \tfrac{7}{2} m^2 e' & \quad ,, \ 2\tau - g' \\
- \tfrac{1}{2} m^2 e' & \quad ,, \ 2\tau + g',
\end{aligned}
$$

$$
\begin{aligned}
v = \quad nt + \epsilon & \\
- 3\ m e' & \quad \sin g' \\
+ \tfrac{11}{8} m^2 & \quad ,, \ 2\tau \\
+ \tfrac{77}{16} m^2 e' & \quad ,, \ 2\tau - g' \\
- \tfrac{11}{8} m^2 e' & \quad ,, \ 2\tau + g'
\end{aligned}
$$

$$(g' = c'mnt + \text{const.}, \quad 2\tau = (2 - 2m)\, nt + \text{const.})$$

For the coordinates of the Sun we have

$$\begin{aligned}\frac{1}{\rho'} &= 1 \\ &\quad + e' \quad \cos g' \\ &\quad + e'^2 \quad ,, \ 2g',\end{aligned}$$

$$\begin{aligned}v' &= n't + \epsilon' \\ &\quad + 2\,e' \quad \sin g' \\ &\quad + \tfrac{5}{4}\,e'^2 \quad ,, \ 2g',\end{aligned}$$

the series being carried as far as e'^2; but the terms in e'^2 are only used for the formation of $\delta\rho'$, $\delta v'$; and, omitting them, we have

$$\begin{aligned}\frac{1}{\rho'} &= 1 \\ &\quad + e' \quad \cos g',\end{aligned}$$

$$\begin{aligned}v' &= n't + \epsilon' \\ &\quad + 2e' \quad \sin g'.\end{aligned}$$

If $e' + f't$ is written for e', then the value of $\delta e'$ is $= f't$; but as only the terms multiplied by the simple power f' are attended to, we may for convenience write $\delta e' = t$, the factor f' being restored in the final results: we thus have

$$\begin{aligned}\delta\frac{1}{\rho'} &= 1 \quad t\cos g' \\ &\quad + 2e' \quad ,, \ 2g',\end{aligned}$$

$$\begin{aligned}\delta v' &= 2 \quad t\sin g' \\ &\quad + \tfrac{5}{2}\,e' \quad ,, \ 2g',\end{aligned}$$

and we may add the equations

$$\begin{aligned}\frac{dv}{dt} &= n \times \overbrace{\quad 1 \quad} \\ &\quad - 3\,m^2e' \quad \cos g' \\ &\quad + \tfrac{11}{4}\,m^2 \quad ,, \ 2\tau \\ &\quad + \tfrac{77}{8}\,m^2e' \quad ,, \ 2\tau - g' \\ &\quad - \tfrac{11}{8}\,m^2e' \quad ,, \ 2\tau + g',\end{aligned}$$

$$\begin{aligned}\delta\frac{1}{\rho'^3} &= 3\ e'\ t \\ &\quad + 3 \quad t\cos g' \\ &\quad + \tfrac{9}{2}\,e' \quad ,, \ 2g',\end{aligned}$$

$$\begin{aligned}\frac{\delta v'}{\rho'^3} &= 2 \quad t\sin g' \\ &\quad + \tfrac{11}{2}\,e' \quad ,, \ 2g',\end{aligned}$$

which will be found useful.

II.

Proceeding now to the development of the solution, we have

$$\begin{aligned}\delta P = &\ \frac{1}{\rho'^3}\left(\tfrac{1}{2} + \tfrac{3}{2}\cos 2v - 2v'\right)\delta\rho \\ &+ \frac{\rho}{\rho'^3}\left(\quad - \tfrac{3}{2}\sin 2v - 2v'\right)\delta v \\ &+ \rho\left[\left(\tfrac{1}{2} + \tfrac{3}{2}\cos 2v - 2v'\right)\delta\frac{1}{\rho'^3} + (3\sin 2v - 2v')\frac{\delta v'}{\rho'^3}\right],\end{aligned}$$

where the terms containing $\delta\rho$ and δv are (see Annex 1)

$$\left.\begin{array}{lll} \tfrac{1}{2} & & \\ +\tfrac{3}{2}e' & \cos & g' \\ +\tfrac{3}{2} & ,, & 2\tau \\ +\tfrac{21}{4}e' & ,, & 2\tau - g' \\ -\tfrac{3}{4}e' & ,, & 2\tau + g' \end{array}\right|\ \delta\rho$$

$$\left.\begin{array}{lll} -3 & \sin & 2\tau \\ -\tfrac{21}{2}e' & ,, & 2\tau - g' \\ +\tfrac{3}{2}e' & ,, & 2\tau + g' \end{array}\right|\ \delta v,$$

and also

$$\begin{aligned}\delta Q = &\ \frac{\rho}{\rho'^3}(-3\sin 2v - 2v')\,\delta\rho \\ &+ \frac{\rho^2}{\rho'^3}(-3\cos 2v - 2v')\,\delta v \\ &+ \rho^2\left[\left(-\tfrac{3}{2}\sin 2v - 2v'\right)\delta\frac{1}{\rho'^3} + (3\cos 2v - 2v')\frac{\delta v'}{\rho'^3}\right],\end{aligned}$$

where the terms containing $\delta\rho$ and δv are (see Annex 2)

$$\left.\begin{array}{lll} -3 & \sin & 2\tau \\ -\tfrac{21}{2}e' & ,, & 2\tau - g' \\ +\tfrac{3}{2}e' & ,, & 2\tau + g' \end{array}\right|\ \delta\rho$$

$$\left.\begin{array}{|lll|} -3 & & \cos 2\tau \\ -\tfrac{21}{2}e' & ,, & 2\tau-g' \\ +\tfrac{3}{2}e' & ,, & 2\tau+g' \end{array}\right. \delta v,$$

but the additional term $\tfrac{171}{4}m^2e'\cos g'$ is ultimately added (see Annex 17) to the coefficient of δv.

Neglecting the terms in $\delta\rho$, δv, we have (see Annex 4)

$$\begin{aligned} \delta P = & \ \tfrac{3}{2}e' & t & \\ & +\tfrac{3}{2} & t\cos g' & \\ & -\tfrac{15}{2}e' & ,, & 2\tau \\ & +\tfrac{21}{4} & ,, & 2\tau-g' \\ & -\tfrac{3}{4} & ,, & 2\tau+g' \\ & +\tfrac{9}{2}e' & ,, & 2g' \\ & +\tfrac{51}{2}e' & ,, & 2\tau-2g', \end{aligned}$$

and similarly (see Annex 5),

$$\begin{aligned} \delta Q = & \ \tfrac{15}{2}e' & t\sin 2\tau & \\ & -\tfrac{21}{4} & ,, & 2\tau-g' \\ & +\tfrac{3}{4} & ,, & 2\tau+g' \\ & -\tfrac{51}{2}e' & ,, & 2\tau-2g'. \end{aligned}$$

But in the foregoing expression of δP the terms belonging to the arguments g', $2g'$ give in $\delta\rho$, terms which rise by integration in δv; and in forming the expressions for δP, δQ, it is proper to take account of these terms. Attending only to the terms in question, we have

$$\frac{d^2\delta\rho}{dt^2}+n^2\delta\rho = m^2n^2\delta P = \overbrace{\begin{array}{ll} \tfrac{3}{2}m^2 & t\cos g' \\ +\tfrac{9}{2}m^2e' & ,,\ 2g'. \end{array}}^{n^2\times}$$

Now in general, if

$$\frac{d^2\delta\rho}{dt^2}+n^2\delta\rho = n^2 \quad t\cos n\alpha t,$$

then

$$\delta\rho = \frac{2\alpha}{(1-\alpha^2)^2}\sin n\alpha t + \frac{1}{1-\alpha^2} \quad t\cos n\alpha t;$$

and hence the foregoing equation gives in $\delta\rho$ the terms

$$\begin{array}{lll} 3m^3 & \sin g' & \\ +18m^3e' & ,, & 2g' \end{array} \qquad \begin{array}{lll} +\tfrac{3}{2} & & t\cos g' \\ +\tfrac{9}{2}m^2e' & ,, & 2g'; \end{array}$$

or, neglecting the terms which contain m^3, the terms of $\delta\rho$ are

$$\begin{array}{ll} \tfrac{3}{2} m^2 & t \cos g' \\ + \tfrac{9}{2} m^2 e' & \;,,\; 2g'. \end{array}$$

Substituting these terms in

$$\frac{d\delta v}{dt} = -2n\delta\rho,$$

we have

$$\frac{d\delta v}{dt} = \overbrace{\begin{array}{ll} -3\,m^2 & t\cos g' \\ -9\,m^2 e' & \;,,\; 2g'; \end{array}}^{n\,\times}$$

and since, in general,

$$\int t \cos nat\, dt = \frac{1}{n^2 a^2} \cos nat + \frac{1}{na} t \sin nat,$$

we obtain in δv the terms

$$\overbrace{\begin{array}{ll} -3 & \cos g' \\ -\tfrac{9}{4} e' & \;,,\; 2g' \end{array}}^{n^{-1}\times} + \overbrace{\begin{array}{ll} -3\,m & t \sin g' \\ -\tfrac{9}{2} me' & \;,,\; 2g', \end{array}}$$

or for the present purpose the terms

$$\overbrace{\begin{array}{ll} -3 & \cos g' \\ -\tfrac{9}{4} e' & \;,,\; 2g'; \end{array}}^{n^{-1}\times}$$

and these give in δP the additional terms (see Annex 6)

$$\overbrace{\begin{array}{ll} \tfrac{27}{2} & \sin 2\tau \\ + \tfrac{9}{2} & \;,,\; 2\tau - g' \\ + \tfrac{9}{2} & \;,,\; 2\tau + g' \\ + \tfrac{153}{8} e' & \;,,\; 2\tau - 2g' \\ + \tfrac{9}{8} e' & \;,,\; 2\tau + 2g', \end{array}}^{n^{-1}\times}$$

and in δQ the additional terms (see Annex 7)

$$\overbrace{\begin{array}{ll} \tfrac{27}{2} & \cos 2\tau \\ + \tfrac{9}{2} & \;,,\; 2\tau - g' \\ + \tfrac{9}{2} & \;,,\; 2\tau + g' \\ + \tfrac{153}{8} e' & \;,,\; 2\tau - 2g' \\ + \tfrac{9}{8} e' & \;,,\; 2\tau + 2g'. \end{array}}^{n^{-1}\times}$$

Combining the foregoing results, we have,

$$
\begin{array}{rllrll}
\delta P = & \tfrac{3}{2}e' & t & + & \multicolumn{2}{c}{\overbrace{\hspace{6em}}^{n^{-1}\times}} \\
& +\tfrac{3}{2} & t\cos g' & & 0 & \sin g' \\
& -\tfrac{15}{2}e' & \text{,,}\ 2\tau & & +\tfrac{27}{2}e' & \text{,,}\ 2\tau \\
& +\tfrac{21}{4} & \text{,,}\ 2\tau - g' & & +\tfrac{9}{2} & \text{,,}\ 2\tau - g' \\
& -\tfrac{3}{4} & \text{,,}\ 2\tau + g' & & +\tfrac{9}{2} & \text{,,}\ 2\tau + g' \\
& +\tfrac{9}{2}e' & \text{,,}\ 2g' & & 0 & \text{,,}\ 2g' \\
& +\tfrac{51}{2}e' & \text{,,}\ 2\tau - 2g' & & +\tfrac{153}{8}e' & \text{,,}\ 2\tau - 2g' \\
& 0 & \text{,,}\ 2\tau + 2g' & & +\tfrac{9}{8}e' & \text{,,}\ 2\tau + 2g',
\end{array}
$$

and

$$
\begin{array}{rllrll}
& & & & \multicolumn{2}{c}{\overbrace{\hspace{6em}}^{n^{-1}\times}} \\
\delta Q = & 0 & t\sin g' & & +0 & \cos g' \\
& +\tfrac{15}{2}e' & \text{,,}\ 2\tau & & +\tfrac{27}{2}e' & \text{,,}\ 2\tau \\
& -\tfrac{21}{4} & \text{,,}\ 2\tau - g' & & +\tfrac{9}{2} & \text{,,}\ 2\tau - g' \\
& +\tfrac{3}{4} & \text{,,}\ 2\tau + g' & & +\tfrac{9}{2} & \text{,,}\ 2\tau + g' \\
& 0 & \text{,,}\ 2g' & & 0 & \text{,,}\ 2g' \\
& -\tfrac{51}{2}e' & \text{,,}\ 2\tau - 2g' & & +\tfrac{153}{8}e' & \text{,,}\ 2\tau - 2g' \\
& 0 & \text{,,}\ 2\tau + 2g' & & +\tfrac{9}{8}e' & \text{,,}\ 2\tau + 2g';
\end{array}
$$

whence also (see Annex 8)

$$
\begin{array}{rllrll}
& & & & \multicolumn{2}{c}{\overbrace{\hspace{6em}}^{n^{-1}\times}} \\
n\int\delta Q dt = & 0 & t\cos g' & & +0 & \sin g' \\
& -\tfrac{15}{4}e' & \text{,,}\ 2\tau & & +\tfrac{69}{8}e' & \text{,,}\ 2\tau \\
& +\tfrac{21}{8} & \text{,,}\ 2\tau - g' & & +\tfrac{15}{16} & \text{,,}\ 2\tau - g' \\
& -\tfrac{3}{8} & \text{,,}\ 2\tau + g' & & +\tfrac{39}{16} & \text{,,}\ 2\tau + g' \\
& 0 & \text{,,}\ 2g' & & 0 & \text{,,}\ 2g' \\
& +\tfrac{51}{4}e' & \text{,,}\ 2\tau - 2g' & & +\tfrac{51}{16}e' & \text{,,}\ 2\tau - 2g' \\
& 0 & \text{,,}\ 2\tau + 2g' & & +\tfrac{9}{16}e' & \text{,,}\ 2\tau + 2g'.
\end{array}
$$

The equation for $\delta\rho$ may be written,

$$\frac{d^2\delta\rho}{dt^2} + n^2\delta\rho = m^2n^2\left(\delta P + 2n\int\delta Q dt\right),$$

and we have, (see Annex 9)

$$\begin{array}{llllll}
\delta P + 2n\int \delta Q dt = & \tfrac{3}{2}\, e' & t & & + \overbrace{\qquad\qquad\qquad}^{n^{-1}\times} & \\
& + \tfrac{3}{2} & t\cos g' & & 0 & \sin g' \\
& - 15\, e' & \text{,,}\ 2\tau & & + \tfrac{123}{4}\, e' & \text{,,}\ 2\tau \\
& + \tfrac{21}{2} & \text{,,}\ 2\tau - g' & & + \tfrac{51}{8} & \text{,,}\ 2\tau - g' \\
& - \tfrac{3}{2} & \text{,,}\ 2\tau + g' & & + \tfrac{75}{8} & \text{,,}\ 2\tau + g' \\
& + \tfrac{9}{2}\, e' & \text{,,}\ 2g' & & 0 & \text{,,}\ 2g' \\
& + 51\, e' & \text{,,}\ 2\tau - 2g' & & + \tfrac{51}{2}\, e' & \text{,,}\ 2\tau - 2g' \\
& 0 & \text{,,}\ 2\tau + 2g' & & + \tfrac{9}{4}\, e' & \text{,,}\ 2\tau + 2g'.
\end{array}$$

Hence observing that a term $n^2 t\cos n\alpha t$ in $\dfrac{d^2\delta\rho}{\delta t^2} + n^2\delta\rho$, gives in $\delta\rho$ the terms

$$\frac{1}{1-\alpha^2}\, t\cos n\alpha t \qquad + \frac{2\alpha}{(1-\alpha^2)^2}\,\frac{1}{n}\,\sin n\alpha t,$$

and a term $n\sin n\alpha t$ in $\dfrac{d^2\delta\rho}{dt^2} + n^2\delta\rho$, gives in $\delta\rho$ the term

$$\frac{1}{1-\alpha^2}\,\frac{1}{n}\,\sin n\alpha t,$$

we have (see Annex 10, but restoring the factor f'),

$\delta\rho = f'$ into as follows, viz.

$$\begin{array}{lllll}
\tfrac{3}{2}\, m^2 e' & t & & + \overbrace{\qquad\qquad\qquad}^{n^{-1}\times} & \\
+ \tfrac{3}{2}\, m^2 & t\cos g' & & 3\ m^3 & \sin g' \\
+ 5\ m^2 e' & \text{,,}\ 2\tau & & - \tfrac{203}{12}\, m^2 e' & \text{,,}\ 2\tau \\
- \tfrac{7}{2}\, m^2 & \text{,,}\ 2\tau - g' & & + \tfrac{61}{24}\, m^2 & \text{,,}\ 2\tau - g' \\
+ \tfrac{1}{2}\, m^2 & \text{,,}\ 2\tau + g' & & - \tfrac{91}{24}\, m^2 & \text{,,}\ 2\tau + g' \\
+ \tfrac{9}{2}\, m^2 e' & \text{,,}\ 2g' & & - 18\ m^3 e' & \text{,,}\ 2g' \\
- 17\ m^2 e' & \text{,,}\ 2\tau - 2g' & & + \tfrac{85}{6}\, m^2 e' & \text{,,}\ 2\tau - 2g' \\
0 & \text{,,}\ 2\tau + 2g' & & - \tfrac{3}{4}\, m^2 e' & \text{,,}\ 2\tau + 2g'.
\end{array}$$

The first column, containing the term in t and the terms $t\cos$ arg., shows that the constant term of ρ, and the terms involving the cosines of the same arguments, as obtained without attending to the variation of the solar eccentricity, are correct as regards the first power of t, when for the constant eccentricity e' we substitute $e'+f't$. In fact, the above-mentioned expression (correct to e'^2) of $\frac{1}{\rho}$ gives

$$\begin{array}{lll}
\rho = 1 - \tfrac{1}{6}m^2 + \tfrac{3}{4}m^2e'^2 & & \\
\quad + \tfrac{3}{2}m^2e'^2 & \cos g' \\
\quad - \quad m^2 + \tfrac{5}{2}m^2e'^2 & ,, \ 2\tau \\
\quad - \tfrac{7}{2}m^2e' & ,, \ 2\tau - g' \\
\quad + \tfrac{1}{2}m^2e' & ,, \ 2\tau + g' \\
\quad + \tfrac{9}{4}m^2e'^2 & ,, \ 2g' \\
\quad - \tfrac{17}{2}m^2e'^2 & ,, \ 2\tau - 2g' \\
\quad\quad 0 & ,, \ 2\tau + 2g',
\end{array}$$

and putting therein $e'+f't$ in the place of e', we have the first column of the foregoing expression of $\delta\rho$.

The second column, involving sin arg., contains the new periodic terms considered in Prof. Adams' Memoir of 1853, and the coefficients for the arguments g', 2τ, $2\tau-g'$, $2\tau+g'$, agree with his values; observing that his terms belong to $\delta\frac{1}{\rho} = -\frac{\delta\rho}{\rho^2} = -\delta\rho$, so that the signs are reversed; those for the remaining arguments $2g'$, $2\tau-2g'$, $2\tau+2g'$, are not given by him.

The equation for δv may be written,

$$\frac{d\delta v}{dt} = -2n\delta\rho + m^2n^2\int \delta Q dt,$$

and we have (see Annex 11)

$$\begin{array}{llll}
\dfrac{d\delta v}{dt} = \overbrace{-\ 3\ m^2e'\ \ t}^{n\,\times} & & & \\
\quad -\ 3\ m^2 & t\cos g' & -\ 6\ m^3 & \sin g \\
\quad + \tfrac{55}{4}m^2e' & ,,\ 2\tau & + \tfrac{1019}{24}m^2e' & ,,\ 2\tau \\
\quad + \tfrac{77}{8}m^2 & ,,\ 2\tau - g' & - \tfrac{199}{48}m^2 & ,,\ 2\tau - g' \\
\quad - \tfrac{11}{8}m^2 & ,,\ 2\tau + g' & + \tfrac{481}{48}m^2 & ,,\ 2\tau + g' \\
\quad -\ 9\ m^2e' & ,,\ 2g' & -\ 36\ m^3e' & ,,\ 2g' \\
\quad + \tfrac{187}{8}m^2e' & ,,\ 2\tau - 2g' & - \tfrac{1207}{48}m^2e' & ,,\ 2\tau - 2g' \\
\quad\quad 0 & ,,\ 2\tau + 2g' & + \tfrac{33}{16}m^2e' & ,,\ 2\tau + 2g',
\end{array}$$

whence, integrating by the formulæ

$$\int t\cos n\alpha t\ \ dt = \frac{1}{n\alpha}t\sin n\alpha t + \frac{1}{n^2\alpha^2}\cos n\alpha t,$$

$$\int \sin n\alpha t\ \ dt = \qquad -\frac{1}{n\alpha}\cos n\alpha t,$$

we have (see Annex 12, but restoring the factor f'),

$\delta v = f'$ into as follows, viz.

$$
\begin{array}{llll}
 & & - \tfrac{3}{2}\ m^2 n e' t^2 & \\
 & & \quad n^{-1}\times & \\
-3\ m & t\sin g' & + \overbrace{-3+6\,m^2} & \cos g' \\
-\tfrac{55}{8}\ m^2 e' & ,,\ 2\tau & -\tfrac{74}{3}\ m^2 e' & ,,\ 2\tau \\
+\tfrac{77}{16}\ m^2 & ,,\ 2\tau - g' & +\tfrac{215}{48}\ m^2 & ,,\ 2\tau - g' \\
-\tfrac{11}{16}\ m^2 & ,,\ 2\tau + g' & -\tfrac{257}{48}\ m^2 & ,,\ 2\tau + g' \\
-\tfrac{9}{2}\ m e' & ,,\ 2g' & (-\tfrac{9}{4}+18\,m^2)\,e' & ,,\ 2g' \\
+\tfrac{187}{8}\ m^2 e' & ,,\ 2\tau - 2g' & +\tfrac{2239}{96}\ m^2 e' & ,,\ 2\tau - 2g' \\
0 & ,,\ 2\tau + 2g' & -\tfrac{33}{32}\ m^2 e' & ,,\ 2\tau + 2g'.
\end{array}
$$

The first column, containing $t \sin$ arg., may be obtained from the before-mentioned expression (accurate to e'^2) of v, by substituting therein $e' + f't$ in the place of e'.

The term $-\tfrac{3}{2} m^2 n e' f' t^2$; or, as it may be written, $-\tfrac{3}{2} m^2 \int n\,[(e' + f't)^2 - e'^2]\,dt$, is the first term of the acceleration; the other terms of the second column are the new periodic terms in δv, considered by Prof. Adams; the coefficients for the arguments g', 2τ, $2\tau - g'$, $2\tau + g'$, agreeing with his values, but those for the remaining arguments $2g'$, $2\tau - 2g'$, $2\tau + 2g'$ not being given by him.

III.

Proceeding now to the calculation of the term in m^4 of the acceleration, we have,

$$\frac{d\delta v}{dt} = -\frac{2}{\rho}\frac{dv}{dt}\,\delta\rho + \frac{m^2 n^2}{\rho^2}\left(C + \int \delta Q dt\right),$$

where the non-periodic part of $\delta\rho$ is of the form,

$$\delta\rho = (\tfrac{3}{2} m^2 + \square m^4)\, e' f' t,$$

and it is in the first place necessary to find the value of the numerical coefficient $\square$, in fact to calculate the secular part of $\delta\rho$ as far as m^4. Reverting to the equation

$$\frac{d^2\delta\rho}{dt^2} + n^2\delta\rho = \left(n^2 + \frac{2n^2}{\rho^3} - 3\left(\frac{dv}{dt}\right)^2\right)\delta\rho + m^2 n^2\left(\delta P + \frac{2}{\rho}\frac{dv}{dt}\left(C + \int \delta Q dt\right)\right),$$

and as before omitting in the process the factor f':

The part $\left(n^2 + \dfrac{2n^2}{\rho^3} - 3\left(\dfrac{dv}{dt}\right)^2\right)\delta\rho$ contains (see Annex 13), a term

$$= \tfrac{381}{8} m^4 n^2 e' t.$$

The part of $m^2n^2\delta P$, which involves $\delta\rho$, contains (see Annex 14) a term

$$= -\tfrac{495}{32}\, m^4n^2e't.$$

The part of $m^2n^2\delta P$, which depends on δv, contains a term

$$= -\tfrac{15}{4}\, m^4n^2e't.$$

The part of $m^2n^2\delta P$, depending on $\delta\rho'$ and $\delta v'$, is found (see Annex 18) to contain, besides the term $\tfrac{3}{2}\, m^2ne't$ in m^2 already obtained, a new term

$$= -\tfrac{647}{32}\, m^4n^2e't.$$

And finally the part $m^2n^2 \dfrac{2}{\rho}\dfrac{dv}{dt}\left(C + \int \delta Q dt\right)$ is found (see Annex 19) to contain a term

$$\left(2 . \tfrac{-1455}{32} + \tfrac{675}{32} =\right) - \tfrac{2235}{32}\, m^4n^2e't,$$

where the component coefficient $-\tfrac{1455}{32}$, which arises from the new periodic terms of $\delta\rho$ and δv is separately calculated (see Annex 21).

Hence $\dfrac{d^2\delta\rho}{dt^2} + n^2\delta\rho$ contains the term

$$\left(\Box = \tfrac{381}{8} - \tfrac{495}{32} - \tfrac{15}{4} - \tfrac{647}{32} - \tfrac{2235}{32} =\right) - \tfrac{1973}{32}\, m^4n^2e't,$$

and this gives in $\delta\rho$ the term

$$-\tfrac{1973}{32}\, m^4e't,$$

so that, restoring the term in m^2, and the common factor f', the complete secular term of $\delta\rho$ is

$$\delta\rho = \left(\tfrac{3}{2}\, m^2 - \tfrac{1973}{32}\, m^4\right) e'f't,$$

which, as will be shown, Art. IV., agrees with Prof. Adams' result.

Resuming now the equation,

$$\frac{d\delta v}{dt} = -\frac{2}{\rho}\frac{dv}{dt}\delta\rho + \frac{m^2n^2}{\rho^2}\left(C + \int \delta Q dt\right),$$

the part $-\dfrac{2}{\rho}\dfrac{dv}{dt}\delta\rho$ contains (see Annex 22) the term

$$\left(\tfrac{1973}{16} + \tfrac{275}{8} =\right) \tfrac{2523}{16}\, m^4ne't,$$

and the part $\dfrac{m^2n^2}{\rho}\left(C + \int \delta Q dt\right)$ contains (see Annex 23) the term

$$\left(\tfrac{45}{8} - \tfrac{1455}{32} =\right) - \tfrac{1275}{32}\, m^4ne't,$$

so that we have in $\dfrac{d\delta v}{dt}$ the term

$$\left(\tfrac{2523}{16} - \tfrac{1275}{32} =\right) \tfrac{3771}{32}\, m^4ne't,$$

giving in δv the term

$$\tfrac{3771}{64} m^4 n e' t^2,$$

or, restoring the term in m^2 and the common factor f', the complete secular term of δv is

$$\delta v = (-\tfrac{3}{2} m^2 + \tfrac{3771}{64} m^4)\, n e' f' t^2,$$

which agrees with the value obtained by Prof. Adams. It is right to remark that the m of Prof. Adams is different from that of the present Memoir; we have in fact,

$$m\,(\text{Adams}) = m\,\{1 + (\tfrac{3}{2} m^2 - \tfrac{3771}{64} m^4)\, 2e'f't\}$$

in the notation of the present Memoir; but as f'^2 is throughout neglected, we may in the foregoing expression of the secular part of δv substitute the m of Prof. Adams. And then if the term be written in the form

$$\delta v = (-\tfrac{3}{2} m^2 + \tfrac{3771}{64} m^2) \int [(e' + f't)^2 - e'^2]\, n dt,$$

the two results are seen to agree together. But as regards the before-mentioned secular term,

$$\delta \rho = (\tfrac{3}{2} m^2 - \tfrac{1973}{32} m^4)\, e'f't,$$

the identification is less easy, and I shall consider it in the following article.

IV.

It will be convenient to write M, N, A, E', in place of the m, n, a, e', of the foregoing part of the present Memoir, and to now use m, n, e', in the significations in which they are employed by Prof. Adams; E' (the constant part of the solar eccentricity) is his E', and his e' is $E' + f't$. As to his symbols a, $a_{,}$, these, I think, ought to have been represented, and I shall here represent them by a, $\mathrm{a}_{,}$ [1]. And I take a such that $n^2 a^3 =$ Sum of the masses of the Earth and Moon; or, taking this to be unity, we have $N^2 A^3 = 1$, $n^2 a^3 = 1$.

The formulæ of Prof. Adams' memoir, which it will be necessary to make use of, may be written

$$\begin{aligned}
\frac{\mathrm{a}}{r} = 1 - \tfrac{11}{8} m^2 - \tfrac{201}{16} m^4 e'^2 & \\
- \tfrac{3}{2} m^2 e' & \quad \cos g' \\
+ \quad m^2 - \tfrac{5}{2} m^2 e'^2 & \quad ,, \ 2\tau \\
+ \tfrac{7}{2} m^2 e' & \quad ,, \ 2\tau - g' \\
- \tfrac{1}{2} m^2 e' & \quad ,, \ 2r + g'
\end{aligned}$$

[1] Plana, in his Lunar Theory, uses the three letters a, $\mathrm{a}_{,}$, a; his a and n being such that $n^2 a^3 =$ Sum of the masses of the Earth and Moon. There is an obvious inconvenience in writing a, $a_{,}$, in the place of his a, $\mathrm{a}_{,}$.

the sine terms being disregarded,

$$n = N\{1 + (-\tfrac{3}{2}m^2 + \tfrac{3771}{64}m^4)\, 2E'f't\},$$

(whence also

$$m = M\{1 + (\ \tfrac{3}{2}m^2 - \tfrac{3771}{64}m^4)\, 2E'f't\}$$

since $m = \dfrac{n'}{n}$, $M = \dfrac{N'}{N}$).

$$\frac{1}{n} = \mathrm{a}_{,}^{\frac{3}{2}}\{1 + \ m^2 - \tfrac{197}{64}m^4 + (\ \tfrac{3}{2}m^2 - \tfrac{3867}{64}m^4)\, e'^2\},$$

$$1 = \frac{\mathrm{a}}{\mathrm{a}_{,}}\{1 - \tfrac{1}{2}m^2 + \tfrac{13}{4}m^4 + (-\tfrac{3}{4}m^2 + \tfrac{3201}{64}m^4)\, e'^2\},$$

which formulæ $\left(\text{observing that } \dfrac{1}{\rho + \delta\rho} = \dfrac{A}{r}\right)$ lead to the value of $\delta\rho$, and we should obtain for the secular part the foregoing expression, which will now be

$$\delta\rho = (\tfrac{3}{2}M^2 - \tfrac{1973}{32}M^4)\, E'f't.$$

We in fact have

$$\frac{1}{n} = a^{\frac{3}{2}} = \mathrm{a}_{,}^{\frac{3}{2}}\{1 + \ m^2 - \tfrac{197}{64}m^4 + (\tfrac{3}{2}m^2 - \tfrac{3867}{64}m^4)\, e'^2\},$$

and thence

$$\begin{aligned} a &= \mathrm{a}_{,}\ \{1 + \tfrac{2}{3}m^2 - \tfrac{197}{96}m^4 + (\ m^2 - \tfrac{3867}{96}m^4)\, e'^2 \\ &\qquad\qquad - \tfrac{1}{9}(m^4 \qquad\qquad + 3\, m^4 e'^2)\} \\ &= \mathrm{a}_{,}\ \{1 + \tfrac{2}{3}m^2 - \tfrac{623}{288}m^4 + (\ m^2 - \tfrac{3899}{96}m^4)\, e'^2\}\,; \end{aligned}$$

but

$$1 = \frac{\mathrm{a}}{\mathrm{a}_{,}}\{1 - \tfrac{1}{2}m^2 + \tfrac{13}{4}m^4 + (-\tfrac{3}{4}m^2 + \tfrac{3201}{64}m^4)\, e'^2\},$$

and therefore

$$\begin{aligned} a = \mathrm{a}\ \{&1 + \tfrac{2}{3}m^2 - \tfrac{623}{288}m^4 + (\qquad m^2 - \tfrac{3899}{96}m^4)\, e'^2 \\ &\quad - \tfrac{1}{2}m^2 + \tfrac{13}{4}m^4 \qquad\qquad - \tfrac{1}{2}\ m^4\ e'^2 \\ &\qquad\qquad - \tfrac{1}{3}\ m^4 + (-\tfrac{3}{4}m^2 + \tfrac{3201}{64}m^4)\, e'^2 \\ &\qquad\qquad\qquad\qquad\qquad - \tfrac{1}{2}\ m^4\ e'^2\} \\ = \mathrm{a}\ \{&1 + \tfrac{1}{6}m^2 + \tfrac{217}{288}m^4 + (\ \tfrac{1}{4}m^2 + \tfrac{1613}{192}m^4)\, e'^2\}\,; \end{aligned}$$

and hence, since for the non-periodic part

$$\frac{\mathrm{a}}{r} = 1 - \tfrac{11}{8}m^2 - \tfrac{201}{16}m^4 e'^2,$$

we find

$$\begin{aligned}\frac{a}{r} &= 1 + \tfrac{1}{6} m^2 + \tfrac{217}{288} m^4 + (\tfrac{1}{4} m^2 + \tfrac{1613}{192} m^4)\, e'^2 \\ &\qquad - \tfrac{11}{8} m^4 \qquad\qquad - \tfrac{201}{16} m^4\, e'^2 \\ &= 1 + \tfrac{1}{6} m^2 - \tfrac{179}{288} m^4 + (\tfrac{1}{4} m^2 - \tfrac{799}{192} m^4)\, e'^2 \\ &= 1 + \tfrac{1}{6} m^2 - \tfrac{179}{288} m^4 + (\tfrac{1}{4} m^2 - \tfrac{799}{192} m^4)\, E'^2 + (\tfrac{1}{4} m^2 - \tfrac{799}{192} m^4)\, 2E'f't.\end{aligned}$$

But

$$\frac{A}{a} = \left(\frac{n}{N}\right)^{\frac{2}{3}} = 1 + (-m^2 + \tfrac{3771}{96} m^4)\, 2E'f't;$$

and therefore

$$\begin{aligned}\frac{A}{r} &= 1 + \tfrac{1}{6} m^2 - \tfrac{179}{288} m^4 + (\tfrac{1}{4} m^2 - \tfrac{799}{192} m^4)\, E'^2 + (\ \tfrac{1}{4} m^2 - \tfrac{799}{192} m^4)\, 2E'f't \\ &\qquad + (-\ m^2 + \tfrac{3771}{96} m^4)\, 2E'f't \\ &\qquad - \tfrac{1}{6} m^4 . 2E'f't \\ &= 1 + \tfrac{1}{6} m^4 - \tfrac{179}{288} m^4 + (\tfrac{1}{4} m^2 - \tfrac{799}{192} m^4)\, E'^2 + (-\tfrac{3}{4} m^2 + \tfrac{2237}{64} m^4)\, 2E'f't.\end{aligned}$$

But we have

$$m = M\,\{1 + (\tfrac{3}{2} M^2 - \tfrac{3771}{64} M^4)\, 2E'f't\};$$

and thence in the foregoing expression

$$m^2 = M^2 + 3M^4 . 2E'f't,$$

$$m^4 = M^4;$$

and therefore

$$\begin{aligned}\frac{A}{r} &= 1 + \tfrac{1}{6} M^2 - \tfrac{179}{288} M^4 + (\tfrac{1}{4} M^2 - \tfrac{799}{192} M^4)\, E'^2 + (-\tfrac{3}{4} M^2 + \tfrac{2237}{64} M^4)\, 2E'f't \\ &\qquad + \ \tfrac{1}{2}\ M^4 . 2E'f't \\ &= 1 + \tfrac{1}{6} M^2 - \tfrac{179}{288} M^4 + (\tfrac{1}{4} M^2 - \tfrac{799}{192} M^4)\, E'^2 + (-\tfrac{3}{4} M^2 + \tfrac{2269}{64} M^4)\, 2E'f't;\end{aligned}$$

and observing that in the periodic terms we may write A, in the place of a, and neglect the sine terms, we have

$$\begin{aligned}\frac{A}{r} = \frac{1}{\rho + \delta\rho} &= 1 + \tfrac{1}{6} M^2 - \tfrac{179}{288} M^4 + (\tfrac{1}{4} M^2 - \tfrac{799}{192} M^4)\, E'^2 \\ &\qquad + (-\tfrac{3}{4} M^2 + \tfrac{2269}{64} M^4)\, 2E'f't \\ &- \tfrac{3}{2} m^2 e' \qquad \cos g' \\ &+ \ m^2 (1 - \tfrac{5}{2} e'^2) \quad \text{,,} \quad 2\tau \\ &+ \tfrac{7}{2} m^2 e' \qquad \text{,,} \quad 2\tau - g' \\ &- \tfrac{1}{2} m^2 e' \qquad \text{,,} \quad 2\tau + g',\end{aligned}$$

say $\dfrac{1}{\rho + \delta\rho} = 1 + X$; and thence $\rho + \delta\rho = \dfrac{1}{1+X} = 1 - X + X^2$, the non-periodic part whereof is

$$\begin{aligned}
1 &- \tfrac{1}{6} M^2 + \tfrac{179}{128} M^4 + (-\tfrac{1}{4} M^2 + \tfrac{179}{192} M^4)\, E'^2 + (\tfrac{3}{4} M^2 - \tfrac{2269}{64} M^4)\, 2E'f't \\
&+ \tfrac{1}{36} M^4 \qquad + \tfrac{1}{12} M^4\, E'^2 \qquad + \tfrac{1}{4} M^4 . 2E'f't \\
&+ \tfrac{1}{2} . \tfrac{9}{4}\, m^4 e'^2 \\
&+ \tfrac{1}{2}\, m^4 (1 - 5e'^2) \\
&+ \tfrac{1}{2} . \tfrac{49}{4}\, m^4 e'^2 \\
&+ \tfrac{1}{2} . \tfrac{1}{4}\, m^4 e'^2,
\end{aligned}$$

where the terms in m^4 and $m^4e'^2$ are

$$= \tfrac{1}{2} m^4 + (\tfrac{9}{8} - \tfrac{5}{2} + \tfrac{49}{8} + \tfrac{1}{8} =)\, \tfrac{39}{8} m^4 e'^2,$$

which are

$$= \tfrac{1}{2} M^4 \qquad + \tfrac{39}{8} M^4 E'^2 \qquad + \tfrac{39}{8} M^4 . 2E'f't;$$

so that the foregoing expression of the non-periodic part of $\rho + \delta\rho$ is

$$\begin{aligned}
= 1 &\quad - \tfrac{1}{6} M^2 + (\ \tfrac{179}{128} + \tfrac{1}{36} + \tfrac{1}{2} =)\ \ \tfrac{2219}{1152} M^4 \\
&+ (-\tfrac{1}{4} M^2 + (\ \tfrac{179}{192} + \tfrac{1}{12} + \tfrac{39}{8} =)\ \ \tfrac{1131}{192} M^4)\, E'^2 \\
&+ (\ \tfrac{3}{4} M^2 + (-\tfrac{2269}{64} - \tfrac{1}{4} + \tfrac{39}{8} =) - \tfrac{1973}{64} M^4)\, 2E'f't;
\end{aligned}$$

or the secular term of $\delta\rho$ is

$$= (\tfrac{3}{4} M^2 - \tfrac{1973}{64} M^4)\, 2E'f't,$$

which is the required formula.

V.

It is interesting to see how the coefficient $\tfrac{3771}{64}$ is made up. In Prof. Adams' Memoir we have

$$\begin{aligned}
\tfrac{3771}{64} = & \\
= & - \tfrac{3}{2} - \tfrac{3}{4} && (= - \tfrac{9}{4}\ ) \\
& + \tfrac{15}{4} + \tfrac{135}{64} - \tfrac{117}{8} + \tfrac{495}{128} - \tfrac{285}{16} - \tfrac{147}{16} - \tfrac{3}{16} - \tfrac{1323}{256} - \tfrac{27}{256} && (= - \tfrac{2391}{64}) \\
& + \tfrac{9}{2} - \tfrac{27}{4} - \tfrac{45}{2} - \tfrac{45}{2} - \tfrac{15}{2} - \tfrac{495}{32} + \tfrac{285}{4} - \tfrac{147}{4} - \tfrac{3}{4} + \tfrac{441}{8} + \tfrac{441}{8} + \tfrac{9}{8} + \tfrac{9}{8} && (= + \tfrac{3153}{32}),
\end{aligned}$$

where it may be remarked that the terms

$$\tfrac{495}{128} - \tfrac{285}{16} = (- \quad \tfrac{1785}{128})$$

and

$$- \tfrac{495}{32} + \tfrac{285}{4} = (+ 4 . \tfrac{1785}{128})$$

make together $3 . \tfrac{1785}{128} = \tfrac{5355}{128}$, and that it is in fact by the addition of these terms that Plana's coefficient $\tfrac{2187}{128}$ is changed into $\tfrac{3771}{64}$.

But in the present Memoir the coefficient $\frac{3771}{64}$ is obtained by means of an entirely different set of component numbers, viz. we have

$$\tfrac{3771}{64} = -\tfrac{381}{8} + \tfrac{495}{32} + \tfrac{15}{4} + \tfrac{647}{32} + \tfrac{1455}{16} - \tfrac{675}{32} + \tfrac{275}{16} + \tfrac{45}{16} - \tfrac{1455}{64}.$$

I had imagined, from the way in which the numbers $\frac{1455}{16} - \frac{1455}{64}$ presented themselves, that, if they were omitted, Plana's value $\frac{2187}{128}$ would have been obtained; but the result shows that this is not so.

As just deduced from the formula of Prof. Adams, the number $\frac{1973}{64}$ is obtained as follows, viz.

$$\begin{aligned}\tfrac{1973}{64} &= (-\tfrac{3771}{96} - 1) - \tfrac{1}{3} - \tfrac{1}{2} + (\tfrac{3153}{64} + \tfrac{3}{4}) - \tfrac{1}{2} - \tfrac{201}{16} + \tfrac{3771}{96} - \tfrac{1}{6} - \tfrac{1}{2} + \tfrac{1}{4} - \tfrac{39}{8} \\ &= -1 - \tfrac{1}{3} - \tfrac{1}{2} + \tfrac{3}{4} - \tfrac{1}{2} - \tfrac{1}{6} - \tfrac{1}{2} + \tfrac{1}{4} \qquad\qquad (=-1) \\ &\quad + \tfrac{3153}{64} - \tfrac{201}{16} - \tfrac{39}{8},\end{aligned}$$

where *ut suprà*

$$\tfrac{3153}{64} = \tfrac{1}{2}\{\tfrac{9}{2} - \tfrac{27}{4} - \tfrac{45}{2} - \tfrac{45}{2} - \tfrac{15}{2} - \tfrac{495}{32} + \tfrac{285}{4} - \tfrac{147}{4} - \tfrac{3}{4} + \tfrac{441}{8} + \tfrac{441}{8} + \tfrac{9}{8} + \tfrac{9}{8}\};$$

$-\frac{201}{16}$ is a number occurring in his Memoir, and which is in effect obtained irrespectively of the new periodic terms, and $-\frac{39}{8}$ is a number obtained as above, irrespectively of the new periodic terms. According to the method of the present Memoir, the number $\frac{1973}{64}$ was obtained in the form

$$\tfrac{1973}{64} = -\tfrac{1}{2}(\tfrac{381}{8} - \tfrac{495}{32} - \tfrac{15}{4} - \tfrac{647}{32} - \tfrac{1455}{16} + \tfrac{675}{32}).$$

VI.

If the investigation were pursued further, a question would arise as to the proper form to be given to the arguments; for in these, $nt + \epsilon$ seems to stand in the place of v, the value whereof is

$$v = nt + \epsilon - (\tfrac{3}{2} m^2 - \tfrac{3771}{64} m^4)\, ne'f't^2,$$

say $v = nt + \epsilon + kne'f't^2$, and it might be considered that in the arguments $nt + \epsilon$ should be changed into $nt + \epsilon + kne'f't^2$, or, what is the same thing, that τ should be changed into $\tau + kne'f't^2$, but that g' should remain unaltered (this assumes that there is not in the Sun's longitude any term corresponding to the acceleration). The arguments, instead of being of the simple form kt, would thus be of the form $kt + k_2 f't^2$. But this would not only increase the difficulty of integration, but would be inconsistent with the general plan of the solution; and it would seem to be the proper course to imagine the cosine or sine of such an argument to be developed $\left(\begin{smallmatrix}\cos\\ \sin\end{smallmatrix}\, kt + k_2 f't^2 = \begin{smallmatrix}\cos\\ \sin\end{smallmatrix}\, kt \mp k_2 f't^2 \begin{smallmatrix}\sin\\ \cos\end{smallmatrix}\, kt\right)$ in such manner as to bring the secular part of the argument outside the cos or sin; this is, in fact, the form which the solution takes when the arguments are left throughout in their original form, for the terms of the form $f't^2 \begin{smallmatrix}\cos\\ \sin\end{smallmatrix}$ arg. would present themselves in the subsequent approximations. But I shall not at present further examine the question.

ANNEXES CONTAINING THE DETAILS OF THE CALCULATION.

Annex 1.

Calculation of part of δP.

$$\delta P = \frac{1}{\rho'^3}(\tfrac{1}{2} + \tfrac{3}{2}\cos 2v - 2v')\,\delta\rho$$

$$+\frac{\rho}{\rho'^3}(\ -3\sin 2v - 2v')\,\delta v.$$

For $\cos 2v - 2v'$, $\sin 2v - 2v'$, see Annex 3.

$\tfrac{1}{2} + \tfrac{3}{2}\cos 2v - 2v' =$			$\dfrac{1}{\rho'^3} =$	
$\tfrac{1}{2}$			1	
$+\tfrac{3}{2}$	cos	2τ	$+3e'$	$\cos g'$.
$+3e'$	,,	$2\tau - g'$		
$-3e'$	,,	$2\tau + g'$		

Product is =

$$\begin{array}{llll} \tfrac{1}{2} & & & +\tfrac{3}{2}e' \cos g' \\ +\tfrac{3}{2} & \cos & 2\tau & +\tfrac{9}{2}e'(\tfrac{1}{2}\cos 2\tau - g' \ +\tfrac{1}{2}\cos 2\tau + g') \\ +3e' & ,, & 2\tau - g' & \\ -3e' & ,, & 2\tau + g' & \end{array}$$

$$\begin{array}{rlll} = & \tfrac{1}{2} & & \\ & +\tfrac{3}{2}e' & \cos & g' \\ & +\tfrac{3}{2} & ,, & 2\tau \\ (+3+\tfrac{9}{4}=) & +\tfrac{21}{4}e' & ,, & 2\tau - g' \\ (-3+\tfrac{9}{4}=) & -\tfrac{3}{4}e' & ,, & 2\tau + g', \end{array}$$

which is the coefficient of $\delta\rho$.

And

$-3\sin 2v - 2v' =$			$\dfrac{1}{\rho'^3} =$	
-3	sin	2τ	1	
$-6e'$	,,	$2\tau - g'$	$+3e'$	$\cos g'$.
$+6e'$	,,	$2\tau + g'$		

Product is =

$$\begin{array}{lll} -3 & \sin & 2\tau \\ -6e' & ,, & 2\tau - g' \\ +6e' & ,, & 2\tau + g' \end{array} \qquad -9e'(\tfrac{1}{2}\sin 2\tau - g' + \tfrac{1}{2}\sin 2\tau + g')$$

$$\begin{array}{llll} = & -3 & \sin & 2\tau \\ (-6-\tfrac{9}{2}=) & -\tfrac{21}{2}e' & ,, & 2\tau - g' \\ (+6-\tfrac{9}{2}=) & +\tfrac{3}{2}e' & ,, & 2\tau + g'; \end{array}$$

or, since $\rho = 1$, this is the coefficient of δv.

Annex 2.

Calculation of part of δQ.

$$\delta Q = \frac{\rho}{\rho'^3}(-3\sin 2v - 2v')\,\delta\rho$$
$$+\frac{\rho^2}{\rho'^3}(-3\cos 2v - 2v')\,\delta v.$$

For $\cos 2v - 2v'$, $\sin 2v - 2v'$, see Annex 3.

$-3\sin 2v - 2v' =$			$\frac{1}{\rho'^3} =$
-3	$\sin$	2τ	1
$-6e'$	,,	$2\tau - g'$	$+3e' \cos g'$.
$-6e'$	,,	$2\tau + g'$	

Product is =

$$\begin{array}{lll} -3 & \sin & 2\tau \\ -6e' & ,, & 2\tau - g' \\ -6e' & ,, & 2\tau + g' \end{array} \qquad -9e'(\tfrac{1}{2}\sin 2\tau - g' + \tfrac{1}{2}\sin 2\tau + g')$$

$$\begin{array}{llll} = & -3 & \sin & 2\tau \\ (-6-\tfrac{9}{2}=) & -\tfrac{21}{2}e' & ,, & 2\tau - g' \\ (+6-\tfrac{9}{2}=) & +\tfrac{3}{2}e' & ,, & 2\tau + g' \end{array}$$

or, since $\rho = 1$, this is the coefficient of $\delta\rho$.

$-3\cos 2v - 2v' =$			$\frac{1}{\rho'^3} =$
-3	$\cos$	2τ	1
$-6e'$	,,	$2\tau - g$	$+3e' \cos g'$.
$+6e'$	,,	$2\tau + g$	

Product is =

$$\begin{array}{lll} -\ 3 & \cos & 2\tau \\ -\ 6\,e' & \ ,, & 2\tau - g' \\ +\ 6\,e' & \ ,, & 2\tau + g' \end{array} \qquad -9e'\left(\tfrac{1}{2}\cos 2\tau - g' + \tfrac{1}{2}\cos 2\tau + g'\right)$$

$$\begin{array}{rlll} = & & & \\ & -\ 3 & \cos & 2\tau \\ (-6-\tfrac{9}{2}=) & -\tfrac{21}{2}\,e' & \ ,, & 2\tau - g' \\ (+6-\tfrac{9}{2}=) & +\ \tfrac{3}{2}\,e' & \ ,, & 2\tau + g'\,; \end{array}$$

or, since $\rho^2 = 1$, this is the coefficient of δv.

Annex 3.

Calculation of $\overset{\cos}{\underset{\sin}{\cdot}}\, 2v - 2v'$.

$$v - \ v' = \tau - 2e' \sin g'$$

$$2v - 2v' = 2\tau - \ e' \sin g'$$

$$\begin{array}{rll} \cos 2v - 2v' = & & \cos\ 2\tau \\ & + & \sin\ 2\tau\,.\,4e' \sin g' \\ = & & \cos\ 2\tau \\ & +2e' & \ ,, \quad 2\tau - g' \\ & -2e' & \ ,, \quad 2\tau \quad g' \end{array}$$

$$\begin{array}{rll} \sin 2v - 2v' = & & \sin\ 2\tau \\ & - & \cos\ 2\tau\,.\,4e' \sin g' \\ = & & \sin\ 2\tau \\ & +2e' & \ ,, \quad 2\tau - g' \\ & -2e' & \ ,, \quad 2\tau + g'. \end{array}$$

The expressions are calculated (*post*, Annex 16) as far as m^2.

Annex 4.

Calculation of a part of δP.

$$\delta P = \rho \left\{ \left(\tfrac{1}{2} + \tfrac{3}{2}\cos 2v - 2v'\right)\delta\,\frac{1}{\rho'^3} + \left(3 \sin 2v - 2v'\right)\frac{\delta v'}{\rho'^3} \right\}.$$

For $\cos 2v - 2v'$, $\sin 2v - 2v'$, see Annex 3.

$$\begin{array}{llll} \multicolumn{4}{l}{\tfrac{1}{2}+\tfrac{3}{2}\cos 2v-2v'=} \\ & \tfrac{1}{2} & & \\ + & \tfrac{3}{2} & \cos & 2\tau \\ + & 3e' & ,, & 2\tau-g' \\ - & 3e' & ,, & 2\tau+g' \end{array} \qquad \begin{array}{llll} \multicolumn{4}{l}{\delta\,\dfrac{1}{\rho'^3}=} \\ & 3e' & t & \\ + & 3 & t\cos & g' \\ + & 9e' & ,, & 2g'. \end{array}$$

Product is =

$$\begin{array}{lllll} & \tfrac{3}{2}e' & t & & \\ + & \tfrac{9}{2}e' & t\cos 2\tau & & \\ + & \tfrac{3}{2} & ,,\ g' & & \\ + & \tfrac{9}{2} & (\tfrac{1}{2}t\cos 2\tau-g' & +\tfrac{1}{2}t\cos 2\tau+g' &) \\ + & 9e' & (\tfrac{1}{2}\ ,,\ 2\tau-2g' & +\tfrac{1}{2}\ ,,\ 2\tau &) \\ - & 9e' & (\tfrac{1}{2}\ ,,\ 2\tau & +\tfrac{1}{2}\ ,,\ 2\tau+2g' &) \\ + & \tfrac{9}{2}e' & t\cos 2g' & & \\ + & \tfrac{27}{2}e' & (\tfrac{1}{2}t\cos 2\tau-2g' & +\tfrac{1}{2}t\cos 2\tau+2g' &), \end{array}$$

which is =

$$\begin{array}{rlll} & \tfrac{3}{2}e' & t & \\ & +\tfrac{3}{2} & t\cos & g' \\ (\tfrac{9}{2}+\tfrac{9}{2}-\tfrac{9}{2}=) & +\tfrac{9}{2}e' & ,, & 2\tau \\ & +\tfrac{9}{4} & ,, & 2\tau-g' \\ & +\tfrac{9}{4} & ,, & 2\tau+g' \\ & +\tfrac{9}{2}e' & ,, & 2g' \\ (\tfrac{9}{2}+\tfrac{27}{4}=) & +\tfrac{45}{4}e' & ,, & 2\tau-2g' \\ (-\tfrac{9}{2}+\tfrac{27}{4}=) & +\tfrac{9}{4}e' & ,, & 2\tau+2g' \end{array}$$

$$\begin{array}{llll} \multicolumn{4}{l}{3\sin 2v-2v'=} \\ & 3 & \sin & 2\tau \\ + & 6e' & ,, & 2\tau-g' \\ - & 6e' & ,, & 2\tau+g' \end{array} \qquad \begin{array}{llll} \multicolumn{4}{l}{\dfrac{\delta v'}{\rho'^3}=} \\ & 2 & t\sin & g' \\ + & \tfrac{11}{2}e' & ,, & 2g'. \end{array}$$

Product is =

$$\begin{array}{lllll} & 6 & (\tfrac{1}{2}t\cos 2\tau-g' & -\tfrac{1}{2}t\cos 2\tau+g' &) \\ + & 12e' & (\tfrac{1}{2}\ ,,\ 2\tau-2g' & -\tfrac{1}{2}\ ,,\ 2\tau &) \\ - & 12e' & (\tfrac{1}{2}\ ,,\ 2\tau & -\tfrac{1}{2}\ ,,\ 2\tau+2g' &) \\ + & \tfrac{33}{2}e' & (\tfrac{1}{2}\ ,,\ 2\tau-2g' & -\tfrac{1}{2}\ ,,\ 2\tau+2g' &), \end{array}$$

which is =

$$\begin{array}{llll}
(-6-6=) & -12e' & t\cos & 2\tau \\
 & +3 & ,, & 2\tau-g' \\
 & -3 & ,, & 2\tau+g' \\
(6+\tfrac{33}{4}=) & +\tfrac{57}{4}e' & ,, & 2\tau-2g' \\
(6-\tfrac{33}{4}=) & -\tfrac{9}{4}e' & ,, & 2\tau+2g',
\end{array}$$

whence, adding the two products, and observing that $\rho^2=1$, the required terms are

$$\begin{array}{llll}
= & \tfrac{3}{2}e' & t & \\
 & +\tfrac{3}{2} & t\cos & g' \\
(\tfrac{9}{2}-12=) & -\tfrac{15}{2}e' & ,, & 2\tau \\
(\tfrac{9}{4}+3=) & +\tfrac{21}{4} & ,, & 2\tau-g' \\
(\tfrac{9}{4}-3=) & -\tfrac{3}{4} & ,, & 2\tau+g' \\
 & +\tfrac{9}{2}e' & ,, & 2g' \\
(\tfrac{45}{4}+\tfrac{57}{4}=) & +\tfrac{51}{2}e' & ,, & 2\tau-2g' \\
(\tfrac{9}{4}-\tfrac{9}{4}=) & 0 & ,, & 2\tau+2g'.
\end{array}$$

Annex 5.

Calculation of a part of δQ, viz.

$$\delta Q=\rho^2\left[(-\tfrac{3}{2}\sin 2v-2v')\,\delta\frac{1}{\rho'^3}+(3\cos 2v-2v')\frac{\delta v'}{\rho'^3}\right],$$

$-\tfrac{3}{2}\sin 2v-2v'=$			$\delta\dfrac{1}{\rho'^3}=$		
$-\tfrac{3}{2}$	sin	2τ	$3e'$	t	
$-3e'$	,,	$2\tau-g'$	$+3$	$t\cos$	g'
$+3e'$	,,	$2\tau+g'$	$+9e'$	,,	$2g'$.

Product is

$$\begin{array}{lllll}
-\tfrac{9}{2}e' & t\sin 2\tau & & & \\
-\tfrac{9}{2} & (\tfrac{1}{2}t\sin & 2\tau+g' & +\tfrac{1}{2}t\sin & 2\tau-g'\ ) \\
-9e'(\tfrac{1}{2} & ,, & 2\tau & +\tfrac{1}{2}\ ,, & 2\tau-2g') \\
+9e'(\tfrac{1}{2} & ,, & 2\tau+2g' & +\tfrac{1}{2}\ ,, & 2\tau\ ) \\
-\tfrac{27}{2}e'(\tfrac{1}{2} & ,, & 2\tau+2g' & +\tfrac{1}{2}\ ,, & 2\tau-2g'),
\end{array}$$

which is =

$$\begin{array}{rlll}
(-\tfrac{9}{2}-\tfrac{9}{2}+\tfrac{9}{2}=) & -\tfrac{9}{2}\,e' & t\ \sin & 2\tau \\
 & -\tfrac{9}{4} & \text{,,} & 2\tau-g' \\
 & -\tfrac{9}{4} & \text{,,} & 2\tau+g' \\
(-\tfrac{9}{2}-\tfrac{27}{4}=) & -\tfrac{45}{4}\,e' & \text{,,} & 2\tau-2g' \\
(+\tfrac{9}{2}-\tfrac{27}{4}=) & -\tfrac{9}{4}\,e' & \text{,,} & 2\tau+2g';
\end{array}$$

and

$$\begin{array}{lll|lll}
\multicolumn{3}{l|}{\cos 2v-2v'=} & \multicolumn{3}{l}{\dfrac{\delta v'}{\rho'^3}=} \\
3 & \cos & 2\tau & 2 & t\ \sin & g' \\
+6\,e' & \text{,,} & 2\tau-g' & & & \\
-6\,e' & \text{,,} & 2\tau+g' & +\tfrac{11}{2}\,e' & \text{,,} & 2g'.
\end{array}$$

Product is

$$\begin{array}{rllll}
6 & (\tfrac{1}{2}\,t\ \sin & 2\tau+2g' & -\tfrac{1}{2}\,t\ \sin & 2\tau-g'\ ) \\
+\,12\,e' & (\tfrac{1}{2}\ \ \text{,,} & 2\tau & -\tfrac{1}{2}\ \ \text{,,} & 2\tau-2g') \\
-\,12\,e' & (\tfrac{1}{2}\ \ \text{,,} & 2\tau+2g' & -\tfrac{1}{2}\ \ \text{,,} & 2\tau\quad\ ) \\
+\,\tfrac{33}{2}\,e' & (\tfrac{1}{2}\ \ \text{,,} & 2\tau+2g' & -\tfrac{1}{2}\ \ \text{,,} & 2\tau-2g'),
\end{array}$$

which is =

$$\begin{array}{rlll}
(6+6=) & 12\,e' & t\ \sin & 2\tau \\
 & -\ 3 & \text{,,} & 2\tau-g' \\
 & +\ 3 & \text{,,} & 2\tau+g' \\
(-6-\tfrac{33}{4}=) & -\tfrac{57}{4}\,e' & \text{,,} & 2\tau-2g' \\
(-6+\tfrac{33}{4}=) & +\tfrac{9}{4}\,e' & \text{,,} & 2\tau+2g'.
\end{array}$$

Adding the two products together, and observing that $\rho^2=1$, the required terms are

$$\begin{array}{rlll}
(-\tfrac{9}{2}+12=) & \tfrac{15}{2}\,e' & t\ \sin & 2\tau \\
(-\tfrac{9}{4}-\ 3=) & -\tfrac{21}{4} & \text{,,} & 2\tau-g' \\
(-\tfrac{9}{4}+\ 3=) & +\tfrac{3}{4} & \text{,,} & 2\tau+g' \\
(-\tfrac{45}{4}-\tfrac{57}{4}=) & -\tfrac{51}{2}\,e' & \text{,,} & 2\tau-2g' \\
(-\tfrac{9}{4}+\tfrac{9}{4}=) & 0 & \text{,,} & 2\tau+2g'.
\end{array}$$

Annex 6.

Calculation of terms in δP, viz.

$$\begin{array}{lll|l}
-\ 3 & \sin & 2\tau & n^{-1}\times\overbrace{\begin{array}{lll}-\ 3 & \cos & g' \\ -\tfrac{9}{4}\,e' & \text{,,} & 2g';\end{array}} \\
-\tfrac{21}{2}\,e' & \text{,,} & 2\tau-g' & \\
+\tfrac{3}{2}\,e' & \text{,,} & 2\tau+g' &
\end{array}$$

the product of which is

$$\begin{array}{llllll} 9 & (\tfrac{1}{2} & \sin & 2\tau+g' & +\tfrac{1}{2} & \sin\ 2\tau-g'\ ) \\ +\tfrac{63}{2}e' & (\tfrac{1}{2} & ,, & 2\tau & +\tfrac{1}{2} & ,,\ \ 2\tau-2g') \\ -\tfrac{9}{2}e' & (\tfrac{1}{2} & ,, & 2\tau+2g' & +\tfrac{1}{2} & ,,\ \ 2\tau\quad) \\ +\tfrac{27}{4}e' & (\tfrac{1}{2} & ,, & 2\tau+2g' & +\tfrac{1}{2} & ,,\ \ 2\tau-2g'), \end{array}$$

which is =

$$\begin{array}{rlll} & \multicolumn{3}{c}{\overbrace{\hphantom{\tfrac{27}{2}e'\quad \sin\ 2\tau}}^{n^{-1}\times}} \\ (\tfrac{63}{4}-\tfrac{9}{4}=) & \tfrac{27}{2}e' & \sin & 2\tau \\ & +\tfrac{9}{2} & ,, & 2\tau-g' \\ & +\tfrac{9}{2} & ,, & 2\tau+g' \\ (\tfrac{63}{4}+\tfrac{27}{8}=) & +\tfrac{153}{8}e' & ,, & 2\tau-2g' \\ (-\tfrac{9}{4}+\tfrac{27}{8}=) & +\tfrac{9}{8}e' & ,, & 2\tau+2g'. \end{array}$$

Annex 7.

Calculation of terms in δQ, viz.

$$\begin{array}{lll|lll} & & & \multicolumn{3}{c}{n^{-1}\times} \\ -3 & \cos & 2\tau & -3 & \cos & g' \\ -\tfrac{21}{2}e' & ,, & 2\tau-g' & -\tfrac{9}{4}e' & ,, & 2g'; \\ +\tfrac{3}{2}e' & ,, & 2\tau+g' & & & \end{array}$$

the product of which is

$$\begin{array}{llllll} 9 & (\tfrac{1}{2} & \cos & 2\tau-g' & +\tfrac{1}{2} & \cos\ 2\tau+g'\ ) \\ +\tfrac{63}{2}e' & (\tfrac{1}{2} & ,, & 2\tau-2g' & +\tfrac{1}{2} & ,,\ \ 2\tau\quad) \\ -\tfrac{9}{2}e' & (\tfrac{1}{2} & ,, & 2\tau & +\tfrac{1}{2} & ,,\ \ 2\tau+2g') \\ +\tfrac{27}{4}e' & (\tfrac{1}{2} & ,, & 2\tau-2g' & +\tfrac{1}{2} & ,,\ \ 2\tau+2g'), \end{array}$$

which is =

$$\begin{array}{rlll} & \multicolumn{3}{c}{\overbrace{\hphantom{\tfrac{27}{2}e'\quad \cos\ 2\tau}}^{n^{-1}\times}} \\ (\ \tfrac{63}{4}-\tfrac{9}{4}=) & \tfrac{27}{2}e' & \cos & 2\tau \\ & +\tfrac{9}{2} & ,, & 2\tau-g' \\ & +\tfrac{9}{2} & ,, & 2\tau+g' \\ (\ \tfrac{63}{4}+\tfrac{27}{8}=) & +\tfrac{155}{8}e' & ,, & 2\tau-2g' \\ (\ \tfrac{9}{4}+\tfrac{27}{8}=) & +\tfrac{9}{8}\ e' & ,, & 2\tau+2g'. \end{array}$$

Annex 8.

Calculation of $n\int \delta Q dt$.

We have

$$n\int \sin nat\, dt = -\frac{1}{a}\, t \cos nat + \frac{1}{na} \sin nat,$$

$$\int \cos nat\, dt = \qquad\qquad \frac{1}{na} \sin nat,$$

and in all the arguments a is taken $= 2$.

$$\begin{array}{lll|llll}
n\int \delta Q dt = -\tfrac{15}{4}e' & t \cos & 2\tau & & \overbrace{\qquad\qquad\qquad}^{n^{-1}\times} & & \\
 & & & (+\tfrac{15}{8}+\tfrac{27}{4}=) & +\tfrac{69}{8}e' & \sin & 2\tau \\
+\tfrac{21}{8} & ,, & 2\tau-g' & (-\tfrac{21}{16}+\tfrac{9}{4}=) & +\tfrac{15}{16} & ,, & 2\tau-g' \\
-\tfrac{3}{8} & ,, & 2\tau+g' & (+\tfrac{3}{16}+\tfrac{9}{4}=) & +\tfrac{39}{16} & ,, & 2\tau+g' \\
+\tfrac{51}{4}e' & ,, & 2\tau-2g' & (-\tfrac{51}{8}+\tfrac{153}{16}=) & +\tfrac{51}{16}e' & ,, & 2\tau-g' \\
0 & ,, & 2\tau+2g' & (\ 0+\tfrac{9}{16}=) & +\tfrac{9}{16} & ,, & 2\tau+2g'.
\end{array}$$

Annex 9.

Calculation of $\delta P + 2n\int \delta Q\, dt$; viz. this is

$$\begin{array}{llll|llll}
 & \tfrac{3}{2}e' & t & & & & & \\
 & \tfrac{3}{2} & t \cos & g' & & \overbrace{\qquad\qquad\qquad}^{n^{-1}\times} & & \\
(-\tfrac{15}{2}-\tfrac{15}{2}=) & -15e' & ,, & 2\tau & (+\tfrac{27}{2}+\tfrac{69}{4}=) & +\tfrac{123}{4}e' & \sin & 2\tau \\
(+\tfrac{21}{4}+\tfrac{21}{4}=) & +\tfrac{21}{2} & ,, & 2\tau-g' & (+\tfrac{9}{2}+\tfrac{15}{8}=) & +\tfrac{51}{8} & ,, & 2\tau-g' \\
(-\tfrac{3}{4}-\tfrac{3}{4}=) & -\tfrac{3}{2} & ,, & 2\tau+g' & (+\tfrac{9}{2}+\tfrac{39}{8}=) & +\tfrac{75}{8} & ,, & 2\tau+g' \\
(+\tfrac{9}{2}+0=) & +\tfrac{9}{2}e' & ,, & 2g' & (\ 0+0=) & 0 & ,, & 2g' \\
(+\tfrac{51}{2}+\tfrac{51}{2}=) & +51e' & ,, & 2\tau-2g' & (+\tfrac{153}{8}+\tfrac{51}{8}=) & +\tfrac{51}{2}e' & ,, & 2\tau-2g' \\
(\ 0+0=) & 0 & ,, & 2\tau+2g' & (+\tfrac{9}{8}+\tfrac{9}{8}=) & +\tfrac{9}{4}e' & ,, & 2\tau+2g'.
\end{array}$$

Annex 10.

Calculation of $\delta\rho$ from the equation

$$\frac{d^2\delta\rho}{dt^2} + n^2 d\rho = m^2 n^2 \left(\delta P + 2n\int \delta Q\, dt\right).$$

In $\dfrac{d^2\delta\rho}{dt^2} + n^2\delta\rho$, a term $n^2 t \cos nat$, gives in $\delta\rho$,

$$\frac{1}{1-\alpha^2}\, t \cos nat + \frac{2\alpha}{(1-\alpha^2)^2}\, \frac{1}{n} \sin nat\,;$$

and a term $n \sin nat$, gives in $\delta\rho$,

$$\frac{1}{1-\alpha^2}\, \frac{1}{n} \sin nat,$$

and $\alpha = 2$, and therefore $\dfrac{1}{1-\alpha^2} = -\tfrac{1}{3}$, for all the args. except g', $2g'$; for these, $\alpha = m$ or $2m$, and therefore $\dfrac{1}{1-\alpha^2} = 1$.

$\delta\rho =$				$n^{-1} \times$	
$\tfrac{3}{2}\, m^2 e'$	t				
$+\tfrac{3}{2}\, m^2$	$t \cos$	g'	$(+3 \quad =)$	$+3\, m^3$	$\sin g'$
$+5\, m^2 e'$	,,	2τ	$(-\tfrac{20}{3} - \tfrac{41}{4} =)$	$-\tfrac{203}{12}\, m^2 e'$	,, 2τ
$-\tfrac{7}{2}\, m^2$	,,	$2\tau - g'$	$(+\tfrac{14}{3} - \tfrac{17}{8} =)$	$+\tfrac{61}{24}\, m^2$	,, $2\tau - g'$
$+\tfrac{1}{2}\, m^2$	,,	$2\tau + g'$	$(-\tfrac{2}{3} - \tfrac{25}{8} =)$	$-\tfrac{91}{24}\, m^2$	,, $2\tau + g'$
$+\tfrac{9}{2}\, m^2 e'$	,,	$2g'$	$(+18 + 0 =)$	$+18\, m^3 e'$	,, $2g'$
$-17\, m^2 e'$	,,	$2\tau - 2g'$	$(+\tfrac{68}{3} + \tfrac{17}{2} =)$	$+\tfrac{85}{6}\, m^2 e'$	,, $2\tau - 2g'$
0	,,	$2\tau + 2g'$	$(\ 0 - \tfrac{3}{4} =)$	$-\tfrac{3}{4}\, m^2 e'$	,, $2\tau + 2g'$.

Annex 11.

Calculation of $\dfrac{d\delta v}{dt}$; viz. this is

$$= -2n\,\delta\rho + m^2 n^2 \int \delta Q\, dt.$$

$\dfrac{d\delta v}{dt} =$

	$n \times$					
$(-3 \quad =)$	$-3\, m^2 e'\, t$					
$(-3 \quad =)$	$-3\, m^2$	$t \cos g'$	$(-6 \quad =)$	$-6\, m^3$	$\sin$	g'
$(-10 - \tfrac{15}{4} =)$	$-\tfrac{55}{4}\, m^2 e'$	,, 2τ	$(+\tfrac{203}{6} + \tfrac{69}{8} =)$	$+\tfrac{1019}{24}\, m^2 e'$	,,	2τ
$(+7 + \tfrac{21}{8} =)$	$+\tfrac{77}{8}\, m^2$	,, $2\tau - g'$	$(-\tfrac{61}{12} + \tfrac{15}{16} =)$	$-\tfrac{199}{48}\, m^2$	,,	$2\tau - g'$
$(-1 - \tfrac{3}{8} =)$	$-\tfrac{11}{8}\, m^2$	,, $2\tau + g'$	$(+\tfrac{91}{12} + \tfrac{39}{16} =)$	$+\tfrac{481}{48}\, m^2$	,,	$2\tau + g'$
$(-9 + 0 =)$	$-9\, m^2 e'$	,, $2g'$	$(-36 + 0 =)$	$-36\, m^3 e'$	,,	$2g'$
$(+34 + \tfrac{51}{4} =)$	$+\tfrac{187}{4}\, m^2 e'$	,, $2\tau - 2g'$	$(-\tfrac{85}{3} + \tfrac{51}{16} =)$	$-\tfrac{1207}{48}\, m^2 e'$	,,	$2\tau - 2g'$
$(\ 0 + 0\ )$	0	,, $2\tau + 2g'$	$(+\tfrac{3}{2} + \tfrac{9}{16} =)$	$+\tfrac{33}{16}\, m^2 e'$	,,	$2\tau + 2g'$.

Annex 12.

Calculation of δv from the foregoing value of $\dfrac{d\delta v}{dt}$.

We have

$$n\int \cos n\alpha t\, dt = \frac{1}{\alpha}\, t \sin n\alpha t + \frac{1}{n\alpha^2}\cos n\alpha t,$$

$$\int \sin n\alpha t\, dt = \qquad -\frac{1}{n\alpha}\cos n\alpha t\,;$$

$\alpha = 2$ for all the arguments, except only $\alpha = m,\ 2m$, for the arguments g', $2g'$, respectively.

$\delta v =$

			$-\ \tfrac{3}{2}\, m^2 n e'\ \ t^2$			
			$+\ n^{-1}\times$			
$-3\, m$	$t \sin g'$	$\left(-\dfrac{3}{m^2} + 6 =\right)$	$-3 + 6\, m^2$		$\cos g'$	
$-\tfrac{55}{8}\, m^2 e'$	„ 2τ	$\left(-\tfrac{55}{16} - \tfrac{1019}{48} =\right)$	$-\tfrac{74}{3}$	$m^2 e'$	„	2τ
$+\tfrac{77}{16}\, m^2$	„ $2\tau - g'$	$\left(+\tfrac{77}{32} + \tfrac{199}{96} =\right)$	$+\tfrac{215}{48}$	m^2	„	$2\tau - g'$
$-\tfrac{11}{16}\, m^2$	„ $2\tau + g'$	$\left(-\tfrac{11}{32} - \tfrac{481}{96} =\right)$	$-\tfrac{257}{48}$	m^2	„	$2\tau + g'$
$-\tfrac{9}{2}\, m e'$	„ $2g'$	$\left(-\dfrac{9}{4m^2} + 18 =\right)$	$(-\tfrac{9}{4} + 18\, m^2)\, e'$		„	$2g'$
$+\tfrac{187}{8}\, m^2 e'$	„ $2\tau - 2g'$	$\left(+\tfrac{187}{16} + \tfrac{1207}{96} =\right)$	$+\tfrac{2329}{96}$	$m^2 e'$	„	$2\tau - 2g'$
0	„ $2\tau + 2g'$	$\left(0 - \tfrac{33}{32} =\right)$	$-\tfrac{33}{32}$	$m^2 e'$	„	$2\tau + 2g'$.

The remaining Annexes relate to the determination of the non-periodic or secular terms of the order m^4, in $\delta\rho$ and δv respectively.

Annex 13.

Calculation of term of $\left(n^2 + \dfrac{2n^2}{\rho^3} - 3\left(\dfrac{dv}{dt}\right)^2\right)\delta\rho$.

We have

$n^2 + \dfrac{2n^2}{\rho^3} - 3\left(\dfrac{dv}{dt}\right)^2 =$ $\qquad\qquad$ $\delta\rho =$

	$n^2\times$			$\delta\rho$	
$(1 + 2 + m^2 - 3 =)$	m^2			$\tfrac{3}{2}\, m^2 e'$	t
$(-9 + 18 =)$	$+\ 9\ m^2 e'$	$\cos g'$		$+\ \tfrac{3}{2}\, m^2$	$t \cos g'$
$(+6 - \tfrac{33}{2} =)$	$-\ \tfrac{21}{2}\ m^2$	„ 2τ		$+\ 5\, m^2 e'$	„ 2τ
$(+21 - \tfrac{231}{4} =)$	$-\ \tfrac{147}{4}\ m^2 e'$	„ $2\tau - g'$		$-\ \tfrac{7}{2}\, m^2$	„ $2\tau - g'$
$(-3 + \tfrac{33}{4} =)$	$+\ \tfrac{21}{4}\ m^2 e'$	„ $2\tau + g'$		$+\ \tfrac{1}{2}\, m^2$	„ $2\tau + g'$
				$+$ &c.	

where, in the second factor, the arguments not occurring in the first factor are omitted, as not giving rise to any non-periodic term; and so in other similar cases. Hence term of product is

$$\begin{array}{llcl} m^4n^2e't & & 1 & .\ \tfrac{3}{2} \\ & +\tfrac{1}{2}\ . & 9 & .\ \tfrac{3}{2} \\ & +\tfrac{1}{2}\ . & -\tfrac{21}{2} & .\ 5 \\ & +\tfrac{1}{2}\ . & -\tfrac{147}{4} & .\ -\tfrac{7}{2} \\ & +\tfrac{1}{2}\ . & \tfrac{21}{4} & .\ \tfrac{1}{2} \end{array}$$

$$=(\tfrac{3}{2}+\tfrac{27}{4}-\tfrac{105}{4}+\tfrac{1029}{16}+\tfrac{21}{16}=)+\tfrac{381}{8}\,m^4n^2e't.$$

Annex 14.

Calculation of term of $m^2n^2\delta P$.

$m^2n^2\delta P$, the part involving δv is

$m^2n^2\times$		$\delta v =$	
-3	$\sin 2\tau$	$-3\,m$	$t\sin g'$
$-\tfrac{21}{2}\,e'$	„ $2\tau-g'$	$-\tfrac{55}{8}\,m^2e'$	„ 2τ
$+\tfrac{3}{2}\,e'$	„ $2\tau+g'$	$+\tfrac{77}{16}\,m^2$	„ $2\tau-g'$
		$-\tfrac{11}{16}\,m^2$	„ $2\tau+g'$
		+ &c.	

and term of the product is

$$\begin{array}{llcl} m^4n^2e't & \tfrac{1}{2}\ . & -3 & .\ -\tfrac{55}{8} \\ & +\tfrac{1}{2}\ . & -\tfrac{21}{2} & .\ \tfrac{77}{16} \\ & +\tfrac{1}{2}\ . & \tfrac{3}{2} & .\ -\tfrac{11}{16} \end{array}$$

$$=(\tfrac{165}{16}-\tfrac{1617}{64}-\tfrac{33}{64}=)-\tfrac{495}{32}\,m^4n^2e't.$$

Annex 15.

Calculation of term of $m^2n^2\delta P$.

$m^2n^2\delta P$, the term involving $\delta\rho$ is

$m^2n^2\times$		$\delta\rho =$	
$\tfrac{1}{2}$		$\tfrac{3}{2}\,m^2e'$	t
$+\tfrac{3}{2}\,e'$	$\cos g'$	$+\tfrac{3}{2}\,m^2$	$t\cos g'$
$+\tfrac{3}{2}$	„ 2τ	$+5\,m^2e'$	„ 2τ
$+\tfrac{21}{4}\,e'$	„ $2\tau-g'$	$-\tfrac{7}{2}\,m^2$	„ $2\tau-g$
$-\tfrac{3}{4}\,e'$	„ $2\tau+g'$	$+\tfrac{1}{2}\,m^2$	„ $2\tau+g'$
		+ &c.	

and term of the product is

$$\begin{array}{rl} m^4n^2e't & \tfrac{1}{2}\,.\quad \tfrac{3}{2} \\ +\tfrac{1}{2}\,. & \tfrac{3}{2}\,.\quad \tfrac{3}{2} \\ +\tfrac{1}{2}\,. & \tfrac{3}{2}\,.\quad 5 \\ +\tfrac{1}{2}\,. & \tfrac{21}{4}\,.-\tfrac{7}{2} \\ +\tfrac{1}{2}\,. & -\tfrac{3}{4}\,.\quad \tfrac{1}{2} \end{array}$$

$$= (\tfrac{3}{4}+\tfrac{9}{8}+\tfrac{15}{4}-\tfrac{147}{16}-\tfrac{3}{16}=)-\tfrac{15}{4}\,m^4n^2e't.$$

Annex 16.

Calculation of $\begin{smallmatrix}\cos\\ \sin\end{smallmatrix}\, 2v-2v'$, as far as m^2.

$$\cos 2v-2v' = \cos 2\tau + X = \begin{array}{l} \cos 2\tau \\ -X\sin 2\tau \\ -\tfrac{1}{2}X^2\cos 2\tau, \end{array}$$

$$\sin 2v-2v' = \sin 2\tau + X = \begin{array}{l} \sin 2\tau \\ +X\cos 2\tau \\ \tfrac{1}{2}X^2\sin 2\tau, \end{array}$$

where

$$\begin{array}{rll} X = -(4+6m)\,e' & \sin & g' \\ +\tfrac{11}{4}m^2 & ,, & 2\tau \\ +\tfrac{77}{8}m^2e' & ,, & 2\tau-g' \\ -\tfrac{11}{8}m^2e' & ,, & 2\tau+g', \end{array}$$

and thence

$$\begin{array}{rlll} X\sin 2\tau = -(4+\ 6\ m)\,e' & (\tfrac{1}{2}\cos 2\tau-g' & -\tfrac{1}{2}\cos 2\tau+g') \\ +\tfrac{11}{4}m^2 & (\tfrac{1}{2} & -\tfrac{1}{2}\ ,,\ 4\tau\qquad) \\ +\tfrac{77}{8}m^2e' & (\tfrac{1}{2}\cos g' & -\tfrac{1}{2}\ ,,\ 4\tau-g') \\ -\tfrac{11}{8}m^2e' & (\tfrac{1}{2}\ ,,\ g' & -\tfrac{1}{2}\ ,,\ 4\tau+g'), \end{array}$$

which is =

$$\begin{array}{rrll} & \tfrac{11}{8}m^2 & & \\ (\tfrac{77}{16}-\tfrac{11}{16}=) & +\tfrac{33}{8}m^2e' & \cos & g' \\ & -(2+3\ m)\,e' & ,, & 2\tau-g' \\ & +(2+3\ m)\,e' & ,, & 2\tau+g' \\ & -\tfrac{11}{8}m^2 & ,, & 4\tau \\ & -\tfrac{77}{16}m^2e' & ,, & 4\tau-g' \\ & +\tfrac{11}{16}m^2e' & ,, & 4\tau+g'; \end{array}$$

$$
\begin{aligned}
X\cos 2\tau = &-(4+6\,m)\,e' \;(\tfrac{1}{2}\sin 2\tau+g' - \tfrac{1}{2}\sin 2\tau-g')\\
&+\tfrac{11}{4}\,m^2 \;(\tfrac{1}{2} \;\text{,,}\; 4\tau \qquad\qquad\qquad)\\
&+\tfrac{77}{8}\,m^2 e' \;(\tfrac{1}{2} \;\text{,,}\; 4\tau-g' - \tfrac{1}{2} \;\text{,,}\; g' \quad)\\
&-\tfrac{11}{8}\,m^2 e' \;(\tfrac{1}{2} \;\text{,,}\; 4\tau+g' - \tfrac{1}{2} \;\text{,,}\; g' \quad),
\end{aligned}
$$

which is =

$$
\begin{array}{rlll}
(-\tfrac{77}{16}-\tfrac{11}{16}=) & \tfrac{11}{2}\,m^2 e' & \sin & g' \\
& +(2+3\,m)\,e' & \text{,,} & 2\tau-g' \\
& -(2+3\,m)\,e' & \text{,,} & 2\tau-g' \\
& +\tfrac{11}{8}\,m^2 & \text{,,} & 4\tau \\
& +\tfrac{77}{16}\,m^2 e' & \text{,,} & 4\tau-g' \\
& -\tfrac{11}{16}\,m^2 e' & \text{,,} & 4\tau+g'\,;
\end{array}
$$

we have, moreover,

$$
\begin{aligned}
X^2 &= -2\,(4+6m)\,e'\sin g'\,.\,\tfrac{11}{4}\,m^2\sin 2\tau\\
&= -22\,m^2e'\,(\tfrac{1}{2}\cos 2\tau-g' - \tfrac{1}{2}\cos 2\tau+g')\\
&= -11\,m^2e' \quad \cos 2\tau-g'\\
&\quad\; +11\,m^2e' \quad \text{,,} \;\; 2\tau-g',
\end{aligned}
$$

and thence

$$
\begin{aligned}
X^2\cos 2\tau = &-11\,m^2e'\,(\tfrac{1}{2}\cos 4\tau-g'+\tfrac{1}{2}\cos g')\\
&+11\,m^2e'\,(\tfrac{1}{2} \;\text{,,}\; 4\tau+g'+\tfrac{1}{2} \;\text{,,}\; g')
\end{aligned}
$$

$$
\begin{array}{rlll}
=(-\tfrac{11}{2}+\tfrac{11}{2}=) & 0 & \cos & g' \\
& -\tfrac{11}{2}\,m^2e' & \text{,,} & 4\tau-g' \\
& +\tfrac{11}{2}\,m^2e' & \text{,,} & 4\tau+g'\,;
\end{array}
$$

and

$$
\begin{aligned}
X^2\sin 2\tau = &-11\,m^2e'\,(\tfrac{1}{2}\sin 4\tau-g'+\tfrac{1}{2}\sin g')\\
&-11\,m^2e'\,(\tfrac{1}{2} \;\text{,,}\; 4\tau+g'+\tfrac{1}{2} \;\text{,,}\; g')
\end{aligned}
$$

$$
\begin{array}{rlll}
=(-\tfrac{11}{2}-\tfrac{11}{2}=) & -11\,m^2e' & \sin & g' \\
& -\tfrac{11}{2}\,m^2e' & \text{,,} & 4\tau-g' \\
& +\tfrac{11}{2}\,m^2e' & \text{,,} & 4\tau+g'\,;
\end{array}
$$

and thence

$$
\begin{array}{rrlll}
\cos 2v-2v' = & & -\tfrac{11}{8}\,m^2 & & \\
& & +1 & \cos & 2\tau \\
& & -\tfrac{33}{8}\,m^2e' & \text{,,} & g' \\
& & +(2+3\,m)\,e' & \text{,,} & 2\tau-g' \\
& & -(2+3\,m)\,e' & \text{,,} & 2\tau+g' \\
& & +\tfrac{11}{8}\,m^2 & \text{,,} & 4\tau \\
& (+\tfrac{77}{16}+\tfrac{11}{4}=) & +\tfrac{121}{16}\,m^2e' & \text{,,} & 4\tau-g' \\
& (-\tfrac{11}{16}-\tfrac{11}{4}=) & -\tfrac{55}{16}\,m^2e' & \text{,,} & 4\tau+g'\,;
\end{array}
$$

and

$\sin 2v - 2v' =$		1	$\sin\ 2\tau$
$(-\tfrac{11}{2}+\tfrac{11}{2}=)$		$0\ m^2 e'$	,, g'
		$+(2+\ 3\ m)e'$	,, $2\tau - g'$
		$-(2+\ 3\ m)e'$	,, $2\tau + g'$
		$+\ \tfrac{11}{8}\ m^2$	,, 4τ
$(\ \tfrac{77}{16}+\tfrac{11}{4}=)$		$+\tfrac{121}{16}\ m^2 e'$	,, $4\tau - g'$
$(-\tfrac{11}{16}-\tfrac{11}{4}=)$		$-\ \tfrac{55}{16}\ m^2 e'$	,, $4\tau + g'$.

Annex 17.

Calculation of term in δQ.

The part of δQ containing δv is $\frac{\rho^2}{\rho'^3}(-3\cos 2v - 2v')\,\delta v$; and it is necessary to find in $\frac{\rho^2}{\rho'^3}(-3\cos 2v - 2v')$ the coefficient of $\cos g'$ as far as m^2; this is in fact required, *post* Annex 21, for the calculation of $m^2n^2\left(C + \int \delta Q\, dt\right)$.

$\rho^2 = 1 - \tfrac{1}{3}\ m^2$		$\frac{1}{\rho'^3} = \ 1$
$+\ 3\ m^2 e'$	$\cos\ g'$	$+\ 3\,e' \quad \cos\ g'$
$-\ 2\ m^2$	,, 2τ	
$-\ 7\ m^2 e'$	,, $2\tau - g'$	
$+\ 1\ m^2 e'$	,, $2\tau + g'$	

and thence

$\frac{\rho^2}{\rho'^3} =$			$-\ 3\ \cos 2v - 2v' =$	
	$1 - \tfrac{1}{3}\ m^2$		$\tfrac{33}{8}\ m^2$	
	$+(3+\ 2\ m^2)\,e'$	$\cos\ g'$	$+\ \tfrac{99}{8}\ m^2\ e'$	$\cos\ g'$
	$-\ 2\ m^2$	,, 2τ	$-\ 3$	,, 2τ
$(-7-\tfrac{1}{2}.2.3=)$	$-\ 10\ m^2\ e'$	,, $2\tau - g'$	$-(6+\ 9\ m)\,e'$	,, $2\tau - g'$
$(+1-\tfrac{1}{2}.2.3=)$	$-\ 2\ m^2\ e'$	,, $2\tau + g'$	$+(6+\ 9\ m)\,e'$	,, $2\tau + g'$
			&c.	

where, in the second column, the omitted terms have arguments containing 4τ, and consequently, do not, by combination with the first column, give rise to any term with the argument g'. The term arg. g' arises from the combinations

$$\begin{array}{llll}
 & \frac{33}{8} m^2 & .\ 3e' & \cos g' \\
+ & \frac{99}{8} m^2 e' & .\ 1 & \quad ,, \ g' \\
- & 3 & .\ -10\, m^2 e' & (\cos 2\tau \cos 2\tau - g' = \frac{1}{2} \cos 4\tau - g' + \frac{1}{2} \cos g') \\
- & 3 & .\ -\ 2\, m^2 e' & (\ ,,\ 2\tau\ ,,\ 2\tau + g' = \frac{1}{2}\ ,,\ 4\tau + g' + \frac{1}{2}\ ,,\ g') \\
- & 6e' & .\ -\ 2\, m^2 & (\ ,,\ 2\tau\ ,,\ 2\tau - g' = \frac{1}{2}\ ,,\ 4\tau - g' + \frac{1}{2}\ ,,\ g') \\
+ & 6e' & .\ -\ 2\, m^2 & (\ ,,\ 2\tau\ ,,\ 2\tau + g' = \frac{1}{2}\ ,,\ 4\tau + g' + \frac{1}{2}\ ,,\ g')
\end{array}$$

so that the required term is

$$(\tfrac{99}{8} + \tfrac{99}{8} + 15 + 3 + 6 - 6 =) + \tfrac{171}{4}\, m^2 e' \cos g',$$

and annexing this to the terms found Annex 2, the part of δQ which contains δv is

$$\left.\begin{array}{llll}
 & \frac{171}{4} & m^2 e' & \cos g' \\
- & 3 & & \quad ,, \ 2\tau \\
- & \frac{21}{2} & e' & \quad ,, \ 2\tau - g' \\
+ & \frac{3}{2} & e' & \quad ,, \ 2\tau + g'
\end{array}\right| \delta v.$$

Annex 18.

Calculation of term of $m^2 n^2 \delta P$.

The part of δP involving $\delta v'$ and $\delta \rho'$ is

$$\delta P = \rho \left[(\tfrac{1}{2} + \tfrac{3}{2} \cos 2v - 2v')\, \delta \frac{1}{\rho'^3} + (3 \sin 2v - 2v') \frac{\delta v'}{\rho'^3} \right];$$

$$\rho = \begin{array}{lll}
1 - \frac{1}{6} m^2 & & \\
+ \frac{3}{2} m^2 e' & \cos g' & \\
- \quad m^2 & \quad ,, & 2\tau \\
- \frac{7}{2} m^2 e' & \quad ,, & 2\tau - g' \\
+ \frac{1}{2} m^2 e' & \quad ,, & 2\tau + g'
\end{array}$$

$$\tfrac{1}{2} + \tfrac{3}{2} \cos 2v - 2v' = \begin{array}{lll}
\frac{1}{2} - \frac{33}{16} m^2 & & \\
\quad - \frac{99}{16} m^2 e' & \cos g' & \\
+ \frac{3}{2} & \quad ,, & 2\tau \\
+ (3 + \frac{9}{2} m)\, e' & \quad ,, & 2\tau - g' \\
- (3 + \frac{9}{2} m)\, e' & \quad ,, & 2\tau + g' \\
+ \ \&c. & &
\end{array}$$

$$\delta \frac{1}{\rho'^3} = \begin{array}{lll}
3\, e' & t & \\
+ 3 & t \cos 2g' & \\
+ 9\, e' & \quad ,, & 2g'
\end{array}$$

where, in the second factor, the terms belonging to the arguments which contain 4τ (i. e. the arguments 4τ, $4\tau - g'$, $4\tau + g'$) are omitted. In fact, the terms in question would, in the product of the second and third factors, give rise to terms with arguments containing 4τ, and as there are no such terms in the first factor, there is no resulting secular term.

The product of the second and third factors is

$$\begin{array}{llll}
(\frac{3}{2} - \frac{99}{16} m^2) e' & t & & \\
+ (\frac{3}{2} - \frac{99}{16} m^2) & t \cos g' & & \\
+ \frac{9}{2} e' & \quad \text{,,} \ 2\tau & & \\
\quad - \frac{297}{16} m^2 e' & (\frac{1}{2} t & + \frac{1}{2} t \cos 2g' &) \\
+ \frac{9}{2} & (\frac{1}{2} t \cos 2\tau - g' & + \frac{1}{2} \quad \text{,,} \ 2\tau + g' &) \\
+ (9 + \frac{27}{2} m) e' & (\frac{1}{2} t \ \text{,,} \ 2\tau - 2g' & + \frac{1}{2} \quad \text{,,} \ 2\tau &) \\
- (9 + \frac{27}{2} m) e' & (\frac{1}{2} \quad \text{,,} \ 2\tau + 2g' & + \frac{1}{2} \quad \text{,,} \ 2\tau &) \\
+ (\frac{9}{2} - \frac{297}{16} m^2) e' & \quad t \ \text{,,} \ 2g' & & \\
+ \frac{27}{2} e' & (\frac{1}{2} t \ \text{,,} \ 2\tau - 2g' & + \frac{1}{2} t \ \text{,,} \ 2\tau + 2g'), &
\end{array}$$

and we may in this product omit the terms with arguments containing $2g'$, since the first factor does not contain any such term. The product then is

$$\begin{array}{rlr}
(-\frac{99}{16} - \frac{297}{32} = -\frac{495}{32}) & (\frac{3}{2} - \frac{495}{32} m^2) e' & t \\
& + (\frac{3}{2} - \frac{99}{16} m^2) & t \cos g' \\
((\frac{9}{2} + \frac{9}{2} - \frac{9}{2}) e' + (\frac{27}{4} - \frac{27}{4}) me' =) & + \frac{9}{2} e' & \text{,,} \ 2\tau \\
& + \frac{9}{4} & \text{,,} \ 2\tau - g' \\
& + \frac{9}{4} & \text{,,} \ 2\tau + g',
\end{array}$$

which is to be multiplied by the first factor, ρ, and the whole by the factor m^2n^2.

The term in the product is

$$\begin{array}{lll}
m^2n^2e't & (\frac{3}{2} - \frac{495}{32} m^2)(1 - \frac{1}{6} m^2) & \\
& + \frac{1}{2} \cdot \frac{3}{2} & \cdot \ \frac{3}{2} m^2 \\
& + \frac{1}{2} \cdot \frac{9}{2} & \cdot - m^2 \\
& + \frac{1}{2} \cdot \frac{9}{4} & \cdot - \frac{7}{2} m^2 \\
& + \frac{1}{2} \cdot \frac{9}{4} & \cdot \ \frac{1}{2} m^2,
\end{array}$$

giving the term $\frac{3}{2} m^2n^2e't$, which was found above, Annex 9, and the new terms

$$(-\tfrac{495}{32} - \tfrac{1}{4} + \tfrac{9}{8} - \tfrac{9}{4} - \tfrac{63}{16} + \tfrac{9}{16} =) - \tfrac{647}{32} m^4n^2 e' t.$$

The term of the part containing $\delta v'$ is found to be $=0$; in fact we have

$\rho =$		$3\sin 2v - 2v' =$		$\dfrac{\delta v'}{\rho'^3} =$	
$1 - \frac{1}{6}m^2$		3	$\sin 2\tau$	2	$t\sin g'$
$+\frac{3}{2}m^2e'$	$\cos g'$	$+(6+9m)e'$	$,,\ 2\tau - g'$	$+\frac{11}{2}e'$	$,,\ 2g'$
$-\ \ m^2$	$,,\ 2\tau$	$-(6+9m)e'$	$,,\ 2\tau + g'$		
$-\frac{7}{2}m^2e'$	$,,\ 2\tau - g'$	&c.			
$+\frac{1}{2}m^2e'$	$,,\ 2\tau + g'$				

where in the second factor the terms with arguments containing 4τ are for the before-mentioned reason omitted.

The product of the second and third factors is

$$\begin{array}{rllll} & 6\ (\frac{1}{2}\ t\cos 2\tau - g' & & -\frac{1}{2}\ t\cos 2\tau + g' &) \\ +(12+18m)e' & (\frac{1}{2} & ,,\ 2\tau - 2g' & -\frac{1}{2} & ,,\ 2\tau\ \ \ ) \\ -(12+18m)e' & (\frac{1}{2} & ,,\ 2\tau & -\frac{1}{2} & ,,\ 2\tau + 2g') \\ +\frac{33}{2}\ \ e' & (\frac{1}{2} & ,,\ 2\tau - 2g' & -\frac{1}{2} & ,,\ 2\tau + 2g') \end{array}$$

which, omitting the terms with arguments containing $2g'$, is

$$\begin{array}{rll} -(6+9m+6+9m) = -(12+18m)e' & t\ \cos & 2\tau \\ +\ 3 & \ \ ,, & 2\tau - g' \\ -\ 3 & \ \ ,, & 2\tau + g' \end{array}$$

which is to be multiplied by the first factor, ρ, and the whole by m^2n^2. The term is

$$\begin{array}{rl} m^4n^2e'\,t. & \ \ \frac{1}{2}\,.\,-12\,.\,-1 \\ & +\frac{1}{2}\,.\ \ \ \ 3\,.\,-\frac{7}{2} \\ & +\frac{1}{2}\,.\,-3\,.\ \ \ \frac{1}{2} \end{array}$$

which is

$$(6 - \tfrac{21}{4} - \tfrac{3}{4} =)\ 0\ m^4n^2e'\,t.$$

Hence the entire term in question is the before-mentioned value

$$-\tfrac{647}{32}\,m^4n^2e'\,t.$$

Annex 19.

Calculation of term in $m^2n^2\dfrac{2}{\rho}\dfrac{dv}{dt}\left(C + \int \delta Q\,dt\right)$.

We have

$$\frac{2}{\rho}\frac{dv}{dt} = n \times \left\{\begin{array}{lll} 2 + \tfrac{1}{3} & m^2 & \\ - 9 & m^2e' & \cos g' \\ + \tfrac{15}{2} & m^2 & \;\;,,\;\; 2\tau \\ + \tfrac{105}{4} & m^2e' & \;\;,,\;\; 2\tau - g' \\ - \tfrac{15}{4} & m^2e' & \;\;,,\;\; 2\tau + g' \end{array}\right.$$

See *post* Annex 20.

$$m^2n^2\left(C + \int \delta Q\, dt\right) = n \times \left\{\begin{array}{lll} -\tfrac{1455}{32} & m^4e'\; t & \\ 0 & & t \cos g' \\ -\tfrac{15}{4} & m^2 & \;\;,,\;\; 2\tau \\ + \tfrac{21}{8} & m^2 & \;\;,,\;\; 2\tau - g' \\ - \tfrac{3}{8} & m^2 & \;\;,,\;\; 2\tau - g' \end{array}\right.$$

For the term $-\tfrac{1455}{32} m^4e'\, t$ see *post* Annex 21.

The term in the product therefore is

$$m^4n^2e'\, t. \quad \begin{array}{l} 2 \;.\; -\tfrac{1455}{32} \\ +\tfrac{1}{2}\;.\;\; \tfrac{15}{2}\;.\; -\tfrac{15}{4} \\ +\tfrac{1}{2}\;.\;\; \tfrac{105}{4}\;.\; -\tfrac{21}{8} \\ +\tfrac{1}{2}\;.\; -\tfrac{15}{4}\;.\;\; \tfrac{3}{8} \end{array}$$

which is

$$= (2\;.\; -\tfrac{1455}{32} - \tfrac{225}{16} + \tfrac{2205}{64} + \tfrac{45}{64})\, m^4n^2e'\, t$$
$$= (2\;.\; -\tfrac{1455}{32} + \tfrac{675}{32} =) \quad -\tfrac{2235}{32}\, m^4n^2e'\, t.$$

Annex 20.

Calculation of $\frac{2}{\rho}\frac{dv}{dt}$.

We have

$$\frac{dv}{dt} = n \times \left\{\begin{array}{lll} 1 & & \\ - 3 & m^2e' & \cos g' \\ + \tfrac{11}{4} & m^2e' & \;\;,,\;\; 2\tau \\ + \tfrac{77}{8} & m^2e' & \;\;,,\;\; 2\tau - g' \\ - \tfrac{11}{8} & m^2e' & \;\;,,\;\; 2\tau + g' \end{array}\right.$$

$$\frac{2}{\rho} = \begin{array}{lll} 2 + \tfrac{1}{3} & m^2 & \\ - 3 & m^2e' & \cos g' \\ + 2 & m^2 & \;\;,,\;\; 2\tau \\ + 7 & m^2e' & \;\;,,\;\; 2\tau - g' \\ - 1 & m^2e' & \;\;,,\;\; 2\tau + g' \end{array}$$

so that the product is

$$\begin{array}{llll} & \overbrace{2 + \tfrac{1}{3}\ m^2}^{n\,\times} & & \\ (-\ 6\ -3 =) & -\ 9\ m^2e' & \cos & g' \\ (+\tfrac{11}{2} + 1 =) & +\tfrac{15}{2}\ m^2 & ,, & 2\tau \\ (+\tfrac{77}{4} + 7 =) & +\tfrac{105}{4}\,m^2e' & ,, & 2\tau - g' \\ (-\tfrac{11}{4}\quad 1 =) & -\ \tfrac{15}{8}\ m^2e' & ,, & 2\tau + g'. \end{array}$$

Annex 21.

Calculation of a term in $m^2n^2\left(C + \int \delta Q\, dt\right)$.

The part $\dfrac{\rho}{\rho'^3}(-3 \sin 2v - 2v')\,\delta\rho$ of δQ gives

$$\frac{\rho}{\rho'^3}(-3\sin 2v - 2v') = \qquad \begin{array}{lll} -\ 3 & \sin & 2\tau \\ \tfrac{21}{2}\ e' & ,, & 2\tau\ \ g \\ +\ \tfrac{3}{2}\ e' & ,, & 2\tau + g' \end{array}$$

$$\delta\rho = \qquad \begin{array}{lll} \overbrace{3\ \ m^3}^{n^{-1}\,\times} & \sin & g' \\ -\ \tfrac{203}{12}\ m^2e' & ,, & 2\tau \\ +\ \tfrac{64}{24}\ m^2 & ,, & 2\tau - g' \\ -\ \tfrac{91}{24}\ m^2 & ,, & 2\tau + g' \end{array}$$

and we have thence in $m^2n^2\,\delta Q$ the term

$$\begin{array}{llll} m^4\,n\,e' & \tfrac{1}{2}\,. & -\ 3\ . & -\ \tfrac{203}{12} \\ & +\ \tfrac{1}{2}\,. & -\ \tfrac{21}{2}\,. & \tfrac{61}{24} \\ & +\ \tfrac{1}{2}\,. & \tfrac{3}{2}\,. & -\ \tfrac{91}{24}\,; \end{array}$$

that is

$$(\tfrac{203}{8} - \tfrac{427}{32} - \tfrac{91}{32} =) + \tfrac{147}{16}\ m^4\,n\,e'.$$

The part $\dfrac{\rho^2}{\rho'^3}(3 \cos 2v - 2v')\,\delta v$ of δQ gives

$$\frac{\rho^2}{\rho'^3}(-3\cos 2v - 2v') = \qquad \begin{array}{llll} \tfrac{171}{4}\ m^2e' & & \cos & g' \\ -\ 3 & & ,, & 2\tau \\ -\ \tfrac{21}{2} & e' & ,, & 2\tau - g' \\ +\ \tfrac{3}{2} & e' & ,, & 2\tau + g' \end{array}$$

See *ante*, Annex 17.

$$\delta v = \qquad \begin{array}{lll} \overbrace{-\ 3\qquad\qquad}^{n^{-1}} & \cos & g' \\ -\ \tfrac{74}{3}\ m^2e' & ,, & 2\tau \\ +\ \tfrac{215}{48}\ m^2 & ,, & 2\tau - g' \\ -\ \tfrac{257}{48}\ m^2 & ,, & 2\tau + g' \end{array}$$

and we have thence in $m^2n^2\,\delta Q$ the term

$$\begin{array}{llll} m^4n\,e' & \tfrac{1}{2}\,. & \tfrac{171}{4}\,. & -\,3 \\ & +\,\tfrac{1}{2}\,. & -\,3\,. & -\,\tfrac{74}{3} \\ & +\,\tfrac{1}{2}\,. & -\,\tfrac{21}{2}\,. & \tfrac{215}{48} \\ & +\,\tfrac{1}{2}\,. & \tfrac{3}{2}\,. & -\,\tfrac{257}{48} \end{array}$$

that is

$$(-\tfrac{513}{8} + 37 - \tfrac{1505}{64} - \tfrac{257}{64} =) - \tfrac{1749}{32}\, m^4n\,e'\,;$$

and, combining this with the other term just obtained, the two together are

$$(\tfrac{147}{16} - \tfrac{1749}{32} =)\, \tfrac{1455}{32}\, m^4n\,e'\,;$$

and this term in $m^2n^2\,\delta Q$ gives in $m^2n^2\left(C + \int \delta Q\,dt\right)$ the term

$$- \tfrac{1455}{32}\, m^4n\,e'\,t.$$

Annex 22.

Calculation of term in $-\dfrac{2}{\rho}\dfrac{dv}{dt}\,\delta\rho$.

We have

$-\dfrac{2}{\rho}\dfrac{dv}{dt}$ (see Annex 19) =

$$\begin{array}{llll} -\,2\,\overbrace{}^{n\,\times} & & & \\ & -\,\tfrac{1}{3} & m^2 & \\ & +\,9 & m^2e' & \cos g' \\ & -\,\tfrac{15}{2} & m^2 & \;\;,,\;\; 2\tau \\ & -\,\tfrac{105}{4} & m^2e' & \;\;,,\;\; 2\tau - g' \\ & +\,\tfrac{15}{4} & m^2e' & \;\;,,\;\; 2\tau + g' \end{array}$$

$\rho =$

$$\begin{array}{llll} (\tfrac{3}{2}\,m^2 - \tfrac{1973}{32}\,m^4)\,e' & & t & \\ +\,\tfrac{3}{2} & m^2 & t\,\cos g' & \\ +\,5 & m^2e' & ,, & 2\tau \\ -\,\tfrac{7}{2} & m^2e' & ,, & 2\tau - g' \\ +\,\tfrac{1}{2} & m^2 & ,, & 2\tau + g' \end{array}$$

giving, besides the term $\tfrac{3}{2}\,m^2n\,e'\,t$ already taken account of, the term

$$\begin{array}{llll} m^4n\,e'\,t\,. & +\,\tfrac{1973}{16}\,. & -\,\tfrac{1}{3}\,. & \tfrac{3}{2} \\ & +\,\tfrac{1}{2}\,. & 9\,. & \tfrac{3}{2} \\ & +\,\tfrac{1}{2}\,. & -\,\tfrac{15}{2}\,. & 5 \\ & +\,\tfrac{1}{2}\,. & -\,\tfrac{105}{4}\,. & -\,\tfrac{7}{2} \\ & +\,\tfrac{1}{2}\,. & \tfrac{15}{4}\,. & \tfrac{1}{2} \end{array}$$

which is

$$= (\tfrac{1973}{16} - \tfrac{1}{2} + \tfrac{27}{4} - \tfrac{75}{4} + \tfrac{735}{16} - \tfrac{15}{16}) = (\tfrac{1973}{16} + \tfrac{275}{8} =)\, \tfrac{2523}{16}\, m^4n^2e'\,t.$$

Annex 23.

Calculation of term in $\dfrac{m^2n^2}{\rho^2}\left(C+\int \delta Q\, dt\right)$.

We have

$\dfrac{1}{\rho^2}=$		$m^2n^2\left(C+\int \delta Q\, dt\right)=$	
		$\overbrace{\qquad\qquad}^{n\,\times}$	
$1+\frac{1}{3}m^2$		$-\frac{1455}{32}\,m^2e'\;\; t$	
$-\,3\,m^2e'$	$\cos g'$		
$+\,2\,m^2$	„ 2τ	$-\,\frac{15}{4}\,m^2e'$	$t\cos 2\tau$
$+\,7\,m^2e'$	„ $2\tau-g'$	$+\,\frac{21}{8}\,m^2e'$	„ $2\tau-g'$
$-\,1\,m^2e'$	„ $2\tau+g$	$-\,\frac{3}{8}\,m^2e'$	„ $2\tau+g'$

giving the term

$$
m^4ne'\,t \qquad \begin{array}{l} -\frac{1455}{32} \\ +\frac{1}{2}\,.\;\;2\,.-\frac{15}{4} \\ +\frac{1}{2}\,.\;\;7\,.\;\;\frac{21}{8} \\ +\frac{1}{2}\,.-1\,.-\,\frac{3}{8} \end{array}
$$

which is

$$
=\left(-\tfrac{1455}{32}-\tfrac{15}{4}+\tfrac{147}{16}\,\tfrac{3}{16}\right)=\left(-\tfrac{1455}{32}+\tfrac{45}{8}=\right)-\tfrac{1275}{32}\,m^4ne'\,t,
$$

and this completes the series of calculations.

222.

ON LAMBERT'S THEOREM FOR ELLIPTIC MOTION.

[From the *Monthly Notices of the Royal Astronomical Society*, vol. XXII. (1862), pp. 238—242.]

THE theorem referred to is that which gives the time of description of an elliptic arc in terms of the radius vectors and the chord. The demonstration given by the author in his "Insigniores Orbitæ Cometarum Proprietates," Augs. 1761, depends upon a series of geometrical propositions of great elegance, which may be thus stated.

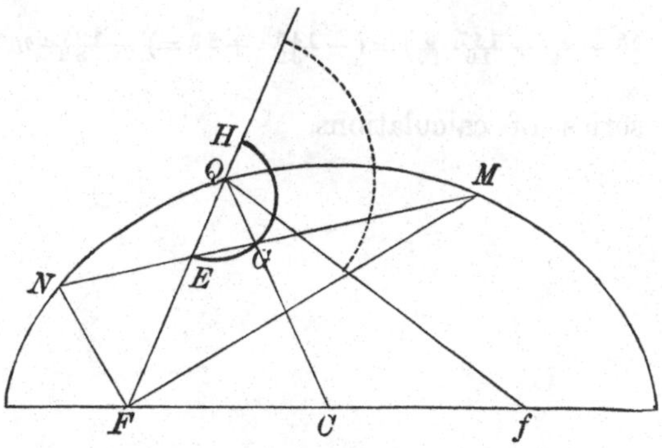

Let FQ be a line given in magnitude and position, E a given point on this line, Qf a line given in magnitude only, the position thereof being determined by assigning a value to its variable inclination to the line FQ. With F, f, as foci describe an ellipse passing through the point Q (the axis major, $=FQ+Qf$, is of course a constant magnitude). Take C, the centre of the ellipse, and join CQ; through E draw a chord, MEN, conjugate to the diameter CQ and meeting it in G. Then treating the inclination as variable,

1°. The locus of G is a circle passing through E, and having its centre on the line FQ.

2°. The semichord GM or GN, and the sum $FM+FN$ of the radius vectors are respectively constant.

3°. The elliptic area NFM, divided by the square root of the latus rectum, is a constant.

It may also be mentioned, that taking 2θ to represent the external inclination (supplement of the angle FQf), and if, moreover, a is the semiaxis major, e the eccentricity, and u, u' the eccentric anomalies of the points M, N, then the square root of the latus rectum, or say $\sqrt{1-e^2}$, $\propto \sin\theta$, and moreover EM, EN, FM, FN, $e\cos u$, $e\cos u'$, $e\sin u$, $e\sin u'$, consists each of them of a constant part, *plus* a part which $\propto \cos\theta$; these expressions give as above $GM=GN=\frac{1}{2}(EM+EN)=$ constant, $FM+FN=$ constant; and they give moreover $e\cos u+e\cos u'=$ constant; $e\sin u-e\sin u'$ $=$ constant; $u-u'=$ const. The expression for the area is

$$\tfrac{1}{2}a^2\sqrt{1-e^2}\,\{u-u'-(e\sin u-e\sin u')\},$$

and consequently the area divided by $\sqrt{1-e^2}$ is a constant; that is, the area is, as stated above, proportional to the square root of the latus rectum.

Hence, assuming the dynamical theorem that for a given central force at F, varying inversely as the square of the distance, the time of describing the elliptic arc is proportional to the area divided by the square root of the latus rectum, the time of describing the elliptic arc is constant. But in the extreme case, where the point f lies in the line FQ produced in the direction from F to Q, the ellipse reduces itself to a finite right line, length $FQ+Qf$, which is considered to be described by a body falling from the extremity with an initial velocity zero; and the arc MN is a portion thereof given in magnitude, and having for its centre the point H (where EH is the diameter of the before-mentioned circle, the locus of G). Hence the time of describing the elliptic arc is equal to the time of describing, under the action of the same central force, a given right line, and as such it is at once obtainable in the form

$$\frac{a^{\frac{3}{2}}}{\sqrt{\mu}}(\phi-\phi'-(\sin\phi-\sin\phi'))$$

where ϕ, ϕ' are functions of the major axis $FQ+Qf$, and of FM, FN, or, what is the same thing, of $FQ+Qf$, and of the chord MN and sum of the radius vectors FM, FN. The preceding is the geometrical mode of getting out the result, without the assistance of any expression for the elliptic area, and latus rectum, and assuming only that we know the formula for rectilineal motion; but, if the expressions for the elliptic area and latus rectum are obtained, then the expression for the time is known, and the problem is solved, without the necessity of passing from the ellipse to the right line.

Writing $FQ=\rho$, $QF=\sigma$, and as before the exterior angle of inclination $=2\theta$, the actual expressions for the various lines of the figure are easily found to be

$$\tfrac{1}{2}(\rho+\sigma) \qquad , = a,$$

$$CF = Cf = \tfrac{1}{2}\sqrt{\rho^2+\sigma^2+2\rho\sigma\cos 2\theta}, = a,$$

$$CQ \qquad = \tfrac{1}{2}\sqrt{\rho^2+\sigma^2-2\rho\sigma\cos 2\theta}, = a',$$

$$CR \qquad = \sqrt{\rho\sigma} \qquad , = b',$$

where CR (not shown in the figure) denotes the semi-diameter conjugate to CQ.

$$1-e^2 = \frac{4\rho\sigma}{(\rho+\sigma)^2}\sin^2\theta,$$

$$\cos F = \frac{\rho+\sigma\cos 2\theta}{2ae}, \quad \cos Q = \frac{\rho-\sigma\cos 2\theta}{2a'},$$

$$\sin F = \frac{\sigma\sin 2\theta}{2ae}, \quad \sin Q = \frac{\sigma\sin 2\theta}{2a'}, \quad \sin C = \frac{\rho\sigma\sin 2\theta}{4a'ae},$$

where F, C, Q, denote the angles of the triangle FCQ, respectively,

$$EG = \frac{2k\sigma}{\rho+\sigma}\cos\theta, \text{ and therefore } EH = \frac{2k\sigma}{\rho+\sigma},$$

$$QG = \frac{k}{\rho+\sigma}2a',$$

and, if for shortness $\Lambda = \sqrt{k\rho\sigma(\rho+\sigma-k)}$, then

$$(EM,\ EN) = \frac{2}{\rho+\sigma}(\Lambda \pm k\sigma\cos\theta),$$

$$(FM,\ FN) = \frac{1}{\rho+\sigma}\{\rho(\sigma+\rho)+k(\sigma-\rho)\pm 2\Lambda\cos\theta\},$$

so that

$$GM = GN = \tfrac{1}{2}(EM+EN) = \frac{2\Lambda}{\rho+\sigma},$$

$$\tfrac{1}{2}(FM+FN) = \frac{1}{\rho+\sigma}\{\rho(\sigma+\rho)+k(\sigma-\rho)\},$$

and moreover

$$(e\cos u,\ e\cos u') = \frac{1}{(\rho+\sigma)^2}\{(\sigma-\rho)(\rho+\sigma-2k)\mp 4\Lambda\cos\theta\},$$

$$(e\sin u,\ e\sin u') = \frac{1}{(\rho+\sigma)^2}\left\{\pm\frac{2(\sigma-\rho)\Lambda}{\sqrt{\rho\sigma}}+2(\rho+\sigma-2k)\sqrt{\rho\sigma}\cos\theta\right\},$$

so that

$$e\cos u + e\cos u' = \frac{2}{(\rho+\sigma)^2}(\rho+\sigma-2k),$$

$$e\sin u - e\sin u' = \frac{4}{(\rho+\sigma)^2}\,\frac{(\sigma-\rho)\,\Lambda}{\sqrt{\rho\sigma}},$$

$$u - u' = 2\tan^{-1}\frac{2\Lambda}{\sqrt{\rho\sigma}\,(\rho+\sigma-2k)} = \sin^{-1}\frac{4\Lambda\,(\rho+\sigma-2k)}{(\rho+\sigma)^2\sqrt{\rho\sigma}},$$

$$u - u' - (e\sin u - e\sin u') = \sin^{-1}\frac{4\,(\rho+\sigma-2k)\,\Lambda}{(\rho+\sigma)^2\sqrt{\rho\sigma}} - \frac{4\,(\sigma-\rho)\,\Lambda}{(\rho+\sigma)^2\sqrt{\rho\sigma}}$$

which is

$$= \phi - \phi' - (\sin\phi - \sin\phi'),$$

if

$$1-\cos\phi = \frac{1}{2a}(FM+FM+MN) = \frac{2}{(\rho+\sigma)^2}\{\rho\,(\sigma+\rho)+k\,(\sigma-\rho)+2\Lambda\},$$

$$1-\cos\phi' = \frac{1}{2a}(FM+FN-MN) = \frac{2}{(\rho+\sigma)^2}\{\rho\,(\sigma+\rho)+k\,(\sigma-\rho)-2\Lambda\}.$$

In fact we then have also

$$1+\cos\phi = \frac{2}{(\rho+\sigma)^2}\{\sigma\,(\sigma+\rho)-k\,(\sigma-\rho)-2\Lambda\},$$

$$1+\cos\phi' = \frac{2}{(\rho+\sigma)^2}\{\sigma\,(\sigma+\rho)-k\,(\sigma-\rho)+2\Lambda\},$$

and thence

$$\sin\tfrac{1}{2}\phi = \frac{1}{\rho+\sigma}\left(\sqrt{\rho\,(\rho+\sigma-k)}+\sqrt{k\sigma}\right),\quad \sin\tfrac{1}{2}\phi' = \left(\sqrt{\rho\,(\rho+\sigma-k)}-\sqrt{k\sigma}\right),$$

$$\cos\tfrac{1}{2}\phi = \frac{1}{\rho+\sigma}\left(\sqrt{\sigma\,(\rho+\sigma-k)}-\sqrt{k\rho}\right),\quad \cos\tfrac{1}{2}\phi' = \left(\sqrt{\sigma\,(\rho+\sigma-k)}+\sqrt{k\rho}\right),$$

whence

$$\sin\phi = \frac{2}{(\rho+\sigma)^2}\left\{\sqrt{\rho\sigma}\,(\rho+\sigma-2k)+\frac{(\sigma-\rho)\sqrt{\Lambda}}{\sqrt{\rho\sigma}}\right\},$$

$$\sin\phi' = \frac{2}{(\rho+\sigma)^2}\left\{\sqrt{\rho\sigma}\,(\rho+\sigma-2k)-\frac{(\sigma-\rho)\sqrt{\Lambda}}{\sqrt{\rho\sigma}}\right\},$$

and therefore
$$\sin\phi - \sin\phi' = \frac{4\,(\sigma-\rho)\sqrt{\Lambda}}{(\rho+\sigma)^2\sqrt{\rho\sigma}},$$

$$\sin\tfrac{1}{2}(\phi-\phi') = \frac{2k\Lambda}{(\rho+\sigma)\sqrt{\rho\sigma}},\quad \cos\tfrac{1}{2}(\phi-\phi') = \frac{\rho+\sigma-2k}{\rho+\sigma},$$

and therefore
$$\sin(\phi-\phi') = \frac{4\,(\rho+\sigma-2k)\,\Lambda}{(\rho+\sigma)^2\sqrt{\rho\sigma}},$$

which verifies the formula.

NOTES AND REFERENCES.

173. I ATTACH some value to this analysis and development of Laplace's Method, showing how it leads to the actual expression for the Potential of an Ellipsoid upon an exterior point in a series of terms of the form $\left(\alpha^2 \frac{d^2}{da^2} + \beta^2 \frac{d^2}{db^2} + \gamma^2 \frac{d^2}{dc^2}\right)^i \frac{1}{\sqrt{a^2+b^2+c^2}}$, being in fact the series deduced by me in the year 1842 from a result of Lagrange's; see vol. I., Notes and References 2 and 3.

191. The theorem obtained at the end of the paper is a very peculiar one; the only paper that I know of in anywise relating to the theory is Donkin, "On an application of the Calculus of Operations to the transformation of Trigonometric Series," *Quart. Math. Jour.* t. III. (1860), pp. 1—15, where (pp. 13—15) my theorem is referred to and a more general theorem involving two arbitrary functions ϕ, F, is arrived at.

194. In connexion herewith see the Memoir, Donkin, "On the Analytical Theory of the Attraction of Solids bounded by Surfaces of a hypothetical Class including the Ellipsoid," *Phil. Trans.* t. 150 (1860), pp 1—11. The author referring to my Note remarks that I there showed that if *two* of the principal theorems of attraction (in the case of the ellipsoid) be given the rest follow very simply and are common to all the surfaces of which these two can be predicated: but that the demonstration of the two assumed theorems constitute the most essential part of the analytical problem, and that it was his present object to show that they and the others connected with them are implied in the two partial differential equations

$$\frac{d^2u}{dx^2} + \frac{d^2u}{dy^2} + \frac{d^2u}{dz^2} = 2\left(\frac{1}{a^2+h} + \frac{1}{b^2+h} + \frac{1}{c^2+h}\right)$$

and

$$\left(\frac{du}{dx}\right)^2 + \left(\frac{du}{dy}\right)^2 + \left(\frac{du}{dz}\right)^2 + 4\frac{du}{dh} = 0,$$

satisfied by the function $\frac{x^2}{a+h} + \frac{y^2}{b+h} + \frac{z^2}{c+h}$: and he accordingly derives the whole theory, and in particular the theorems V. and VI. (equivalent to my assumed theorems) from these two partial differential equations.

221. It is well known that Plana, developing the explanation given by Laplace for the secular variation of the moon's mean motion, obtained in the expression of the true longitude the terms $-(\frac{3}{2}m^2 - \frac{2187}{128}m^4)\int(e'^2 - E'^2)\,ndt$, and that Prof. Adams in his memoir "On the Secular Variation of the Moon's Mean Motion," *Phil. Trans.* t. 143 (1853), pp. 397—406, corrected this into $-(\frac{3}{2}m^2 - \frac{3771}{64}m^4)\int(e'^2 - E'^2)\,ndt$. The validity of the correction was a good deal discussed, and it was interesting to establish the result by an entirely independent method.

END OF VOL. III.

CAMBRIDGE:
PRINTED BY C. J. CLAY, M.A. AND SONS,
AT THE UNIVERSITY PRESS.

Printed in the United States